Microbial Factories

Vipin Chandra Kalia
Editor

Microbial Factories

Biodiversity, Biopolymers, Bioactive Molecules: Volume 2

Editor
Vipin Chandra Kalia
Microbial Biotechnology and Genomics
CSIR-Institute of Genomics
and Integrative Biology
Delhi University Campus
Delhi, India

ISBN 978-81-322-2594-2 ISBN 978-81-322-2595-9 (eBook)
DOI 10.1007/978-81-322-2595-9

Library of Congress Control Number: 2015957418

Springer New Delhi Heidelberg New York Dordrecht London

Printed on acid-free paper

Springer (India) Pvt. Ltd. is part of Springer Science+Business Media (www.springer.com)

Dedicated to my family and friends

Preface

Human beings are an integral part of the environment. Biological activities have a strong influence on physical and chemical components of the ecosystem. Plants are the major contributors as producers of bioproducts, to be used by animals and microbes. In the animal kingdom, human beings are the most aggressive consumers, and their needs are increasing geometrically with time. Unlike animals, human needs extend beyond food and shelter. The innovative nature of man has led to discoveries and inventions, which apparently are for the benefit of human beings. However, these developments are a big drain on the available natural resources with a cascade effect. At the base of this chain reaction, the most adversely affected is the energy sector. The demand for energy is increasing rapidly because of the needs and attitudes of humans, who are thus transforming to a society of high-end consumers. Since fossil fuels are the major source of energy, their consumption is the root cause of irreparable damage to the environment. Another factor which adds to the ever-increasing environmental pollution is the unmanageable quantities of wastes. The conventional means of disposal of wastes and waste waters, adopted in most parts of the world, pollutes the land, atmosphere, and the water bodies. Here, we may need to approach the most efficient organisms on the planet Earth. These efficient organisms are the microbes, which can metabolize organic matter content of the biowastes, especially those produced due to human activities. These bioproducts are eco-friendly, biodegradable, and highly energy efficient. Microbes can be exploited as factories for producing energy (biofuels), biopolymers (bioplastics), and bioactive molecules (antimicrobial, anticancer, antidiabetic, antioxidants, etc.). There has been a vigorous scientific pursuit to exploit microbes for the welfare of human beings. The most exciting are the possibilities of generating clean fuels (biohydrogen, biodiesel, etc.) and biodegradable plastics as an alternative to nondegradable plastics. Apart from these, the most curiosity-driven activities have been to learn about those microbes which are yet to be cultured. During the last 2–3 decades, many scientific activities have been demonstrated and published in scientific journals of repute; however, it is yet to reach the curious young minds – the graduate and postgraduate students – of our future scientists. This compilation, contributed by the experts in these research domains, speaks a lot about the present status of microbial factories and their future potential for the welfare of human beings. In principle, experts exist in all domains; however, most of the times, they are too busy in their pursuits to spare time for such activities. The young, curious, and tender minds are eager to learn, but those who know what and how to say do not get the right platform and access. I am extremely thankful to all those who readily agreed to share their expertise for the *Ignited Minds*, to whom the book is dedicated. Although it is impossible to acknowledge the reality and true worth of the efforts of the contributing authors, however, I am still indebted to their prompt responses and dedicated efforts. My inspiration to learn well and transmit the knowledge to the next generation burgeons from the tireless efforts and constant support of my close ones – Mrs. Kanta Kalia and Mr. R.B. Kalia (parents); Amita (wife); Sunita and Sangeeta (sisters); Ravi, Vinod, and Satyendra (brothers); Daksh and Bhrigu (sons); and my teachers and friends Rup, Hemant, Yogendra, Rakesh, Atya, Jyoti, Malabika, Neeru, and Ritushree – to write this book. I must also acknowledge the selfless and dedicated support of my next-generation colleagues – Prasun, Sanjay, Subhasree, Shikha, Anjali, and Jyotsana.

Delhi, India Vipin Chandra Kalia

Contents

About the Editor

Vipin Chandra Kalia is presently working as chief scientist at Microbial Biotechnology and Genomics, CSIR-Institute of Genomics and Integrative Biology, Delhi. He is a professor at the Academy of Scientific and Innovative Research (AcSIR), Delhi. He obtained his M.Sc. and Ph.D. degrees in genetics from the Indian Agricultural Research Institute, Delhi. He has been elected as (1) Fellow of the Association of Microbiologists of India (FAMI) and (2) Fellow of the National Academy of Sciences (FNASc). His main areas of research are microbial biodiversity, genomics, and evolution, bioenergy, biopolymers, antimicrobials, quorum sensing, and quorum quenching. He has published 85 papers in scientific journals such as (1) *Nature Biotechnology*, (2) *Biotechnology Advances*, (3) *Trends in Biotechnology*, (4) *Critical Reviews in Microbiology*, (5) *Bioresource Technology*, (6) *International Journal of Hydrogen Energy*, (7) *PLoS ONE*, (8) *BMC Genomics*, and (9) *Gene*. His works have been cited 2080 times with an *h* index of 25 and an i10 index of 39. He has edited a book: *Quorum Sensing Versus Quorum Quenching: A Battle with No End in Sight* (2015, Springer India). He is presently the editor in chief of the Indian *Journal of Microbiology* and editor of (1) *PLoS ONE*, (2) *Journal of Microbiology and Biotechnology* (Korea), (3) *Applied Biochemistry and Biotechnology* (USA), (4) *International Scholarly Research Notices* (energy), (5) *Dataset Papers in Science* (microbiology), and (6) *Journal of Molecular and Genetic Medicine*. He is a life member of the following scientific societies: (1) Society of Biological Chemists of India; (2) Society for Plant Biochemistry and Biotechnology, India; (3) Association of Microbiologists of India; (4) Indian Science Congress Association; (5) BioEnergy Society of India, and (6) the Biotech Research Society of India (BRSI). He is also a member of the American Society for Microbiology. He can be contacted at vckalia@igib.res.in; vc_kalia@yahoo.co.in

1 Biopolymers and Their Application as Biodegradable Plastics

Scott Lambert

Abstract

Plastics have gained widespread use because of their plasticity in form and function, and their benefits are wide-ranging. The drawbacks to petrochemical plastics are that they are considered to be biologically inert. This in turn represents a waste management problem, especially for plastic materials with a short use phase. Presently, the development of biodegradable plastics as an alternative to petrochemical plastics is seen as an important waste management option. Polyhydroxyalkanoates (PHAs) are polyesters produced by the bacterial fermentation of sugars and lipids. PHAs have attracted commercial and academic interest because they are considered to be highly biodegradable. One of the most promising areas of application for PHAs is in the production of thin film materials for use as packaging materials. Presently, about 40 % of the plastics produced worldwide are utilised for packaging purposes. The disposable nature of packaging materials makes the use of PHA plastics an attractive alternative. As such PHAs may help to solve some of the waste management issues associated with single-use plastic items.

1.1 Introduction

The majority of plastic products in use today are made from petrochemical-based polymers. These materials have gained widespread use in a variety of industries such as packaging, transport and agriculture, because of their plasticity in form and function. It is clear that plastic materials play an important role in our everyday lives. However, the problems associated with modern petrochemical plastics from a waste management and environmental perspective come from the large volumes of plastic waste materials that are generated. These waste materials do not readily degrade in landfills and composting facilities. Indeed, littered plastic materials are one of the most ubiquitous and conspicuous environmental pollution problems and are present in one form or another globally. In addition to the obvious visual impacts of littering, there are also many broader environ-

S. Lambert (✉)
Goethe University Frankfurt am Main,
Max-von-Laue-Str. 13, 60438 Frankfurt, Germany
e-mail: scottl210@hotmail.co.uk

V.C. Kalia (ed.), *Microbial Factories*, DOI 10.1007/978-81-322-2595-9_1

mental issues as they present a persistent pollution problem that has hazardous consequences for organisms and environmental degradation.

Presently, society is working towards the goals of sustainable development. Among others, research and development increasingly focuses on waste and pollution prevention. However, both the generation of waste materials and pollution cannot always be prevented, and biodegradation is seen as an important management option. To this end, there is strong interest in the development of biodegradable materials made from biological resources, commonly termed 'biopolymers' or 'bio-based polymers', as a substitution for petrochemical-based polymers, and as a potential solution to waste management and environmental issues. Bio-based polymers can be produced from various resources; these include carbohydrate-rich products from the agricultural sector (natural polymers) such as polyhydroxyalkanoates, starch and cellulose (Lambert et al. 2014). Next are the polymers produced by bacteria via fermentation of sugars and lipids, which form a class composed of polyhydroxyalkanoates (PHA), polylactides (PLA), aliphatic polyesters, polysaccharides and their copolymers and/or blends (Lambert et al. 2014; Reddy et al. 2003). The two most talked about bio-based polymers in the scientific literature are PHAs and PLA. This chapter focuses mainly on the microbially formed PHAs as an alternative to traditional petrochemical plastics and their potential utilisation as materials for bioplastic applications.

1.2 Worldwide Production of Polymers

The present-day usage of plastics can be traced back to the development of rubber technology in the nineteenth century. the discovery of vulcanisation of natural rubber by Charles Goodyear was a major breakthrough in this area. Subsequently, efforts were focused on the development of synthetic polymers. Polystyrene (PS) and polyvinyl chloride (PVC) were discovered during this era, but could not be commercialised due to their brittle nature. Bakelite, developed by Leo Baekeland, in 1909 was among the first synthetic polymers to be produced on a mass scale. The modern forms of PVC, polyurethane (PUR) and a more processable PS were created during 1926–1938. The early 1950s saw the development of high-density polyethylene and polypropylene. The advances in the material sciences through the 1960s have seen the development of plastic materials produced from other natural resources, such as PHA, PLAs, polysaccharides and aliphatic polyesters (Reddy et al. 2003).

Today world synthetic and natural polymer production is approximately 327 million t annually (Fig. 1.1) and is dominated by synthetic polymers (315 million t), for which polyolefins (114 million t) are the most important. The polyolefins and other thermoplastics, as listed in Fig. 1.1, are the primary commodity plastics because they have the desired properties and are inexpensive to produce. World natural rubber latex production is approximately 10.4 million t annually (NRS 2011), and the global production of biopolymers is estimated to be 2.33 million t (Shen et al. 2009).

Packaging represents one of the most important applications of plastics. It accounts for around 40.1 % of the total production. The other usages of plastics are in building and construction (20.4 %), where as automotive electrical and electronic equipment account for 5.6–7.0 %. Apart from these major areas, markets including leisure and agriculture account for 26.9 %. The benefits of plastic materials are wide ranging. They are inexpensive to produces when compared to alternative materials, are long lasting and offer thermal and electrical insulation properties, and many are resistant to chemicals and water. Plastics are flexible and very versatile materials, meaning that they are malleable for design purposes. They are also strong and lightweight with valuable properties for the automotive and aviation industries.

1.3 Plastics and Their Waste Management Issues

As outlined in the above section, the vast majority of plastics are made from petrochemical polymers. Petrochemical-based plastics are generally considered bioinert and do not degrade, or at

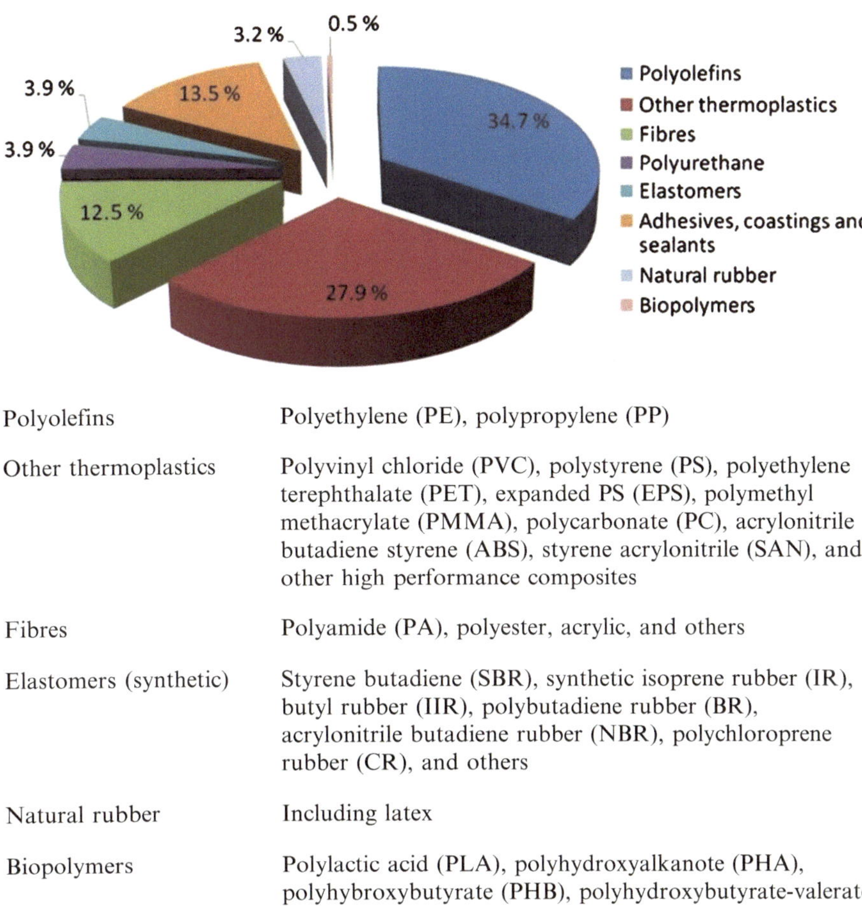

Polyolefins	Polyethylene (PE), polypropylene (PP)
Other thermoplastics	Polyvinyl chloride (PVC), polystyrene (PS), polyethylene terephthalate (PET), expanded PS (EPS), polymethyl methacrylate (PMMA), polycarbonate (PC), acrylonitrile butadiene styrene (ABS), styrene acrylonitrile (SAN), and other high performance composites
Fibres	Polyamide (PA), polyester, acrylic, and others
Elastomers (synthetic)	Styrene butadiene (SBR), synthetic isoprene rubber (IR), butyl rubber (IIR), polybutadiene rubber (BR), acrylonitrile butadiene rubber (NBR), polychloroprene rubber (CR), and others
Natural rubber	Including latex
Biopolymers	Polylactic acid (PLA), polyhydroxyalkanote (PHA), polyhybroxybutyrate (PHB), polyhydroxybutyrate-valerate (PHBV), cellulose, and others.

Fig. 1.1 Annual world polymer production

least degrade very slowly, in landfill and under composting conditions.

There are several waste streams that waste plastic materials are likely to enter, i.e. landfill, incineration, wastewater treatment plants and recycling. The suitable treatment is a key question of waste management. Different countries have a wide range of waste management prioritisations for solid waste materials, ranging from countries that heavily favour landfill to those that favour recycling and incineration. Landfill is still a major end-of-life disposal system employed for municipal and industrial solid waste in many countries. However, waste disposal in landfill is becoming increasing undesirable in many countries due to legislative pressure, rising costs and hazards associated with landfill leachate contaminating groundwater. Incineration is an important alternative to landfill. Chinese administration expects to expand its capacity for waste treatment via incineration from 3 % in 2011 to 30 % by 2020 (Cheng and Hu 2010). Japan, Denmark and Sweden also have extensive incinerator infrastructure in place for dealing with solid waste. Incineration is seen as having a preferred waste management option in terms of energy recovery and reducing the need for landfill.

The utilisation of biodegradation processes as a waste management option is today viewed as a method to improve sustainability of managing societies' waste. A plastic material labelled as either compostable or biodegradable should, in theory, undergo biological decomposition within industrial composting sites. This will require the development of plastic technologies that are able to fully degrade under such conditions.

Table 1.1 Examples of PHA production from different media types

Organism	Carbon source	Production level [g/L]	References
Azotobacter vinelandii	Molasses	23	Page et al. (1992)
Pseudomonas fluorescens	Molasses	22	Jiang et al. (2008)
Ralstonia eutropha	Potato starch	94	Haas et al. (2008)
Cupriavidus necator	Glycerol	38	Cavalheiro et al. (2009)
Cupriavidus necator	Glycerol	16	Koller et al. (2005)
Escherichia coli	Whey	96.2	Ahn et al. (2000)
Bacillus megaterium	Sugarcane molasses	1.27	Kulpreecha et al. (2009)
Bacillus megaterium	Sugarcane molasses	0.45	Kulpreecha et al. (2009)

1.4 Biopolymers: PHA and PLA

Although biopolymer, at present, represents only a small portion of the current market share, their production costs in the future are expected to decrease making them more competitive. The following section provides an overview of the biopolymers PHAs and PLA. For an in-depth review of biodegradable plastics, the reader is pointed towards recent reviews by Reddy et al. (2013) and Soroudi and Jakubowicz (2013).

1.4.1 Polyhydroxyalkanoates

PHAs are polyesters produced by the bacterial fermentation of sugars and lipids. These simple macromolecules are synthesised by a very wide range of microorganisms. PHAs are accumulated intracellularly and function as carbon and energy reserves. Microorganisms capable of accumulating PHAs include the genera *Alcaligenes*, *Bacillus* and *Pseudomonas*; for a detailed list, the reader is encouraged to see Koller et al. (2010). PHA molecules, extracted from the bacterial cell, have sufficiently high molecular mass, which exhibit characteristics quite similar to some common petrochemical plastics, for example, polyethylene and polypropylene. It is thought that there are over 150 different types of PHAs that can be combined to provide a wide range of different properties. The majority of PHAs are identified as primarily linear polyesters composed of 3-hydroxy fatty acid monomers (Madison and Huisman 1999). In these PHAs, the carboxyl group and the hydroxyl group of neighbouring monomers form an ester bond (Madison and Huisman 1999). As a general rule, PHAs are classified based on the number of carbon atoms in the polymer chain. Those PHAs consisting of three to five carbon atoms in the polymer chain are classified as short-chain length PHAs (scl-PHA) and can be synthesised by numerous microorganisms including *Alcaligenes latus* and *Ralstonia eutropha* (Madison and Huisman 1999). PHAs consisting of 6–14 carbon atoms are classified as medium-chain length PHAs (mcl-PHA) and can be synthesised by *Pseudomonas oleovorans* and *Pseudomonas putida* (Kim et al. 2000). Some microorganisms such as *Aeromonas hydrophila* and *Thiococcus pfennigii* are capable of synthesising scl-PHA and mcl-PHA copolyesters (Kim do et al. 2007). In addition, like all polymers, PHAs can also be classified as either homopolymers or heteropolymers depending on the number of different hydroxyalkanoate monomer types which makes up the monomeric unit. In terms of their material properties, the scl-PHAs display thermoplastic properties, while mcl-PHAs have more elastic properties. In addition, like all polymers, PHAs can also be classified as either homopolymers or heteropolymers depending on the number of different hydroxyalkanoate monomer types which makes up the monomeric unit.

The production of PHAs can be undertaken using various materials including sugar-based media (Jiang et al. 2008; Kulpreecha et al. 2009; Page 1992), starch-based media (Haas et al. 2008), whey-based media (Ahn et al. 2000), and

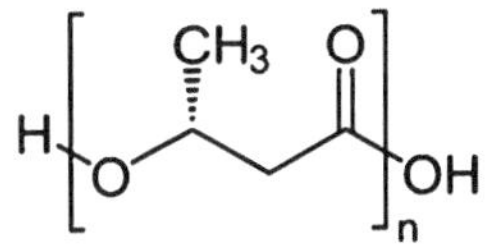

Fig. 1.2 Structure of polyhydroxybutyrate

glycerol-based media (Cavalheiro et al. 2009; Koller et al. 2005) (Table 1.1). The data presented in Table 1.1 identifies starch-based and whey-based materials as easily utilised by microorganisms, as shown by their high production levels. Table 1.1 also highlights differences in the species of microorganism used. *Azotobacter vinelandii* and *Pseudomonas fluorescens* are shown to demonstrate high production levels compared to *Bacillus megaterium* when using sugar-based media.

The most common member of this family of polymers and often the most talked about in the scientific literature, is polyhydroxybutyrate (PHB; Fig. 1.2) or more accurately poly(3-hdroxybutyrate). PHB when within the cell exists in a fluid, amorphous state, but after extraction, it becomes highly crystalline (Madison and Huisman 1999) and is often described as producing material films that are stiff but brittle (Woolnough et al. 2008). The brittleness of PHB means that it is not very stress resistant. In addition, PHB has a melting temperature in the region of 170 °C; this is close to the temperature where PHB thermally decomposes; therefore, PHBs ability to be processed as a homopolymer is limited (Madison and Huisman 1999). To improve the material characteristics of PHB, copolymers can be formed that contain 3-hydroxyvalerate or 3-hydroxybutyrate monomers. Poly(hydroxybutyrate-*co*-hydroxyvalerate) (PHBV) is a copolymer and is made through the copolymerisation of 3-hydroxybutyrate with units of 3-hydroxyvalerate to yield PHBV, a material with modified physicochemical properties making it a more flexible polymer. In addition, the prosperities of PHBV are reported to be easily tailored for the required application by varying the valerate content of the copolymer. Increasing the hydroxyvalerate content is shown to increase impact strength of PHBV copolymers. At the same time, copolymers of PHAs are reported to display decreased water permeability and tensile strength (Madison and Huisman 1999).

PHAs are considered to be highly biodegradable (Álvarez-Chávez et al. 2012) and are reported to be biodegradable in a range of natural environments. The literature consists of studies reporting their biodegradation potential in soils (Kunioka et al. 1989; Mergaert et al. 1993; Woolnough et al. 2010), aerobic and anaerobic sludge and compost (Mergaert et al. 1996; Luo and Netravali 2003) and freshwater and marine water environments (Mergaert et al. 1995). PHAs have been demonstrated to be degradable under laboratory conditions of soil burial, at temperatures of 25–30 °C, when samples have undergone pre-ultraviolet exposure (Saad et al. 2010; Sadi et al. 2010). Overall, the biodegradability of PHA materials depends on (i) the ability of degrading bacteria to secrete specific extracellular PHA depolymerases, (ii) the structure of the resulting material, and (iii) the polymer type as PHA copolymers are shown to degrade faster than homopolymers (Kunioka et al. 1989). In addition, the major advantage of PHAs is that they are believed to produce no toxic biodegradation residues.

The diverse ranging properties that PHAs display make them suitable for a wide range of application. PBHV, because of its greater flexibility compared to PHA, makes it a suitable material for packaging. PHAs have also generated interest as biomedical devices used in tissue repair and as bone graft substitutes (Chen and Wu 2005). The continuing development of PHAs, including PHA blends and composites, promises the potential for a broader range of applications (Madison and Huisman 1999; Reddy et al. 2013).

1.4.2 Polylactic Acid (PLA)

PLA is a thermoplastic aliphatic polyester (Fig. 1.3). It can be produced by the polymerisation of lactic acid, a metabolic product of microbial fermentation of carbohydrates present in corn or sugarcane. The PLA monomer is a chiral molecule and exhibits two isomeric forms, D-lactic acid and L-lactic acid. PLA is considered

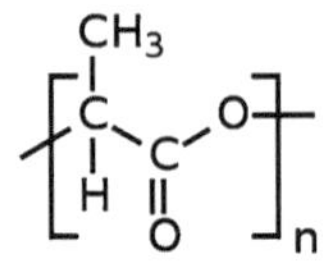

Fig. 1.3 Structure of polylactic acid

to be biodegradable and microorganisms capable of degrading its L-isoform have been identified. However, these microbial communities are thought to not be widely distributed in soil, and soil burial studies using L-PLA over a 6-week time period resulted in no weight loss (Ohkita and Lee 2006). Therefore, the degradation of PLAs is generally favoured under conditions that will support chemical hydrolysis of ester bonds in the first step (Garlotta 2001). This process does not require the presence of enzymes to catalyse the hydrolysis, and the rate at which this process happens depends on the temperature of the degrading environment. This process then leads to the microbial degradation of the lower molecular weight compounds that are produced (Garlotta 2001). On the other hand, D-PLA is considered hydrolysable in water, but not biodegradable (Shogren et al. 2003; Tokiwa and Calabia 2006; Kim et al. 2006). The hydrolysis of this isomeric form through the backbone ester groups is considered slow, and a process that under ambient moisture and temperature conditions can take several years. This process is possible to accelerate by subjecting it to temperatures above 50 °C (Tokiwa et al. 2009).

PLA has good mechanical properties, processability and good thermal properties and can be moulded into bottles and containers or extruded into fibres, films and sheets (Soroudi and Jakubowicz 2013). PLA has also been approved by the regulatory agencies of many countries for medical applications such as suture threads, implantable scaffolds and bone fixation devices. Although PLA is considered a biomaterial with excellent properties comparable to conventional plastics, drawbacks have been identified when compared to the requirements for certain applications. PLA is a very brittle material and without the addition of additives has been shown to have less than 10 % elongation at break; this means that PLA is potentially not suitable for applications that demand a high mechanical performance unless it is modified (Soroudi and Jakubowicz 2013). PLA is not quite suitable for industrial sectors such as packaging because of its limited gas barrier properties (Singh and Sharma 2008). There is currently considerable interest in the development of modified PLA, through the development of coploymers and composites, to improve material characteristics such as stiffness, permeability, crystallinity, and thermal stability (Reddy et al. 2013; Soroudi and Jakubowicz 2013).

1.5 Life Cycle Analysis of Biopolymers

Life cycle analysis (LCA) is a standardised method that is used to assesses a range of impact categories (e.g. acidification, carcinogens, etc.), with the intention of minimising total environmental impacts (Askham 2012). The LCA method has been formalised by the International Standards Organization (ISO) and has been applied, in the literature, to evaluate the environmental aspects of biopolymers derived from renewable resources. There are a number of studies that have compared the LCA for biopolymers to their petrochemical counterparts. When considering the production process only, PHAs have been identified to deliver better global warming potential and energy consumption scores than petrochemical polymers, as well as ranking as environmentally superior to petrochemical polymers over almost all impact categories (Harding et al. 2007). Álvarez-Chávez et al. (2012) also showed that production consumes 30–50 % less fossil energy and emits 50–70 % less CO_2 in comparison to petrochemical polymers.

Predicting the environmental impacts of plastics made from biological resources and ranking them alongside those predicted for their petrochemical counterparts are not easy tasks. At present, bioplastics produced from PHAs seem to be the preferred choice. However, one of the key questions is the use of suitable feedstocks. The preference is to use locally available and renewable carbon sources that do not compete for land with food production. Of economic interest are agricultural cellulose waste materials that are leftover from, for example, rice, corn and sugarcane plants, after harvesting (Koller et al. 2010; Álvarez-Chávez et al. 2012). In addition, the surplus whey from the dairy industries is available in large amounts in Europe and North America and is considered as a suitable feedstock (Koller et al. 2010). From a life cycle perspective, the use of agricultural by-products as a feedstock for bioplastic production is expected to improve their sustainability ranking compared to petrochemical plastics (Álvarez-Chávez et al. 2012). As always the end-of-life management will also have an important role to play. Questions that remain to be answered are as follows: (i) Can the recycling of bio-based plastics be as an efficient option as waste disposal through, for example, composting? (ii) To what extent will combining the use of bio-based plastics with today's collection systems affect the recycling of petrochemical plastics? Overall, biopolymers are likely to rank favourably for environmental impacts during LCA, because they represent a decrease in fossil fuel use and global warming potential (Tabone et al. 2010).

1.6 Applications for Biopolymers

In principle, polymers derived from bio-based resources could potentially find utilisation in all areas. However, the shift from traditional petrochemical polymers to biopolymers is mostly likely to play an important role for applications that have a relatively short use phase, for example, the production of thin film materials for use as packaging materials and disposable plastic bags. Presently, about 40 % of the plastics produced worldwide are utilised for packaging purposes. The disposable nature of packaging materials makes the use of biopolymers, such as these made from PHAs, an attractive alternative.

Packaging materials are needed to pack and contain a wide variety of products, such as liquids, powders and solids. Polyethylene is a successful material for the use of thin films that are used to make carrier bags, cling film, freezer and sandwich bags. In the food industry, packaging has an important role in providing protection from environmental, chemical and physical damage (during transport). Packaging materials provide a barrier to block light and prevent oxygen and moisture from deteriorating foodstuffs. PHAs because of their unique characteristics such as good tensile strength, printability, flavour and odour barriers, resistance to grease and oil and high stability to temperatures are highly preferred in food packaging industry (Tabone et al. 2010).

PHAs can also be processed into fibres for nonwoven fabrics, as well as diaper materials and other compostable personal hygiene products (Madison and Huisman 1999). The medical and pharmaceutical industries also offer areas for PHA application (Box 1.1) because of their biodegradability. Biodegradable polymers also have the potential to offer specific advantages in the agriculture and horticulture industries. Plastics have various agricultural applications. These include irrigation piping, greenhouses, low grow tunnels, mulches and storage (e.g. silage bails). They are desirable because they conserve moisture thereby reducing irrigation; reduce weed growth and increase soil temperature which reduces competition for soil nutrients and reduces

Box 1.1: PHA Applications Specific to the Medical Industry

- Sutures and suture fasteners
- Meniscus repair and regeneration devices
- Rivets, tacks, staples and screws
- Bone plates and bone plating systems
- Surgical mesh, repair patches and cardiovascular patches
- Vein valves and bone marrow scaffolds
- Ligament and tendon grafts
- Ocular cell implants
- Skin substitutes, bone graft substitutes and wound dressings

Information adapted from the text in Chen and Wu (2005); Madison and Huisman (1999) fertiliser costs thereby improving crop yields; and protect against adverse weather conditions. The development of biodegradable mulching films for use in crop production offers a practical alternative to petrochemical plastics, because their collection and removal from the field are time-consuming and costly.

1.7 Concluding Thoughts

Traditionally, plastic materials have been designed in the past to resist degradation. For plastic items that have a short use phase, such as packaging materials, this has created a waste management problem. The use of biological systems and the creation of microbial factories for the production of monomers to produce biodegradable materials for uses that have a relatively short use phase would seem likely to be a logical strategy. The challenge, therefore, is to develop biopolymers that have the necessary functionality during use, but are able to undergo full mineralisation after use, and leave behind no toxic residues. At present, the biopolymer market remains small compared to their petrochemical-based counterparts. However, this market has experienced fast growth in the past decade and global capacity is expected to reach 3.45 million metric tons by 2020 (Shen et al. 2009). For now, PLAs and PHAs are expected to be the major types of biopolymers in the future (Shen et al. 2009; Peelman et al. 2013).

Acknowledgements The author would like to thank Dr. V. C. Kalia for his critical reading of the manuscript. The author received no financial support for the research, authorship and/or publication of this article.

References

Ahn WS, Park SJ, Lee SY (2000) Production of Poly(3-Hydroxybutyrate) by fed-batch culture of recombinant Escherichia coli with a highly concentrated whey solution. Appl Environ Microbiol 66(8):3624–3627

Álvarez-Chávez CR, Edwards S, Moure-Eraso R, Geiser K (2012) Sustainability of bio-based plastics: general comparative analysis and recommendations for improvement. J Cleaner Prod 23(1):47–56, doi:http://dx.doi.org/10.1016/j.jclepro.2011.10.003

Askham C (2012) REACH and LCA—methodological approaches and challenges. Int J Life Cycle Assess 17(1):43–57. doi:10.1007/s11367-011-0329-z

Cavalheiro JMBT, de Almeida MCMD, Grandfils C, da Fonseca MMR (2009) Poly(3-hydroxybutyrate) production by Cupriavidus necator using waste glycerol. Process Biochem 44(5):509–515, doi:http://dx.doi.org/10.1016/j.procbio.2009.01.008

Chen G-Q, Wu Q (2005) The application of polyhydroxyalkanoates as tissue engineering materials. Biomaterials 26(33):6565–6578, doi:http://dx.doi.org/10.1016/j.biomaterials.2005.04.036

Cheng H, Hu Y (2010) Municipal solid waste (MSW) as a renewable source of energy: current and future practices in China. Bioresour Technol 101(11):3816–3824

Garlotta D (2001) A literature review of Poly(lactic acid). J Polym Environ 9(2):63–84. doi:10.1023/a:1020200822435

Haas R, Jin B, Zepf FT (2008) Production of Poly(3-hydroxybutyrate) from waste potato starch. Biosci Biotechnol Biochem 72(1):253–256. doi:10.1271/bbb.70503

Harding KG, Dennis JS, von Blottnitz H, Harrison STL (2007) Environmental analysis of plastic production processes: comparing petroleum-based polypropylene and polyethylene with biologically-based poly-β-hydroxybutyric acid using life cycle analysis. J Biotechnol 130(1):57–66, doi:http://dx.doi.org/10.1016/j.jbiotec.2007.02.012

Jiang Y, Song X, Gong L, Li P, Dai C, Shao W (2008) High poly(β-hydroxybutyrate) production by Pseudomonas fluorescens A2a5 from inexpensive substrates. Enzym Microb Technol 42:167–172. doi:10.1016/j.enzmictec.2007.09.003

Kim DY, Kim HW, Chung MG, Rhee YH (2007) Biosynthesis, modification, and biodegradation of bacterial medium-chain-length polyhydroxyalkanoates. J Microbiol 45(2):87–97

Kim DY, Kim YB, Rhee YH (2000) Evaluation of various carbon substrates for the biosynthesis of polyhydroxyalkanoates bearing functional groups by Pseudomonas putida. Int J Biol Macromol 28(1):23–29

Kim EY, Lee JK, Lee WK (2006) Hydrolytic kinetics of langmuir monolayers of enantiorneric poly(lactide)s. Curr Appl Phys 6:735–738. doi: 10.1016/j.cap.2005.04.029

Koller M, Bona R, Braunegg G, Hermann C, Horvat P, Kroutil M, Martinz J, Neto J, Pereira L, Varila P (2005) Production of polyhydroxyalkanoates from agricultural waste and surplus materials†. Biomacromolecules 6(2):561–565. doi:10.1021/bm049478b

Koller M, Atlić A, Dias M, Reiterer A, Braunegg G (2010) Microbial PHA production from waste raw materials. In: Chen GG-Q (ed) Plastics from bacteria, vol 14, Microbiology monographs. Springer, Berlin/Heidelberg, pp 85–119. doi:10.1007/978-3-642-03287-5_5

Kulpreecha S, Boonruangthavorn A, Meksiriporn B, Thongchul N (2009) Inexpensive fed-batch cultivation for high poly(3-hydroxybutyrate) production by a new isolate of Bacillus megaterium. J Biosci Bioeng 107(3):240–245

Kunioka M, Kawaguchi Y, Doi Y (1989) Production of biodegradable copolymers of 3-hydroxybutyrate and 4-hydroxybutyrate by Alcaligenes-eutrophus. Appl Microbiol Biotechnol 30(6):569–573

Lambert S, Sinclair CJ, Boxall ABA (2014) Occurrence, degradation and effects of polymer-based materials in the environment. Rev Environ Contam Toxicol 227:1–53. doi:10.1007/978-3-319-01327-5_1

Luo S, Netravali AN (2003) A study of physical and mechanical properties of poly(hydroxybutyrate-co-hydroxyvalerate) during composting. Polym Degrad Stab 80(1):59–66. doi:10.1016/s0141-3910(02)00383-x

Madison LL, Huisman GW (1999) Metabolic engineering of poly(3-hydroxyalkanoates): from DNA to plastic. Microbiol Mol Biol Rev 63(1):21–53

Mergaert J, Webb A, Anderson C, Wouters A, Swings J (1993) Microbial-degradation of poly(3-hydroxybutyrate) and poly(3-hydroxybutyrate-co-3-hydroxyvalerate) in soils. Appl Environ Microbiol 59(10):3233–3238

Mergaert J, Wouters A, Anderson C, Swings J (1995) In-situ biodegradation of poly(3-hydroxybutyrate) and poly(3-hydroxybutyrate-co-3-hydroxyvalerate) in natural-waters. Can J Microbiol 41:154–159

Mergaert J, Glorieux G, Hauben L, Storms V, Mau M, Swings J (1996) Biodegradation of Poly (3 -hYdroxYalkanoates) in Anaerobic Sludge and Characterization of a Poly (3 -hYdroxYalkanoates) Degrading Anaerobic Bacterium. Systematic and Applied Microbiology 19(3):407–413, doi:http://dx.doi.org/10.1016/S0723-2020(96)80070-1

NRS (2011) Natural rubber statistics 2011. http://www.lgm.gov.my/nrstat/nrstats.pdf. Accessed 30 July 2012

Ohkita T, Lee S-H (2006) Thermal degradation and biodegradability of poly (lactic acid)/corn starch biocomposites. Journal of Applied Polymer Science 100 (4):3009–3017. doi:10.1002/app.23425

Page WJ (1992) Production of polyhydroxyalkanoates by Azotobacter vinelandii UWD in beet molasses culture. FEMS Microbiol Lett 103(2–4):149–157, doi:http://dx.doi.org/10.1016/0378-1097(92)90304-7

Page WJ, Manchak J, Rudy B (1992) Formation of poly(hydroxybutyrate-co-hydroxyvalerate) by Azotobacter vinelandii UWD. Appl Environ Microbiol 58(9):2866–2873

Peelman N, Ragaert P, De Meulenaer B, Adons D, Peeters R, Cardon L, Van Impe F, Devlieghere F (2013) Application of bioplastics for food packaging. Trends Food Sci Technol 32(2):128–141, doi:http://dx.doi.org/10.1016/j.tifs.2013.06.003

Reddy CSK, Ghai R, Rashmi KVC (2003) Polyhydroxyalkanoates: an overview. Bioresour Technol 87(2): 137–146

Reddy MM, Vivekanandhan S, Misra M, Bhatia SK, Mohanty AK (2013) Biobased plastics and bionanocomposites: current status and future opportunities. Prog Polym Sci 38(10–11):1653–1689. doi:10.1016/j.progpolymsci.2013.05.006

Saad GR, Khalil TM, Sabaa MW (2010) Photo- and biodegradation of poly(ester-urethane)s films based on poly[(R)-3-Hydroxybutyrate] and poly(epsilon-Caprolactone) blocks. J Polym Res 17(1):33–42

Sadi RK, Fechine GJM, Demarquette NR (2010) Photodegradation of poly(3-hydroxybutyrate). Polym Degrad Stab 95(12):2318–2327

Shen L, Haufe J, Patel M (2009) Product overview and market projection of emerging biobased plastics. Report no: NWS-E-2009-32. Universiteit Utrecht, The Netherlands

Shogren RL, Doane WM, Garlotta D, Lawton JW, Willett JL (2003) Biodegradation of starch/polylactic acid/poly(hydroxyester-ether) composite bars in soil. Polym Degrad Stab 79(3):405–411, doi:http://dx.doi.org/10.1016/S0141-3910(02)00356-7

Singh B, Sharma N (2008) Mechanistic implications of plastic degradation. Polym Degrad Stab 93(3):561–584

Soroudi A, Jakubowicz I (2013) Recycling of bioplastics, their blends and biocomposites: a review. Eur Polym J 49(10):2839–2858, doi:http://dx.doi.org/10.1016/j.eurpolymj.2013.07.025

Tabone MD, Cregg JJ, Beckman EJ, Landis AE (2010) Sustainability metrics: life cycle assessment and green design in polymers. Environ Sci Technol 44(21):8264–8269. doi:10.1021/es101640n

Tokiwa Y, Calabia BP (2006) Biodegradability and biodegradation of poly(lactide). Appl Microbiol Biotechnol 72(2):244–251. doi:10.1007/s00253-006-0488-1

Tokiwa Y, Calabia BP, Ugwu CU, Aiba S (2009) Biodegradability of plastics. Int J Mol Sci 10(9):3722–3742. doi:10.3390/ijms10093722

Woolnough CA, Charlton T, Yee LH, Sarris M, Foster LJR (2008) Surface changes in polyhydroxyalkanoate films during biodegradation and biofouling. Polym Int 57(9):1042–1051. doi:10.1002/pi.2444

Woolnough CA, Yee LH, Charlton T, Foster LJR (2010) Environmental degradation and biofouling of 'green' plastics including short and medium chain length polyhydroxyalkanoates. Polym Int 59(5):658–667. doi:10.1002/pi.2746

Scott Lambert received his Ph.D. degree from the University of York, UK. He is currently working as a Research Fellow at Goethe University Frankfurt am Main. His main research interests are the biodegradability of plastic materials for better waste management and the environmental effects of plastics and their degradation products.

Approaches for the Synthesis of Tailor-Made Polyhydroxyalkanoates

2

Carlos F. Peña Malacara, Andrés García Romero, Modesto Millán Ponce, and Tania Castillo Marenco

Abstract

Polyhydroxyalkanoates (PHAs) are biodegradable and biocompatible thermoplastics. These have been proposed for a wide range of biotechnological applications, especially in the field of the medicine and chemistry. PHAs are produced by more than 300 bacterial species, the most efficient being *Cupriavidus necator* (*formerly Ralstonia eutropha*), *Alcaligenes latus*, and recombinant strains of *Escherichia coli*. PHAs are produced by fermentation using different culture systems, from batch culture to exponentially fed-batch cultures, and it is known that culture conditions, such as pH, aeration, and nutritional conditions, influence the chemical characteristic PHAs synthesized by microorganisms; because of that, it has been proposed that by manipulating the microbial metabolism and culture conditions, it is possible to design biopolymers with specific chemical properties. This paper describes four cases of PHAs production: the copolymers of poly-3-hydroxybutyrate-co-poly-3-hydroxyvalerate [P(3HB-co-3HV)] and poly-3-hydroxybutyrate-co-poly-3-hydroxyhexanoate [P(3HB-co-3HHx)], the medium-chain-length PHAs, the P3HB of ultrahigh molecular mass, and finally, the production of other short-chain-length PHAs, with a special emphasis on the species that have been reported for their production as well as the molecular and fermentation strategies evaluated in order to modify the chemical composition of PHAs.

C.F.P. Malacara (✉) • A.G. Romero • M.M. Ponce
Institute of Biotechnology, National University of Mexico (UNAM), 62250 Cuernavaca, Morelos, Mexico
e-mail: carlosf@ibt.unam.mx; agromero@ibt.unam.mx; mmillanp@ibt.unam.mx

T.C. Marenco
Center of Research in Biotechnology (CEIB), Autonomous University of Morelos State (UAEM), 62250 Cuernavaca, Morelos, Mexico
e-mail: tania.castillo.marenco@gmail.com

2.1 Introduction

Commercial interest in bioplastics has increased due to the possibility of replacing synthetic materials, which have disadvantages from the environmental perspective. In this regard, polyhydroxyalkanoates (PHAs) are a suitable option to substitute the plastics derived from petroleum.

V.C. Kalia (ed.), *Microbial Factories*, DOI 10.1007/978-81-322-2595-9_2

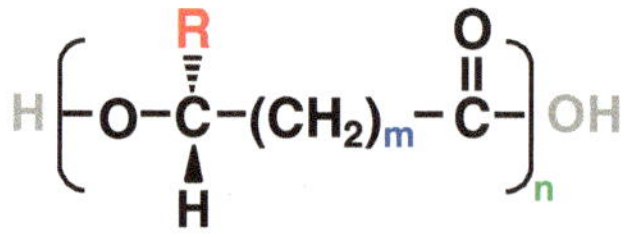

Polyhydroxyalkanoate

PHA	R	m	Name of the monomer
SCL	H	1	3-hydroxypropionate (3HP)
	CH_3	1	3-hydroxybutyrate (3HB)
	H	2	4-hydroxybutyrate (4HB)
	C_2H_5	1	3-hydroxyvalerate (3HV)
	H	3	5-hydroxyvalerate (5HV)
MCL	C_3H_7	1	3-hydroxyhexanoate (3HHx)
	C_5H_{11}	1	3-hydroxyoctanoate (3HO)
	C_7H_{15}	1	3-hydroxydecanoate (3HD)
	C_9H_{19}	1	3-hydroxydodecanoate (3HDD)
	C_9H_{18}	1	3-hydroxydodecenoate (3HDDe)
	$C_{11}H_{23}$	1	3-hydroxytetradecanoate (3HTD)

n = 1 - 230,000 monomers.

Fig. 2.1 Chemical structure of polyhydroxyalkanoates. *R* alkyl group, *m* length of carbon chains, *n* number of monomers

PHAs are polyesters composed of 3-hydroxy fatty acid monomers (Fig. 2.1) (Chen 2010; Peña et al. 2014a). The main advantages of these biopolymers are their biodegradability and biocompatibility, making them suitable for a wide range of applications, from the traditional plastic industry to their use as materials in the biomedical field, with emphasis on chemical composition and purity of the product (Peña et al. 2014a; Leong et al. 2014). PHAs are synthesized by many microorganisms as energy reserve material. In general, it has been well documented that these polymers are produced under nutrient limitation, mainly nitrogen, phosphorus, or oxygen (Anderson and Dawes 1990; Peña et al. 2011, 2014a; Ienczak et al. 2013). There is a wide variability in PHA composition that includes homopolymers, heteropolymers, and up to 150 different types of monomers (Steinbüchel and Lütke-Eversloh 2003). The thermomechanical properties of PHAs and therefore their specific applications will depend on their chemical structure, specifically monomer composition (type, ratio, and distribution) and, in the case of homopolymers, their mean molecular mass (MMM). In this line, several attempts, that include genetic manipulation of microorganisms as well as changes on the culture conditions in which the cells are grown, have been evaluated in order to obtain materials of specific characteristics (Reddy et al. 2003; Peña et al. 2014a; Leong et al. 2014). The subjects covered by this chapter include properties of PHAs and their applications, bacterial sources and PHAs biosynthesis, as well as the influence of culture conditions (i.e., medium composition, temperature, pH, etc.), which determine the composition of PHAs, including specific examples regarding the production of PHAs with different chemical compositions.

2.2 Chemical Structure, Physicochemical Properties, and Applications of PHAs

Polyhydroxyalkanoates (PHAs) are polyesters produced and accumulated by several bacteria as a carbon and energy reservoir. These polymers protect organisms against starvation and may enable them to survive under adverse conditions. The PHA accumulation occurs mainly under conditions of excess of carbon and limitation of other nutrients (Anderson and Dawes 1990; Peña et al. 2011, 2014a). These polymers are water-insoluble and they are stored in the cytoplasm as granules (Legat et al. 2010). The monomeric composition of PHAs depends primarily on the microbe and the type of the carbon used for growth. Based on their monomeric chemical structure, three PHA groups can be defined: the short-chain-length PHAs (SCL-PHAs), with monomers from 3 to 5 carbon atoms, the medium-chain-length PHAs (MCL-PHAs) composed of units from 6 to 15 carbon atoms, and, finally, the long-chain-length PHAs (LCL-PHAs) with monomers of more than 15 carbon atoms (Peña et al.2014a; Leong et al. 2014; Fig. 2.1). On the other hand, PHAs could also be classified as homopolymers, such as P3HB, or copolymers that could be found as SCL copolymers, MCL copolymers, and SCL-MCL-PHA copolymers. The thermoelastic properties of the PHAs will be influenced by the type, ratio, and distribution of the monomer units (Leong et al. 2014; Table 2.1); homopolymers of SCL-PHAs such as the poly-3-hydroxybutyrate P3HB are brittle and stiff materials, while copolymers of MCL-PHAs have improved elastomeric properties (Reddy et al. 2003). It must be emphasized that thermo-processibility, biodegradability, and biocompatibility of PHAs make them of great interest for biomedical applications such as the emerging field of tissue engineering (Hazer et al. 2012; Peña et al. 2014a; Leong et al. 2014).

2.3 Bacterial Sources of Polyhydroxyalkanoates

PHAs are produced by several bacterial and archaea species (Olivera et al. 2001; Chanprateep 2010; Peña et al. 2014a). It is noteworthy that species able to produce and accumulate these biopolymers could be found in diverse environments, from marine sediments with genera such as *Vibrio*, *Beneckea*, and *Paracoccus* (López-Cortés et al. 2010) to soil environments, where species such as *Azotobacter vinelandii*, *Bacillus* spp., *Cupriavidus necator*, and *Pseudomonas* spp. are natural producers of PHAs, or even species which can be found on extreme hypersaline environments, in which the *haloarchaeal* genera *Haloferax*, *Halococcus*, *Halobacterium*, *Halorubrum*, and *Haloarcula* are an interesting group for PHA production (Legat et al. 2010; Poli et al. 2011; Kumar et al. 2013). Until now only two species have been successfully used for PHA production at a commercial scale: *C. necator* and *Azohydromonas lata* (Chen 2009; Ienczak et al. 2013); however, some of the genera and species listed above could bring advantages for the tailor-made production of these biopolymers, such as *Haloferax mediterranei*, which produces the poly(3-hydroxybutyrate-co-3-hydroxyvalerate) [P(3HB-co-3HV)] copolymer in high-cell-density cultures, reaching PHA concentrations of up to 77.8 g L^{-1} (Huang et al. 2006). Furthermore, production of PHAs using this archaea has additional advantages. For example, some archaea have the ability to grow on hypersaline conditions, using inexpensive carbon sources, and the feasibility to lyse cells using distilled water, which could be of great impact for PHA recovery costs (Hezayen et al. 2000; Huang et al. 2006; Poli et al. 2011). In contrast, *Pseudomonas* species are able to produce a wide range of MCL-PHAs from cheap and renewable substrates, such as plant oils. In addition, various *Bacillus* spp. are able to accumulate homopolymer P3HB and copolymer P(3HB-co-3HV) using different carbon sources as glycerol, carbohydrates, and biowaste (pea shells) and also produce MCL-PHAs, when various carbon sources are co-fed or when this bacterium was employed as host for overexpression of the biosynthetic operon *phaCAB* from *P. aeruginosa* and *C. necator* (Singh et al. 2009; Kumar et al. 2009, 2013, 2015). Another interesting case is *E. coli*, a nonnatural PHA producer; however, recombinant strains of this bacterium harboring PHA biosynthetic genes from *C. necator, A. lata, A. vinelandii*, or *Pseudomonas oleovorans* are important alternatives for the production of a wide range of PHAs, which *E. coli* can synthesize using a wide range of substrates. *E. coli* can grow in high-cell-density cultures and does not have PHA depolymerases, unlike natural PHA producers (Lee 1996; Olivera et al. 2001; Reddy et al. 2003; Chen 2009; Centeno-Leíja et al. 2014; Leong et al. 2014). In addition to the production of PHA by bacteria and archaea in pure cultures, the use of microbial mixed cultures for the production of these polymers is an attractive alternative. The mixed culture of several microbial species in one single process allows the use of very low-cost complex substrates, or mixtures of substrates, such as those present in waste materials (derived from agro-industry or other waste sources), with no sterilization requirements and with the possibility of a continuous process (Kleerebezem and Loosdrecht

Table 2.1 Mechanical properties and applications of PHAs

PHA	Tensile strength (MPa)	Elongation at break (%)	Young's modulus (GPa)	Crystallinity (%)	Applications	References
P3HB (1400 kDa)	161	45	2.8	71	P3HB with Mw: 1143 for scaffolds for nerve cells	Domínguez-Díaz et al. (2015); Chan et al. (2014)
P3HB (230 kDa)	43	5	–	46	Construction of cast film for the growth of human embryonic cells (HEK293) without any cytotoxic effect	Domínguez-Díaz et al. (2015)
P(3HB-4HB) (62–38 mol%)	–	48	0.65	–	Biomaterials for human dermal fibroblasts and orthopedic support. Furthermore, P(3HB-4HB) is a potential temporary substrate that can be used in transplantation to replace damaged bone or skin	Chanprateep et al. (2010)
P(3HB/3HB-co-3HHx) (100/70–30 mol%)	2.0	10	0.19	53.7	Construction of matrices as cell growth supporting materials for applications in skin engineering and in nerve regeneration	Li et al. (2008)
P(3HB/3HB-co-3HHx) (100/60–40 mol%)	3.5	17	0.32	50.8	–	–
P(3HO-co-3HHx) (88–12 mol %)	9.0	380	0.008	–	Scaffolds for tissue engineering of cardiac valves; suitable for engineering both soft and hard tissues	Jana et al. (2014)
P(3HB-co-3HV) (80–20 mol %)	32	50	1.2	–		

2007). In this type of cultures, it is very difficult to determine the species composition; however, some studies have started to identify some of the PHA-producing species present. Particularly interesting are those works reporting *Alcaligenes, Azoarcus, Amaricoccus, Comamonas, Achromobacter, Pseudomonas, Kluyvera, Acine-tobacter, Paracoccus, Xanthobacter, Curto-bacterium, Flavobacterium*, and *Thauera* (Dionisi et al. 2005, 2006, 2007; Serafim et al. 2006; Lemos et al. 2008) as the dominant genera present in mixed cultures promoting high PHA accumulation.

2.4 PHA Biosynthesis Pathways

PHAs biosynthesis and its regulation has been well documented (Anderson and Dawes 1990; Slater et al. 1992; Steinbuchel and Schlegel, 1991; Peña et al. 2011, 2014a). The synthesis of P3HB, the simplest SCL-PHA, involves three enzymatic reactions; the first reaction involved the condensation of two molecules of acetyl-CoA, mainly from the tricarboxylic acid (TCA) cycle, into acetoacetyl-CoA by the β-ketothiolase (encoded by *phaA*). Then, acetoacetyl-CoA gets reduced to 3-hydroxybutyryl-CoA (3HB-CoA) with the help of the enzyme acetoacetyl-CoA reductase (encoded by *phaB*). Finally, the PHA synthase (encoded by *phaC*) polymerizes the 3-hydroxybutyryl-CoA monomers to P3HB, with the subsequent liberation of CoA (Stubbe et al. 2005; Peña et al. 2011, 2014a). However, biosynthesis of PHAs with different monomeric compositions involved biosynthetic pathways of hydroxyacyl-CoA thioester precursors (Fig. 2.2). In the case of the biosynthesis of SCL-copolymers such as P(3HB-co-3HV), two pathways are involved, leading to C4 monomer (3-hydroxybutyryl-CoA) or to C5 monomer

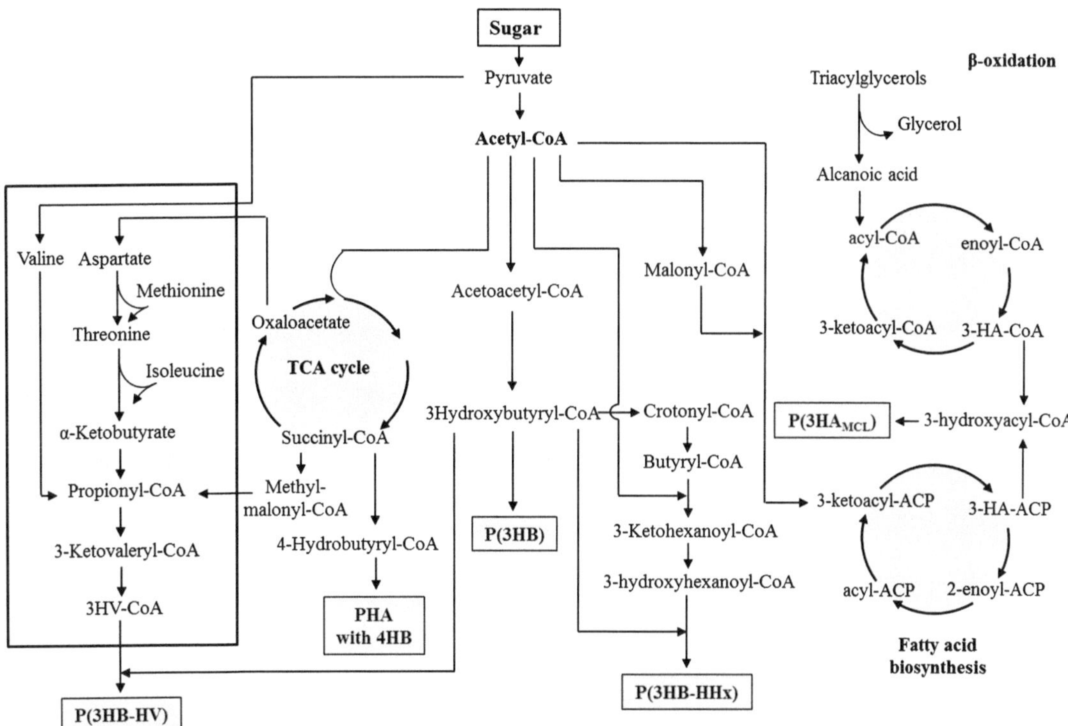

Fig. 2.2 Pathways involved in the biosynthesis of polyhydroxyalkanoates. Amino acid metabolic pathways, the tricarboxylic acids cycle, butyrate metabolism, fatty acid biosynthesis, and β-oxidation pathways (from left to right) are shown. Abbreviations: *ACP* acyl-carrier protein, *3HB* 3-hydroxybutyric acid, *3HA* 3-hydroxyalkanoic acid, *HV* hydroxyvaleric acid, *4HB* 4-hydroxybutyric acid, *HHx* hydroxyhexanoic acid, *MCL* medium chain length

(3-hydroxyvaleryl-CoA) (Steinbuchel and Schlegel 1991). As it has been previously described, the synthesis of the monomer of 3-hydroxybutyryl-CoA involves the condensation of two molecules of acetyl-CoA and its further reduction to 3-hydroxybutyryl-CoA which will be available for its incorporation into the copolymer by the PHA synthase (*phaC*). On the other hand, formation of the 3-hydroxyvaleryl-CoA involves the condensation of acetyl-CoA and propionyl-CoA into 3-ketovaleryl-CoA, a reaction catalyzed by the β-ketothiolase (*phaA*). Afterward 3-ketovaleryl-CoA is reduced by the acetoacetyl-CoA reductase (*phaB*) into the monomer 3-hydroxyvaleryl-CoA which could be incorporated into the growing polymer chain by the PHA synthase or polymerase (*phaC*). In this case, the propionyl-CoA, precursor of the 3-ketovaleryl-CoA, could be the result of the amino acid metabolism, from threonine, which could be converted in α-ketobutyrate and then reduced to propionyl-CoA with the help of the enzyme pyruvate dehydrogenase (Slater et al. 1998; Fig. 2.2), or it could be synthesized through β-oxidation during the growth of bacteria on fatty acids, amino acids, and other substrates that can first be converted into fatty acids (Steinbuchel and Schlegel 1991).

For the biosynthesis of the MCL-PHAs, which are composed of C_6 to C_{15}, two pathways are involved: one of these is the biosynthesis and degradation of fatty acids (*β*-oxidation pathway), wherein a wide variety of substrates are available for the polymer production (Lageveen et al. 1988; Timm and Steinbüchel 1990). From the fatty acid metabolism, precursors such as enoyl-CoA, hydroxyacyl-CoA, and ketoacyl-CoA could be used as substrates for the PHA polymerase for their further conversion into MCL-PHAs (Kraak et al. 1997; Lageveen et al. 1988; Fig. 2.2). Fatty acid biosynthesis is built by adding two carbons through intermediates linked to acyl-carrier protein (ACP), whereas in the *β*-oxidation pathway, two carbons are reduced from the fatty acyl substrates, the whole process liberates a molecule of acetyl-CoA in each cycle and their intermediates are linked to CoA (Fig. 2.2). Although, both fatty acid metabolic pathways are present in all organisms, carbon sources can vary and affect MCL-PHA production. The capability of incorporating different hydroxyacyl-CoA units will be dependent on the PHA synthase (*phaC*). There are two types of these enzymes: the Type I which is harbored by organisms such as *C. necator* and synthesizes SCL-PHAs and the Type II which is present mainly in *Pseudomonas* and is able to polymerize MCL-PHAs.

2.5 Effect of the Culture Conditions on the PHAs Synthesized by Native and Recombinant Bacteria

It is known that the culture conditions affect the chemical characteristics of PHAs synthesized by microorganisms, and those chemical properties have an important effect on the mechanical properties and therefore the final applications of PHAs. In this section, some cases regarding the manipulation of the chemical composition of PHAs by the manipulation of strains and the culture conditions will be discussed (Tables 2.2 and 2.3).

Case 1: Production of the Heteropolymers P(3HB-co-3HV) and P(3HB-co-3HHx)

The copolymers poly(3-hydroxybutyrate-co-3-hydroxyvalerate) [P(3HB-co-3HV)] and poly(3-hydroxybutyrate-co-3-hydroxyhexanoate) [P(3HB-co-3HHx)] are conformed by monomers of 3-hydroxybutyrate (3HB) and 3-hydroxyvalerate (3HV) or 3-hydroxyhexanoate (3HHx), respectively. Both copolymers are interesting candidates as alternative materials for replacement of petrochemical plastics. In the case of P(3HB-co-3HV), this was manufactured and commercialized by ICI (Biopol), Zeneca BioProducts, Biomer Inc. (Biomer), and Tianan Biologic (Enmat) (Braunegg et al. 1998; Chanprateep 2010).

Due to the presence of 3HV or 3HHx residues, polymer crystallinity is reduced, and these residues contribute to an increase of flexibility, elasticity, and melting temperature as compared with the homopolymer P3HB (Feng et al. 2002; Zhuang et al. 2014). In this line, when the molar ratio of 3HV is of only 20 mol %, the copolymer

Table 2.2 Production of polyhydroxyalkanoates under different culture conditions

PHA	Organism	Production scale	Culture conditions	Biomass (g L^{-1})	PHA content (%)	References
P3HP	Recombinant *E. coli*	Fed-batch fermentation 2 L	Crude glycerol Pure glycerol For both: 300 mM, 37°C and 400 rpm for 92 h	5.2 12.0	5.2 11.8	Andreeβen et al. (2010)
P3HB, P3HHx, P3HO, P3HD	Recombinant *E. coli*	Shake flasks	Decanoate Glucose and decanoate Cultures kept at 37 °C and 200 rpm	2.5–3.0 4.3–5.1	1.8–9.3 5.9–36.4	Li et al. (2011)
P4HB	Recombinant *E. coli*	Bioreactor 1 L	Glycerol (20 g L^{-1}) with K-4HB (4 g L^{-1}), 32 °C, pH 7, 800 rpm, 30 % DO_2, 1 L air/min, for 48 h Glucose (20 g L^{-1}) with K-4HB (4 g L^{-1}), propionic acid, and NZ-amines (1 g L^{-1}), 32 °C, pH 7, 800 rpm, 30 % DO_2, 1 L air/min. for 48 h	N.S. 6.5–6.7	2.0–61.0 63.0–65.0	Kämpf et al. (2014)
P(3HB-co-3HHx) (90–10 mol %)	*E. coli* LS5218 (pBBJPC)	Shake flasks	M9 medium supplied with glucose (20 g L^{-1}), 37 °C, 200 rpm, for 48 h	7.8	14.1	Wang et al. (2015)
P3HP	*E. coli* Q1911 (harboring both pHP302 and pHP513)	Baffled shake flasks 500 mL	100 mL of minimal medium with glycerol (20 g L^{-1}) and glucose (3 g L^{-1})	4.9	10.2	Wang et al. (2014)
P(3HB-3HHx-3HO-3HD-3HDD-3HTD) (21.2, 6.1, 45.8, 11.0, 9.2, 6.8 mol %)	*E. coli* LS5218	Shake flasks batch 300 mL	50 mLof medium supplemented with glucose (30 g L^{-1}), 30 °C, 250 rpm, for 72 h	6.5	12.1	Zhuang et al. (2014)
P(3HB-co-3HV) (85–15 mol %)	*E. coli* XL10	Fed-batch (first stage, glucose; second stage, propionate)	Continuous feeding of glucose (20 g L^{-1}) and propionic acid (2 g L^{-1})	39.8	60.5	Liu et al. (2009)
P(3HB-co-3HHx) (60–40 mol %)	*C. necator Re2133/pCB81*	Shake flasks 250 mL	Butyrate (0.5 %)	0.6	65	Jeon et al. (2014)

N.S. not specified

has an excellent strength and flexibility (Luzier 1992). Besides, in some cases, these copolymers have better biocompatibility compared to either P3HB or polylactic acid, which makes them promising materials for medical fields, for example, in cardiovascular problems, wound-healing process, orthopedic issues, drug delivery, and tissue engineering (Yang et al. 2002; Chen and Wu 2005; Table 2.1).

The ability to produce these copolymers is directly attributed to the specificity of polymerase synthase, which has been characterized only at preliminary level in *Bacillus* sp. (Lee et al. 2008). The P(3HB-co-3HV) is synthesized by several

Table 2.3 Culture conditions for the P3HB production with different molecular masses using recombinant *E. coli* strains and *Azotobacter* species

Organism	Mean molecular mass Mw or Mn* (kDa)	Production scale	Culture conditions	Biomass (g L^{-1})	PHB content (%)	References
E. coli XL-1 Blue (pSYL105)	20,000*	Bioreactor 2.6 L	LB medium with glucose 20 g L^{-1}, pH 6, 37 °C	7.4	48	Kusaka et al. (1997)
E. coli JM109 (pGEM-phaC$_{Re}$AB)	1800*	Shake flasks 500 mL with 100 mL medium	LB medium with glucose 20 g L^{-1}, 37 °C, 14 h culture time	3.2	33	Agus et al. (2006)
E. coli JM109 (pGEM-phaC$_{Da}$AB)	4000*			5.8	51	
E. coli JM109 (pGEM-phaC$_{Ac}$AB)	380*			2.7	24	
E. coli JM109 (pGEM-phaRCBspAB)	170–48*	Shake flasks	LB medium with glucose 20 g L^{-1}, 37 °C at 14 and 60 h of culture time	7.9–9.2	54–61	Agus et al. (2010)
E. coli JM109 (pGEM-phaRCBspAB)	1800*	Shake flasks	LB medium with glucose 20 g L^{-1}, 25 °C	5	24	Agus et al. (2010)
E. coli DH5α (pGETS109-pha) with order gene phaABC, phaACB, phaBAC, phaBCA, phaCAB, and *phaCBA*	2000–6200	Shake flasks	LB medium with glucose 20 g L^{-1}, 30 °C, 130 rpm at 72 h	8.0–11.2	31–57	Hiroe et al. (2012)
A. vinelandii UWD	4100	Shake flasks	5 % (w/v) beet molasses at 24 h	N.S.	N.S.	Chen and Page, (1994)
A. vinelandii OPN (ptsIIA^{Ntr}-)	3670	Shake flasks	Low aeration conditions (200 mL PY medium)	N.S.	62	Peña et al. (2014b)
A. chroococcum 7B	2215	Shake flasks	Microaerophilic conditions	2.87	61.3	Myshkina et al. (2008)

N.S. not specified

bacteria such as *C. necator*, some species of *Bacillus*, *Azotobacter*, recombinant strains of *E. coli*, and *Haloferax mediterranei.* This last is a natural P(3HB-co-3HV) producer (Don et al. 2006). The composition of P(3HB-co-3HV) produced by bacterial sources could be manipulated by the kind of carbon sources employed. Several studies have shown that the supply of propionyl-CoA in cells is the key factor for the production of the 3HV fraction during the synthesis of P(3HB-co-3HV) (Aldor et al. 2002). Therefore, most attempts aimed to produce P(3HB-co-3HV) copolymer or increasing the 3HV fraction are based on the strategies to improve propionate utilization. On the other hand, the P(3HB-co-3HHx) is generally produced from plant oil and fatty acids by several wild-type and recombinant bacteria. Previous reports have shown that it is possible to produce P(3HB-co-3HHx), containing greater than 20 % content of 3HHx monomer, using plant oils as carbon source (Kahara et al. 2004; Budde et al. 2011; Riedel et al. 2012). More recently, Jeon et al. (2014) demonstrated that engineered

C. necator can produce P(3HB-co-3HHx), when this strain was grown on mixed acids or on butyrate as carbon source. This strain produced a polymer containing up to 40 wt % of 3HHx monomer. It is important to point that this was the first report for the production of P(3HB-co-3HHx) copolymer in *C. necator* using butyrate. In this section the more recent attempts to improve composition and production of the copolymers P(3HB-co-3HV) and P(3HB-co-3HHx) will be discussed.

Choi and Lee (1999) described a strategy for production of copolymer P(3HB-co-3HV) at high concentration using a recombinant strain of *E. coli* with different feeding solutions containing propionic acid and glucose. In that study, a maximal copolymer concentration of 141.9 g L^{-1} with a P(3HB-co-3HV) up to 62.1 wt % and a 3HV component of 15.3 mol % was reached. It has been reported that the copolymer composition can be manipulated by adding propionate in the feed (Fidler and Dennis 1992; Slater et al. 1992, 1998; Yim et al. 1996; Choi and Lee 1999). However, industrial production of propionate is more expensive than glucose (Poirier et al. 1995; Aldor et al. 2002) making difficult the scale-up process for P(3HB-co-3HV) production. In addition, propionate being toxic must be fed at relatively low concentrations (Steinbüchel and Lütke-Eversloh 2003). An alternative strategy has been to design genetically modified strains in which it is possible to induce the expression of a critical gene in the polymer-producing pathway (Aldor and Keasling 2001). Some examples of genetic modifications that increased P(3HB-co-3HV) synthesis have been reported in different bacteria. For example, Yang et al. (2012), by introducing the genes of propionyl-CoA transferase (*pct*), β-ketothiolase *(bktB)*, acetoacetyl-CoA reductase *(phaB)* and PHA synthase *(phaC)* from *C. necator* into *E. coli* strain YH090, were able to produce P(3HB-co-3HV) with an ultra-high 3HV monomer composition reaching over 80 wt %. More recently, Yang et al. (2014) reported the *E. coli* strain (XL-1) harboring *E. coli poxB L253F V380A* gene along *C. necator prpE* (propionyl-CoA synthase) and *phaCAB* genes, which was able to produce propionyl-CoA via citramalate pathway. When this strain was cultured in a defined medium having 20 g L^{-1} of glucose as carbon source, P(3HB-co-3HV) was produced up to polymer content of 61.7 % based on dry weight. Furthermore, the 3HV monomer fraction in P(3HB-co-3HV) increased up to 5.5 mol % by additional deletion of the genes responsible for the metabolism of propionyl-CoA (*prpC* and *scpC*). Another interesting case is that of *Salmonella enterica* serovar *Typhimurium* that by expression of the *E. coli* (2R)-methylmalonyl-CoA mutase (YliK) and (2R)-methylmalonyl-CoA decarboxylase (YgfG) was able to biosynthesize P(3HB-co-3HV) from a single carbon source through the generation of propionyl-CoA from succinyl-CoA (Aldor et al. 2002).

It is important to point out that the production cost of these polymers can be significantly reduced by using activated sludge instead of pure substrates. This step enables easy operation, since this does not require sterile conditions and uses renewable substrates as carbon sources (Bosco and Chiampo 2010). Others have focused on the use of dairy waste (Pandian et al. 2010), sewage water (Hu et al. 1997; Wong et al. 2000), waste from food processing industry (Wong et al. 2000), as well as agricultural feed stocks (Solaiman et al. 2006). More recently, Narayanan et al. (2014) reported that the culture of *B. mycoides* DFC1 in rice husk hydrolyzate in combination with gluten hydrolyzate resulted in maximum synthesis of P(3HB-co-3HV) in the presence of valeric acid as co-substrate at different induction intervals and concentrations.

On the other hand, focus has been on the production of P(3HB-co-3HHx), which take advantage of the microbial fatty acid degradation pathways (Khanna and Srivastava 2005; Budde et al. 2011). A recent report of P(3HB-co-3HHx) production by *C. necator* has showed that the engineered *C. necator* accumulated P(3HB-co-3HHx) from fructose via the inverted β-oxidation pathway (Insomphun et al. 2015). Since the metabolic flux from acetyl-CoA to 3HB-CoA was too high in the natural PHA producer *C. necator*, a cellular content of 48 wt % P(3HB-co-3HHx) composed of 22 mol % 3HHx was obtained. In this context, Wang et al. (2015) with the purpose to produce P(3HB-co-3HHx) from glucose as carbon source designed an *E. coli* recombinant strain, where they combined the BktB-dependent condensation pathway with the inverted β-oxidation cycle pathway, by cloning five exogenous genes (*bktB, phaB1, phaJ, ter*, and *phaC*). The resultant recombinant strain was able to produce a copolymer with a 3HHx fraction of 10 mol %. On the other hand, the biosynthesis of P(3HB-co-3HHx) from sugars involves an artificial pathway that allowed to build up the C6-monomer from three acetyl-CoA molecules, which is a challenge from metabolic engineering point of view. Based on this strategy, the recombinant *E. coli* strain PHB4 designed by Fukui et al. (2002) was able to accumulate P(3HB-co-3HHx) up to 48 wt % of dry weight from fructose, although the (3HHx) monomer composition in the copolymer was lower than 1.5 mol %.

Case 2: Production of Medium-Chain-Length PHAs (MCL-PHAs)

The MCL-PHAs may be used in diverse applications due to their better physical and mechanical properties as compared with the SCL-PHAs (Table 2.1). The MCL-PHAs are characterized due to their low degree of crystallinity, low melting point, and glass-transition temperatures combined with their improved flexibility, elasticity, and sticky properties that are required for applications in certain biomedical areas (Abe et al. 2012; Chen et al. 2014; Table 2.1). *Pseudomonas* spp. are able to produce MCL-PHAs, and their composition is directly related to the carbon source used as growth substrate. This is because the former monomers, as it was previously described (Fig. 2.2), are derived from intermediates of fatty acid biosynthesis or β-oxidation pathways; therefore, the nature of the PHA monomers produced by *Pseudomonas* species will depend on the metabolism of the specific carbon sources. When the carbon sources are carbohydrates, *P. aeruginosa* accumulates C10 (3-hydroxydecanoate; 3HD) from the biosynthetic fatty acid pathway as the predominant monomer. However, when the carbon source used are fatty acids, the precursors for the PHA synthesis are produced by β-oxidation pathway, and the predominant monomers are C8 (3-hydroxyoctanoate; 3HO), C10, and C12 (3-hydroxydodecanoate; 3HDD) (Madison and Huisman 1999; Nitschke et al. 2011). In addition, *Pseudomonas* spp. produce MCL-PHAs due to their PHA synthases (type II), which are able to polymerize hydroxy acids of short and medium chain length ($3HA_{SCL}$ and $3HA_{MCL}$), covalently linked within the same polyester molecules (Steinbüchel and Lütke-Eversloh 2003; Chen et al. 2014). MCL-PHAs such as poly-3-hydroxydodecanoate (P3HDD) and poly-3-hydroxyoctanoate (P3HO) are of commercial interest, because these polymers exhibit a considerable interval of thermomechanical properties and elastomeric behavior.

Simon-Colin et al. (2008) showed that *P. guezennei* was able to produce MCL-PHA copolymers with a great diversity in their structures and properties. The carbon sources include saturated and unsaturated monomers, from C4 to C14, but preferably C8 and C10 monomers. Furthermore, this strain was able to use a broad range of carbon sources as carbohydrates or fatty acids. For example, when the strain was grown in glucose, cells accumulated 3-hydroxybutyrate (3HB: 1.3 mol %), 3-hydroxyhexanoate (3HHx: 0.9 mol %), 3-hydroxyoctanoate (3HO: 22 mol %), 3-hydroxydecanoate (3HD: 62.8 mol %), 3-hydroxydodecanoate (3HDD: 6.2 mol %), 3-hydroxydodecenoate (3HDDE: 5.6 mol %), and 3-hydroxytetradecanoate (3HTD: 1.2 mol %). In contrast, when the bacterium was cultivated using oleic acid as carbon source, the MCL-PHAs included 3HTD (13.8 mol %) with less fraction mol of monomer 3HD (35.6 mol %). In another study, Simon-Colin et al. (2012) using sodium octanoate as sole carbon source observed that *P. guezennei* synthesized MCL-PHAs mainly composed of 3HO accounting for up to 94 mol % and lower amounts of 3HHx and 3HD. Recently, in cultures of the *P. fulva* strain TY16 grown on petrochemical wastes as carbon source, the production of MCL-PHAs was reported (Ni et al. 2010). Interestingly, when this strain was grown in glucose, toluene, benzene, ethylbenzene, gluconic acid, and acetic acid, it was able to synthesize MCL-PHA copolyesters, containing saturated and unsaturated units of 3HDD, 3HHx, 3HO, and 3HD. On the other hand, copolyesters – MCL-PHAs synthesized by *P. fulva* strain TY16 from octanoic and decanoic acids – were composed of repeating units of 3HHx, 3HO, and 3HD with a mean molecular mass (MMM) between 42 and 43 kDa (Ni et al. 2010).

Another interesting case was reported using the strain of *P. putida* KT2440, grown with nonanoic acid and glucose as carbon sources at a 1:1–1.5 (w/w) ratio for the PHA production. Under such conditions, this strain accumulated a biopolyester with the following composition: 3-hydroxynonanoate (66 mol %), 3-hydroxyheptanoate (32 mol %), and 3HV (1 mol %). The terpolymer produced exhibited a MMM of 11 kDa, with a polydispersion index of 1.8 (Sun et al. 2009). Chan et al. (2014) evaluated the PHA production employing the strain of *P. mosselii* TO7 through utilization of plant oils such as soybean and palm kernel oil as carbon sources. These authors demonstrated that this strain accumulated up to 50 % (cell dry weight) of poly-3-hydroxyoctanoate (P3HO) achieving a productivity of 2.05 g_{PHA} L^{-1} h^{-1}.

Hori et al. (2011) evaluated the effect of temperature (within a range from 15 to 30 °C), in the biosynthesis of the PHAs produced by *P. aeruginosa* IFO3924. The results indicated that the MCL-PHA composition was closely dependent on the temperature and the culture time in which the biopolymer accumulation is carried out. At the beginning of the culture, 3HD and 3-hydroxydodecenoate (C12:1) units were found in the PHA samples at all temperatures evaluated (15, 20, and 25 °C). In contrast, the 3HO was detected only at 30 °C. On the other hand, when the maximum cellular content of PHA was achieved, 3HO and 3HD were the major monomer units present at all the temperatures tested. Haba et al. (2007) studied the effect of temperature (in the range of 18–42 °C) on PHA composition in cultures of *P. aeruginosa* 47 T2. In this study, *P. aeruginosa* was grown in mineral medium supplemented with urea as nitrogen source and 2 % of waste cooking oil. The results obtained indicated that the most abundant monomer was 3HD, except for PHAs produced at 42 °C in which 3HO was the monomer present in greater proportion (43.2 %). At elevated temperatures, long chain monomers such as 3HD, 3HDD, and C14:1 decreased, whereas at 37 °C the content of unsaturated monomers (C12:1, C14:2, C14:1) increased. Later on a significant decrease was observed at 42 °C (Haba et al. 2007). Another interesting example is the production of MCL-PHAs by cultures of *P. mediterranea* using reagent-grade or partially refined glycerol (Pappalardo et al. 2014). The gas chromatography analysis indicated that the biopolymer structure was composed by six monomers: 3HHx, 3HO, 3HD, 3HDD, *cis* 3-hydroxydodec-5-enoate (C12:1Δ^5), and *cis* 3-hydroxydodec-6-enoate (C12:1Δ^6).

Besides *Pseudomonas*, recombinant strains of *E. coli* are also an alternative for production of MCL-PHAs. For example, Li et al. (2011) reported an *E. coli* recombinant strain harboring *phaA* and *phaB* genes from *C. necator* and the *phaC2* gene from *P. stutzeri* (pCJY02). When this strain was cultivated in medium with decanoate, the bacterium accumulated SCL-MCL-PHAs with a monomer composition of 3HB, 3HHx, 3HO, and 3HD in mol ratios of 43.2:12.8:10.3:33.6. However, when this strain was grown on decanoate and glucose, the recombinant strain synthesized the same biopolymer, but the mol ratios were 3HB (83.4), 3HHx (4.0), 3HO (5.6), and 3HD (7.0). These results indicated that it is possible to modulate the monomer content and type of the PHA accumulated by adding different carbon sources and manipulating metabolic pathways of the host. In the same line, Zhuang et al. (2014) designed in its *E. coli* host metabolic pathways to synthesize MCL-PHAs directly from glucose. Engineering the reversed fatty acid β-oxidation cycle, Zhuang et al. (2014) employed this route to generate the key intermediates for the production of MCL-PHAs in *E. coli*. By using a PHA synthase with broad substrate specificity and using glucose as carbon source, recombinant *E. coli* was able to produce MCL-PHA copolymers with monomer composition ranging from 4 to 14 carbons. The PHA compositions in mol %

were 3HB:21.2, 3HHx:6.1, 3HO:45.8, 3HD:11.0, 3HDD:9.2, and 3HTD:6.8 (Table 2.2).

Case 3: Production of Poly-3-hydroxybutyrate of High Molecular Mass (HMM-P3HB)

The third case of microbial PHAs is the poly-3-hydroxybutyrate (P3HB), the common homopolymer of SCL; it is composed by monomers of 3-hydroxybutyrate (3HB), which are linked by ester bonds between the hydroxyl group and the carbonyl groups of the two adjacent monomers (Fig. 2.1). This polymer has similar thermomechanical properties to those found in conventional petrochemical plastics (Chanprateep et al. 2010), and these properties of P3HB are highly dependent on the mean molecular mass (MMM) of the polymer (Peña et al. 2014a; Domínguez-Díaz et al. 2015). Previous reviews have pointed out two interesting bacterial sources for the production of P3HB of high molecular mass (HMM-P3HB): one of them belongs to the genus *Azotobacter*, which is able to accumulate P3HB that exhibits a high molecular mass (>1000 kDa), and the other are recombinant strains of *E. coli* (Peña et al. 2014a; Leong et al. 2014). Some of the most relevant cases are discussed here.

Several authors have studied the culture parameters that could affect the MMM of the P3HB synthesized by *Azotobacter*, finding that the composition of the culture medium, the oxygen availability, and temperature are some of the factors that could have an important effect on the MMM. For example, Chen and Page (1994) observed that the UWD strain of *A. vinelandii* accumulated P3HB with a high MMM (4100 kDa), when it was cultivated using beet molasses of 5 % (w/v) in contrast with cultures without this substrate. These authors suggested that the nitrogen compounds of the beet molasses such as organic acids and salts stimulate the synthesis of P3HB of very high molecular mass (Chen and Page 1994). More recently, Peña et al. (2014b), in shaken flask cultivations of mutant strain OPN, reported a polymer with an MMM of 3670 ± 270 kDa in cultures conducted under low aeration conditions (conventional shaken flasks) as compared with cultures under high aeration. They proposed that by manipulating the aeration conditions of the culture and therefore the oxygen availability, it is possible to modify the MMM of the P3HB. Similar results were reported by Myshkina et al. (2008); these authors evaluated the P3HB production by *A. chroococcum* 7B, under microaerophilic conditions. Under low aeration condition, this strain was able to synthesize P3HB with a MMM of 2215 kDa and an increase on aeration negatively affected the MMM of P3HB. They also found that the optimal temperature for P3HB production with a high molecular mass was 30 ° C; in contrast, at low (20 °C) or high (37 °C) temperatures, the MMM decreased.

On the other hand, an interesting example for production of P3HB with a high molecular mass is that reported for *E. coli* recombinant strains harboring the *C. necator* biosynthesis *phbCAB* genes (Kusaka et al. 1997). These authors reported, for first time, the production of a polymer with ultrahigh molecular mass (20,000 kDa) during the stationary phase of growth by culturing *E. coli* XL-1 Blue (pSYL105), in a bioreactor of 2.6 L, under controlled pH conditions at 6.0 with LB medium supplemented with glucose (20 g L^{-1}). Interestingly, when this *E. coli* strain was grown at pH within a range of 7.0–8.0, the molecular mass of P3HB decreased to values below 5000 kDa after 12 h of culture.

Another interesting case was reported by Agus et al. (2006), who demonstrated that MMM of the P3HB accumulated by recombinant strains of *E. coli* depends on the specific PHA synthase (type and organism of origin) employed. Their results indicated that P3HB with high number-average molecular mass (Mn: 1500-4000 kDa) were synthesized by PHA synthases from *C. necator* (type I), *Delftia acidovorans* (type I), and *Allochromatium vinosum* (type III). P3HB with the lowest Mn (170–790 kDa) were accumulated by PHA synthases from *Aeromonas caviae* (type I), *Pseudomonas* sp. (type III), and *Bacillus* sp. (type IV). On the contrary, these authors found out that the highest MMM were obtained using the PHA synthase from *D. acidovorans* (4000 kDa). They also observed that for the strain harboring the PHA synthase from *D. acidovorans*, an acid pH (4.8) favored the P3HB production with high Mn (2100 kDa) as compared with the biopolymer produced under basic conditions (pH 7.4–7.8), where the Mn was 1500 kDa, and these results were similar to those observed by Kusaka et al. 1997. When they investigated the effect of temperature on the Mn of P3HB using *E. coli* recombinant strains with PHA synthase *D. acidovorans*, they found that at 37 °C the biopolymer exhibited a higher Mn (4300 kDa) than that accumulated in the condition of 30 °C (580 kDa). In contrast, when Agus et al. (2006) used a recombinant *E. coli* strain with PHA synthase from *Bacillus* sp., they found a different behavior to that observed with the strain containing PHA synthase from *D. acidovorans.* They found that the Mn of P3HB produced by the strain with PHA synthase from *Bacillus* sp. in the condition of 37 °C was lower (440 kDa) than the Mn of the P3HB synthesized at 25 °C (nearly 1900 kDa).

On the other hand, through rearrangement of gene order of *phbCAB* operon biosynthesis (*phaABC*, *phaACB*, *phaBAC*, *phaBCA*, *phaCAB*, and *phaCBA*) in recombinant *E. coli* DH5α (pGES109-pha), Hiroe et al. (2012) found that it was possible to produce P3HB with different MMM in the range between 2000 and 6200 kDa. The results indicate that the MMM of P3HB accumulated by the six strains was higher during the exponential growth phase (12 h of cultivation) as compared with the biopolymer produced at the stationary phase (72 h). They also found an inverse correlation between MMM and P3HB synthase activity, in contrast to the accumulation percentage (quantified as dry weight), which increased as the synthase activity increased (Hiroe et al. 2012; Table 2.3).

Case 4: Production of Homopolymers of Short Chain Length: Poly-4-hydroxybutyrate (P4HB) and Poly-3-hydroxypropionate (P3HP)

Another example of SCL-homopolymers is the poly-4-hydroxybutyrate (P4HB; Fig. 2.1), which is one of the most promising PHA for biomedical applications, because of its unique properties, which include biodegradability, biocompatibility, nontoxicity, and superior mechanical properties. It must be emphasized that synthesis of P4HB requires precursor like 4-hydroxybutyric acid, 1.4-butanediol, or γ-butyrolactone (Valappil et al. 2007). Recently, Kämpf et al. (2014) investigated the production of P4HB using recombinant *E. coli* JM109 that harbors a 4-hydroxybutyric acid CoA transferase gene (*orfZ*) from *Clostridium kluyveri*, using glycerol and propionic acid. They found that biopolymer accumulation in the cells was of 80 % (dry weight) achieving 3.7 gL^{-1} (Table 2.2). On the other hand, Le Meur et al. (2013) reported that recombinant *E. coli* JM109 was able to produce P4HB using xylose as carbon source and sodium-4-hydroxybutyrate (Na-4HB) as biopolymer precursor. The highest P4HB concentration achieved was 4.33 g L^{-1} with a yield ($Y_{P4HB/Na\text{-}P4HB}$) of 92 % g g^{-1}. Also, Le Meur et al. (2014) using fed-batch high-density bacterial mass using glycerol as the sole carbon source along with precursor 4HB for biopolymer synthesis achieved a concentration of 15 g L^{-1} of P4HB.

The last example is the poly-3-hydroxypropionate (P3HP), which combines the properties of P3HB and poly-2-hydroxypropionate (known as polylactic acid). Andreeβen et al. (2010) reported the conversion of glycerol to P3HP in an *E. coli* recombinant strain, harboring genes encoding for glycerol dehydratase (*dhaB1*) of *Clostridium butyricum*, the propionaldehyde dehydrogenase (*pduP*) of *Salmonella enterica*, and the PHA polymerase (*phaC1*) of *C. necator.* After 92 h of incubation at 37 °C with 300 mM of pure glycerol, 1.42 gL^{-1} of P3HP were achieved with a yield of 17.5 mmol $_{P3HP}$ mol $_{glycerol}^{-1}$ consumed, with the drawback production of ethanol (8.04 gL^{-1}), succinate (48.92 gL^{-1}), and acetate (0.26 gL^{-1}) as by-products. Another case was reported by Wang et al. (2014). They built a recombinant strain of *E. coli* (with *panM, panD, pp0596, ydfG, prpE*, and *phaC1*) for the P3HP production. This strain was able to produce 0.5 g L^{-1} of biopolymer when it was cultivated in shaken flasks, using glycerol and glucose as carbon sources and without any addition of precursors. In cultures of the same strain in stirred bioreactors in fed-batch aerobic cultures, they obtained up to 10.1 g L^{-1} of P3HP (Wang et al. 2013). In the same line, Gao et al. (2014) designed a recombinant stable *E. coli* strain harboring seven exogenous genes of P3HP synthesis pathway. This strain in aerobic fed-batch cultures was able to produce 25.7 g L^{-1} of biopolymer from glycerol.

2.6 Perspectives

PHAs are biomaterials of great importance not only due to their biodegradability and thermomechanical capabilities similar to those of the plastics derived from the petrochemical industry but also due to their biocompatibility, which is a characteristic required in medical and biomedical fields. In addition, the success of application of these biopolymers will depend on their chemical nature, mainly the monomer composition and mean molecular mass, and other properties which influence the mechanical properties, biodegradability, and biocompatibility of PHAs. Current advances in fermentation, purification technology, as well as the design of mutant strains by recombinant DNA technology would allow the tailor-made production of new PHAs. These tailor-made PHAs can be used as materials for biomedical uses, such as tissue engineering. From the economic viewpoint, the efforts are now focusing on the design of new strains, which can use complex substrates of very low cost, such as those present in waste materials, and having the versatility to produce PHAs with a wide chemical variety and molecular mass.

Acknowledgements The authors gratefully thank the financial support of DGAPA-UNAM (grant **IT100513**) and Conacyt (grants **131851** and **238535**).

References

Abe H, Ishii N, Sato S, Tsuge T (2012) Thermal properties and crystallization behaviors of medium-chain-length poly(3-hydroxyalkanoate)s. Polymer 53:3026–3034. doi:10.1016/j.polymer.2012.04.043

Agus J, Kahar P, Abe H, Doi Y, Tsuge T (2006) Molecular weight characterization of poly[(R)-3-hydroxybutyrate] synthesized by genetically engineered strains of *Escherichia coli.* Polym Degrad Stab 91:1138–1146. doi:10.1016/j.polymdegradstab.2005.07.006

Agus J, Kahar P, Hyakutake M, Tomizawa S, Abe H, Tsuge T, Satoh Y, Tajima K (2010) Unusual change in molecular weight of polyhydroxyalkanoate (PHA) during cultivation of PHA-accumulating *Escherichia coli*. Polym Degrad Stab 95:2250–2254. doi:10.1016/j.polymdegradstab.2010.09.009

Aldor I, Keasling JD (2001) Metabolic engineering of poly(3-hydroxybutyrate-*co*-3-hydroxyvalerate) in recombinant *Salmonella enterica* serovar Typhimurium. Biotechnol Bioeng 76:108–114. doi:10.1002/bit.1150

Aldor I, Kim SW, Prather KL, Keasling JD (2002) Metabolic engineering of a novel propionate-independent pathway for the production of poly(3-hydroxybutyrate-co- 3-hydroxyvalerate) in recombinant *Salmonella enterica* serovar typhimurium. Appl Environ Microbiol 68:3848–3854. doi:10.1128/AEM.68.8

Anderson AJ, Dawes EA (1990) Occurrence, metabolism, metabolic role, and industrial uses of bacterial polyhydroxyalkanoates. Microbial Rev 54:450–472, doi:0146-0749/90/040450-23

Andreeβen B, Lange AB, Robenek H, Steinbüchel A (2010) Conversion of glycerol to poly(3-hydroxypropionate) in recombinant *Escherichia coli.* Appl Environ Microbiol 76:622–626. doi:10.1128/AEM.02097-09

Bosco F, Chiampo F (2010) Production of polyhydroxylcanoates (PHAs) using milk whey and dairy wastewater activated sludge production of bioplastics using dairy residues. J Biosci Bioeng 109:418–421. doi:10.1016/j.jbiosc.2009.10.012

Braunegg G, Lefebvre G, Genser KF (1998) Polyhydroxyalkanoates, biopolyesters from renewable resources: physiological and engineering aspects. J Biotechnol 65:127–161. doi:10.1016/S0168-1656(98)00126-6

Budde CF, Riedel SL, Hubner F, Risch S, Popovic MK, Rha C, Sinskey AJ (2011) Growth and polyhydroxybutyrate production by *Ralstonia eutropha* in emulsified plant oil medium. Appl Microbiol Biotechnol 89(5):1611–1619. doi:10.1007/s00253-011-3102-0

Centeno-Leíja S, Huerta-Beristain G, Giles-Gomez M, Bolivar F, Gosset G, Martinez A (2014) Improving poly-3-hydroxybutyrate production in *Escherichia coli* by combining the increase in the NADPH pool and acetyl-CoA availability. Antonie Van Leeuwenhoek 105:687–696. doi:10.1007/s10482-014-0124-5

Chan RTH, Rusell RA, Marcal H, Lee TH, Holden PJ, Foster JR (2014) BioPEGylation of polyhydroxybutyrate promotes nerve cell health and migration. Biomacromolecules 15:339–349. doi:10.1021/bm401572a

Chanprateep S (2010) Current trends in biodegradable polyhydroxyalkanoates. J Biosci Bioeng 110(6):621–32. doi:10.1016/j.jbiosc.2010.07.014

Chanprateep S, Buasri K, Muangwong A, Utiswannakul P (2010) Biosynthesis and biocompatibility of biodegradable poly (3-hydroxybutyrate-co-4-hydroxybutyrate). Polym Degrad Stabil 95:2003–2012. doi:10.1016/j.polymdegradstab.2010.07.014

Chen G (2009) A microbial polyhydroxyalkanoates (PHA) based bio-and materials industry. Chem Soc Rev 38:2434–2446. doi:10.1039/b812677c

Chen G (2010) Plastics completely synthesized by bacteria: polyhydroxyalkanoates. Microbiol Monogr 14:17–37. doi:10.1007/978-3-642-03287_5_2

Chen GQ, Page W (1994) The effect of substrate on the molecular weight of poly-β-hydroxybutyrate produced by *Azotobacter vinelandii* UWD. Biotechnol Lett 16:155–160. doi:10.1007/BF01021663

Chen GQ, Wu Q (2005) Polyhydroxyalkanoates as tissue engineering materials. Biomaterials 26:6565–6578. doi:10.1016/j.biomaterials.2005.04.036

Chen YJ, Huang YC, Lee CY (2014) Production and characterization of medium-chain-length polyhydroxyalkanoates by *Pseudomonas mosselii* TO7. J Biosci Bioeng 118(2):145–152. doi:10.1016/j.jbiosc.2014.01.012

Choi J, Lee SY (1999) High-level production of poly(3-hydroxybutyrate-co-3- hydroxyvalerate) by fed-batch culture of recombinant *Escherichia coli*. Appl Environ Microbiol 65:4363–4368

Dionisi D, Beccari M, Di Gregorio S, Majone M, Petrangeli Papini M, Vallini G (2005) Storage of biodegradable polymers by an enriched microbial community in a sequencing batch reactor operated at high organic load rate. J Chem Tech Biotech 80:1306–1318. doi:10.1002/jctb.1331

Dionisi D, Majone M, Vallini G, Di Gregorio S, Beccari M (2006) Effect of applied organic load rate on biodegradable polymer production by mixed microbial cultures in a sequencing batch reactor. Biotechnol Bioeng 93:76–88. doi:10.1002/bit.20683

Dionisi D, Majone M, Vallini G, Di Gregorio S, Beccari M (2007) Effect of the length of the cycle on biodegradable polymer production and microbial community selection in a sequencing batch reactor. Biotechnol Prog 23:1064–1073. doi:10.1021/bp060370c

Domínguez-Díaz M, Meneses-Acosta A, Romo-Uribe A, Peña C, Segura D, Espín G (2015) Thermo-mechanical properties, microstructure and biocompatibility in poly-β-hydroxybutyrates (PHB) produced by OP and OPN strains of *Azotobacter vinelandii*. Eur Polym J 63:101–112. doi:10.1016/j.eurpolymj.2014.12.002

Don TM, Chen PC, Shang WW, Chiu HJ (2006) Studies on spherulitic morphology and crystallization kinetics of poly (3-hydroxybutyrate) blended with a medium molecular Poly(ethyl eneoxide). Tamkang J Sci Eng 9:279–290

Feng L, Watanabe T, Wang Y, Kichise T, Fukuchi T, Chen GQ, Doi Y, Inoue Y (2002) Studies on comonomer compositional distribution of bacterial poly(3-hydroxybutyrate co-3-hydroxyhexanoate)s and thermal characteristics of their factions. Biomacromolecules 3:1071–1077. doi:10.1021/bm0200581

Fidler S, Dennis D (1992) Production of polyhydroxyalkanoates in recombinant *E. coli* strains. FEMS Microbiol Rev 103:231–236

Fukui T, Abe H, Doi Y (2002) Engineering of *Ralstonia eutropha* for the production of poly(3-hydroxybutyrate-co-hydroxyhexanoate) from fructose and solid-state properties of the copolymer. Biomacromolecules 3:618–624. doi:10.1021/bm0255084

Gao Y, Liu C, Ding Y, Sun C, Zhang R, Xian M, Zhao G (2014) Development of genetically stable Escherichia coli strains for poly(3-hydroxypropionate) production. PLoS One 9:1–8. doi:10.1371/journal.pone.0097845

Haba E, Vidal-Mas J, Bassas M, Espuny MJ, Llorens J, Manresa A (2007) Poly 3-(hydroxyalkanoates) produced from oily substrates by *Pseudomonas aeruginosa* 47T2 (NCBIM 40044): effect of nutrients and incubation temperature on polymer composition. Biochem Eng J 35:99–106. doi:10.1016/j.bej.2006.11.021

Hazer DB, Kılıçay E, Hazer B (2012) Poly(3-hydroxyalkanoate)s: diversification and biomedical applications a state of the art review. Mater Sci Eng C 32:637–647. doi:10.1016/j.msec.2012.01.021

Hezayen FF, Rehm BHA, Eberhardt R, Steinbüchel A (2000) Polymer production by two newly isolated extremely halophilic archaea: application of a novel corrosion-resistant bioreactor. Appl Microbiol Biotechnol 54:319–325. doi:10.1007/s002530000394

Hiroe A, Tsuge K, Nomura CT, Itaya M, Tsuge T (2012) Rearrangement of gene order in the *phaCAB* operon leads to effective production of ultra-high molecular-weight poly[(R)-3-hydroxybutyrate] in genetically engineered *Escherichia coli*. Appl Environ Microbiol 78:3177–3184. doi:10.1128/AEM.07715-11

Hori K, Ichinohe R, Unno H, Marsudi S (2011) Simultaneous syntheses of polyhydroxyalkanoates and rhamnolipids by *Pseudomonas aeruginosa* IFO3924 at various temperatures and from various fatty acids. Biochem Eng J 53:196–202. doi:10.1016/j.bej.2010.10.011

Hu WF, Chua H, Yu PHF (1997) Synthesis of poly (3-hydroxybutyrate-co-3-hydroxyvalerate) from activated sludge. Biotech Lett 19:695–698. doi:10.1023/A:1018307402891

Huang TY, Duan KJ, Huang SY, Chen CW (2006) Production of polyhydroxyalkanoates from inexpensive extruded rice bran and starch by *Haloferax mediterranei*. J Ind Microbiol Biotechnol 33:701–706. doi:10.1007/s10295-006-0098-z

Ienczak JL, Schmidell W, Falcão de Aragão GM (2013) High-cell-density culture strategies for polyhydroxyalkanoate production: a review. J Ind Microbiol 40:275–286. doi:10.1007/s10295-013-1236-z

Insomphun C, Xie H, Mifune J, Kawashima Y, Orita I, Nakamura S, Fukui T (2015) Improved artificial pathway for biosynthesis of poly(3-hydroxybutyrate-co-3-hydroxyhexanoate) with high Cmonomer composition from fructose in Ralstonia eutropha. Metab Eng 27C:38–45. doi:10.1016/j.ymben.2014.10.006

Jana S, Tefft BJ, Spoon DB, Simari RD (2014) Scaffolds for tissue engineering of cardiac valves. Acta Biomater 10:2877–2893. doi:10.1016/j.actbio.2014.03.014

Jeon JM, Brigham CJ, Kim YH, Kim HJ, Yi DH, Kim H, Rha C, Sinskey AJ, Yang YH (2014) Biosynthesis of poly(3-hydroxybutyrate-*co*-3-hydroxyhexanoate) (P(HB-*co*-HHx)) from butyrate using engineered *Ralstonia eutropha* Appl. Microbial Biotechnol 98:5461–5469. doi:10.1007/s00253-014-5617-7

Kahara P, Tsuge T, Taguchi K, Doi Y (2004) High yield production of polyhydroxyalkanoates from soybean oil by *Ralstonia eutropha* and its recombinant strain. Polym Degrad Stab 83(1):79–86. doi:10.1016/S0141-3910(03)00227-1

Kämpf MM, Thöny-Meyer L, Ren Q (2014) Biosynthesis of poly(4-hydroxybutyrate) in recombinant *Escherichia coli* grown on glycerol is stimulated by propionic acid. Int J Biol Macromol 71:8–13. doi:10.1016/j.ijbiomac.2014.04.023

Khanna S, Srivastava AK (2005) Recent advances in microbial polyhydroxyalkanoates. Process Biochem 40(2):607–619. doi:10.1016/j.procbio.2004.01.053

Kleerebezem R, Loosdrecht MCV (2007) Mixed culture biotechnology for bioenergy production. Curr Opin Biotechnol 18:207–12. doi:10.1016/j.copbio.2007.05.001

Kraak MN, Smits THM, Kessler B, Witholt B (1997) Polymerase C1 levels and poly(R-3- hydroxyalkanoate) synthesis in wild-type and recombinant *Pseudomonas* strains. J Bacteriol 179:4985–4991

Kumar T, Singh M, Purohit HJ, Kalia VC (2009) Potential of *Bacillus* sp. to produce polyhydroxybutyrate from biowaste. J Appl Microbiol 106:2017–2023. doi:10.1111/j.1365-2672.2009.04160.x

Kumar P, Patel SKS, Lee JK, Kalia VC (2013) Extending the limits of *Bacillus* for novel biotechnological applications. Biotechnol Adv 31:1543–61. doi:10.1016/j.biotechadv.2013.08

Kumar P, Mehariya S, Ray S, Mishra A, Kalia VC (2015) Biodiesel industry waste: a potential source of bioenergy and biopolymers. Indian J Microbiol 55:1–7. doi:10.1007/s12088-014-0509-1

Kusaka S, Abe H, Lee SY (1997) Molecular mass of poly[(R)-3-hydroxybutyric acid] produced in a recombinant *Escherichia coli*. Appl Microbiol Biotechnol 47:140–143. doi:10.1007/s002530050902

Lageveen RG, Huisman GW, Preusting H, Ketalaar P, Eggink G, Witholt B (1988) Formation of polyesters by *Pseudomonas oleovorans*: effect of substrates on formation and composition of poly-(R)-3-hydroxyalkanoates and poly-(R)-3-hydroxyalkenoates. Appl Environ Microbiol 54:2924–2932

Lee SY (1996) Bacterial polyhydroxyalkanoates. Biotechnol Bioeng 49:1–14. doi:10.1002/(SICI)1097-0290

Lee WH, Loo CY, Nomura CT, Sudesh K (2008) Biosynthesis of polyhydroxyalkanoate copolymers from mixtures of plant oils and 3-hydroxyvalerate precursors. Bioresour Technol 99:6844–6851. doi:10.1016/j.biortech.2008.01.051

Legat A, Gruber C, Zangger K, Wanner G, Stan-Lotter H (2010) Identification of polyhydroxyalkanoates in *Halococcus* and other *haloarchaeal* species. Appl Microbiol Biotechnol 87(3):1119–27. doi:10.1007/s00253-010-2611-6

Lemos PC, Levantesi C, Serafim LS, Rossetti S, Reis MAM, Tandoi V (2008) Microbial characterisation of polyhydroxyalkanoates storing populations selected under different operating conditions using a cell-sorting RT-PCR approach. Appl Microbiol Biotechnol 78:351–360. doi:10.1007/s00253-007-1301-5

Le Meur S, Zinn M, Egli T, Thöny-Meyer L, Ren Q (2013) Poly(4-hydroxybutyrate) (P4HB) production in recombinant Escherichia coli: P4HB synthesis is uncoupled with cell growth. Microb Cell Fact 12:123. doi:10.1186/1475-2859-12-123

Le Meur S, Zinn M, Egli T, Thöny-Meyer L, Ren Q (2014) Improved productivity of poly (4-hydroxybutyrate) (P4HB) in recombinant Escherichia coli using glycerol as the growth substrate with fed-batch culture. Microb Cell Fact 13:131. doi:10.1186/s12934-014-0131-2

Leong YK, Show PL, Ooi CW, Ling TC, Lan JCW (2014) Current trends in polyhydroxyalkanoates (PHAs) biosynthesis: insights from the recombinant *Escherichia coli*. J Biotechnol 180:52–65. doi:10.1016/j.jbiotec.2014.03.020

Li XT, Zhang Y, Chen GQ (2008) Nanofibrous polyhydroxyalkanoate matrices as cell growth supporting materials. Biomaterials 29:3720–3728. doi:10.1016/j.biomaterials.2008.06.004

Li Q, Chen Q, Li MJ, Wang FS, Qi QS (2011) Pathway engineering results the altered polyhydroxyalkanoates composition in recombinant *Escherichia coli*. N Biotechnol 28:92–95. doi:10.1016/j.nbt.2010.08.007

Liu XW, Wang HH, Chen JY, Li XT, Chen GQ (2009) Biosynthesis of poly(3-hydroxybutyrate-co-3-hydroxyvalerate) by recombinant *Escherichia coli* harboring propionyl-CoA synthase gene (prpE) or propionate permease gene (*prp*P). Biochem Eng J 43:72–77. doi:10.1016/j.bej.2008.09.001

López-Cortés A, Rodríguez-Fernández O, Latisnere-Barragán H, Mejía-Ruíz HC, González-Gutiérrez G, Lomelí-Ortega C (2010) Characterization of polyhydroxyalkanoate and the phaC gene of *Paracoccus seriniphilus* E71 strain isolated from a polluted marine microbial mat. World J Microbiol Biotechnol 26:109–118. doi:10.1007/s11274-009-0149-5

Luzier WD (1992) Materials derived from biomass/biodegradable materials. Proc Natl Acad Sci U S A 89:839–842. doi:10.1073/pnas.89.3.839

Madison LL, Huisman GW (1999) Metabolic engineering of poly(3-hydroxyalkanoates): from DNA to plastic. Microbiol Mol Biol Rev 63:21–53

Myshkina VL, Nikolaeva DA, Makhina TK, Bonartsev AP, Filatova EV, Ruzhitsky AO, Bonartseva GA (2008) Effect of growth conditions on the molecular weight of Poly-3-hydroxybutyrate produced by *Azotobacter chroococcum* 7B. Appl Biochem Microbiol 44:482–486. doi:10.1134/S0003683808050050

Narayanan A, Sajeev Kumar VA, Ramana KV (2014) Production and Characterization of Poly (3-Hydroxybutyrateco-3-Hydroxyvalerate) from *Bacillus mycoides* DFC1 Using Rice Husk Hydrolyzate. Waste Biomass Valor 5:109–118. doi:10.1007/s12649-013-9213-3

Ni YY, Kim DY, Chung MG, Lee SH, Park HY, Rhee YH (2010) Biosynthesis of medium-chain-length poly (3-hydroxyalkanoates) by volatile aromatic hydrocarbons-degrading *Pseudomonas fulva* TY16. Bioresour Technol 101:8485–8488. doi:10.1016/j.biortech.2010.06.033

Nitschke M, Costa S, Contiero J (2011) Rhamnolipids and PHAs: Recent reports on *Pseudomonas*-derived molecules of increasing industrial interest. Process Biochem 46:621–630. doi:10.1016/j.procbio.2010.12.012

Olivera E, Carnicero D, Jodra R, Miñambres B, García B, Abraham G (2001) Genetically engineered *Pseudomonas*: a factory of new bioplastics with broad applications. Environ Microbiol 3:612–618. doi:10.1046/j.1462-2920.2001.00224.x

Pandian SR, Deepa KV, Kalishwaralal K, Rameshkumar N, Jeyaraj M, Gurunathan S (2010) Optimization and fed-batch production of PHB utilizing dairy waste and sea water as nutrient sources by *Bacillus megaterium* SRKP-3. Bioresour Technol 101:705–711. doi:10.1016/j.biortech.2009.08.040

Pappalardo F, Fragalà M, Mineo PG, Damigella A, Catara AF, Palmeri R, Rescifina A (2014) Production of filmable medium-chain-length polyhydroxyalkanoates produced from glycerol by *Pseudomonas mediterranea*. Int J Biol Macromol 65:89–96. doi:10.1016/j.ijbiomac.2014.01.014

Peña C, Castillo T, Nuñez C, Segura D (2011) Bioprocess design: fermentation strategies for improving the production of alginate and poly-β-hydroxyalkanoates (PHAs) by *Azotobacter vinelandii*. INTECH- Open Access Publisher, London, pp 217–242. doi:10.5772/20393

Peña C, Castillo T, García A, Millán M, Segura D (2014a) Biotechnological strategies to improve production of microbial poly-(3-hydroxybutyrate): a review of recent research work. Microb Biotechnol 7(4):278–293. doi:10.1111/1751-7915.12129

Peña C, López S, García A, Espín G, Romo-Uribe A, Segura D (2014b) Biosynthesis of poly-β-hydroxybutyrate (PHB) with a high molecular mass by a mutant strain of *Azotobacter vinelandii* (OPN). Ann Microbiol 64:39–47. doi:10.1007/s13213-013-0630-0

Poirier Y, Nawrath C, Somerville C (1995) Production of polyhydroxyalkanoates, a family of Biodegradable plastics and elastomers, in bacterial and plant. Biotechnol 13:142–150. doi:10.1038/nbt0295-142

Poli A, Di Donato P, Abbamondi GR, Nicolaus B (2011) Synthesis, production, and biotechnological applications of exopolysaccharides and polyhydroxyalkanoates by archaea. Archaea. doi:10.1155/2011/693253

Reddy CSK, Ghai R, Kalia RVC (2003) Polyhydroxyalkanoates: an overview. Bioresour Technol 87:137–46. doi:10.1016/S0960-8524(02)00212-2

Riedel SL, Bader J, Brigham CJ, Budde CF, Yusof ZA, Rha C, Sinskey AJ (2012) Production of poly(3-hydroxybutyrate-co-3-hydroxyhexanoate) by *Ralstonia eutropha* in high cell density palm oil fermentations. Biotechnol Bioeng 109(1):74–83. doi:10.1002/bit.23283

Serafim LS, Lemos T, Rosetti S, Levantesi C, Tandoi V, Reis MA (2006) Microbial community analysis with a high PHA storage capacity. Wat Sci Technol 54:183–188. doi:10.2166/wst.2006.386

Simon-Colin C, Raguénès G, Costa B, Guezennec J (2008) Biosynthesis of medium chain length poly-3-hydroxyalkanoates by *Pseudomonas guezennei* from various carbon sources. React Funct Polym 68:1534–1541. doi:10.1016/j.reactfunctpolym.2008.08.005

Simon-Colin C, Gouin C, Pierre Lemechko P, Schmitt S, Senant A, Kervarec N, Guezennec J (2012) Biosynthesis and characterization of polyhydroxyalkanoates by *Pseudomonas guezennei* from alkanoates and glucose. Int J Biol Macromol 51:1063–1069. doi:10.1016/j.ijbiomac.2012.08.018

Singh M, Patel SKS, Kalia VC (2009) *Bacillus subtilis* as potential producer for polyhydroxyalkanoates. Microb Cell Fact 8:38. doi:10.1186/1475-2859-8-38

Slater S, Gallaher T, Dennis D (1992) Production of poly-(3- hydroxybutyrate-Co-3-hydroxyvalerate) in a recombinant *Escherichia coli* strain. Appl Environ Microbiol 58:1089–1094

Slater S, Houmiel KL, Tran M, Mitsky TA, Taylor NB, Padgette SR, Gruys KJ (1998) Multiple -ketothiolases mediate poly (–hydroxyalkanoate) copolymer synthesis in *Ralstonia eutropha*. J Bacteriol 180:1979–1987

Solaiman DKY, Ashby RD, Foglia TA, Marmer WN (2006) Conversion of agricultural feedstock and coproducts into poly(hydroxyalkanoates). Appl Microbiol Biotechnol 71:783–789. doi:10.1007/s00253-006-0451-1

Steinbüchel A, Lütke-Eversloh T (2003) Metabolic engineering and pathway construction for biotechnological production of relevant polyhydroxyalkanoates in microorganisms. Biochem Eng J 16:81–96. doi:10.1016/S1369-703X(03)00036-6

Steinbüchel A, Schlegel HG (1991) Physiology and molecular genetics of poly(beta-hydroxyalkanoic acid) synthesis in Alcaligenes eutrophus. Mol Microbiol 5(3):535–542. doi:10.1111/j.1365-2958.1991.tb00725.x

Stubbe J, Tian J, He A, Sinskey AJ, Lawrence AG, Liu P (2005) Nontemplate-dependent polymerization processes: polyhydroxyalkanoate synthases as a paradigm. Annu Rev Biochem 74:433–480. doi: 10.1146/annurev.biochem.74.082803.133013

Sun Z, Ramsay J, Guay M, Ramsay B (2009) Enhanced yield of medium-chain-length polyhydroxyalkanoates from nonanoic acid by co-feeding glucose in carbon-limited, fed-batch culture. J Biotechnol 143:262–267. doi:10.1016/j.jbiotec.2009.07.014

Timm A, Steinbüchel A (1990) Formation of polyesters consisting of medium-chain-length 3 hydroxyalkanoic acids from gluconate by *Pseudomonas aeruginosa* and other fluorescent pseudomonads. Appl Environ Microbiol 56:3360–3367

Valappil SP, Peiris D, Langley GJ, Herniman JM, Boccaccini AR, Bucke C, Roy I (2007) Polyhydroxyalkanoates (PHA) biosynthesis from structurally unrelated carbon sources by a newly characterised *Bacillus spp*. J Biotechnol 127:475–487. doi:10.1016/j.jbiotec.2006.07.015

Wang Q, Yang P, Liu C, Xue Y, Xian M, Zhao G (2013) Biosynthesis of poly(3-hydroxypropionate) from glycerol by recombinant Escherichia coli. Bioresour Technol 131:548–551. doi:10.1016/j.biortech.2013.01.096

Wang Q, Yang P, Xian M, Feng L, Wang J, Zhao G (2014) Metabolic engineering of *Escherichia coli* for poly(3-hydroxypropionate) production from glycerol and glucose. Biotechnol Lett 36:2257–2262. doi:10.1007/s10529-014-1600-8

Wang Q, Luan Y, Cheng X, Zhuang Q, Qi Q (2015) Engineering of *Escherichia coli* for the biosynthesis of poly(3-hydroxybutyrate-co-3-hydroxyhexanoate) from glucose. Appl Microbiol Biotechnol. doi:10.1007/s00253-015-6380-0

Wong AL, Chua H, Lo WH, Yu PHF (2000) Synthesis of bioplastics from food industry wastes with activated sludge biomass. Water Sci Technol 41:55–59

Yang XS, Zhao K, Chen GQ (2002) Effect of surface treatment on the biocompatibility of microbial polyhydroxyalkanoates. Biomaterials 23:1391–1397. doi:10.1016/S0142-9612(01)00260-5

Yang YH, Brigham CJ, Song E, Jeon JM, Rha CK, Sinskey AJ (2012) Biosynthesis of poly(3-hydroxybutyrate-co-3-hydroxyvalerate) containing a predominant amount of 3-hydroxyvalerate by engineered *Escherichia coli* expressing propionate-CoA transferase. J Appl Microbiol 113:815–23. doi:10.1111/j.1365-2672.2012.05391.x.

Yang JE, Choi YJ, Lee SJ, Kang KH, Lee H, Oh YH, Lee SH, Park SJ, Lee SY (2014) Metabolic engineering of *Escherichia coli* for biosynthesis of poly(3-hydroxybutyrate-co-3-hydroxyvalerate) from glucose. Appl Microbiol Biotechnol 98:95–104. doi:10.1007/s00253-013-5285-z

Yim KS, Lee SY, Chang HW (1996) Synthesis of poly-(3-hydroxybutyrate-co-3-hydroxyvalerate) by recombinant *Escherichia coli*. Biotechnol Bioeng 49:495–503. doi:10.1002/(SICI)1097-0290(19960305)

Zhuang Q, Wang Q, Liang Q, Qi Q (2014) Synthesis of polyhydroxyalkanoates from glucose that contain medium-chain-length monomers via the reversed fatty acid β-oxidation cycle in *Escherichia coli*. Metab Eng 24:78–86. doi:10.1016/j.ymben.2014.05.004

Carlos Peña received his M.Sc. and Ph.D. degree in Biotechnology from UNAM, México. Dr. Peña is currently working as Senior Research at Institute of Biotechnology, México. He is the member of the Editorial Board of the Electronic Journal of Biotechnology and fellow of the Society of Chemical Industry (SCI). His interests are the biopolymer production and the bioprocess engineering.

Andrés García studied Biology at the UAEM, Mexico. He later studied a Master's in Biotechnology from the Institute of Biotechnology at UNAM. He is currently a Ph.D. student of Biochemical Sciences at UNAM. His work is focused on understanding the metabolic pathways involved in the synthesis of alginate and PHB in *Azotobacter vinelandii*.

Modesto Millán graduated in Chemical Engineer in 2006 from Universidad Autónoma del Estado de Morelos in México. He is currently enrolled in the program of Doctorate of Biomedical Sciences in the UNAM and works at Institute of Biotechnology. He is studying the processes of synthesis and degradation of poly-3-hydroxybutyrate (P3HB) as these processes affect the molecular weight of the polymer in cultures of *A. vinelandii*.

Tania Castillo received her M.Sc. degree in Marine Biotechnology at CIBNOR, México, and her Ph.D. degree in Biochemical Sciences at the UNAM, México. Dr. Castillo is currently a postdoc researcher in Systems Biology at the CEIB/UAEM. Her main interests are the metabolic engineering for biopolymer production.

Biodegradable Polymers: Renewable Nature, Life Cycle, and Applications

3

Manjusha Dake

Abstract

Biopolymers are superior to synthetic polymer due to their eco-friendly nature. Microbial biopolymers being a good substitute for conventional plastics causing a waste management problem. Polyhydroxyalkanoates (PHAs) produced as microbial polyesters can provide promising prospects for food and allied industries due to their versatile properties assisting viscosifying, gelling, and film-forming ability. Microbial and biocatalytic production of functionalized polyhydroxyalkanoates with novel monomer structure and tailor-made properties can be feasible by manipulating the metabolic network in host microbes by genetic modification enabling them to utilize a diverse range of low-cost substrate as unsaturated fatty acid constituents. But expensive technology associated extraction and isolation of PHAs is a major hindrance for their commercial applications. A collective knowledge about PHAs as microbial biopolymers, their production from cheap and renewable resources, metabolic pathways involved in their production, economics of PHA production, and decisive factors involved could possibly assist their effective utilization as a substitute to synthetic polymers.

3.1 Introduction

Biopolymers are natural polymers originating from plants and microbes or synthesized chemically from biological building blocks. Production of biopolymers from renewable resources is a promising remedy for their sustainable development (Reddy et al. 2003; Porwal et al. 2008; Kumar et al. 2009; Patel et al. 2011; Singh et al. 2015). They can be degraded using natural entity as microorganisms and their integrated enzyme system to simpler molecular assembly recycled in

M. Dake (✉)
Department of Biotechnology, Dr. D.Y. Patil Biotechnology and Bioinformatics Institute, 411033 Pune, India
e-mail: manjusha.dake@dpu.edu.in; man1d_2@rediffmail.com

V.C. Kalia (ed.), *Microbial Factories*, DOI 10.1007/978-81-322-2595-9_3

the environment itself. With a low environmental footprint and eco-friendly nature, biopolymers could also prove as asset to waste processing and contribute to a sustainable society (Kalia et al. 2000). In nature, biopolymers often play structural role and other important roles in maintaining cell viability by conserving genetic information, by storing carbon-based macromolecules for producing either energy or reducing power, and by defending an organism against attack from hazardous environmental factors (Steinbuchel 2001). Biocompatible and biodegradable nature of biopolymers makes them superior than petrochemical-derived polymers. Synthetic polymers as plastics derived from nonrenewable fossil (petrochemical) in spite of their desirable properties as suitability and durability. Strength, lightness, and cost are causing environmental threats due to the short-term convenience of using and throwing these conventional plastics. Fossil fuels are considered nonrenewable where their utilization from earth's reserves without replacing them has led to the exploitation by human society and as a result created discrepancy in the nature carbon cycle by emission of carbon dioxide.

Synthetic plastics primarily derived from nonrenewable fossil (petrochemical) surpassed the market by replacing the glass, wood, and other construction material. But lack of natural enzymes and biological processes for degradation of synthetic plastic has created alarming situation for environment as well as human society. So there is a worldwide demand for biopolymers originating from renewable raw material (Chen and Martin 2012). Persistence of such synthetic plastics in the environment is dangerous for the human community and wildlife. Thus, nonbiodegradable nature makes synthetic plastic waste management problems. This created an urgent need in implementing the usage of biodegradable plastics for an increasing awareness among consumers regarding the use of plastic-based materials. Thus, deleterious effects of synthetic plastic derived from nonrenewable fossil fuel have created environmental awareness by one and all for making a transition insisting demand for use of biodegradable material. Biopolymers available on a sustainable basis have several economic and environmental advantages. Microbial degradation of biopolymers to CO_2 and water serves as a potential remedy to waste processing where conventional plastics are environmentally unfriendly in the public perception.

Synthesis of biopolymers and their degradation by composting governed by microbial system has proved their biodegradable nature and user-friendly and environmental-friendly attribute. Thus, biopolymers as renewable feedstock fit nicely within the principles underlying green chemistry. Plant-derived biopolymers include starch, cellulose, oils, polyesters, or PHAs, while those from animal source are proteins, fats, and hydrocarbons. Biopolymers being of natural origin and derived from renewable resources are constantly restored through natural processes in spite of their constant utility for human life. Biopolymers provide energy for microbial life which in turn causes the synthesis of bio-based polymeric compounds, establishing a symbiotic association. Biopolymers are superior to synthetic polymers due to their biodegradability and both environmental and human compatibilities which cannot be emulated by synthetic polymers. Biopolymers can be classified according to the monomers that constitute them, as polysaccharides, polyamides (proteins and poly γ-glutamic acid (γ-PGA)), nucleic acids (DNA and RNA), polyesters (polyhydroxyalkanoates, PHAs), polyphosphates, and polyisoprenoids (natural rubber). Biopolymers synthesized by bacteria are applied in food, pharma, and agriculture sector. Natural rubber, cellulosics, and nylon-11 are the most used biopolymers, while newer biopolymers include polylactic acid, polyhydroxyalkanoate, and bio-based thermoplastic polyurethane (Jogdand 2014). Biopolymers from plants serve as energy carrier, carbon storage compounds, and protective elements against pathogens. Biopolymers derived from plants encompass diverse group of polysaccharides as starch, cellulose, hemicellulose, pectin, galactomannans (guar, locust bean gum), exudates gum (gum Arabic), and β-glucans, while seaweed polysaccharides include alginates and carrageenans. Plant polysaccharides are applicable in textile, leather, pharmaceutical (drug formulation and drug delivery), cosmetics (moisturizing ingredient; hair shampoos), agriculture (plant growth regulators, soil fertilizers, antifungals, and anti-nematodals), and food (thickeners, gelling agent). Starch-based plant sources and cellulose derivatives contribute to almost 80 % of the bioplastic market (Rajendran et al. 2012). Starch-based films blended with thermoplastic polyesters are used as packaging materials, in hygiene products and agriculture where 50 % starch combination can be used. Polylactic acid, the most widely used biodegradable aliphatic polyester, is derived from cane sugar. But plant-produced biopolymers present certain limitations as limited availability, seasonal price fluctuations, expensive technology, and the absence of rheological properties for specific applications (Sutherland 2005).

Other limitations for plant-produced biopolymers include lower and delayed biomass production and adverse effect on human food chain. Compared to natural polymers originating from plant sources and those synthesized from fossil fuel, microbial biopolymers are more favorable and advantageous on the basis of social and economic forecast. Compared to plant-produced polysaccharides, commercial production of bioactive polysaccharides through the process of microbial fermentation is always preferred. A wide range of polysaccharides synthesized by microorganisms are xanthan, gellan, curdlan, pullulan, chitosan, and bacterial cellulose. Exopolysaccharides (EPPs) from bacterial and fungal source are comprised of monomeric sugars such as glucose, fructose, rhamnose, glucuronic acid, mannuronic acid, and acetyl glucosamine. Microbial polysaccharides are valuable and cost-effective tools as thickeners, bioadhesives, stabilizers, probiotics, and gelling agents in food and cosmetics (Nicolaus et al. 2010; Freitas et al. 2011) and as emulsifier, biosorbent, and bioflocculant in the environmental sector (Sam et al. 2011). In bacterial polysaccharides applicable in food industry, clinical research medicines include levan, alginate, dextran, xanthan, and scleroglucan. Microbial biopolymers also encompass biodegradable polyesters such as polyhydroxyalkanoates (PHAs), polylactides, and aliphatic polyesters that serve as intracellular reservoir for carbon and energy. Such biodegradable polymers derived from microbial source have the potential to help in waste management and an alternative to conventional plastics. Intrinsic properties of biopolymers assure their promising prospects making them suitable as a packaging material in dairy industry and flight catering products, protecting the product from moisture with increase in its shelf life. Biopolymers have specialized applications in fabrics, adsorbents, biosensors, and data storage elements.

Nanoscale biopolymers produced from natural polymers such as starch and chitin are useful in therapeutics, coatings, packaging materials, and bioremediation of toxic heavy metals. Microbial polyhydroxyalkanoate (PHA) reserve polymers are interesting polyesters sustainably produced from renewable material and as a result can be an effective substitute for synthetic plastic material. The present chapter reveals the biopolymers from microbial origin and their environmental and biomedical applications. Other members of biopolymers include proteins (silk, collagen, elastin, poly amino acids, soy, wheat gluten, zein, and casein), lipids, polyphenols, and natural rubber.

3.2 Microbial Biopolymers

Microorganisms as bacteria, archaea, fungi, and algae produce a diverse group of intracellular and extracellular biological macromolecules such as polysaccharides, polyesters, and polyamides termed as microbial biopolymers. Biopolyesters (polyhydroxyalkanoates) are intracellular and spherical organic inclusion synthesized by microorganisms (Singh et al. 2015). Microbial polysaccharides may be capsular polysaccharides associated with the cell surface, or they may be exopolysaccharides loosely connected with cell surface (Cuthbertson et al. 2009). Microbial polysaccharides are composed of various monosaccharides such as D forms of glucose, galactose, mannose, arabinose, ribose, xylose, and uronic acids. Exopolysaccharides like pullulan, kefiran, bacterial cellulose (BC), gellan, and levan produced by microorganisms exhibit film-forming ability. The physicochemical and rheological properties of exopolysaccharides vary with the type of microbial strain. Microbial biopolymers are applicable in food and allied industries due to their viscosifying, gelling, and film-forming properties. Microbial biopolymers play important roles as energy reserve materials, protective agents, aid in cell functioning, the establishment of symbiosis, and osmotic adaptation in response to changing environmental applications (Vijayendra and Shamala 2014).

Improvement in properties of existing polysaccharides for generation of novel biopolymers of great commercial interest and value with desired properties can be done by proper designing and creating of new property based on requirements through controlled synthesis. The microbial biopolymers can be successfully produced on industrial scale accompanied with the use of efficient producer strain, cheaper fermentation substrates as renewable feedstock, process design along with optimized fermentation parameters like pH and temperature, and superior technology like genetic engineering. Production of microbial polysaccharides using cheap biomass resources like syrups and molasses, olive mill wastewater, cheese whey, vegetable and fruit pomace, pulp and kernels, lignocellulosic biomass like rice hull and bran, as well as carbon dioxide has been reported with a suitable pretreatment method (Oner 2013). Production of biopolymers with customized properties by genetic manipulation of microorganisms is carried out for tissue engineering and drug delivery (Rehm 2009). PHAs are potential microbial polymers for drug delivery (Shrivastav et al. 2013). Pullulan could be used as a reducing as well as a capping agent for AgNPs synthesized using pullulan act as strong antimicrobial agents (Kanmani and Lim 2013).

3.2.1 Alginate

Alginic acid is a linear, negatively charged, viscous or gel-like water-soluble polysaccharide with a molecular formula ($C_6H_8O_6$) n with a molar mass of~ $1–3 \times 10^6$ g mol^{-1}. It is a linear chain of copolymers of β-D-mannuronate ($1 \rightarrow 4$) and α-L-glucuronate (Wong et al. 2000). Brown algae produce 40 % by weight of alginate (Matsubara et al. 2000) (Fig. 3.1). Commercial alginates are produced. Marine microalgae producing alginate are *Ascophyllum nodosum*, *Durvillea antarctica*, *Ecklonia maxima*, *Lessonia nigrescens*, etc. Bacteria including *Azotobacter* sp. and *Pseudomonas* sp. are producing alginates as extracellular polysaccharides during late exponential growth phase. Alginate in *Azotobacter vinelandii* prevents desiccation under adverse environmental conditions (Remminghorst and Rehm 2006). Bacterial yield of alginate is affected by different cultivation conditions as pH, aeration, and oxygen supply. The batch cultivation process is designed for commercial production of alginate on an industrial scale. Alginate is applied in biomedical science and engineering, in cell and tissue immobilization (Wang et al. 2003), and as food additives (Draget et al. 2005). Alginate is also used as a thickener in ice creams and as a chelator for removal of metals and in dye printing due to its unreactive nature.

3.2.2 Xanthan

Xanthan is an extracellular polysaccharide produced by microorganisms belonging to genus Xanthomonas like *Xanthomonas campestris*, *Xanthomonas malvacearum*, and *Xanthomonas axonopodis* through the process of microbial fermentation. Xanthan has molecular weight ranging from 2×10^6 to 20×10^6 Da mol^{-1}. Xanthan, a polyanionic heteropolysaccharide, is composed of glucose, mannose, and glucuronic acid forming a repeated pentasaccharide unit (Fig. 3.2) where noncarbohydrate components as O-acetate and pyruvate substitute on the terminal mannose. Acetylation and pyruvylation of mannose produce modified xanthan. It shows rigid rod-like conformation. Xanthan shows desirable properties like viscosity, pseudoplasticity, thixotropy, and gellation applicable in food industry. Bacteria producing xanthan gum are

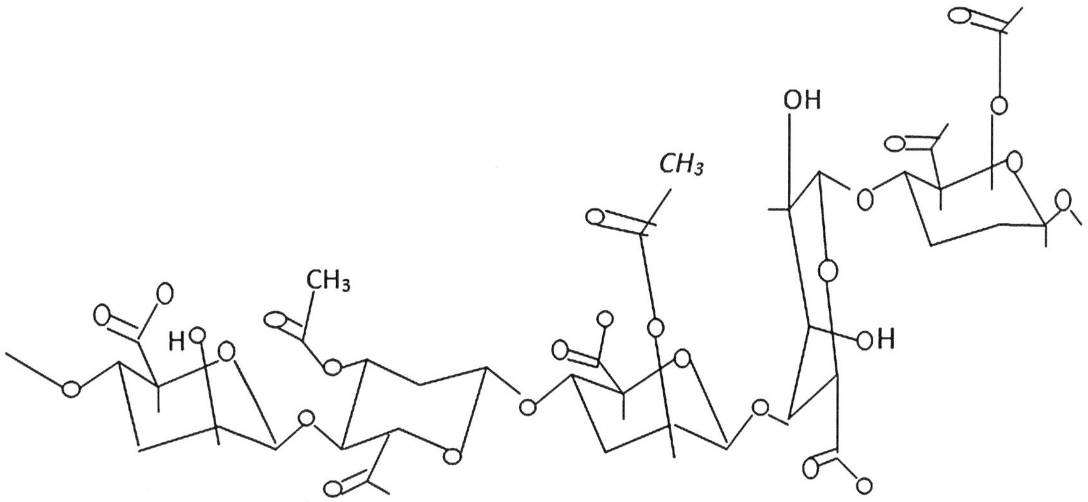

Fig. 3.1 Structure of alginate

Fig. 3.2 Structure of xanthan

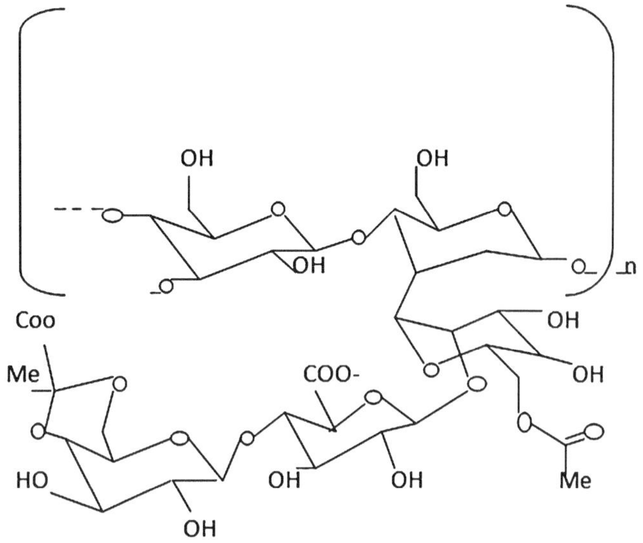

X. campestris, *X. vasculorum*, *X. juglandis*, *X. fragaria*, etc. Xanthan formed by Xanthomonas species serves the bacterium in acting as plant pathogen by allowing its penetration and formation of local lesions, soft roots, scabs, and cancers. Production of xanthan can be achieved using cheaper carbon sources like sugarcane molasses, whey along with yeast extract, soymeal peptone as a complex nitrogen source, and organic acids as a stimulator through the process of aerobic fermentation in batch culture or continuous culture. Production medium with higher carbon to nitrogen ratio shows a better yield of xanthan. *Xanthomonas campestris* is responsible to produce xanthan by anaerobic fermentation at a temperature of 28 °C.

3.2.3 Pullulan

Aureobasidium pullulans is a dimorphic fungi producing pullulan, a homopolysaccharide of natural glucan polymer with average molecular weight ranging from 1.5×10^4 to 1.0×10^7. It is also known as α-1,4- and α-1,6-glucan consisting of 0.6 % maltotetraose and a major component of maltotriose (Fig. 3.3). Structure of pullulan composed of repeating maltotriose units linked through (1–6) linkage using terminal glucose residue of trisaccharide, while each maltotriose unit is made up of three α-1,4-linked glucose residues (Carolan et al. 1983). Production of pullulan varies with culture conditions and type of microbial strain that get decreased in the late stationary phase due to pullulanase production in medium (Pollock et al. 1992). Pullulan has adhesive properties applied during formation of fibers, moldings, and oxygen-impermeable films used for preservation purpose (Sakata and Otsuka 2009) and used for coating of food containers for preservation of perishable fruits and vegetables. It has the molecular formula (C6H10O5) n and is considered as a composite of panose or isopanose subunits. Pullulan due to its high solubility in water is applied during controlled drug delivery. Consequently, pullulan can be used as food additive, flocculant, and wound healing (Cheng et al. 2011). Pullulan shows pH tolerance from 3 to 8 pH and it carbonizes at 25–280 °C. The production of pullulan varies with fermentation parameters like pH and temperature, composition of production medium, and physiological state of microbial strain. Commercially, pullulan can be produced using coconut by-products, beet molasses, and agro-industrial waste. Pullulan is produced by various microbes including *Aureobasidium pullulans* (Leathers 2003), *Teloschistes flavicans* (Reis et al. 2002), and *Cryphonectria parasitica* (Delben et al. 2006). Anionic or amphiphilic pullulan microparticles play an important role in controlled drug delivery system (Jeong et al. 2006).

3.2.4 Levan

Levan a neutral, biodegradable, and water-soluble homopolysaccharide is a polyfructan which consists of D-fructofuranosyl residues joined by β-(2,6) linkages forming a core along with β-(2,1) branching points (Fig. 3.4) (Tanaka et al. 1990). Levans are produced by plants termed as phelins. Levan is produced by bacterial sp. like Bacillus and Pseudomonas (Oliveira et al. 2007; Shih et al. 2005). Low fermentation temperature (25 °C) supports optimal production of levan, while high temperature (35–40 °C) inhibits its production. Levan has high solubility in oil, low viscosity, high water-holding capacity, good biocompatibility with surfactants, and stability to heat (Sima et al. 2011). Levans have a high molecular weight of 2–100 million Da and the density 1.4 g/cm^3. Levan is a nontoxic and a soluble dietary fiber where the hydrolysates resulting from levan particularly help to improve gut function.

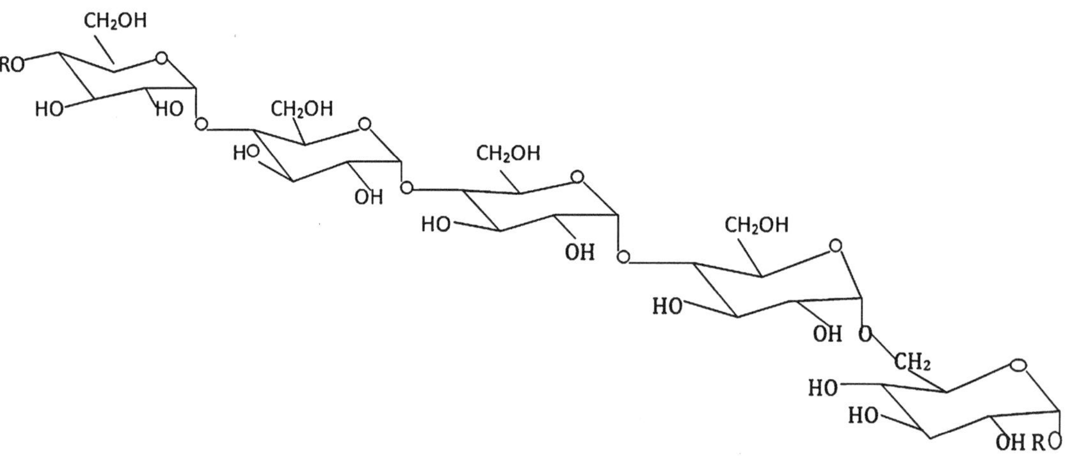

Fig. 3.3 Structure of pullulan

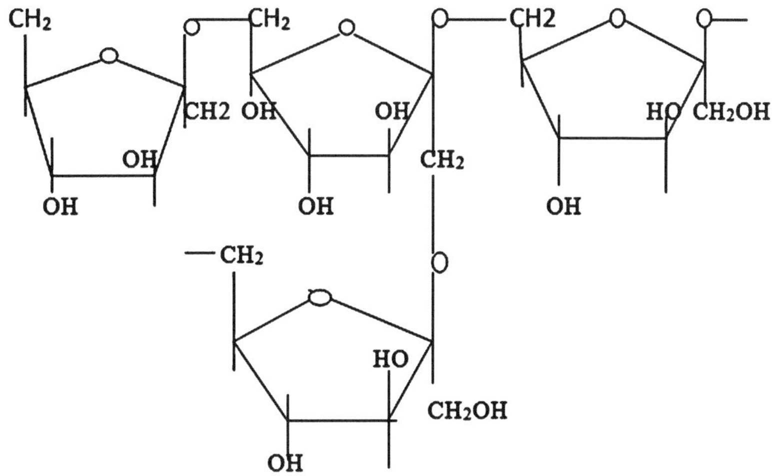

Fig. 3.4 Structure of levan

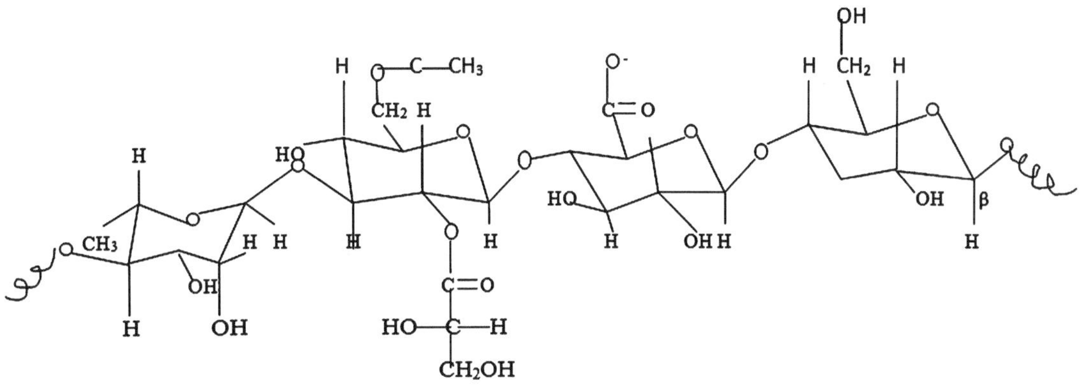

Fig. 3.5 Structure of gellan

Levans can be used in foods, feeds, medicines, and cosmetics, in the chemical and biotech industry (Rairakhwada et al. 2007; Gupta et al. 2008). It is useful as encapsulating agent, thickener, emulsifier, and carrier of colors and flavors for food (Jang et al. 2001). Levan from *Microbacterium laevaniformans*, *Rahnella aquatilis*, and *Zymomonas mobilis* was found to exhibit in vitro antitumor activity (Yoo et al. 2004). It also shows cell proliferative and anti-inflammatory effects (Ki Ho Kim et al. 2005). Levan provides a better substitute for gum Arabic in a variety of foods, pharmaceuticals, medicines, cosmetics, adhesives, paints, ink, lithography, textile, and others. Production of microbial levan is generally carried out using sucrose-based substrates like fructose, molasses, and sugarcane juice.

3.2.5 Gellan

Gellan gum is a linear anionic polysaccharide with a molecular mass approximately 1×10^5 Da due to the presence of tetracyclic units (Gong et al. 2009). It is produced by the bacterium *Sphingomonas elodea* using a fermentative process through immersion method. It is a linear polymer composed of a repeating tetrasaccharide unit consisting of glucose, glucuronic acid, and rhamnose in 2:1:1 ratio (Fig. 3.5). The structure of gellan contains about 1.5 % acyl group as acetyl and glycerate molecules per unit. With respect to acyl groups, it is classified into high-acyl and low-acyl gellan where the natural form of biopolymer is high-acyl gellan and has shorter length than low-acyl gellan. Gellan with high-acyl groups forms soft, elastic, and non-brittle gels

(Sworn 2000). Gellan inhibits sedimentation or suspension of particles in fruit juices with pulps, cacao, and non-soluble minerals where the suspended particles are trapped in liquid gel structures (Young 2002). Gellan is an appropriate biopolymer for coating and protecting the probiotics by resisting acidic conditions and protecting cells from acidic damages. The gellan gum shows plastic flow behavior above 1.0 % concentration at 25 °C, while formation of gel requires polysaccharide concentration over 0.8 % (Tako and Nakamura 1989). Gellan reduces oil uptake due to its thermo-gelling property (Bajaj and Singhal 2007). It is produced by *Sphingomonas elodea* through the process of aerobic fermentation using a carbohydrate as a substrate. Gellan gum is used in dairy products and sugar confectionary and to modify traditional gelatin dessert jellies. Gellan is sold in the market with trade names Kelcogel, Gelrite, Phytagel, and Gel-Gro. The thickness gellan gum can be controlled by the addition of sodium and potassium salts. Plasticizers like glycerol are added to make gellan films less brittle. Gellan is applied as a novel drug delivery vehicle and in fruit products due to heat and acid stability.

3.2.6 Kefiran

Kefiran is a water-soluble exopolysaccharide present in kefir grains. It is a glucogalactan with a molecular weight corresponding to 10^7 Da. The heteropolysaccharide structure of kefiran mainly consists of glucose and galactose in equal amounts (Fig. 3.6). Yeasts, bacteria producing lactic and acetic acid, and mycelial fungi are microbial constituents of kefir grains (Witthuhn et al. 2005). Kefiran is produced by Lactobacillus sp. like *Lactobacillus kefiranofaciens* (Mitsue et al. 1999). Exocellular kefiran recovered from culture supernatant by centrifugation and recovered by precipitation of supernatant with an equal amount of cold ethanol. Capsular kefiran recovered from cells with water at 95 °C and precipitated with ethanol. Kefiran has several commercial applications ranging from being an emulsifier, gelling agent, stabilizer, thickener, etc. (Cheirsilp and Radchabut 2011). Kefiran in kefir grains can form edible transparent films at concentrations ranging from 5 to 10 g/kg. The addition of glycerol as plasticizer improved flexibility of kefiran film matrix compared to low density polyethylene film and reduced water vapor permeability (Piermaria et al. 2009). Kefiran films are transparent showing pseudoplastic behavior, partially soluble in water at 25–37 °C while completely solubilized at 100 °C.

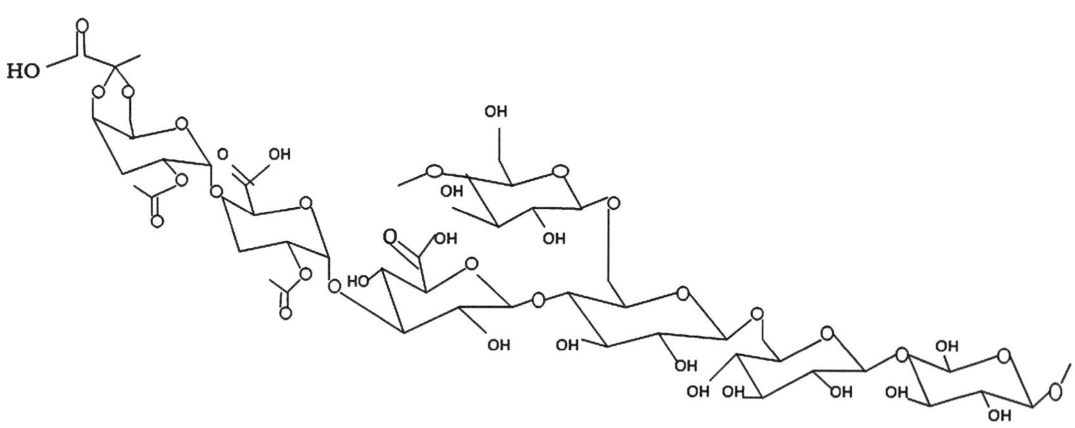

Fig. 3.6 Structure of kefiran

Kefiran exhibits good potential as an edible film in the food and packaging industries (Ghasemlou et al. 2011a, b). Coculture of *L. kefiranofaciens* and *S. cerevisiae* consumes lactic acid, thereby lowering its concentration, which enhances cell growth and kefiran producing capacity of *L. kefiranofaciens* (Cheirsilp et al. 2003). Production of kefiran by these bacteria depends upon the cultural conditions like pH, temperature, agitation as well as carbon and nitrogen source. The kefiran represents a very important option as natural additive to the food and industry that can be used as thickeners and stabilizers in food. It also shows antimicrobial properties.

3.2.7 Cellulose

Bacterial cellulose is produced extracellularly in the form of nanofibers by *Acetobacter*, *Aerobacter*, *Alcaligenes*, *Rhizobium*, *Agrobacterium*, etc. (El-Saied et al. 2004). Microbial cellulose consists of bundles of cellulose microfibrils 2–4 nm diameter (Nakagaito et al. 2005). Cellulose nanofibers are arranged in a ribbon form with 100 μm length and 100 nm diameters. Such network of cellulose nanofibers exhibits biocompatibility, bioadaptability, biodegradability, and chemical stability (Moreira et al. 2009). Microbial cellulose consists of microfibrils which are made up of hydrogen bond linked glucan chains. Microbial cellulose is a linear and unbranched polymer comprised of D-glucopyranose units with β-1,4 linkages (Fig. 3.7). It has a molecular formula $(C_6H_{10}O_5)$ n. Bacterial cellulose is preferred than plant cellulose due to its versatile properties like higher purity, tensile strength, water-holding capacity, and crystallinity index making it more suitable raw material for paper and dessert foods (Shoda and Sugano 2005). Unlike plant cellulose, bacterial cellulose does not require processing to remove contaminating lignin and hemicelluloses components (Nishi et al. 1990). Cellulose accounts for its tensile strength and elastic modulus due to its fine network of cellulose microfibrils. Bacterial cellulose is applicable diet food, paper additive, skin tissue repair, vascular grafts, and for regenerative medicines (Lin et al. 2013). The processed cellulose membrane is highly applicable as in food packaging due to its hydrophilic nature and continuous moisture removal. Nutrient media with high carbon content and limitation of nutrients favors cellulose production (George et al. 2005). Various carbon sources used during cellulose production include sucrose, mannitol, glucose, and xylose.

3.2.8 *Haloferax* Exopolysaccharide

Bacterial spp. of the genera such as *Methanosarcina*, *Thermococcus*, *Sulfolobus*, and *Archaeoglobus* represent a valuable source of exopolysaccharides. Molecular weight of bacterial exopolysaccharides ranges from 10 to 30 KDa (Singha 2012). EPSs have heterogeneous,

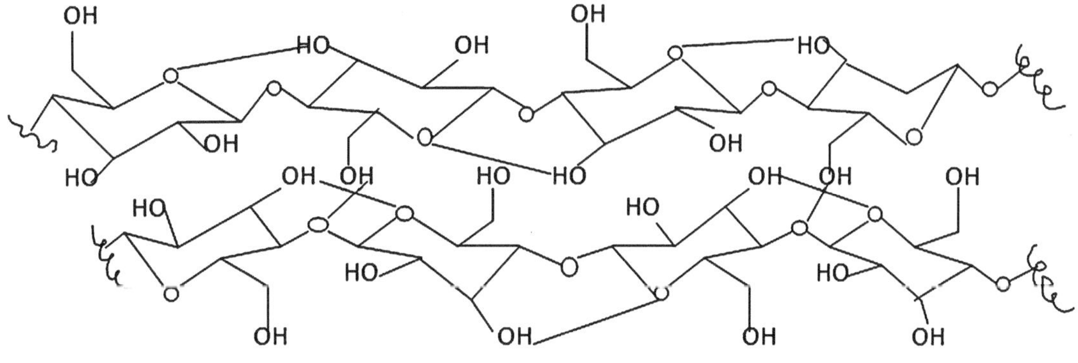

Fig. 3.7 Structure of microbial cellulose

neutral, or polyanionic nature due to the presence of phosphate, sulfate, and uronic acid residues. The polysaccharide produced by halobacterium *Haloferax mediterranei* is well studied due to its film-forming property. This exopolysaccharide consists of sugars such as glucose, mannose, small amount of galactose, and ribose, along with amino sugars and uronic acids (Fig. 3.8). Halophilic Archaea producing EPS are *Haloferax*, *Haloarcula*, *Halococcus*, and *Natronococcus*. EPSs are produced by extremophiles in response to biotic and abiotic stress factors (Donot et al. 2012) and serve for protection against desiccation and predation. Exopolysaccharide from *Haloferax gibbonsii* (ATCC 33959) consists of neutral sugars like D-mannose, D-glucose, D-galactose, and L-rhamnose. EPS produced by *H. mediterranei* has certain rheological properties like pseudoplastic and viscoelastic behavior and flow behavior, maintaining the structure at high temperature (Mironescu and Mironescu 2011) and also shows higher salt resistance. Due to such remarkable resistance, EPS from *H. mediterranei* is applicable as a thickening agent and in oil recovery process. *Haloferax denitrificans* grows with salt concentration of 1.5–4.5 M (Parolis et al. 1999). Production of biopolymer is related to nutritional composition of media, pH, temperature, physiological state, and the genetics of organism (Finore et al. 2014). The biofilm matrix prepared with exopolysaccharides exhibits interesting properties such as antimicrobial action, preserving activity, and biodegradability (Mironescu and Mironescu 2006).

3.3 Polyhydroxyalkanoates (PHAs)

Biodegradable plastics can be chemically synthesized polymer, starch-based biodegradable plastics, and polyhydroxyalkanoates (PHAs) (Kalia et al. 2003; Khanna and Srivastava 2005; Singh et al. 2009). Biodegradable plastics produced from bacterial fermentation include polyhydroxyalkanoates (PHAs). Polyhydroxyalkanoates are biocompatible and branched heteropolyesters composed of 150 dif-

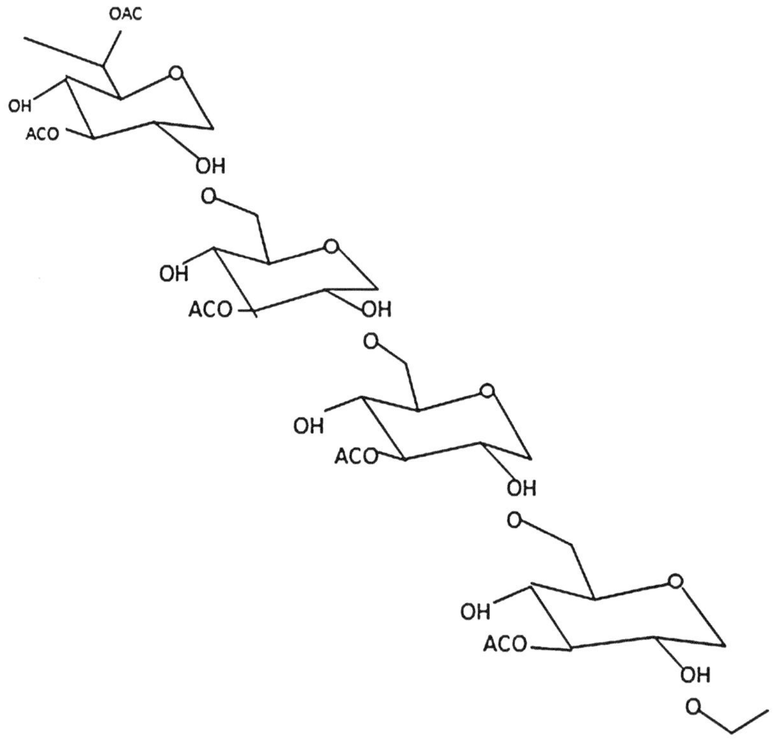

Fig. 3.8 Structure of Haloferax exopolysaccharide

ferent (*R*)-3-hydroxyalkanoic acid monomers containing oxoester bonds joining hydroxyl and carboxyl groups. PHAs are synthesized as storage polymers by a wide variety of bacterial species, *Archaea*, and recombinant bacteria by metabolic transformation (Table 3.1). PHAs possess unique properties as biocompatibility, biodegradability both in aerobic and anaerobic conditions including aquatic environments, biobased and renewable origin, and structural diversity providing a remedy to reduce pollution of the environment (Tortajada et al. 2013). PHAs can be also produced from CO_2 (Snell and Peoples 2002). The production of various types of PHA has been using several transgenic crops, including corn, sugarcane, and cotton (Snell and Peoples 2009). Microbial degradation of PHAs to CO_2 and H_2O under aerobic conditions in soil, freshwater, and marine environments (Mergaert et al. 1994) or to methane and water under anaerobic conditions is also reported (Volova et al. 2010). Their degradation occurs more rapidly than degradation of synthetic polyesters and lignocelluloses. PHAs consisting of various hydroxyalkanoic acids are developed through biotechnological routes and show thermal properties similar to polypropylene.

PHAs are storage biopolymers in prokaryotic cells that occur as granular, hydrophobic inclusion bodies (PHA granules) within the cytoplasm. PHAs' granules stored as insoluble inclusion bodies are coated with lipids and proteins termed as phasins on their surface which prevents coalescence where lipids involved are phospholipid monolayer. All enzymes necessary for synthesis of PHB from acetyl-CoA associated with PHA granule as observed in *R. eutropha* (Uchino et al. 2007). Polymers stored in granules remain in the amorphous state in vivo, while in isolated state, the polymer rapidly crystallizes. Intracellular utilization of PHA occurs by PHA synthesizing organisms, while extracellular utilization of PHA released from lysed cells takes place into the environment.

PHAs are accumulated in cell cytoplasm when carbon is in excess and inorganic nitrogen and phosphate are limiting (Koller et al. 2010; Singh et al. 2015). Thus, the bacterial cells switch carbon pool to Krebs cycle for synthesis of PHA under such growth-limiting conditions. PHAs are biocompatible and responsible to regulate intracellular energy flow by directing the carbon pool toward metabolic pathways and protecting the cell against stress conditions (Koller et al. 2011).

Table 3.1 Microbial strains producing PHA on industrial scale

Microbial strain	Polymer produced	References
Cupriavidus necator (Alcaligenes eutrophus)	PHB and its copolymers	Volova and Kalacheva 2005
Cupriavidus necator H16	PHB	Pohlmann et al. 2006
Azohydromonas latus DSM 1124	PHA	Yu et al. 1999
Rhizobium meliloti, R. viciae, Bradyrhizobium japonicum	PHA	Mercan and Beyatli 2005
Recombinant *Escherichia coli*	(UHMW) PHB	Nikel et al. 2006
Alcaligenes latus, Staphylococcus epidermidis	PHB	Wong et al. 2005
Ralstonia eutropha (*Cupriavidus necator)*	PHB, PHBV, P_3HB_4HB	Kahar et al. 2004
Aeromonas hydrophila	mcl-PHAs	Lee et al. 2000
Pseudomonas aeruginosa	mcl-PHAs	Hoffmann and Rehm 2004
Pseudomonas putida, P. fluorescens, P. jessenii	Aromatic polymers	Wang et al. 2005
Bacillus sp.	PHB, PHBV copolymers	Full et al. 2006
Burkholderia cepacia	PHB, PHBV	Nakas et al. 2004
Halomonas boliviensis	PHB	Quillaguaman et al. 2006
Microlunatus phosphovorus	PHB	Akar et al. 2006
Spirulina platensis (cyanobacterium)	PHB	Jau et al. 2005
Halophilic archaeal species	*Natrialba*	Han et al. 2007
	Haloquadratum	Legault et al. 2006

They exhibit nonlinear optical activity. The PHA types are polyhydroxybutyrate (PHB), polyhydroxyvalerate (PHV), polyhydroxyhexanoate (PHH), and polyhydroxyoctanoate (PHO) where PHB is the main biodegradable polymer (Chen 2009; Singh et al. 2013) (Fig. 3.9). The best-known PHAs are PHB, poly-3-hydroxybutyrate-co-3-hydroxy valerate P(HB-co-HV), poly-3-hydroxoctanoate-co-polyhydroxyhexanoate P(HO-co-HH) (Table 3.2).

Pendant R group in PHAs varies from C3 to C14 carbon atoms (Doi and Abe 1990). PHA has molecular weight varying from 2×10^5 to 3×10^6 Da. It is synthesized by 30 % soil bacteria as well as other bacterial sp. inhabiting activated sludge, high seas, extreme environments, and soil. *Cupriavidus necator* is the most commonly studied strain for PHB production (Atlic et al. 2011). PHB is widely found in *Bacillus*; *Pseudomonas*; plant symbionts, *Rhizobium*; nitrogen-fixing *Azotobacter* sp.; *Azohydromonas lata*; and recombinant *Escherichia coli* (Volova 2004; Reddy et al. 2003; Kumar et al. 2013, 2014) (Table 3.2). Microorganisms living in extreme environmental habitats accumulate endocellular polyhydroxyalkanoates (PHAs) during stationary phase. Several halophilic archaeal species like *Haloferax*, *Haloarcula*, and *Halococcus* have been also reported to produce PHAs like PHB and PHBV.

Polyhydroxyalkanoates are categorized as SCL-PHA (3–5 carbons), MCL-PHA (6–14 carbon), and LCL-PHA (17–18 carbon). Copolymers of scl-PHA like hydroxybutyrate (HB) and 3-hydroxyhexanoate improve their mechanical property making them comparable with conventional plastics such as polypropylene and polystyrene. Mcl-PHAs such as poly (hydroxyoctanoate-co-hydroxydecanoate) [P (HO-co-HD)] behave as a complete elastic substance. PHB is highly crystalline due to its stereospecific nature with a melting temperature (Tm) 170–180 °C (Matko et al. 2005) and Tg around 5 °C. Unfavorable aging of PHB homopolymer can be prevented by annealing for changing its lamellar morphology and improving mechanical properties. Increasing porosity and surface area of PHA products enhance their degradation rate in the environment. The control of PHA monomer composition along with new processing and compounding technologies and their novel properties like moisture resistance

Fig. 3.9 General structure of polyhydroxyalkanoates (PHAs) and their structural derivatives

Poly(3-hydroxyalkanoates) [PHA]

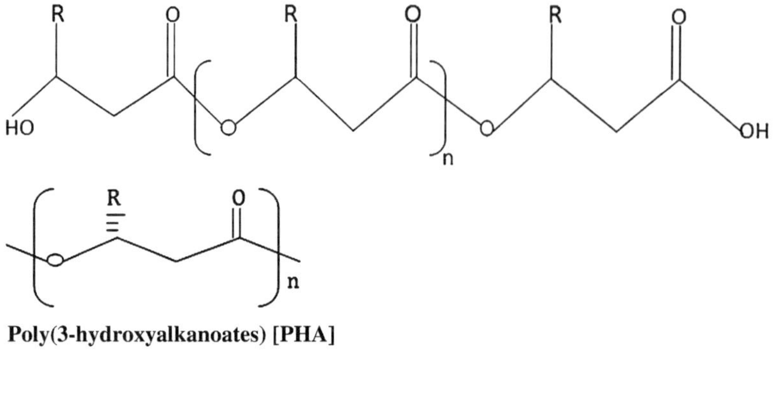

Table 3.2 Polyhydroxyalkanoates (PHAs) and their structural derivatives based upon "R" groups

R group	Polyhydroxyalkanoate	Abbreviation
----CH_3	Poly(3-hydroxyalkanoates)	PHA
----- CH_3	Poly(3-hydroxyvalerate)	PHV
------$(CH)_2$.........CH_3	Poly(3-hydroxyhexanoate)	PHHex
------$(CH)_4$.........CH_3	Poly(3-hydroxyoctaonate)	PHO
------$(CH)_6$.........CH_3	Poly(3-hydroxydecanoate)	PHD
—— **(CH_2)** ——	Poly(3-hydroxy-5-phenylvalerate)	PHPV

and oxygen impermeability will make PHAs more functional and expand the number of applications for which these plastics can be used.

Processing PHAs using certain methods can alter their biodegradability and improve their mechanical properties. Properties of polymers can be altered through the use of additives such as fillers, plasticizers, stabilizers, reinforcers, and pigments. The composition used in bioplastic film includes P(HB-co-HV) with acetyl tributyl citrate as a plasticizer and cyclohexylphosphoric acid/zinc stearate as a nucleating agent and thermal stabilizer (Asrar and Pierre 2000). Drawing PHA fibers and films can change the physical structure of the polymer and alter its mechanical properties. Tanaka and coworkers developed melt-spinning technique that produced highly crystalline and P (HB-co-HV) fibers with greater strength (Tanaka et al. 2007). Gel-spinning techniques have also been developed for preparation of high-strength PHA fiber (Antipov et al. 2006) increased the toughness of P(HB-co-HV) films by cross-linking the polymer and drying the films under uniaxial strain. High degree of crystallinity of PHB leads to a plastic that is strong but very stiff high Young's modulus and brittle that limits the commercial potential of PHB. PHA copolymers have more favorable properties than PHB where the addition of HV units to the polymer chain in case of PHB-co-HV resulted in a material with a lower Young's modulus and a greater extension to break (Sudesh et al. 2000). P (HB-co-$HA_{MCL)}$ copolymers are weaker than PHB but are tougher and more flexible thermoplastics. Strength of PHBV can be increased by lowering Tm and Tg based upon its variable HV content. Higher HV content increases strength and results in reduction in Tm and Tg values (Amass et al. 1998) and crystallinity. PHB serves as a promising source of biodegradable thermoplastic material useful for packaging industries to solve environmental pollution problem for waste management strategies (Mona et al. 2001). *Haloferax mediterranea*, a denitrifying halophile, might present advantages for production of PHB and PBV as their culture requirements in terms of salinity and temperature provide little opportunity for growth of contaminants. Functionalized PHAs with tailor-made properties can be generated by controlling monomer composition where the side chains in mcl-PHAs are reactive groups and a potential target for post-biosynthetic modification. Higher fraction of unsaturated side chains in PHA monomers cause inhibition of crystallization and subsequent lowering of melting and glass transition temperature. Cross-linking of unsaturated PHAs can transform them into rubbers (Bassas et al. 2008).

3.3.1 PHA Production from Cheap and Renewable Resources: Renewable Nature and Life Cycle

Biodegradable nature of PHA resulting formation of carbon dioxide and water indicates their positive ecological impact making them an effective substitute to synthetic plastic. PHAs can be produced from renewable resources through the process of fermentation using agricultural feeds as a source of sugars and fatty acids. Carbon source plays a predominant role in PHA production reducing the cost by 50 %. The greatest challenge in PHA production is the use of waste and biowaste, mostly because of their substrate and contaminant contents (Fig. 3.10). Biowastes, namely, glycerol, pea shells, rice chaff, coconut oil cake, cottonseed cake, wafer residue, and citrus pulp waste, have been tested as substrates to produce PHA (Patel et al. 2012, 2015; Kumar et al. 2015a,b; Singh et al. 2015). The use of defined media with carbohydrates (glucose, sucrose, fructose), alcohols as methanol, alkanes (C_6–C_{12}), and organic acids (butyrate, valerate) is expensive for industrial scale synthesis of PHA. High-volume industrial PHA production can be possible using cheap renewable resources as starch, lactose, fats, and oils (Braunegg et al. 2002).

PHB production is carried using a wide variety of carbon sources including beet molasses, ethanol, methanol, wheat and casein hydrolysate, corn steep liquor, etc. *Ralstonia eutropha* H16, a promising producer of PHA showing higher yield of SCL-PHA (up to 80–90 %), utilize H_2-CO_2

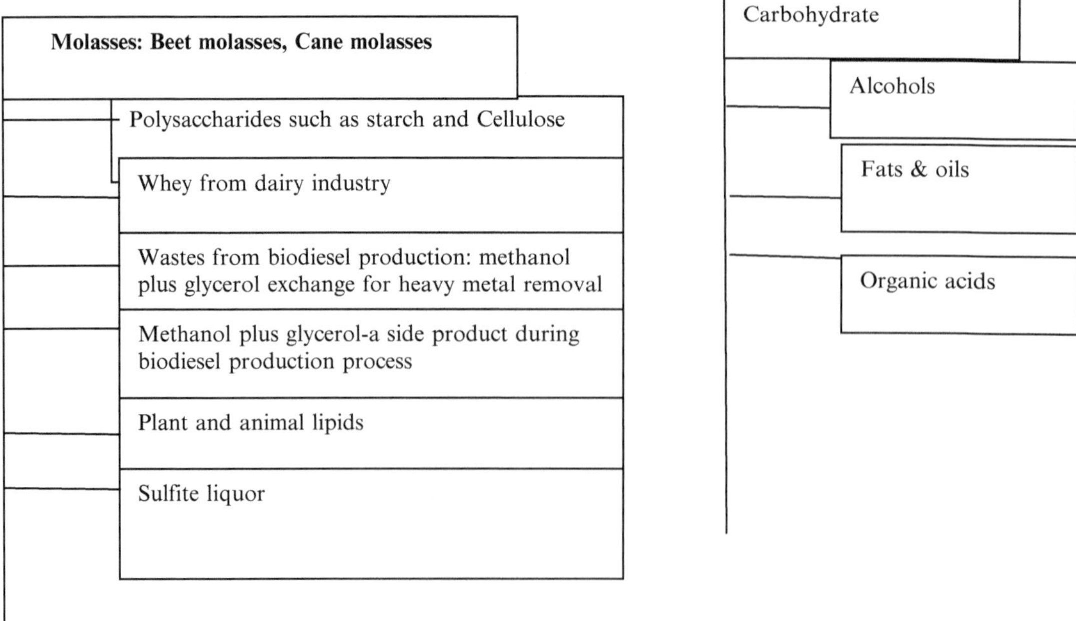

Fig. 3.10 PHA production from cheap and renewable resources

mixture as a sole carbon and energy source (Pohlmann et al. 2006) for production. The production of polyhydroxybutyrate (PHB) in *Bacillus megaterium* ATCC 6748 was carried out using renewable carbon and nitrogen source (Chaijamrus and Udpuay 2008). Synthesis of multicomponent PHAs is a very complicated biotechnological task. Bacterial cultures cannot be grown in the presence of mixed carbon substrates (CO_2 + valerate, hexanoate, etc.) as monomers with different number of carbon atoms that cannot be incorporated into the polymer at the same rate during their synthesis. The fatty acid salts added to the culture as co-substrates were found to be toxic to bacteria. Production of PHB by a recombinant *Halomonas campaniensis* LS21 using energy-saving, seawater-based, continuous open process, and cheaper substrates (cellulose, proteins, fat, and starch) was established (Yue et al. 2014). Thus, it is essential to determine maximal permissible concentrations of every acid for every PHA producer. Bacterial strains of *Wautersia eutropha*, B5786 and H16 grown with CO_2-heptanoic acid mixture, synthesize HV and HB. The addition of octanoate inhibited both culture growth and polymer synthesis (Volova et al. 2007). The high cost of industrial production and recovery of bioplastics can be improved by optimization of fermentation process, by the use of recombinant organisms utilizing cheap carbon sources. This will facilitate the commercialization of bioplastic production. Zero-cost feedstocks like industrial vegetable wastes were used by extremophiles for the production of biopolymers like PHAs (Donato et al. 2011). *Haloarcula* sp. IRU1 effectively used petrochemical waste water as a carbon source to produce PHB (Taran 2011). Industrial production of PHB was successfully carried out from cardboard industry waste water as a sole carbon source using *Enterococcus* sp. NAP11 and *Brevundimonas* sp. NAC1 isolates (Bhuwal et al. 2013). PHAs are valuable biopolymers due to their renewable nature and life cycle (Fig. 3.11).

3.3.2 Metabolic Pathway to PHB

PHB synthase catalyzes synthesis of PHB from (R)-3-hydroxybutyryl-CoA (Fig. 3.12). Under limiting conditions of nitrogen acetyl-CoA undergoes condensation with another acyl-CoA

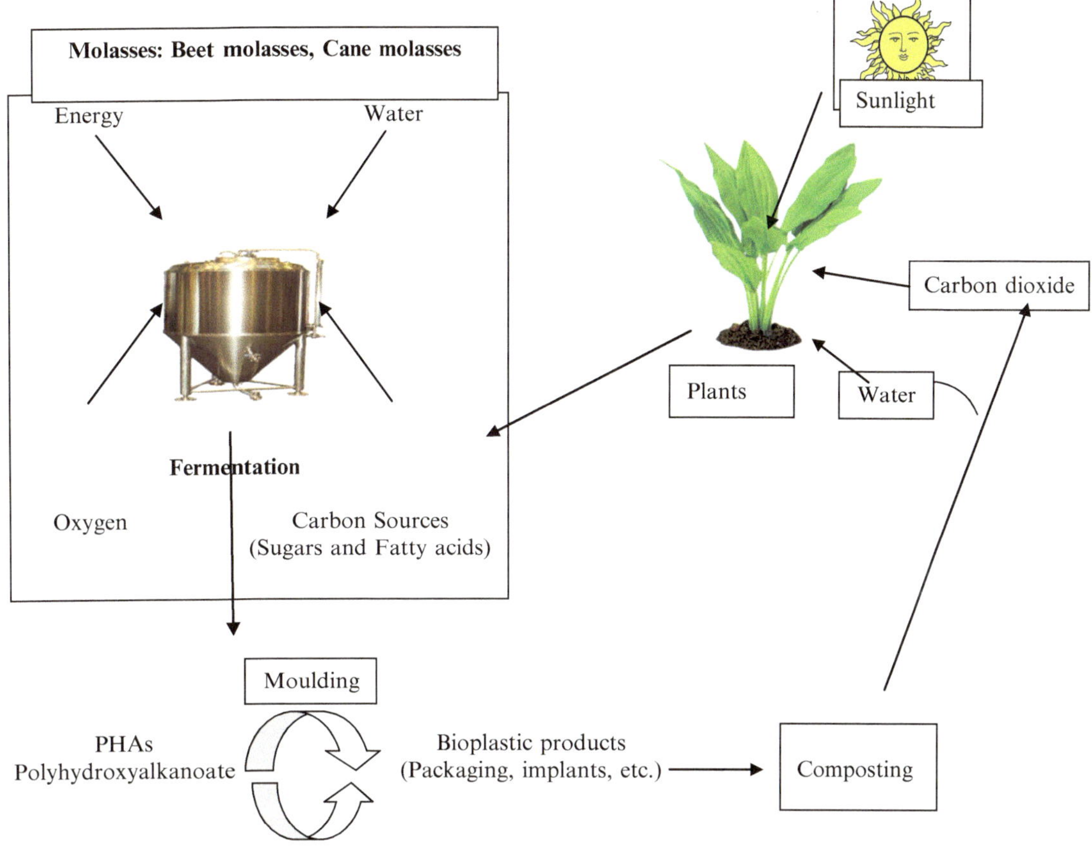

Fig. 3.11 Bacterial synthesis of PHAs: renewable nature and life cycle

derivative using 3-ketothiolase resulting formation of acetoacetyl-CoA that is stereoselectively converted to (R)-3-hydroxybutyryl-CoA with the help of NADPH-dependent acetoacetyl-CoA reductase (Kalia et al. 2007). Polymerization of (R)-3-hydroxybutyryl-CoA to form the homopolyester (poly(R)-3-hydroxybutyrate) or copolyesters (poly-3-hydroxybutyrate-co-3-hydroxyvalerate) is executed by PHA synthase catalyzes. These polymers are stored by the bacterial cell as globular granules. PHAs are depolymerized to monomeric components that are metabolized to CO_2 and water along with the production of ATP under growth promoting conditions in the presence of assimilable nitrogen. When PHA production process aims to produce copolyesters, the addition of various precursors like propionate, γ-butyrolactone, 1,4-butanediol, and 4-hydroxybutyrate to the production medium is required.

Functionalized PHAs are synthesized by feeding structurally related substrates like (*R*)-3-hydroxyacyl-CoAs, processed through β-oxidation pathway by bacterial sp. like *pseudomonads* (Figs. 3.13 and 3.14). Nonfatty acid precursors such as carbohydrate like fructose, glucose, glycerol, acetate, and ethanol are channelized to PHA by the de novo pathway via (*R*)-3-hydroxyacyl-acyl carrier protein intermediates (Fig. 3.15). A plethora of tailor-designed mcl-PHAs, with highly diverse structures that include acetylthioester, acetoxy, alkoxy, amino, cyano, cyclohexyl, epoxy, halogenated, hydroxy, or propylthiol groups, can be synthesized (Tables 3.3, 3.4, and 3.5). Synthesis of mcl-PHAs with desired properties can be possible by using low-cost substrates as unsaturated fatty acid con-

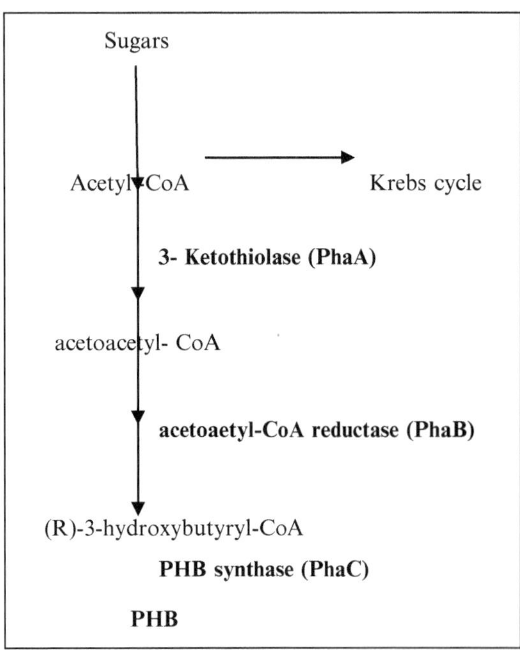

Fig. 3.12 General pathway for the synthesis of PHB

stituents present in oils or fats providing an ideal opportunity for insertion of required functional groups in PHA. But a major challenge in mcl-PHA production is a more complex metabolic network needed for diverting mixture of fatty acids in fats and oils toward PHA synthesis. Isolation, identification, and genetic manipulation of microbes which produce bioplastics with different structures, properties, and applications indicate a promising future for the industrialization of bioplastics.

3.3.3 Various PHA Recovery Methods

The extraction procedure applied for the recovery of PHAs can affect the molecular mass of PHB (Nuti et al. 1972; Senior and Dawes 1973). Various different extraction methods have been reported for PHAs (Table 3.6). Cells grown in fermentation media were harvested by centrifugation or flocculation, and the PHA can be recovered by the use of surfactants or hypochlorite to lyse the cells for release of the intracellular product. But hypochlorite can cause some degradation of the product so that alternative solvent extraction method is used which removes bacterial endotoxins and causes negligible degradation of PHAs. Solvent extraction is the widely accepted method to isolate PHA from cellular mass due to its simplicity and rapidity. In this method, cell mass is first dried by spray drying or by lyophilisation followed by addition of large amounts of solvent since PHA solutions are concentrated and highly viscous. The first step in solvent extraction method is solubilization of PHA followed by nonsolvent precipitation using methanol and ethanol. Most commonly used solvents for extraction of PHA are chloroform, 1,2-dichloroethane, ethylene carbonate, 1,2-propylene carbonate, and acetone. The use of solvents involves high capital and operational cost. Alternative extraction method involves chemical digestion method using sodium hypochlorite and surfactants such as Triton X-100 and SDS that reduced the cost by 50 % (Yu and Stahl 2008; Dong and Sun 2000) and facilitates rapid PHA recovery. Isolation of PHA from *S. meliloti* using treatment with surfactant-chelate (EDTA) showed 90 % purity (Lakshman and Shamala 2006). The action of protease secreted by *Microbispora* sp. induced hydrolysis of *S. meliloti* cells. In other methods, PHA is recovered by enzymatic digestion in which cell lysis is carried out by the action of proteases (Yasotha et al. 2006). Other methods applied for extraction of PHA includes bead milling and high pressure (Bury et al. 2001), the use of supercritical-carbon dioxide showing 89 % recovery in *C. necator* (Hejazi et al. 2003), and the use of gamma irradiation as in *B. flexus* (Divyashree and Shamala 2009). Apart from the solvents used, parameters applied during separation of PHAs such as pH and temperature minimize degradation of PHAs.

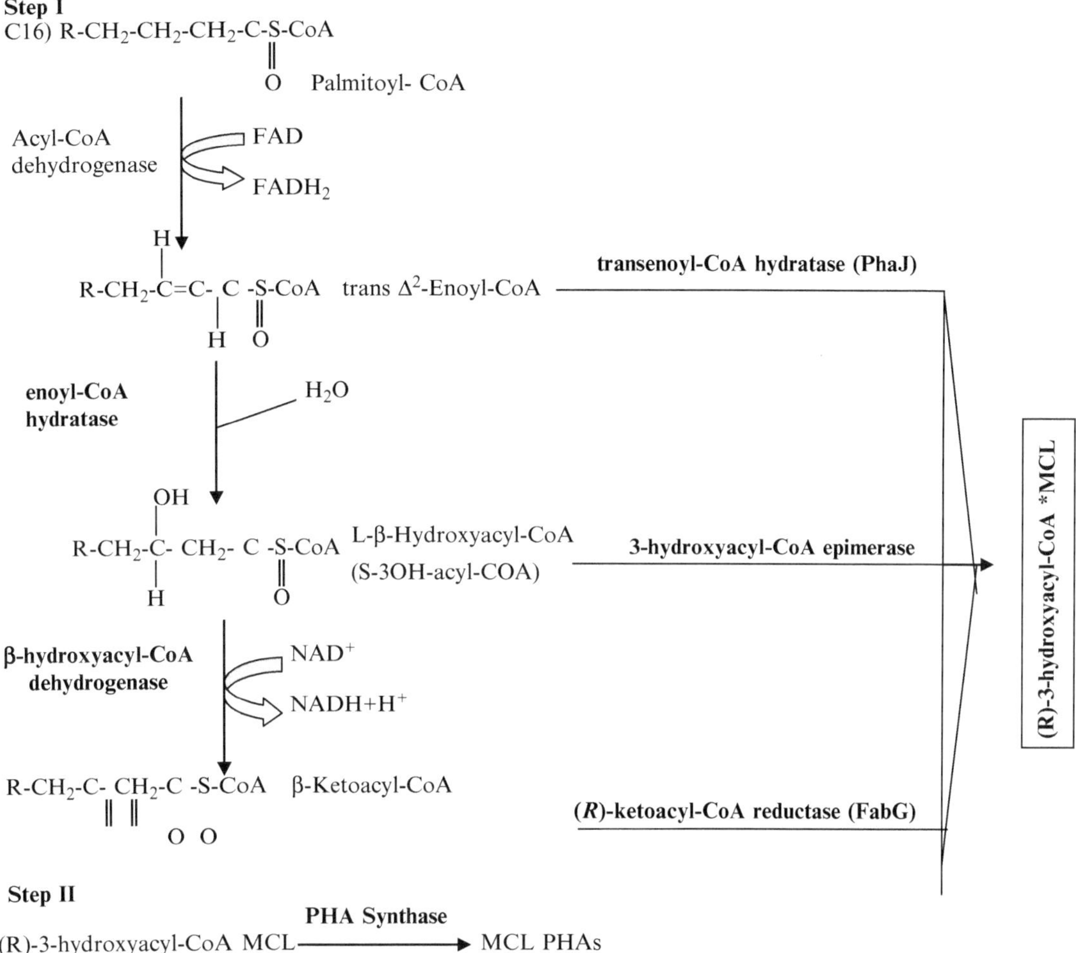

Fig. 3.13 Oxidation of fatty acids with even number of carbon atoms

3.3.4 Applications of Polyhydroxyalkanoates (PHAs)

PHAs are natural polymers produced by numerous microorganisms like bacteria and algae under depletion of nitrogen and phosphorous with excess of carbon source. Polyhydroxyalkanoates can be an effective substitute for industrial thermoplastics and are useful in compostable packaging, molding, veterinary practice, agriculture, and biodegradable rubbers and as paint binder and resorbable materials for medical applications (Fig. 3.16) (Bucci and Tavares 2005; Noda 2001; Chen 2009). PHA monomers can be useful synthons in the pharmaceutical industry (De Eugenio et al. 2010). Small molecular weight PHAs with tertiary helical structure play an important role like ion channels (Brandl et al. 1995). Functionalized mcl-PHAs with longer side chains are promising and versatile candidates for high added-value applications.

Bioplastics are normally highly crystalline and optically active, possess piezoelectric properties, and can be used as an alternative for conventional synthetic plastic. PHAs remain as a resource of chiral hydroxy acid. PHA can be

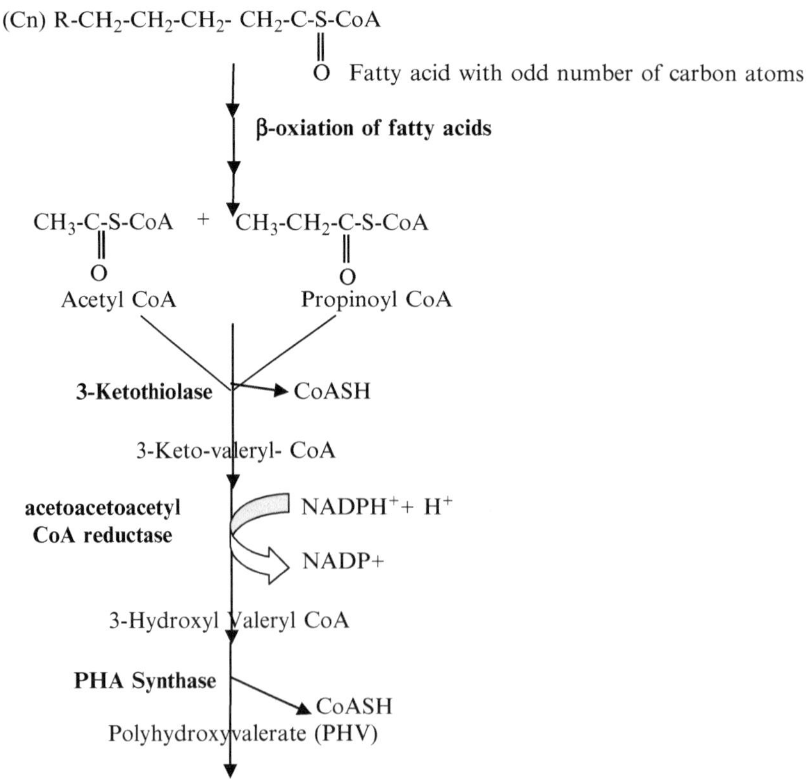

Fig. 3.14 Oxidation of fatty acids with odd number of carbon atoms

a promising candidate in microsphere- or microcapsule-based drug delivery systems due to their unique physiochemical and mechanical properties. PHA microsphere or microcapsule can transport antibiotics, vaccines, and steroids (Orts et al. 2008). Application of PHA as a carrier of drug in anticancer study is reported (Lu et al. 2011). PHAs are applied in medical tools such as sutures, meshes, implants, repair patches, cardiovascular patches, orthopedic pins, nerve guides, and tendon repair devices (Chen and Wu 2005; Valappil et al. 2006; Hazer 2010). Board and paper coated with biobased material as PHB are useful for dry foods, as these materials have relatively high water vapor barrier capacity as well as confer paper or board mechanical strength, thereby protecting products from breakage. Composite of PHA and hydroxyapatite (HA) has the potential to be used in hard tissue replacement and regeneration.

3.3.5 Economics of the PHA Production: Decisive Factors

For successful commercialization of bioplastics, production cost for PHAs should be reduced by optimization of fermentation parameters and purification procedure, cheaper and renewable carbon source, and the use of recombinant microbes (Verlinden et al. 2007). Using sheer diversity of microbial world to screen the microbes capable of producing large amount of PHB using cheaper nutrient sources is also an important aspect. Almost 50 % cost for the production of PHA is due to the feed carbon source used. Waste renewable resources like lignocellulosics, carbohydrates, waste lipids, or alcohols are beneficial in PHAs' production (Braunegg et al. 1998; Solaiman et al. 2006; Khardenavis et al. 2007). Productivity of PHAs can be enhanced by improvements in downstream processing where fermentation by using the continu-

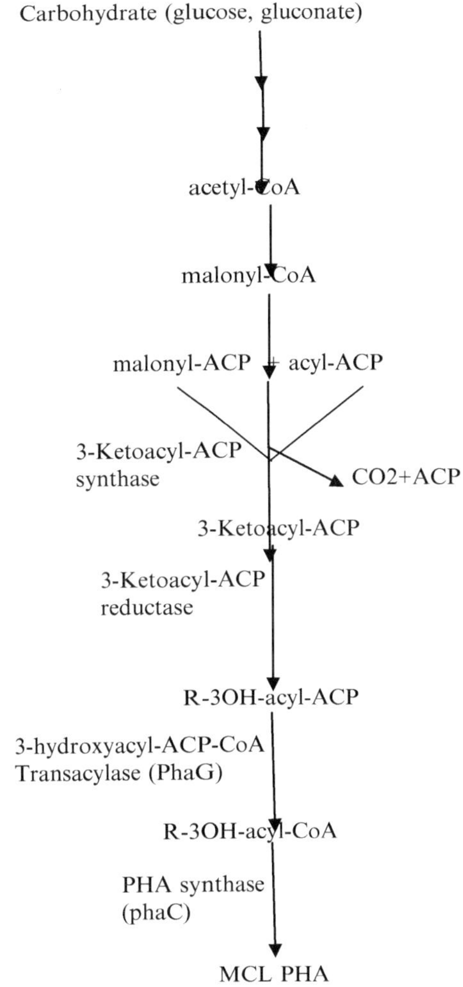

Fig. 3.15 De novo synthesis pathway, acyl carrier protein

ous production mode is beneficial for high productivity in microbial sp. like *Cupriavidus necator* DSM 545 (Horvat et al. 2013).

Application of inexpensive growth additive to enhance rate of biomass production and genetic engineering aspects involving inactivation or modification of enzymes causing intracellular PHA degradation will be new criteria for enhancing volumetric productivity of the process. Alteration of metabolic pathway or introducing the new one by recombinant DNA techniques can also improve the efficiency of microbes to produce PHA enabling them to utilize a wide range of renewable carbon sources as substrate. Genetically modified *Aeromonas hydrophila* had enhanced ability to produce poly(3-hydroxybutyrate-co-3-hydroxyhexanoate) (PHBHHx) (Han et al. 2004).

Recombinant *E. coli* strains developed by inserting the genes for PHA biosynthesis from *R. eutropha,* yielded PHB with mass of $3–11 \times 10^6$ Da. Genes for PHA synthesis were reported to be cloned from *Alcaligenes eutrophus* that were successfully expressed in *E. coli* with 80 % PHB content (Kato et al. 1996) where the same microbial strain is currently utilized for producing PHA on commercial scale by the company Metabolix in the USA (Ren et al. 2000). Other companies producing PHA on industrial scale are Biomatera, PolyFerm Canada, Tianan, Bio-On, Biomer, Kaneka, Tianzhu, etc. Thus, bioplastics will capture 30 % of the total plastics market within the next decade and will decrease the reliance on nonrenewable fossil fuel.

3.4 Perspectives

Polyhydroxyalkanoates representing a renewable source for bioplastics can overcome the petrochemical-derived synthetic biopolymers. Cost-effective production of PHA can be possible by bioprocess design through development of mutant microbial strain by genetic manipulation capable of utilizing low-cost renewable substrates as a carbon source. The major challenge in PHA production is the recovery process that can be optimized using low-cost eco-friendly extraction method.

Acknowledgments The author is thankful to Springer and the editor Dr. V. C. Kalia for giving the opportunity to contribute a book chapter. The support from Dr. D. Y. Patil Vidyapeeth, Pune, is also gratefully acknowledged.

Table 3.3 Precursors used in the literature to produce functionalized mcl-PHA (branched alkyl, cyclohexyl, halogenated)

mcl-PHAs	Precursor	Microbial strain
Branched alkyl mcl-PHA	Citronellol	*P. citronellolis* ATCC 13674
	Alkylhydroxyoctanoates	*P. putida* GPo1
	Methyloctanoates	*P. putida* GPo1
Cyclohexyl mcl-PHA	Cyclohexylbutyric acid	*P. cichorii* YN2
	Cyclohexylvaleric/butyric acid	*P. putida* GPo1
Unsaturated mcl-PHAS	Alkenes (C7–C9)	*P. putida* GPo1
	Undecenoic acid	*P. putida* KCTC 2407
	Undecenoic acid	*P. putida* GPo1
	Hydroxyoctanoic acids	*P. putida* GPo1
	Dicarboxylic acids (C4–C10)	*P. citronellolis* ATCC 13.674
	Undecynoic acid	*P. putida* GPo1
	Undecynoic acid	*P. putida* KCTC 2407
Halogens mcl-PHA	Bromoalkanoic acids (C6–C11)	*P. putida* GPo1
	Chlorooctane	*P. putida* GPo1
	Fluorohexanoic/nonanoic acids	*P. putida* GPo1
	Fluorohexanoic/nonanoic acids	*P. putida* KT2440
	Fluorophenoxyundecanoic acid	*P. putida* 27 N01

Table 3.4 Precursors used in the literature to produce functionalized mcl-PHA (acetoxy, ester, alkoxy, epoxy, thio, cyano, nitro)

mcl-PHAs	Precursor	Microbial strain
Acetoxy mcl-PHA	Octanone, octylacetate	*P. putida* GPo1
Ester/alkoxy/epoxy mcl-PHA	Alkylheptanoate	*P. putida* GPo1
	Alkxyhexanoic/octanoic/undecanoic acids	*P. putida* GPo1
	10-Epoxyundecanoic acid	*P. putida* GPo1
	C7–C12 alkenes	*P. cichorii* YN2
	Soybean oil	*P. stutzeri* 1317
Thio, sulfanyl mcl-PHA	Acetylthiohexanoic acid	*P. putida* KT2442, KT24FadB
	Propylthiohexanoic acid	*Ralstonia eutropha* DSM541
	Propylthioundecanoic acid	*P. putida* KT2440
	Methylsulfanylphenoxyvaleric acid	*P. cichorii* H45, YN2
	Methylsulfanylphenoxyvaleric acid	*P. jessenii* P161
	Thiophenoxyundecanoic acid	*P. putida* 27 N01
Cyano, nitro mcl-PHA	Cyanoundecanoic acid	*P. putida* GPo1
	Cyanophenoxyhexanoic acid	
	Nitrophenoxyhexanoic acid	
	Dinitrophenylvaleric acid	
	Cyanophenoxyhexanoic acid	*P. putida* KT2440
	Nitrophenoxyhexanoic acid	

Table 3.5 Precursors used in the literature to produce functionalized mcl-PHA (aromatics) group at mcl-PHA side chain: aromatics (benzoyl, methylphenoxy, phenoxy, and phenyl)

mcl-PHAs	Precursor	Microbial strain
Aromatics mcl-PHA	Benzoylalkanoic acids (C4–C8)	*P. cichorii* YN2
	Methylphenoxyalkanoic acids (C6, C8)	*P. putida* KCTC 2407
	Methylphenoxyalkanoic acids (C6, C8)	*P. putida* GPo1
	Methylphenoxyalkanoic acids (C6, C8)	*P. putida* KCTC 2407
	Phenoxyundecanoic acid	*P. putida* GPo1
	Phenoxyalkanoic acids (C6, C8, C11)	*P. putida* GPo1
	Phenoxyundecanoic acid	*P. putida* BM01
	Phenylvaleric acid	*P. putida* BM01
	Phenylvaleric acid	*P. putida* GPo1
	Phenyl, tolylvaleric/octanoic acids	*P. putida* GPo1
	Phenylalkanoic acids (C4–C8)	*P. jessenii* C8
	Phenylalkanoic acids (C4–C8)	*P. putida* S12, CA-1, H4, F6, D5
	Phenylalkanoic acids (C6–C11)	*P. putida* U fadA-, ΔFadBA-PhaZ
	Phenylvaleric acid	*P. putida* GPo1
	Phenylvaleric acid	*P. putida* GPo1

Table 3.6 PHA recovery: solvent extraction methods

Extraction method	Solvent used	Strain used	Yield
Solvent extraction	Chloroform	*Bacillus cereus* SPV	31 %
		Cupriavidus necator DSM54	96 %
	Acetone-water process		80–85 %
	Methylene chloride	*C. necator*	98 %
	Acetone, room temperature	*P. putida* GPo1	94 %
	Nonhalogenated solvents: isoamyl propionate, propyl butyrate, isoamyl valerate	*C. necator*	–
Surfactant	SDS	Rec. *Escherichia coli*	89 %
Sodium hypochlorite	Sodium hypochlorite	*C. necator* DSM 545, Rec. *coli*	–
Surfactant-sodium hypochlorite	SDS-sodium hypochlorite	*Azotobacter chroococcum* G-3	87 %
Surfactant-chelate	Triton X-100-EDTA	*Sinorhizobium meliloti*	–
Dispersion of sodium hypochlorite and chloroform	Chloroform-sodium hypochlorite	*B. cereus* SPV	30 %
	Chloroform-sodium hypochlorite	*C. necator*, Rec. *E. coli*	–
Selective dissolution by protons	Sulfuric acid	*C. necator*	95 %
Enzymatic digestion	Microbispora sp. culture-chloroform	*S. meliloti*	–
	Enzyme combined with SDS-EDTA	*P. putida*	–
	Bromelain; pancreatin	*C. necator*	–
Mechanical disruption	Bead mill	*A. latus*	–
	High-pressure homogenization	*A. latus*	–
	SDS-high-pressure homogenization	*Methylobacterium* sp. V49	98 %
	Sonication	*Bacillus flexus*	20 %

(continued)

Table 3.6 (continued)

Extraction method	Solvent used	Strain used	Yield
Supercritical fluid	SC-CO2	*C. necator*	89 %
Cell fragility	Chloroform	*B. flexus*	43 %
	Sodium hypochlorite	*B. flexus*	50 %
	Alkaline hydrolysis	*B. flexus*	50 %
Gamma irradiation	Radiation-chloroform	*B. flexus*	45–54 %
Air classification	–	*E. coli*	90 %
		C. necator	85 %
Spontaneous liberation	–	*E. coli*	80 % (of autolysis)

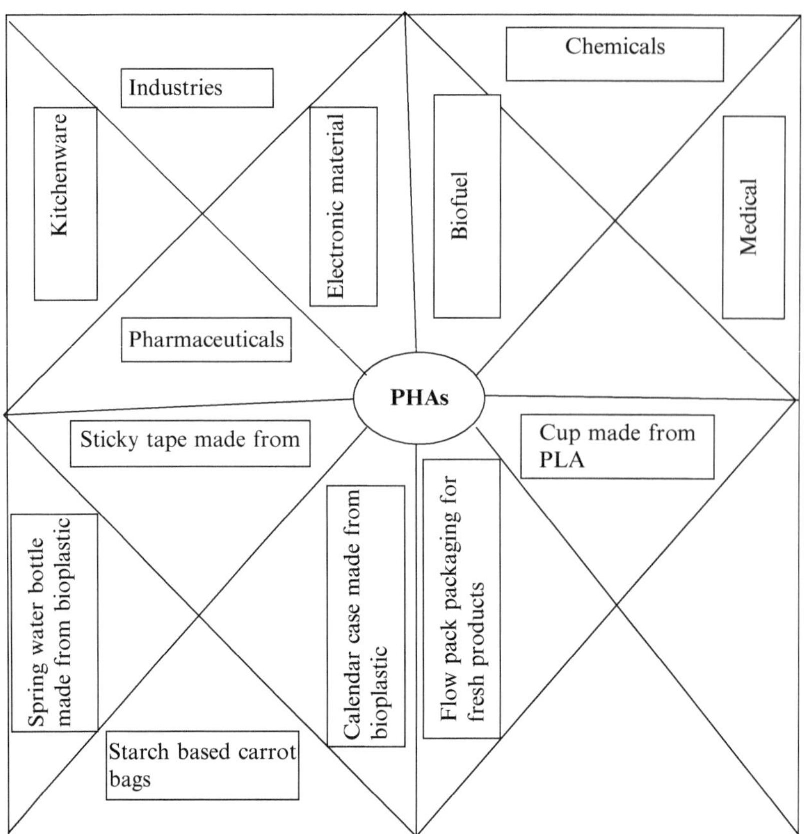

Fig. 3.16 Applications of PHAs in various fields

References

Akar A, Akkaya EU, Yesiladali SKC, Celikyilmaz G, Cokgor EU, Tamerler C, Orhon D, Cakar ZP (2006) Accumulation of polyhydroxyalkanoates by Microlunatus phosphovorus under various growth conditions. J Ind Microbiol Biotechnol 33:215–220. doi:10.1007/s10295-004-0201-2

Amass W, Amass A, Tighe B (1998) A review of biodegradable polymers: uses, current developments in the synthesis and characterization of biodegradable polyesters, blends of biodegradable polymers and recent advances in biodegradation studies. Polym Int 47:89–144. doi:10.1002/(SICI)1097-0126(1998100)47:2<89::AID-PI86>3.0.CO;2-F

Antipov EM, Dubinsky VA, Rubrov AV, Nekrasov YP, Gordeev SA, Ungar G (2006) Strain induced mesophase and hard- elastic behaviour of biodegradable polyhydroxyalkanoates fibers. Polymer 47:5678–5690. doi:10.1016/j.polymer.2005.04.111

Asrar J, Pierre JR (2000) Plasticized polyhydroxyalkanoate compositions and methods for their use in the production of shaped polymer articles. US patent 6127512

Atlic A, Koller M, Scherzer D, Kutschera C, Grillo-Fernandes E, Horvat P, Chiellini E, Braunegg G (2011) Continuous production of poly([R]-3-hydroxybutyrate) by *Cupriavidus necator* in a multistage bioreactor cascade. Appl Microbiol Biotechnol 91:295–304. doi:10.1007/s00253-011-3260-0

Bajaj I, Singhal R (2007) Gellan gum for reducing oil uptake in sev, a legume based product during deep-fat frying. Food Chem 104:1472–1477. doi:10.1016/j.foodchem.2007.02.011

Bassas M, Marques AM, Manresa A (2008) Study of the crosslinking reaction (natural and UV induced) in polyunsaturated PHA from linseed oil. Biochem Eng J 40:275–283. doi:10.1016/j.bej.2007.12.022

Bhuwal AK, Singh G, Aggarwal NK, Goyal V, Yadav A (2013) Isolation and screening of polyhydroxyalkanoates producing bacteria from pulp, paper, and cardboard industry wastes. Int J Biomater 2013:1–10. doi:10.1155/2013/752821

Brandl H, Aeberli B, Bachofen R, Schwegler I, Muller HM, Burger MH, Hoffmann T, Lengweiler UD, Seebach D (1995) Biodegradation of cyclic and substituted linear oligomers of poly (3-hydroxybutyrate). Can J Microbiol 41:180–186. doi:10.1139/m95-185

Braunegg G, Lefebvre G, Genser KF (1998) Polyhydroxyalkanoates, biopolyesters from renewable resources: physiological and engineering aspects. J Biotechnol 65:127–161. doi:10.1016/S0168-1656(98)00126-6

Braunegg G, Bona R, Koller M, Martinz J (2002) Production of polyhydroxyalkanoates: a contribution of biotechnology to sustainable development. In: Proceedings of sustainable development and environmentally degradable plastics in China, Beijing, pp 56–71

Bucci DZ, Tavares LBB (2005) PHB packaging for the storage of food products. J Polym Test 24:564–571. doi:10.1016/j.polymertesting.2005.02.008

Bury D, Jelen P, Kalab M (2001) Disruption of *Lactobacillus delbrueckii* ssp. bulgaricus 11842 cells for lactose hydrolysis in dairy products: a comparison of sonication, high-pressure homogenization and bead milling. Innovat Food Sci Emerg Tech 2:23–29. doi:10.1016/S1466-8564(00)00039-4

Carolan G, Catley BJ, McDougal FJ (1983) The location of tetrasaccharide units in pullulan. Carbohydr Res 114:237–243. doi:10.1016/0008-6215(83)88190-7

Chaijamrus S, Udpuay N (2008) Production and Characterization of polyhydroxybutyrate from molasses and corn steep liquor produced by *Bacillus megaterium* ATCC 6748. Agric Eng Int Cigr J 10:1–12

Cheirsilp B, Radchabut S (2011) Use of whey lactose from dairy industry for economical kefiran production by *Lactobacillus kefiranofaciens* in mixed cultures with yeasts. N Biotechnol 28:574–580. doi:10.1016/j.nbt.2011.01.009

Cheirsilp B, Shimizu H, Shioya S (2003) Enhanced kefiran production by mixed culture of *Lactobacillus kefiranofaciens* and *Saccharomyces cerevisiae*. J Biotechnol 100:43–53. doi:10.1016/s1389-1723(03)80194-9

Chen GQ (2009) A microbial polyhydroxyalkanoates (PHA) based bio- and materials industry. Chem Soc Rev 38:2434–2446. doi:10.1039/B812677C

Chen GQ, Martin KP (2012) Plastics derived from biological sources: present and future: a technical and environmental review. Chem Rev 112:2082–2099. doi:10.1021/cr200162d

Chen GQ, Wu Q (2005) The application of polyhydroxyalkanoates as tissue engineering materials. Biomaterials 26:6565–6578. doi:10.1016/j.biomaterials.2005.04.036

Cheng KC, Catchmark JM, Demicri A (2011) Effect of CMC addition on bacterial cellulose production in a biofilm reactor and its paper sheets analysis. Biomacromolecules 12:730–736. doi:10.1021/bm101363t

Cuthbertson L, Mainprize IL, Naismith JH, Whitfield C (2009) Pivotal roles of the outer membrane polysaccharide export and polysaccharide copolymerase protein families in export of extracellular polysaccharides in Gram-negative bacteria. Microbiol Mol Biol Rev 73:155–177. doi:10.1128/MMBR.00024-08

De Eugenio LI, Escapa IF, Morales V, Dinjaski N, Galan B, Garcia JL, Prieto MA (2010) The turnover of medium-chain-length polyhydroxyalkanoates in *Pseudomonas putida* KT2442 and the fundamental role of PhaZ depolymerase for the metabolic balance. Environ Microbiol 12:207–221. doi:10.1111/j.1462-2920.2009.02061.x

Delben F, Forabosco A, Guerrini M, Liut G, Torri G (2006) Pullulans produced by strains of Cryphonectria parasitica-II. Nuclear magnetic resonance evidence. Carbohydr Polym 63:545–554. doi:10.1016/j.carbpol.2005.11.012

Di Donato P, Fiorentino G, Anzelmo G, Tommonaro G, Nicolaus B, Poli A (2011) Re-use of vegetable wastes as cheap substrates for extremophile biomass production. Waste Biomass Valoriz 2:103–111. doi:10.1007/s12649-011-9062-x

Divyashree MS, Shamala TR (2009) Effect of gamma irradiation on cell lysis and polyhydroxyalkanoate produced by *Bacillus flexus*. Radiat Phys Chem 78:147–152. doi:10.1016/j.radphyschem.2008.08.010

Doi Y, Abe C (1990) Biosynthesis and characterization of new bacterial copolyester of 3-hydroxyalkanoates and 3-hydroxy-ω-chloroalkanoates. Macromolecules 23:3705–3707. doi:10.1021/ma00217a027

Dong Z, Sun X (2000) A new method of recovering polyhydroxyalkanoate from *Azotobacter chroococcum*. Chin Sci Bull 45:252–256. doi:10.1007/BF02884685

Donot F, Fontana A, Baccou JC, Schorr-Galindo S (2012) Microbial exopolysaccharides: main examples of synthesis, excretion, genetics and extraction. Carbohydr Polym 87:951–962. doi:10.1016/j.carbpol.2011.08.083

Draget KI, Smidsrod O, Skjak-Braek G (2005) Alginates from algae. In: Steinbüchel EA, Rhee SK (eds) Polysaccharides and polyamides in the food industry: properties, production, and patents. Wiley-VCH/Verlach GmbH, Weinheim/Germany, pp 1–30

El-Saied H, Basta AH, Gobran RH (2004) Research progress in friendly environmental technology for the production of cellulose products (bacterial cellulose and its application). Polym-Plast Technol Eng 43:797–820. doi:10.1081/ppt-120038065

Finore I, Donato PD, Mastascusa V, Nicolaus B, Poli A (2014) Fermentation technologies for the optimization of marine microbial exopolysaccharide production. Mar Drugs 12:3005–3024. doi:10.3390/md12053005

Freitas F, Alves VD, Reis MA (2011) Advances in bacterial exopolysaccharides: from production to biotechnological applications. Trends Biotechnol 29:388–398. doi:10.1016/j.tibtech.2011.03.008

Full TD, Jung DO, Madigan MT (2006) Production of poly-beta-hydroxyalkanoates from soy molasses oligosaccharides by new, rapidly growing *Bacillus species*. Lett Appl Microbiol 43:377–384. doi:10.1111/j.1472-765x.2006.01981.x

George J, Ramana KV, Sabapathy SN, Jambur HJ, Bawa AS (2005) Characterization of chemically treated bacterial (*Acetobacter xylinum*) biopolymer: some thermo-mechanical Properties. Int J Biol Macromol 37:189–194. doi:10.1016/j.ijbiomac.2005.10.007

Ghasemlou M, Khodaiyan F, Oromiehie A, Yarmand MS (2011a) Development and characterisation of a new biodegradable edible film made from kefiran, an exopolysaccharide obtained from kefir grains. Food Chem 127:1496–1502. doi:10.1016/j.foodchem.2011.02.003

Ghasemlou M, Khodaiyan F, Oromiehie A, Yarmand MS (2011b) Characterization of edible emulsified films with low affinity to water based on kefiran and oleic acid. Int J Biol Macromol 49:378–384. doi:10.1016/j.ijbiomac.2011.05.013

Gong Y, Wang C, Lai RC, Su K, Zhang F, Wang D (2009) An improved injectable polysaccharide hydrogel; modified gellan gum for long term cartilage regeneration in vitro. J Mater Sci 19:1968–1977. doi:10.1039/b818090c

Gouda MK, Swellam AE, Omar SH (2001) Production of PHB by a *Bacillus megaterium* strain using sugarcane molasses and corn steep liquor as sole carbon and nitrogen source. Microbiol Res 156:201–220. doi:10.1078/0944-5013-00104

Gupta S, Pal A, Sahu N, Dalvi R, Kumar V, Mukherjee S (2008) Microbial levan in the diet of *Labeo rohita* Hamilton juveniles: effect on nonspecific immunity and histopathological changes after challenge with *Aeromonas hydrophila*. J Fish Dis 3:649–657. doi:10.1111/j.1365-2761.2008.00939.x

Han J, Qiu YZ, Liu DC, Chen GQ (2004) Engineered Aeromonas hydrophila for enhanced production of poly(3-hydroxybutyrate-co-3-hydroxyhexanoate) with alterable monomers composition. FEMS Microbiol Lett 239:195–201. doi:10.1016/j.femsle.2004.08.044

Han J, Lu Q, Zhou L, Zhou J, Xiang H (2007) Molecular characterization of the *phaECHm* genes, required for biosynthesis of poly(3-hydroxybutyrate) in the extremely halophilic archaeon *Haloarcula marismortui*. Appl Environ Microbiol 73:6058–6065. doi:10.1128/AEM.00953-07

Hazer B (2010) Amphiphilic poly(3-hydroxy alkanoate)s: potential candidates for medical applications. Int J Polym Sci 2010:1–8. doi:10.1155/2010/423460

Hejazi P, Vasheghani-Farahani E, Yamini Y (2003) Supercritical fluid disruption of *Ralstonia eutropha* for poly(beta-hydroxybutyrate) recovery. Biotechnol Prog 19:1519–1523. doi:10.1021/bp034010q

Hoffmann N, Rehm BHA (2004) Regulation of polyhydroxyalkanoate biosynthesis in *Pseudomonas putida* and *Pseudomonas aeruginosa*. FEMS Microbiol Lett 237:1–7. doi:10.1111/j.1574-6968.2004.tb09671.x

Horvat P, Vrana Spoljaric I, Lopar M, Atlic A, Koller M, Braunegg G (2013) Mathematical modelling and process optimization of a continuous 5-stage bioreactor cascade for production of poly[–(R)-3-hydroxybutyrate] by *Cupriavidus necator*. Bioprocess Biosyst Eng 36:1235–1250. doi:10.1007/s00449-012-0852-8

Jang KH, Song KB, Park BS, Kim CH, Chung BH, Choue RW, Lee KS, Lee C, Chun UH, Rhee SK (2001) Levan production by use of the recombinant levansucrase immobilized on titanium-activated magnetite. Process Biochem 37:339–342. doi:10.1016/s0032-9592(01)00215-1

Jau MH, Yew SP, Toh PSY, Chong ASC, Chu WL, Phang SM, Najimudin N, Sudesh K (2005) Biosynthesis and mobilization of poly(3-hydroxybutyrate) P(3HB) by *Spirulina platensis*. Int J Biol Macromol 36:144–151. doi:10.1016/j.ijbiomac.2005.05.002

Jeong YI, Na HS, Oh JS, Choi KC, Song CE, Lee HC (2006) Adriamycin release from self-assembling nanospheres of poly (dl-lactide-co-glycolide)-grafted pullulan. Int J Pharm 322:154–160. doi:10.1016/j.ijpharm.2006.05.020

Jogdand SN (2014) Biopolymers, 408, Archana Building, Sector-17. Vashi, Navi Mumbai

Kahar P, Tsuge T, Taguchi K, Doi Y (2004) High yield production of polyhydroxyalkanoates from soybean oil by *Ralstonia eutropha* and its recombinant strain. Mater Res Innov 83:79–86. doi:10.1016/s0141-3910(03)00227-1

Kalia VC, Raizada N, Sonakya V (2000) Bioplastics. J Sci Ind Res 59:433–445

Kalia VC, Chauhan A, Bhattacharyya G (2003) Genomic databases yield novel bioplastic producers. Nat Biotechnol 21:845–846. doi:10.1038/nbt0803-845

Kalia VC, Lal S, Cheema S (2007) Insight in to the phylogeny of polyhydroxyalkanoate biosynthesis: horizontal gene transfer. Gene 389:19–26. doi:10.1016/j.gene.2006.09.010

Kanmani P, Lim ST (2013) Synthesis and structural characterization of silver nanoparticles using bacterial exopolysaccharide and its antimicrobial activity against food and multidrug resistant pathogens. Process Biochem 47:1099–1106. doi:10.1016/j.procbio.2013.05.011

Kato M, Fukui T, Doi Y (1996) Biosynthesis of polyester blends by *Pseudomonas* sp. 61–3 from alkanoic acids. Bull Chem Soc Jpn 69:515–520, doi:http://dx.doi.org/10.1246/bcsj.69.515

Khanna S, Srivastava AK (2005) Recent advances in microbial polyhydroxyalkanoates. Process Biochem 40:607–619. doi:10.1016/j.procbio.2004.01.053

Khardenavis AA, Suresh Kumar M, Mudliar SN, Chakrabarti T (2007) Biotechnological conversion of agro-industrial wastewaters into biodegradable plastic, poly-beta-hydroxybutyrate. Bioresour Technol 98:3579–3584. doi:10.1016/j.biortech.2006.11.024

Ki Ho Kim, Chan Bok Chung, Young Heui Kim, Ki Soo Kim, Chang Sung Han, Chul Ho Kim RD (2005) Cosmeceutical properties of levan produced by *Zymomonas mobilis*. J Cosmet Sci 56:395–406

Koller M, Salerno A, Miranda de SousaDias M, Reiterer A, Braunegg G (2010) Modern biotechnological polymer synthesis: a review. Food Technol Biotechnol 48:255–269. doi:10.1016/j.biombioe.2010.10.008

Koller M, Gasser I, Schmid F, Berg G (2011) Linking ecology with economy: insights into polyhydroxyalkanoate-producing microorganisms. Eng Life Sci 11:222–237. doi:10.1002/elsc.201000190

Kumar T, Singh M, Purohit HJ, Kalia VC (2009) Potential of *Bacillus* sp. to produce polyhydroxybutyrate from biowaste. J Appl Microbiol 106:2017–2023. doi:10.1111/j.1365-2672.2009.04160.x

Kumar P, Patel SKS, Lee JK, Kalia VC (2013) Extending the limits of *Bacillus* for novel biotechnological applications. Biotechnol Adv 31:1543–1561. doi:10.1016/j.biotechadv.2013.08.007

Kumar P, Singh M, Mehariya S, Patel SKS, Lee JK, Kalia VC (2014) Ecobiotechnological approach for exploiting the abilities of *Bacillus* to produce co-polymer of polyhydroxyalkanoate. Indian J Microbiol 54:151–157. doi:10.1007/s12088-014-0457-9

Kumar P, Mehariya S, Ray S, Mishra A, Kalia VC (2015a) Biodiesel industry waste: a potential source of bioenergy and biopolymers. Indian J Microbiol 55:1–7. doi:10.1007/s12088-014-0509-1

Kumar P, Ray S, Patel SKS, Lee JK, Kalia VC (2015b) Bioconversion of crude glycerol to polyhydroxyalkanoate by *Bacillus thuringiensis* under non-limiting nitrogen conditions. Int J Biol Macromol. doi:10.1016/j.ijbiomac.2015.03.046

Lakshman K, Shamala TR (2006) Extraction of polyhydroxyalkanoate from *Sinorhizobium meliloti* cells using *Microbispora* sp. culture and its enzymes. Enzym Microb Technol 39:1471–1475. doi:10.1016/j.enzmictec.2006.03.037

Leathers TD (2003) Biotechnological production and applications of pullulan. Appl Microbiol Biotechnol 62:468–473. doi:10.1007/s00253-003-1386-4

Lee SH, Oh DH, Ahn WS, Lee Y, Choi JI, Lee SY (2000) Production of poly(3-hydroxybutyrate-co-3-hydroxyhexanoate) by high-cell-density cultivation of Aeromonas hydrophila. Biotechnol Bioeng 67:240–244. doi:10.1002/(sici)1097-0290(20000120)67:2<240::aid-bit14>3.0.co;2-f

Legault BA, Lopez-Lopez A, Alba-Casado JC, Doolittle WF, Bolhuis H, Rodriguex-Valera F, Papke RT (2006) Environmental genomics of "*Haloquadratum walsbyi*" in a saltern crystallizer indicates a large pool of accessory genes in an otherwise coherent species. BMC Genomics 7:171. doi:10.1186/1471-2164-7-171

Lin S-P, Calvar IL, Catchmark JM, Liu J-R, Demirci A, Cheng K-C (2013) Biosynthesis, production and applications of bacterial cellulose. Cellulose 20:2191–2219. doi:10.1007/s10570-013-9994-3

Lu XY, Ciraolo E, Stefenia R, Chen GQ, Zhang Y, Hirsch E (2011) Sustained release of PI3K inhibitor from PHA nanoparticles and in vitro growth inhibition of cancer cell lines. Appl Microbiol Biotechnol 89:1423–1433. doi:10.1007/s00253-011-3101-1

Matko E, Tonka K, Ivka K (2005) Thermal degradation of poly(3- hydroxybutyrate) plasticized with acetyl tributyl citrate. Polym Degrad Stab 90:313–318. doi:10.1016/j.polymdegradstab.2005.04.048

Matsubara Y, Kawada R, Iwasaki K, Kimura Y, Oda T, Muramatsu T (2000) Cloning and sequence analysis of a gene (aly PG) encoding poly-(α-L-guluronate) lyase from *Corynebacterium* sp. strain ALY-1. J Biosci Bioeng 89:199–202. doi:10.1016/s1389-1723(00)88738-1

Mercan N, Beyatli Y (2005) Production of polybetahydroxybutyrate (PHB) by *Rhizobium meliloti*, *R. viciae,* and *Bradyrhizobium japonicum* with different carbon and nitrogen sources, and inexpensive substrates. Zuckerindustrie 130:410–415

Mergaert J, Webb A, Anderson C, Wouters A, Swings J (1994) Microbial degradation of poly(3-hydroxybutyrate) and poly(3-hydroxybutyrate-co-3-hydroxyvalerate) in compost. J Environ Polym Degrad 2:177–183. doi:10.1007/bf02067443

Mironescu M, Mironescu V (2006) New concept for the obtention of biopolymers-based food biofilms. J Agroaliment Proc Technol 12:219–216

Mironescu M, Mironescu ID (2011) Rheological behaviour of a novel microbial polysaccharide. Rom Biotechnol Lett 16:6105–6114. doi:10.1016/j.profoo.2011.09.050c

Mitsue T, Tachibana K, Fujio J (1999) Efficient kefiran production by a mixed culture of *Lacto bacillus kefiranofaciens* KF-75 and yeast strains. Seibutsu Kogaku Kaishi 76:93–103. doi:10.1016/s1389-1723(99)80057-7

Moreira S, Silva NB, Almeida-Lima J, Rocha HAO, Medeiros SRB, Alves C Jr, Gama FM (2009) BC nanofibres, in vitro study of genotoxicity and cell proliferation. Toxicol Lett 189:235–241. doi:10.1016/j.toxlet.2009.06.849

Nakagaito AN, Iwamoto S, Yano H (2005) Bacterial cellulose: the ultimate nano-scalar cellulose morphology for the production of high-strength composites. Appl Phys A Mater Sci Process 80:93–97. doi:10.1007/s00339-004-2932-3

Nakas JP, Keenan TJ, Stipanovic AJ, Tanenbaum SW (2004) Bioconversion of xylose and levulinic acid to polyhydroxyalkanoate (PHA) copolymers. Abstr Pap Am Chem Soc 227:U299

Nicolaus B, Kambourova M, Oner ET (2010) Exopolysaccharides from extremophiles: from fundamentals to biotechnology. Environ Technol 31:1145–1158. doi:10.1080/09593330903552094

Nikel PI, Pettinari MJ, Galvagno MA, Mendez BS (2006) Poly(3-hydroxybutyrate) synthesis by recombinant *Escherichia coli* arcA mutants in microaerobiosis. Appl Environ Microbiol 72:2614–2620. doi:10.1128/aem.72.4.2614-2620.2006

Nishi Y, Uryu M, Yamanaka S, Watanabe K, Kitamura N, Iguchi M, Mitsuhashi S (1990) The structure and mechanical-properties of sheets prepared from bacterial cellulose. 2. Improvement of the mechanical-properties of sheets and their applicability to diaphragms of electroacoustic transducers. J Mater 25:2997–3001. doi:10.1007/bf00584917

Noda I (2001) Films comprising biodegradable PHA copolymers. US patent 6174990, 16 Jan 2001

Nuti MP, De Bertoldi M, Lepidi AA (1972) Influence of phenylacetic acid on poly-β-hydroxybutyrate (PHB) polymerization and cell elongation in Azotobacter chroococcum Beij. Can J Microbiol 18:1257–1261. doi:10.1139/m72-194

Oliveira MR, Silva RSSF, Buzato JB, Celligoi MAPC (2007) Study of levan production by *Zymomonas mobilis* using regional low-cost carbohydrate sources. Appl Microbiol Biotechnol 37:177–183. doi:10.1016/j.bej.2007.04.009

Oner ET (2013) Microbial production of extracellular polysaccharides from Biomass. In: Fang Z (ed) Pretreatment techniques biofuels and biorefineries, Green energy and technology. Springer, Berlin, pp 35–56. doi:10.1007/978-3-642-32735-3_2

Orts WJ, Nobes GAR, Kawada J, Nguyen S, Yu GE, Ravenelle F (2008) Poly(hydroxyalkanoates): biorefinery polymers with a whole range of applications. The work of Robert H. Marchessault. Can J Chem 86:628–640. doi:10.1139/v08-050

Parolis LAS, Parolis H, Paramonov NA, Boan IF, Anton J, Rodriguez-Valera F (1999) Structural studies on the acidic exopolysaccharide from Haloferax denitrificans ATCC 35960. Carbohydr Res 319:133–140. doi:10.1016/s0008-6215(99)00111-1

Patel SKS, Singh M, Kalia VC (2011) Hydrogen and polyhydroxybutyrate producing abilities of *Bacillus* spp. from glucose in two stage system. Indian J Microbiol 51:418–423. doi:10.1007/s12088-011-0236-9

Patel SKS, Singh M, Kumar P, Purohit HJ, Kalia VC (2012) Exploitation of defined bacterial cultures for production of hydrogen and polyhydroxybutyrate from pea-shells. Biomass Bioenergy 36:218–225. doi:10.1016/j.biombioe.2011.10.027

Patel SKS, Kumar P, Singh M, Lee JK, Kalia VC (2015) Integrative approach to produce hydrogen and polyhydroxybutyrate from biowaste using defined bacterial cultures. Bioresour Technol 176:136–141. doi:10.1016/j.biortech.2014.11.029

Piermaria JA, Pinotti A, Garcia MA, Abraham AG (2009) Films based on kefiran, an exopolysaccharide obtained from kefir grain: development and characterization. Food Hydrocoll 23:684–690. doi:10.1016/j.foodhyd.2008.05.003

Pohlmann A, Fricke WF, Reinecke F, Kusian B, Liesegang H, Cramm R, Eitinger T, Ewering C, Potter M, Schwartz E, Strittmatter A, Voss I, Gottschalk G, Steinbuchel A, Friedrich B, Bowien B (2006) Genome sequence of the bioplastic-producing 'Knallgas' bacterium *Ralstonia eutropha* H16. Nat Biotechnol 24:1257–1262. doi:10.1038/nbt1244

Pollock TJ, Throne L, Armentrout RW (1992) Isolation of new *Aureobasidium* strains that produce high-molecular-weight pullulan with reduced pigmentation. Appl Environ Microbiol 58:877–883

Porwal S, Kumar T, Lal S, Rani A, Kumar S, Cheema S, Purohit HJ, Sharma R, Patel SKS, Kalia VC (2008) Hydrogen and polyhydroxybutyrate producing abilities of microbes from diverse habitats by dark fermentative process. Bioresour Technol 99:5444–5451. doi:10.1016/j.biortech.2007.11.011

Quillaguaman J, Delgado O, Mattiasson B, Hatti-Kaul R (2006) Poly(beta-hydroxybutyrate) production by a moderate halophile, *Halomonas boliviensis* LC1. Enzyme Microb Technol 38:148–154. doi:10.1016/j.enzmictec.2005.05.013

Rairakhwada D, Pal A, Bhathena Z, Sahu N, Jha A, Mukherjee S (2007) Dietary microbial levan enhances cellular non-specific immunity and survival of common carp (*Cyprinus carpio*) juveniles. Fish Shellfish Immunol 22:477–486. doi:10.1016/j.fsi.2006.06.005

Rajendran N, Sharanya P, Sneha RM, Ruth AB, Rajam C (2012) Seaweeds can be a new source for bioplastics.

J Pharm Res 5:1476–1479. doi:10.1007/s00436-010-2219-x

Reddy CSK, Ghai R, Kalia V (2003) Polyhydroxyalkanoates: an overview. Bioresour Technol 87:137–146. doi:10.1016/S0960-8524(02)00212-2

Rehm B (2009) Microbial exopolysaccharides: variety and potential applications. In: Microbial production of biopolymers and polymer precursors: applications and perspectives. Caister Academic Press, Norfolk, pp 229–254

Reis RA, Tischer CA, Gorrin PAJ, Iacomini M (2002) A new pullulan and a branched (1–3)-, (1–6)- linked β-glucan from the lichenised ascomycete *Teloschistes flavicans*. FEMS Microbiol Lett 210:1–5. doi:10.1111/j.1574-6968.2002.tb11152.x

Remminghorst U, Rehm BHA (2006) Bacterial alginates: from biosynthesis to applications. Biotechnol Lett 28:1701–1712. doi:10.1007/s10529-006-9156-x

Ren Q, Sierro N, Kellerhals M, Kessler B, Witholt B (2000) Properties of engineered poly-3-hydroxyalkanoates produced in recombinant *Escherichia coli* strains. Appl Environ Microbiol 66:1311–1320. doi:10.1128/AEM.66.4.1311-1320.2000

Sakata Y, Otsuka M (2009) Evaluation of relationship between molecular behaviour and mechanical strength of pullulan films. Int J Pharm 374:33–38. doi:10.1016/j.ijpharm.2009.02.019

Sam S, Kucukasik F, Yenigun O, Nicolaus B, Oner ET, Yukselen MA (2011) Flocculating performances of exopolysaccharides produced by a halophilic bacterial strain cultivated on agro-industrial waste. Bioresour Technol 102:1788–1794. doi:10.1016/j.biortech.2010.09.020

Senior PJ, Dawes EA (1973) The regulation of poly-β-hydroxybutyrate metabolism in *Azotobacter beijerinckii*. J Biochem 134:225–238

Shih IL, Yu YT, Shieh CJ, Hsieh CY (2005) Selective production and characterization of levan by *Bacillus subtilis (natto)* Takahashi. J Agric Food Chem 53:8211–8215. doi:10.1021/jf058084o

Shoda M, Sugano Y (2005) Recent advances in bacterial cellulose production. Biotechnol Bioprocess Eng 10:1–8. doi:10.1007/bf02931175

Shrivastav A, Kim HY, Kim TR (2013) Advances in the applications of polyhydroxyalkanoate nanoparticles for novel drug delivery system. BiomMed Res Int 12:1–12. doi:10.1155/2013/581684

Sima F, Mutlu EC, Eroglu MS, Sima LE, Serban N, Ristoscu C, Petrescu SM, Oner ET, Mihailescu IN (2011) Levan nanostructured thin films by MAPLE assembling. Int J Biol Macromol 12:2251–2256. doi:10.1021/bm200340b

Singh M, Patel SKS, Kalia VC (2009) *Bacillus subtilis* as potential producer for polyhydroxyalkanoates. Microb Cell Fact 8:38. doi:10.1186/1475-2859-8-38

Singh M, Kumar P, Patel SKS, Kalia VC (2013) Production of polyhydroxyalkanoate co-polymer by *Bacillus thuringiensis*. Indian J Microbiol 53:77–83. doi:10.1007/s12088-012-0294-7

Singh M, Kumar P, Ray S, Kalia VC (2015) Challenges and opportunities for customizing polyhydroxyalkanoates. Indian J Microbiol 55:234–249. doi:10.1007/s12088-015-0528-6

Singha TK (2012) Microbial extracellular polymeric substances: production, isolation and applications. IOSR J Pharm 2:276–281. doi:10.9790/3013-0220276281

Snell K, Peoples O (2002) Polyhydroxyalkanoate polymers and their production in transgenic plants. Metab Eng 4:29–40. doi:10.1006/mben.2001.0214

Snell KD, Peoples OP (2009) PHA bioplastic: a value-added coproduct for biomass biorefineries. Biofuels Bioprod Bioref 3:456–467. doi:10.1111/j.1467-7652.2008.00350.x

Solaiman SKY, Ashby RD, Foglia TA, Marmer WN (2006) Conversion of agricultural feedstock and co-products into poly(hydroxyalkanoates). Appl Microbiol Biotechnol 71:783–789. doi:10.1007/s00253-006-0451-1

Steinbuchel A (2001) Perspectives for biotechnological production and utilization of biopolymers: metabolic engineering of polyhydroxyalkanoate biosynthesis pathways as a successful example. Macromol Biosci 1:1–24. doi:10.1002/1616-5195(200101)1:1<1::AID-MABI1>3.0.CO;2-B

Sudesh K, Abe H, Doi Y (2000) Synthesis, structure and properties of polyhydroxyalkanoates: biological polyesters. Prog Polym Sci 25:1503–1555. doi:10.1016/S0079-6700(00)00035-6

Sutherland IW (2005) Polysaccharides from microorganisms, plants and animals. In: Steinbuchel A (ed) Bipopolymers online. Wiley, New York. doi:10.1002/3527600035.bpo15001

Sworn G (2000) Gellan gum. In: Phillips GO, Williams PA (eds) Handbook of hydrocolloids. Woodhead Publishing Ltd, Cambridge, UK, pp 117–135

Tako M, Nakamura S (1989) Evidence for intramolecular associations in xanthan molecules in aqueous media. Agric Biol Chem 53:1941–1946. doi:10.1271/bbb1961.53.1941

Tanaka K, Shimonishi M, Kitagaki M, Ikunaga M (1990) Action of levan fructotransferase of Arthrobacter ureafaciens on three oligosaccharides containing a bifurcose residue. Agric Biol Chem 54:815–817

Tanaka T, Yabe T, Teramachi S, Iwata T (2007) Mechanical properties and enzymatic degradation of poly[(R)-3-hydroxybutyrate] fibers stretched after isothermal crystallization near Tg. Polym Degrad Stab 92:1016–1024. doi:10.1016/j.polymdegradstab.2007.02.017

Taran M (2011) Utilization of petrochemical wastewater for the production of poly (3-hydroxybutyrate) by *Haloarcula* sp. IRU1. J Hazard Mater 188:26–28. doi:10.1016/j.jhazmat.2011.01.036

Tortajada M, da Silva LF, Prieto MA (2013) Second-generation functionalized medium chain-length polyhydroxyalkanoates: the gateway to high-value bioplastic applications. Int Microbiol 16:1–15. doi:10.2436/20.1501.01.175

Uchino K, Saito T, Gebauer B, Jendrossek D (2007) Isolated poly (3-hydroxybutyrate) (PHB) granules complex bacterial organelles catalyzing formation of PHB from acetyl Coenzyme A (CoA) and degradation of PHB to acetyl CoA. J Bacteriol 189:8250–8256. doi:10.1128/jb.00752-07

Valappil SP, Misra SK, Boccaccini A, Roy I (2006) Biomedical applications of polyhydroxyalkanoates, an overview of animal testing and in vivo responses. Expert Rev Med Devices 3:853–868. doi:10.1586/17434440.3.6.853

Verlinden RAJ, Hill DJ, Kenward MA, William CD, Radeckal I (2007) Bacterial synthesis of biodegradable polyhydroxyalkanoates. J Appl Microbiol Rev 102:1437–1449.doi:10.1111/j.1365-2672.2007.03335.x

Vijayendra SV, Shamala TR (2014) Film forming microbial biopolymers for commercial applications-a review. Crit Rev Biotechnol 34:338–357. doi:10.3109/07388551.2013.798254

Volova TG (2004) Properties in polyhydroxyalkanoates-plastic materials of the 21st century: production, properties, applications. Nova Science Publishers, New York, p 282

Volova TG, Kalacheva GS (2005) The synthesis of hydroxybutyrate and hydroxyvalerate copolymers by the bacterium *Ralstonia eutropha*. Microbiology 74:54–59. doi:10.1007/s11021-005-0028-5

Volova TG, Kalachevaa GS, Steinbuchel A (2007) Biosynthesis of multi-component polyhydroxyalkanoates by the bacterium *Wautersia eutropha*. Microbiology 76:704–711. doi:10.1134/S0026261707060082

Volova TG, Boyandin AN, Vasiliev AD, Karpov VA, Prudnikova SV, Mishukova OV, Boyarskikh UA, Filipenko ML, Rudnev VP, Xuan BB, Dung VV, Gitelson II (2010) Biodegradation of polyhydroxyalkanoates (PHAs) in tropical coastal waters and identification of PHA-degrading bacteria. Polym Degrad Stab 95:2350–2359. doi:10.1016/j.polymdegradstab.2010.08.023

Wang L, Shelton RM, Cooper PR, Lawson M, Triffitt JT, Barralet JE (2003) Evaluation of sodium alginate for bone marrow cell tissue engineering. Biomaterials 24:3475–3481. doi:10.1016/s0142-9612(03)00167-4

Wang YW, Wu Q, Chen GQ (2005) Gelatin blending improves the performance of poly(3-hydroxybutyrate-co-3-hydroxyhexanoate)films for biomedical application. Biomacromolecules 6:566–571. doi:10.1021/bm049342d

Witthuhn RC, Schoeman T, Britz TJ (2005) Characterisation of the microbial population at different stages of Kefir production and Kefir grain mass cultivation. Int Dairy J 15:383–389. doi:10.1016/j.idairyj.2004.07.016

Wong TY, Preston LA, Schiller NL (2000) Alginate lyase: review of major sources and enzyme characteristics, structure-function analysis, biological roles, and applications. Annu Rev Microbiol 54:289–340. doi:10.1146/annurev.micro.54.1.289

Wong PAL, Cheung MK, Lo WL, Chua H, Yu PHF (2005) Effects of types of food waste as carbon source on the molecular weight distributions and thermal properties of the biopolymer (polyhydroxybutyrate) produced by two strains of microorganisms. Mater Res Innov 9:4–5. doi:10.1515/epoly.2004.4.1.324

Yasotha K, Aroua MK, Ramachandran KB, Tan IKP (2006) Recovery of medium-chain-length polyhydroxyalkanoates (PHAs) through enzymatic digestion treatments and ultrafiltration. Biochem Eng J 30:260–268. doi:10.1016/j.bej.2006.05.008

Yoo SH, Yoon EJ, Cha J, Lee HG (2004) Antitumor activity of levan polysaccharides from selected microorganisms. Int J Biol Macromol 34:37–41. doi:10.1016/j.ijbiomac.2004.01.002

Young NWG (2002) The yield stress phenomenon and related issues – an industrial view. In: Williams PA, Phillips GO (eds) Gums and stabilizers for the food industry. The Royal Society of Chemistry, Cambridge, UK, pp 226–234

Yu J, Stahl H (2008) Microbial utilization and biopolyester synthesis of bagasse hydrolysates. Bioresour Technol 99:8042–8048. doi:10.1016/j.biortech.2008.03.071

Yu P, Chua H, Huang AL, Ho KP (1999) Conversion of industrial food wastes by *Alcaligenes latus* into polyhydroxyalkanoates. Appl Biochem Biotechnol 78:445–454. doi:10.1385/abab:78:1-3:445

Yue H, Ling C, Yang T, Chen X, Chen Y, Deng H, Wu Q, Chen J, Chen GQ (2014) A seawater-based open and continuous process for polyhydroxyalkanoates production by recombinant *Halomonas campaniensis* LS21 grown in mixed substrates. Biotechnol Biofuels 7:108–119. doi:10.1186/1754-6834-7-108

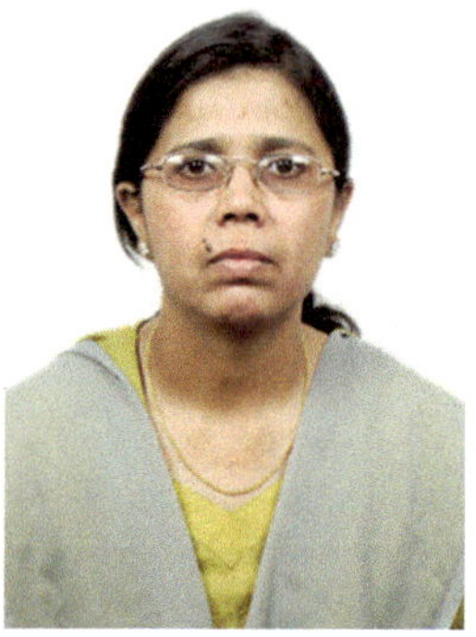

Manjusha Dake received her M.Sc. and Ph.D. degree in Biochemistry from Shivaji University, Kolhapur. She is presently working as an Assistant Professor in Biochemistry at Dr. D.Y. Patil Biotechnology and Bioinformatics Institute, Pune. Her current interests are biofuel and degradative enzymes. She was also a member of the Reviewer Board of the Journal of Microbial and Biochemical Technology. Currently, she is working on applications of enzymes in bioremediation.

Phylogenetic Affiliation of *Pseudomonas* sp. MO2, a Novel Polyhydroxyalkanoate-Synthesizing Bacterium

4

Parveen Kumar Sharma, Jilagamazhi Fu, Xiang Zhang, Richard Sparling, and David B. Levin

Abstract

A bacterium, isolated from wastewater after enrichment with waste canola fryer oil, was found to synthesize 12–20 % of cell dry weight (cdw) medium chain length polyhydroxyalkanoates (mcl-PHAs) using different carbon substrates. On the basis of partial 16S rDNA sequence analysis, this bacterium was first identified as *Pseudomonas putida* and designated as *P. putida* strain MO2. However, the full 16S rDNA gene sequence from a whole genome sequence analysis of this strain showed 100 % identity to 16S rDNA of *Pseudomonas monteilii* SB3101 and *Pseudomonas monteilii* SB3078. The comparison of the *cpn60* gene sequence of *Pseudomonas* sp. strain MO2 with strains of *P. putida*, *P. aeruginosa*, *P. fluorescens*, *P. stutzeri*, *P. syringae*, and *P. monteilii* indicated that this strain is more closely related to *P. monteilii* than *P. putida* type strain KT2440. Based on gene sequence similarity index and phylogenetic analyses, some *P. putida* strains, which were earlier classified as *P. putida*, are also more closely related to the *P. monteilii* cluster. Our analyses show that *P. putida* is a very diverse group with divergent strains, and many strains of *P. putida* that cluster with *P. monteilii* species may need to be reclassified.

P.K. Sharma • J. Fu • D.B. Levin (✉)
Department of Biosystems Engineering, University of Manitoba, R3T 2N2 Winnipeg, Manitoba, Canada
e-mail: pksharmajan11@yahoo.com; Parveen.Sharma@umanitoba.ca; umfu23@myumanitoba.ca; fujilagamazhi@gmail.com; david.levin@umanitoba.ca

X. Zhang
Department of Plant Science, University of Manitoba, R3T 2N2 Winnipeg, Manitoba, Canada
e-mail: justinzhang.xl@gmail.com

R. Sparling
Department of Microbiology, University of Manitoba, R3T 2N2 Winnipeg, Manitoba, Canada
e-mail: richard.sparling@umanitoba.ca

4.1 Introduction

Polyhydroxyalkanoates (PHAs) are intracellular polymers synthesized by bacteria under carbon excess and nitrogen or phosphorus limiting different pathways (Steinbüchel and Füchtenbusch 1998; Lee et al. 1999; Steinbüchel 2001; Rehm and Steinbüchel 2001). PHAs are classified into two classes, short chain length (scl) and medium chain length (mcl), depending on the carbon chain length of the repeat units. The scl-PHAs have 3–5 carbons in the polymer subunits, while mcl-PHAs have 6–14 carbons in the polymer subunits. The

V.C. Kalia (ed.), *Microbial Factories*, DOI 10.1007/978-81-322-2595-9_4

monomer composition of PHAs determines their properties as thermoplastics, elastomers, and/or adhesives (Lee 1996). PHA production in *P. putida* (formerly *P. oleovorans*) was first report by Smet et al. (1983) when this organism was grown in n-octane. Since then, a large number of *Pseudomonas* species have been identified as PHA producers and PHA production appears to be a general feature among pseudomonads.

Pseudomonas spp. are known to produce medium chain length polyhydroxyalkanoates (mcl-PHAs) from a variety of carbon sources, such as glucose, glycerol, and fatty acids, in batch and fed batch cultures (Sun et al. 2007; Sharma et al. 2012). Daird et al. (2002) demonstrated mcl-PHA synthesis by different *Pseudomonas* species belonging to the pseudomonad rRNA homology group. PHA synthesis has been demonstrated and confirmed in *P. putida* (Sharma et al. 2012, 2014), *P. aeruginosa* (Chan et al. 2006; Pham et al. 2004), *P. fluorescens* (Gamal et al. 2013), *P. stutzeri* (Chen et al. 2004), *P. mendocina* (Daird et al. 2002; Guo et al. 2011), *P. resinovorans* (Solaiman et al. 2000), and *P. entomophila* (Chung et al. 2011, 2013). Recently *P. mosselii* TK07 was isolated from wastewater from a vegetable oil processing plant, and this strain was shown to synthesize mcl-PHAs from palm kernel and soybean oil (Chen et al. 2014).

Pseudomonas putida KT2440 (type strain of *P. putida*) and *P. putida* Gpo1 are under development for industrial PHA production. However, although several companies have commercially produced scl-PHAs, mcl-PHAs have yet to be produced in large scale due to their low mcl-PHA yield and the high cost of production (Elbahloul and Steinbüchel 2009). There are only few reports on production of PHAs using low-cost renewable carbon sources like animal fats and vegetable oils (Ashby and Foglia 1998; Cromwick et al. 1996; Füchtenbusch et al. 2000; Tan et al. 1997) and dairy waste like whey (Park et al. 2002). However, production of PHAs from vegetable oils by *Pseudomonas putida* has not been reported.

The genus *Pseudomonas* is a very heterogeneous taxon containing many Gram-negative, rod-shaped, non-sporulating, polar-flagellated, aerobic bacteria (Stanier et al. 1966; Palleroni 1984; 2005; Regenhardt et al. 2002). *Pseudomonas* species have been differentiated on the basis of biochemical tests, fatty acid composition, siderophores, 16S rDNA sequence, and protein profiles (Stanier et al. 1966; Janse et al. 1992; Stead 1992; Vancanneyt et al. 1996; Anzai et al. 2000; Meyer et al. 2008; Santos and Ochman 2004; Goris et al. 2007; Mulet et al. 2009). The type I chaperonin, encoded by *cpn60* (also known as GroEL or Hsp60), is a highly conserved protein found in bacteria that has also been widely used for phylogenetic studies as well as for identification of bacteria (Hill et al. 2004; Lazarovits et al. 2013).

The gene encoding 16S rRNA has been widely used as a molecular sequence to reconstruct inferred phylogenies because it was assumed that the intraspecific variation and horizontal transfer of this gene between organisms were low. Pseudomonads belonging to rRNA homology group I include both fluorescent and nonfluorescent species. Yamamoto et al. (2000) identified two major clusters, designated as intrageneric cluster I (IGC I) and intrageneric cluster II (IGC II). *P. aeruginosa* and *P. stutzeri* formed two subclusters within the IGC I cluster, while *P. putida*, *P. syringae*, and *P. fluorescens* represent three subclusters within IGC II. Pseudomonads belonging to these three branches are able to produce medium chain length PHAs from octanoate (Daird et al. 2002).

Phylogenies based on 16S rDNA intrageneric cluster analyses, however, lack resolution due to their slow rate of evolution, so other molecular targets have been investigated to elucidate the phylogenetic relationships among *Pseudomonas* species (Moore et al. 1996; Yamamoto et al. 2000; Anzai et al. 2000). Genes like *atpD* (ATP synthase), *carA* (carbamoyl phosphate synthase, subunit A), *recA* (recombinase A), *gyrB* (gyrase B), and *rpoD* (RNA polymerase sigma 70) have been employed for phylogenetic analyses (Hilario et al. 2004; Tayeb et al. 2005; Martınez-Murcia et al. 2009). Mulet et al. (2010) demonstrated that three concatenated genes (16S rRNA, *gyrB*, and *rpoD*) were successfully used for phylogenetic analysis of pseudomonads. Konstantinidis et al. (2006) pointed out that concatenation of these three genes could also be employed for the analyses of other bacteria groups, including *Escherichia coli*, *Salmonella*, *Burkholderia*, and *Shewanella*, when compared with their whole genomes.

On the basis of phenotypic heterogeneity, *P. putida* has been divided into three groups: Biovar A, Biovar B, and Biovar C (Palleroni 1984; Stanier et al. 1966; Barrett et al. 1986; Palleroni 2005). Numerical analyses (Barrett et al.1986; Champion et al. 1980) and 16S rRNA gene restriction patterns (Elomari et al. 1994) identified several subgroups within these Biovars and further suggested that *P. putida* consisted of a number of strains, some of which could constitute new species. *P. putida* phylogenetic studies, based on 16S rRNA sequence comparisons, confirmed that strains of Biovars A and B were different (Moore et al.1996; Mulet et al.2013). Multilocus sequence analyses (MLSA) using genes like *cpn60*, *recA*, *rpoA*, *rpoD*, and *dnaJ* have also been employed to identify intraspecific diversity and to confirm the phylogenetic relationships among *P. putida* strains. Consequently, some strains previously described as *P. putida* were separated into other species groups, like *Pseudomonas mosselii* (Dabboussi et al. 2002) or *Pseudomonas monteilii* (Elomari et al. 1997). Using unique features of 16S rRNA genes, Bhushan et al. (2013) developed a strategy to identify *Pseudomonas* strains to the species level. It was concluded that majority of *P. aeruginosa*, *P. stutzeri*, and *P. syringae* strains were identified and classified properly. However, *P. putida* and *P. fluorescens* strains may require reclassification (Bossis et al. 2000).

We isolated a mcl-PHA-synthesizing pseudomonad bacterium from wastewater after enrichment with waste fryer oil and, on the basis of 16S rDNA sequence analysis, identified the bacterium a member of the *Pseudomonas putida* group (99 % nucleotide homology to other *P. putida* strains). However, upon further analyses, the 16S rDNA sequence showed 100 % nucleotide sequence homology to two newly sequenced *Pseudomonas monteilii* strains, SB3101 and SB3078. Identification of the newly isolated bacterium as a strain of *P. monteilii* was further confirmed by phylogenetic analyses and gene similarity indices (GSI) using the *cpn60* gene, phylogenetic analyses using the PHA synthase genes (*phaC1* and *phaC2*), and phylogenetic analyses with several individual or concatenated housekeeping genes (*dnaJ*, *gyrA*, *gyrB*, *rpoA*, and *rpoD*). The pseudomonads are a diverse group of saprophytic bacteria which are present in soil, water, plants, and animals. *Pseudomonas putida* strains are well known for degradation of xenobiotics (Timmis 2002; Nakazawa 2003; Timmis and Pieper 1999; Nelson et al. 2002).

4.2 Isolation and Identification of a Novel PHA-Producing Bacterium

Pseudomonas sp. MO2 was isolated from wastewater using Ramsay's minimal medium with waste fryer oil as sole carbon source. The culture was purified by streaking on Luria Bertani plates. Genomic DNA was isolated using ChargeSwitch® gDNA Mini Bacteria Kit (Life Technologies, Burlington, ON). The whole genome was sequenced and assembled by Genome Quebec, Montreal, using Pacific Biosciences (PacBio) RS II Sequencing Technology. The genome was annotated using NCBI pipeline (Acc. No. JFBC00000000). *Pseudomonas* strains used in present study are listed in Table 4.1. BLASTn analysis of the complete 16S rDNA of *Pseudomonas* MO2 (1538 bp) showed 100 % identity to *P. monteilii* SB3078, *P. monteilii* SB3101, and *Pseudomonas* sp. VLB120 (Fig. 4.2), but only 99 % identity to *Pseudomonas* FGI182, *P. putida* S16, *P. putida* H8234, *P. putida* HB3267, and other *P. putida* strains. *P. monteilii* SB3078 and SB3101 are benzene-, toluene-, and ethylbenzene-degrading bacteria, while *Pseudomonas* sp. VLB120 is a solvent-tolerant, styrene-degrading bacterium isolated from soil (Dueholm et al. 2014; Otto et al. 2004). All the three bacteria have recently been sequenced after December 2013. *P. putida* S16 is a solvent-tolerant nicotine-degrading bacterium (Wang et al. 2007; Tang et al. 2012). *Pseudomononas putida*

Table 4.1 *Pseudomonas* strains, genome accession numbers, and genes (locus tags) used in present study

Genome name	Genome acc. no	Gene locus tag						
		16S rRNA	*cpn60*	*rpoA*	*rpoD*	*gyrA*	*gyrB*	*dnaJ*
P. putida S16	NC_015733	PPS_r01	PPS_4332	PPS_0475	PPS_0383	PPS_1408	PPS_0012	PPS_0403
P. putida 14164	AP013070	PP4_r0010	PP4_08720	PP4_05120	PP4_04200	PP4_40040	PP4_00040	PP4_04400
P. putida SJTE-1	AKCL00000000	A210DRAFT_00860	A210DRAFT_01900	A210DRAFT_03068	A210DRAFT_02963	A210DRAFT_01435	A210DRAFT_00004	A210DRAFT_02143
P. putida CSV86	NZ_ AMWJ00000000	CSV86_r20517	CSV86_04192	CSV86_08943	CSV86_22371	CSV86_27714	CSV86_19158	CSV86_22598
P. putida PC9	CP003738	B479_r26783	B479_21790	B479_02880	B479_02415	B479_06830	B479_00265	B479_02515
P. putida H8234	CP005976	L483_00455	L483_26850	L483_02465	L483_01975	L483_06360	L483_32300	L483_02075
P. putida MTCC 5279	AMZE00000000	D095DRAFT_04923	D095DRAFT_03746	D095DRAFT_02128	D095DRAFT_01341	D095DRAFT_00193	D095DRAFT_04077	D095DRAFT_02038
P. putida LS46	ALPV02000000	PPUTLS46_0r25527	PPUTLS46_013648	PPUTLS46_023893	PPUTLS46_014694	PPUTLS46_015444	PPUTLS46_016579	PPUTLS46_007106
P. putida F1	NC_009512	Pput_R0001	Pput_4363	Pput_0512	Pput_0421	Pput_3947	Pput_0004	Pput_0441
P. putida B6-2	AGCS00000000	KKKDRAFT_00141	KKKDRAFT_04728	KKKDRAFT_00530	KKKDRAFT_04519	KKKDRAFT_04318	KKKDRAFT_00025	KKKDRAFT_05822
P. putida W619	NC_010501	PputW619_R0006	PputW619_0968	PputW619_4724	PputW619_4815	PputW619_1374	PputW619_0004	PputW619_0706
P. putida B001	NZ_ CAED01000000	PPB1DRAFT_04917	PPB1DRAFT_03185	PPB1DRAFT_02238	PPB1DRAFT_04775	PPB1DRAFT_01264	PPB1DRAFT_04934	PPB1DRAFT_03026
P. putida GB-1	NC_010322	PputGB1_R0001	PputGB1_4488	PputGB1_0508	PputGB1_0418	PputGB1_1358	PputGB1_0006	PputGB1_4727
P. putida TRO1	APBQ0100001	C206_r13460	C206_04422	C206_22169	C206_03629	C206_11184	C206_04924	C206_10802
P. putida KT2440	NC_002947	PP_16SA	PP_1361	PP0479	PP_0387	PP_1767	PP_0013	PP_0407
P. putida S12	NZ_ AYKV01000000	B478DRAFT_00819	B478DRAFT_0315	B478DRAFT_0570	B478DRAFT_0025	B478DRAFT_0562	B478DRAFT_0004	B478DRAFT_05215
P. putida DOT-T1E	CP003734	T1E_5788	T1E_4381	T1E_3477	T1E_2166	T1E_3016	T1E_0874	T1E_0653
P. putida Idaho	AGFJ00000000	KICDRAFT_04973	KICDRAFT_05289	KICDRAFT_01684	KICDRAFT_06289	KICDRAFT_04345	KICDRAFT_04591	KICDRAFT_06260
P. putida BIRD-1	CP002290	PPUBIRD1_r0001	PPUBIRD1_4201	PPUBIRD1_0516	PPUBIRD1_0424	PPUBIRD1_3846	PPUBIRD1_0077	PPUBIRD1_0445
P. putida UW4	CP003880	PputUW4_R0030	PputUW4_04386	PputUW4_04859	PputUW4_04952	PputUW4_01358	PputUW4_00004	PputUW4_00691
P. putida ND6	CP003588	YSA_11387	YSA_02940	YSA_05974	YSA_05784	YSA_02244	YSA_05066	YSA_03420
P. putida NB2011	ASJX01000001	L321_r02742	L321_23936	L321_12624	L321_11605	L321_00737	L321_17452	L321_03911
P. monteilii SB3101	CP006979	X970_00550	X970_20960	X970_00800	X970_00290	X970_04940	X970_25675	X970_00400
P. monteilii SB3078	CP006978	X969_00560	X969_10100	X969_00810	X969_00300	X969_04965	X969_26040	X969_00410
Pseudomonas FGI182	CP007012	C163_00600	C163_21760	C163_02435	C163_01930	C163_00020	C163_06545	C163_02030
Pseudomonas VLB120	CP003961.1	PVLB_r25916	PVLB_04630	PVLB_22805	PVLB_23270	PVLB_02625	PVLB_00020	PVLB_03450

P. aeruginosa PA7	NC_009656	PSPA7_0811	PSPA7_4956	PSPA7_0862	PSPA7_0718	PSPA7_1961	PSPA7_0004	PSPA7_5480
P. aeruginosa PAO1	CP006831	PA0668.1	PA4385	PA4238	PA0576	PA3168	PA0004	PA4760
P. stutzeri A1501	CP000304	PST_0759	PST_3145	PST_0809	PST_0712	PST_2343	PST_0004	PST_3326
P. fluorescens Pf5	CP000076	PFL_0119	PFL_4838	PFL_5558	PFL_5663	PFL_4314	PFL_0004	PFL_0828
P. mendocina NK-01	CP002620	MDS_r003	MDS_3990	MDS_4268	MDS_4366	MDS_1954	MDS_0004	MDS_3928
P. mendocina ymp	CP000680	Pmen_R0008	Pmen_3688	Pmen_3884	Pmen_4028	Pmen_1848	Pmen_0004	Pmen_3623
P. entomophila L48	CT573326	PSEEN_16s_1	PSEEN4460	PSEEN0514	PSEEN0413	PSEEN1487	PSEEN0004	PSEEN0434
P. putida H3267	CP003738	B479_r26783	B479_21790	B479_02880	B479_02415	B479_06830	B479_00265	B479_22975
Pseudomonas MO2	JFBC00000000	BC89_30710	BC89_13485	BC89_07790	BC89_22730	BC89_07010	BC89_23970	BC89_15680
P. resinovorans 106553	AP013069	PCA10_r0010	PCA10_46990	PCA10_06160	PCA10_05250	PCA10_39190	PCA10_00040	PCA10_49260
P. fulva 12-X	CP002727	Psefu_R0001	Psefu_3538	Psefu_0665	Psefu_3960	Psefu_2016	Psefu_0004	Psefu_3600
P. syringae DC3000	AE016853	PSPTOimg_006280	PSPTOimg_045270	PSPTOimg_006790	PSPTOimg_005530	PSPTOimg_018060	PSPTOimg_000040	PSPTOimg_0046590

HB3267 was isolated from a hospital in Besançon, France, from an inpatient and is known to kill insects as well as carry antibiotic resistance genes (Molina et al. 2014).

4.3 PHA Production by *Pseudomonas* sp. MO2

Three waste substrates (waste fryer oil from McCain Food, Portage La Prairie, Manitoba, Canada) and two by-products of biodiesel industries (REG glycerol and REG fatty acids from REG, Danville, Illinois, USA) were used for PHAs production. *Pseudomonas* sp. MO2 was grown in Ramsay's minimal medium (Ramsay et al. 1991) with REG glycerol (2 % V/V), REG fatty acids (1 % V/V), and waste canola fryer oil (1 % V/V). Cell dry weight, PHA production, and PHA monomer composition were determined as described earlier (Braunegg et al. 1978; Sharma et al. 2012). The maximum cell mass (measured as cell dry weight=cdw) of *Pseudomonas* sp. MO2 varied with the substrate used. REG fatty acid-grown cultures accumulated maximum cdw (2.77 g/L) after 48 h post-inoculation (h pi) followed by glucose (2.27 g/L) after 24 h pi (Table 4.2). The mcl-PHAs were detected after 12 h pi; however, the production was very low in medium containing glucose (4.8 % of cdw) and glycerol (4.13 % of cdw) compared with medium containing REG fatty acids (13.37 % of cdw) or waste fryer oil (12.37 % of cdw). In glucose and glycerol medium, maximum PHA synthesis was observed at 48 h pi, while maximum PHA synthesis was observed after 36 h pi in medium containing free fatty acids or waste fryer oil medium. The PHAs production decreased significantly after 48 h pi in medium containing glucose, glycerol, or waste fryer oil, but not in medium containing REG free fatty acids. Like other *Pseudomonas* strains, *Pseudomonas* sp. MO2 had six *pha* genes (*phaC1ZC2DFI*) which are highly conserved (unpublished data). The *pha* genes of *Pseudomonas* sp. MO2 had the same size and organization as *Pseudomonas putida* KT2440 and *Pseudomonas putida* LS46 (Sharma et al. 2014).

The monomer compositions of mcl-PHAs synthesized (Table 4.2) in medium containing glucose versus glycerol were very similar, and the major component of polymers was 3-hydroxydecanoate (52–64 mol %). In medium containing free fatty acids, 3-hydroxyoctanoate and 3-hydroxytetradecanoate were the major monomers (≠30 mol % each), but 3-hydroxydecanoate and 3-hydroxydodecanoate were present in lesser amounts (17.5–19.0 mol %). Finally, the major constituents of mcl-PHAs synthesized in medium containing waste fryer oil and REG fatty acids were 3-hydroxyoctanoate (39.5 mol %) and 3-hydroxydecanoate (32.9 mol %). Not only did the monomer composition of the polymers vary with substrate, they also varied with the growth state of the cells, as related to the time of incubation before samples were taken. In the case of glucose-grown cells, 3-hydroxyhexanoate was more than 39 mol % at 12 h pi, but decreased with further incubation, with a corresponding increase in 3-hydroxydecanoate, while the 3-hydroxyoctanoate content remained more or less constant. In glycerol medium, a decrease in the 3-hydroxyhexanoate content of the polymer between 12 and 96 h was accompanied by an increase in the 3-hydroxyoctanoate and 3-hydroxydecanoate content. In fatty acid medium, the 3-hydrohexanoate and 3-hydroxydecanoate content remained the same from 12 to 96 h pi, but the 3-hydroxyoctanoate and 3-hydroxytetradecanoate contents of the polymer changed. Interestingly, there was no significant change in monomer composition of the mcl-PHAs in waste fryer oil medium from 12 to 72 h pi, but at 96 h pi, the 3-hydroxydodecanoate and 3-hydroxytetradecanoate content increased as the mol % of 3-hydroxyhexanoate, 3-hydroxyoctanoate, and 3-hydroxydodecanoate decreased.

The monomer composition of mcl-PHAs synthesized by *Pseudomonas* sp. MO2 indicates that precursors for PHA synthesis are provided by de novo fatty acid synthesis pathway when the cells are cultured in medium containing glucose or glycerol or by the β-oxidation pathway when the cells are grown in medium containing fatty acids or waste fryer oil. The role of de novo synthesis of fatty acids in the synthesis of mcl-PHAs has

Table 4.2 Cell biomass and PHA production *Pseudomonas* MO2 in RMM with glucose, REG glycerol, REG fatty acids, waste fryer oil, and the corresponding mcl-PHA polymer monomer composition

Substrate	Incubation time (h)	CDW (g/L)	% PHA production	Monomer composition (% mol)[a]				
				HHx	HO	HD	HDD	HTD
Waste fryer oil (1 %)	12	1.48±0.04	12.78±0.69	9.25±1.22	37.66±5.17	28.05±3.83	11.52±3.27	13.55±1.71
	24	2.91±0.17	15.22±0.91	8.36±1.14	40.52±0.85	33.95±1.44	10.36±1.07	6.79±0.89
	36	2.77±0.07	15.05±1.39	9.04±0.76	39.50±0.48	32.92±3.28	11.47±2.33	7.05±3.29
	48	2.74±0.18	12.60±0.92	9.10±0.33	39.78±0.88	33.02±1.05	12.37±1.84	5.70±0.53
	72	2.64±0.09	10.43±0.78	7.80±1.13	36.73±1.23	29.11±1.68	17.59±2.82	8.75±1.06
	96	2.54±0.03	8.48±1.78	3.73±2.63	39.08±2.45	26.44±3.21	18.76±1.89	11.96±2.31
Glucose (2 %)	12	0.75±0.06	4.80±2.64	15.95±2.45	2.35±3.68	1.74±3.32	55.99±3.31	23.94±4.89
	24	2.27±0.38	6.29±0.34	17.62±2.12	8.58±0.81	32.31±3.76	25.56±4.18	11.90±2.18
	36	1.99±0.08	16.68±2.65	12.11±3.75	13.74±1.04	52.12±8.23	18.06±2.26	3.95±1.22
	48	2.05±0.05	16.94±4.37	7.53±2.10	16.39±0.28	54.89±4.52	15.76±3.68	5.40±2.78
	72	1.87±0.08	16.30±3.17	8.54±1.81	19.57±1.06	44.92±6.13	19.18±2.10	7.77±1.14
	96	0.92±0.13	17.21±2.87	8.28±1.88	19.25±1.78	35.76±4.23	28.76±5.12	7.93±2.26
REG glycerol (2 %)	12	0.74±0.02	4.13±0.37	15.95±2.51	2.35±0.98	1.74±0.89	55.99±0.72	23.94±5.57
	24	2.01±0.06	5.37±0.53	17.72±1.87	8.58±0.58	32.31±5.56	29.56±3.29	11.90±0.93
	36	2.14±0.07	11.32±1.24	12.11±3.04	13.74±1.71	52.12±3.04	18.06±2.22	3.95±1.22
	48	1.97±0.12	12.24±0.88	7.53±1.46	16.39±1.04	54.89±2.13	15.76±1.67	5.40±3.07
	72	1.92±0.09	11.71±1.32	8.54±0.43	19.57±0.71	44.92±3.15	19.18±1.89	7.77±1.59
	96	1.48±0.09	9.02±1.42	8.28±1.40	19.25±2.75	35.76±4.23	28.76±4.53	7.93±1.82
REG fatty acids (1 %)	12	0.71±0.02	13.37±2.72	4.92±1.69	37.62±8.2	33.34±1.48	14.91±0.52	9.18±0.91
	24	1.80±0.02	17.81±2.72	4.89±1.27	45.49±5.43	17.36±5.32	15.58±3.15	16.66±3.22
	36	1.96±0.11	20.37±1.57	3.33±0.45	31.26±5.44	19.02±3.06	17.50±1.91	28.87±7.04
	48	1.58±0.08	20.70±0.59	3.04±0.30	29.46±2.86	16.76±3.21	17.82±0.81	32.89±4.87
	72	1.12±0.08	20.77±0.51	3.63±0.15	36.86±1.48	22.45±4.17	16.77±2.26	20.26±4.65
	96	1.04±0.04	20.22±1.10	5.28±1.12	45.12±7.97	20.11±3.23	13.86±3.21	15.62±7.85

[a]Mean of three independent replicate samples±standard deviation

been established (Rehm et al. 2001). During fatty acid synthesis, R-3-hydroxyl fatty acid-ACP intermediates are synthesized and then converted to R-3-hydroxyl fatty acid by the PhaG enzyme (Matsumoto et al. 2001). In addition to this, an enzyme, 3-hydroxyacyl-CoA ligase, has also been hypothesized to convert R-3-hydroxyl fatty acid to (R)-3-hydroxyl fatty acid-CoA (Wang et al. 2012). The (R)-specific enoyl-CoA hydratase, PhaJ, is an enzyme that could potentially provide (R)-3 hydroxyacyl-CoA precursors for mcl-PHA synthesis derived via the fatty acid β-oxidation pathway when the bacterial cells are grown on fatty acids (Fiedler et al. 2002). Both these enzymes are the connecting link between fatty acid metabolism and PHA synthesis. Changes in monomer composition of mcl-PHA polymers occur during the exponential growth phase, but the final polymer subunit composition is determined during stationary phase when cell growth is arrested by nutrient limitation (e.g., nitrogen) in the presence of excess carbon. Cells under carbon excess and nutrient limitation conditions produce PHA polymers employing PHA synthase (PhaC). Under carbon-limited (starvation) conditions, however, PHA depolymerase (PhaZ) degrades the PHA polymers to release R-hydroxyalkanoic acids, which are then used as a carbon and energy source for cell growth (de Eugenio et al. 2010).

4.4 Phylogenetic Affiliation of *Pseudomonas* sp. MO2

Nucleotide sequences were aligned using Bioedit (Thompson et al. 1994) and the ends of the sequences were trimmed to equal lengths. Phylogenetic analyses were conducted using MEGA6 (Tamura et al. 2013). A series of individual neighbor-joining trees based on 16S rDNA, *cpn60, dnaJ, gyrA, gyrB, rpoA, and rpoD* were generated. Five genes (*cpn60, dnaJ, gyrA, rpoA*, and *rpoD*) were concatenated using DnaSP software (Librado and Rozas 2009; Rozas 2009) and phylogenetic trees were constructed using the Jukes and Cantor method (Jukes and Cantor 1969; Saitou and Nei 1987). Bootstrap values greater than 50 % (from 1000) are indicated at the branch nodes of the corresponding trees. Nucleotide polymorphic sites, the average number of nucleotide differences, nucleotide diversity (per site), synonymous site PI (JC), and non-synonymous site PI (JC) were calculated using DnaSP software (Rozas 2009).

Nucleotide sequences of the *Pseudomonas* sp. MO2 *cpn60* (locus tag BC89_13485), *dnaJ* (locus tag BC89_15680), *gyrA* (locus tag BC89_07010), *gyrB* (locus tag BC89_23970), *rpoA* (locus tag BC89_07790), *rpoD* (locus tag BC89_22730), *phaC1* (locus tag BC89_22160), and *phaC2* (locus tag BC89_22170) genes were retrieved from the annotated genome sequence and compared with genes from other *Pseudomonas* strains. Other *P. putida* (belonging to Biovar A and Biovar B), as selected by Mulet et al. (2013), were also included in this study. For comparison, other strains of *Pseudomonas* species were also included in the present study, including *P. aeruginosa* PA7 and PAO1, *P. fluorescens* Pf0-1, *P. stutzeri* A1501, *P. syringae* pv. *tomato* DC3000, *P. entomophila* L48, *P. mendocina* NK-01 and ymp, *P. monteilii* SB3101 and SB3078, *P. putida* H3267, *Pseudomonas* sp. VL120, and *Pseudomonas* FG1182. In all, 38 *Pseudomonas* strains were included. All of the gene sequences mentioned above were obtained from the Integrated Microbial Genomes (IMG) database (www.img.jgi.doe.gov). The genome accession numbers and locus tags for all the genes from all the *Pseudomonas* species included in this study are listed in (Table 4.1).

Phylogenetic analysis based on partial 16S rDNA sequences (including sequences from Mulet et al. 2013) revealed two major clades (Fig. 4.1). One clade contained *P. mendocina*, *P. aeruginosa*, and *P. stutzeri*, while the second clade contained several subgroups consisting of mostly *P. putida* strains (Fig. 4.1), a mixture of *P. monteilii* strains, *P. putida* strains, and other *Pseudomonas* species (group 2b), a distinct subgroup of *P. putida* consisting of strains W619 and NBRC 14612 (group 3), and a mixture of other *Pseudomonas* species that includes *P. putida* UW4, *P. fluorescens*, and *P. syringae* (group 4). *Pseudomonas* sp. MO2 formed a separate subgroup (Fig. 4.1) with *P. monteilii* SB3101 and SB3078, *P. putida* S16, and *Pseudomonas* sp. VLB120. On the basis of 16S

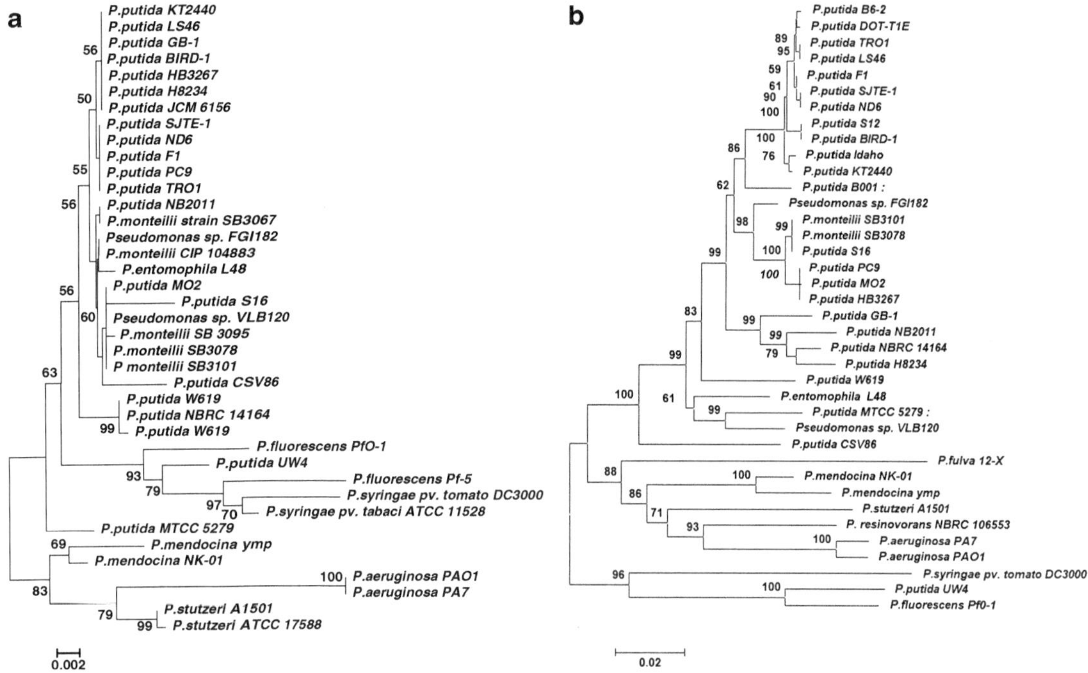

Fig. 4.1 Molecular phylogenetic analysis of *Pseudomonas* species by maximum likelihood method based on Juke and Cantor model. (**a**) *cpn60* and (**b**) 16S rDNA. The tree is drawn to scale, with branch lengths measured in the number of substitutions per site

rDNA phylogeny, it was evident that *Pseudomonas* sp. MO2 was more closely related to *P. monteilii* than the *P. putida* type strains ATCC 12633 and *P. putida* JCM6156, from which the *P. putida* strain KT2440 was derived. On the basis of 16S rRNA, 9 strains (*P. putida* S16, *Pseudomonas* MO2, *P. monteilii* SB3101, *P. monteilii* 3078, *P. monteilii* SB3095, *P. entomophila* L48, *P. putida* CFBP5934, and *P. mosselii* CFML90-83) formed a separate subgroup. These data suggest that *Pseudomonas* MO2 is closer to *P. monteilii* than to *P. putida*. To further resolve this relationship, phylogenic analyses were conducted with *cpn60* sequences and concatenated sequences.

4.4.1 Phylogenetic Analysis Based on *cpn60* Genes

Chaperonin 60 (encoded by the *cpn60* gene) is a highly conserved protein found in bacteria, and the analysis of this gene from wide variety of bacteria has indicated that a 549–567 bp region of the *cpn60* gene (the universal target or UT) can be used for the identification and classification of bacteria. We used the *cpn60* UT sequence to conduct phylogenetic analyses for 38 *Pseudomonas* strains. Phylogenetic tree based on *cpn60* gene sequence analysis separated the *P. putida* type strain KT2440 from *P. fluorescens*, *P. aeruginosa*, *P. syringae*, *P. resinovorans*, *P. stutzeri*, *P. mendocina*, and *P. entomophila* and confirmed previous phylogenetic analyses based on 16S rRNA gene sequences. The *cpn60* phylogenetic analysis further separated *P. putida* strains into five subgroups. Ten *P. putida* strains (B6-2, DOT-T1E, TRO1, LS46, F1, SJTE-1, ND6, S12, BIRD-1, and Idaho) clustered with the *P. putida* type strain KT2440. Our recently isolated *P. putida* strain LS46 also clustered with *P. putida* KT2440. Four *P. putida* strains, NB2011, GB-1, NBRC14164, and H8234, formed another subgroup, and six *Pseudomonas* strains, *P. monteilii* SB3101, SB3078, *P. putida* S16, *P. putida* PC2, *Pseudomonas* sp. FGI182, and *Pseudomonas* MO2, formed a separate subgroup. Four *P. putida*

MTCC5279, *P. putida* CSV86, *Pseudomonas* spp. VLB120, and *P. entomophila* L48 were clustered in one subgroup, while *P. putida* W619 did not cluster with any of *P. putida* strains. The *cpn60* gene of *P. putida* type strain KT2440 showed 99.2–99.6 % similarity indices relative to ten other *P. putida* strains (B6-2, DOT-T1E, TRO1, LS46, F1, SJTE-1, ND6, S12, BIRD-1, and Idaho), but this group had only 96.4–96.6 % gene similarity indices relative to *P. monteilii* group (SB3101, SB3078, S16, PC2, FGI182, and MO2). The *cpn60* gene of both *P. putida* KT2440 group and *P. monteilii* SB3101 group had low (91.6–95.7 %) similarity indices relative to other *P. putida* strains (W619, GB-1, NBRC14164, H8234, MTCC5279, CSV86, and UW4). The similarity index of *cpn60* genes among 38 *Pseudomonas* strains supports the idea of reassigning some species designated as *P. putida* strains into different *Pseudomonas* species (Fig. 4.1 and Table 4.3).

4.4.2 Phylogeny Based on *dnaJ* and *gyrA*

Other constitutively expressed ("housekeeping") genes, such as *dnaJ*, the gene encoding the molecular chaperone DnaJ, and *gyrA*, the gene encoding DNA gyrase subunit A, have also been used as phylogenetic markers. The ubiquitous distribution of *dnaJ* and *gyrA* among bacteria and their high degree of amino acid sequence conservation are key features that make these genes suitable for inferring phylogenetic relationships. *dnaJ* has been used to differentiate species and subspecies, even within genera where 16S rRNA gene sequences have low resolution, as in pathogenic bacteria such as *Mycobacterium* (Morita et al. 2004), *Streptococcus* (Itoh et al. 2006), and more recently *Staphylococcus* (Shah et al. 2007). Phylogenetic analyses of pseudomonads based on *dnaJ* and *gyrA* genes confirmed our observations, based on 16S rRNA gene sequences and *cpn60* genes (Fig. 4.2), that *Pseudomonas* sp. MO2 always grouped with *P. monteilii* SB3038 and SB3101.

4.4.3 Phylogenetic Analysis Based on *phaC1* and *phaC2* Genes

There are six genes in the polyhydroxyalkanoate (PHA) synthesis operon: *phaC1*, *phaZ*, *phaC2*, *phaD*, *phaF*, and *phaI*. The *phaC1* and *phaC2* encode distinct PHA synthases (their gene sequences are divergent with only 60–70 % nucleotide sequence identity). *phaZ* encodes the PHA depolymerase, and the other three genes (*phaD*, *phaF*, and *phaI*) are regulatory genes (Sharma et al. 2014). All the six *pha* genes were present in an operon and their organization is identical to other *P. putida* strains (Sharma et al. 2013, 2014). Phylogenetic analyses based on the nucleotide sequences of *phaC1* and *phaC2* revealed that *phaC1* and *phaC2* form distinct clusters (Fig. 4.3). Analyses of the *phaC1* and *phaC2* of *Pseudomonas* sp. MO2 placed these strains in *phaC1* and *phaC2* clusters with *P. monteilii* SB3101, *P. monteilii* SB3078, *P. putida* PC9, *P. putida* H3267, and *P. putida* S16, while the genes from *P. putida* KT2440, *P. putida* F1, *P. putida* LS46, and *P. putida* BIRD1 resulted in separate *phaC1* and *phaC2* clusters. Moreover, the phylogenetic trees generated with the *phaC1* and *phaC2* genes contained associations that were similar to those formed using *cpn60* gene sequences.

4.4.4 Phylogenetic Analysis of *Pseudomonas* MO2 Based on Concatenated Genes

Phylogenetic analyses based on concatenation of five genes, *cpn60*, *dnaJ*, *gyrA*, *rpoA*, and *rpoD*, indicated that *Pseudomonas* MO2 was closely associated with *P. monteilii* SB3101, *P. monteilii* SB3078, *P. putida* S16, and *P. putida* PC9 (Fig. 4.4). This phylogenetic analysis confirmed the relationships determined with 16S RNA, *cpn60*, or the *phaC* genes. The sequence similarity index of ten *P. putida* strains (F1, BIRD-1, DOT-T1E, TRO1, Idaho, LS46, ND6, SJTE-1, S12, B6-2) with the type strain, *P. putida* KT2440, was 0.99. However, *P. monteilii* group strains (*P. monteilii*

Table 4.3 Genome similarity indices of different *Pseudomonas* species based on *cpn60* gene sequences

Strains	KT2440	BIRD-1	LS46	MO2	H3267	3101	3078	S16
P. putida S12	0.992	0.993	0.992	0.962	0.962	0.964	0.964	0.964
P. putida Idaho	0.996	0.992	0.992	0.963	0.964	0.966	0.966	0.966
P. putida B6-2	0.991	0.991	0.997	0.961	0.962	0.964	0.964	0.964
P. putida SJTE-1	0.992	0.991	0.996	0.961	0.962	0.964	0.964	0.964
P. putida BIRD-1	0.992	1.000	0.992	0.962	0.962	0.964	0.964	0.964
P. putida TRO1	0.992	0.992	1.000	0.962	0.963	0.965	0.965	0.965
P. putida ND6	0.992	0.991	0.996	0.961	0.962	0.964	0.964	0.964
P. putida KT2440	1.000	0.992	0.992	0.965	0.966	0.968	0.968	0.968
P. putida F1	0.994	0.992	0.997	0.962	0.963	0.965	0.965	0.965
P. putida LS46	0.992	0.992	1.000	0.962	0.963	0.965	0.965	0.965
P. putida DOT-T1E	0.992	0.991	0.997	0.962	0.962	0.964	0.964	0.964
P. putida MO2	0.965	0.962	0.962	1.000	0.999	0.993	0.993	0.993
P. putida H3267	0.966	0.962	0.963	0.999	1.000	0.994	0.994	0.994
P. putida S16	0.968	0.964	0.965	0.999	0.994	1.000	1.000	1.000
P. monteilii SB3101	0.968	0.964	0.965	0.993	0.994	1.000	1.000	1.000
Pseudomonas sp. FGI182	0.971	0.968	0.968	0.978	0.979	0.982	0.982	0.982
P. putida B001	0.973	0.969	0.971	0.962	0.962	0.964	0.964	0.964
P. putida GB-1	0.957	0.957	0.957	0.95	0.951	0.953	0.953	0.953
P. putida NBRC 14164	0.953	0.949	0.952	0.949	0.95	0.951	0.951	0.951
P. putida H8234	0.951	0.946	0.948	0.947	0.947	0.949	0.949	0.949
Pseudomonas sp. VLB120	0.942	0.939	0.94	0.947	0.947	0.949	0.949	0.949
P. putida W619	0.947	0.946	0.946	0.944	0.944	0.947	0.947	0.947
P. putida NB2011	0.947	0.942	0.944	0.949	0.95	0.951	0.951	0.951
P. entomophila L48	0.945	0.943	0.943	0.953	0.953	0.956	0.956	0.956
P. putida MTCC 5279	0.937	0.936	0.936	0.94	0.94	0.944	0.944	0.944
P. putida CSV86	0.92	0.916	0.916	0.92	0.92	0.923	0.923	0.923
P. mendocina NK-01	0.892	0.894	0.891	0.887	0.888	0.891	0.891	0.891
P. stutzeri A1501	0.875	0.875	0.872	0.878	0.879	0.881	0.881	0.881
P. mendocina ymp	0.88	0.88	0.879	0.877	0.877	0.881	0.881	0.881
P. aeruginosa PA7	0.867	0.869	0.866	0.87	0.871	0.873	0.873	0.873
P. aeruginosa PAO1	0.869	0.869	0.866	0.87	0.871	0.875	0.875	0.875
P. fluorescens Pf0-1	0.861	0.858	0.861	0.863	0.864	0.864	0.864	0.864
P. fulva 12-X	0.854	0.856	0.855	0.855	0.856	0.858	0.858	0.858
P. resinovorans 106553	0.877	0.878	0.875	0.877	0.877	0.881	0.881	0.881
P. syringae pv. DC3000	0.846	0.844	0.846	0.851	0.852	0.85	0.85	0.85
P. putida UW4	0.858	0.856	0.859	0.86	0.86	0.862	0.862	0.862

SB3101, *P. monteilii* SB3078, *P. putida* S16, *P. putida* 3267, and *Pseudomonas* sp. MO2) had a low sequence similarity index (0.95). A number of other *P. putida* strains also showed low sequence similarity indices with the type strain, *P. putida* KT2440 (Table 4.3). The association of *Pseudomonas MO2* with the *P. monteilii* strains was also reinforced by the very high sequence

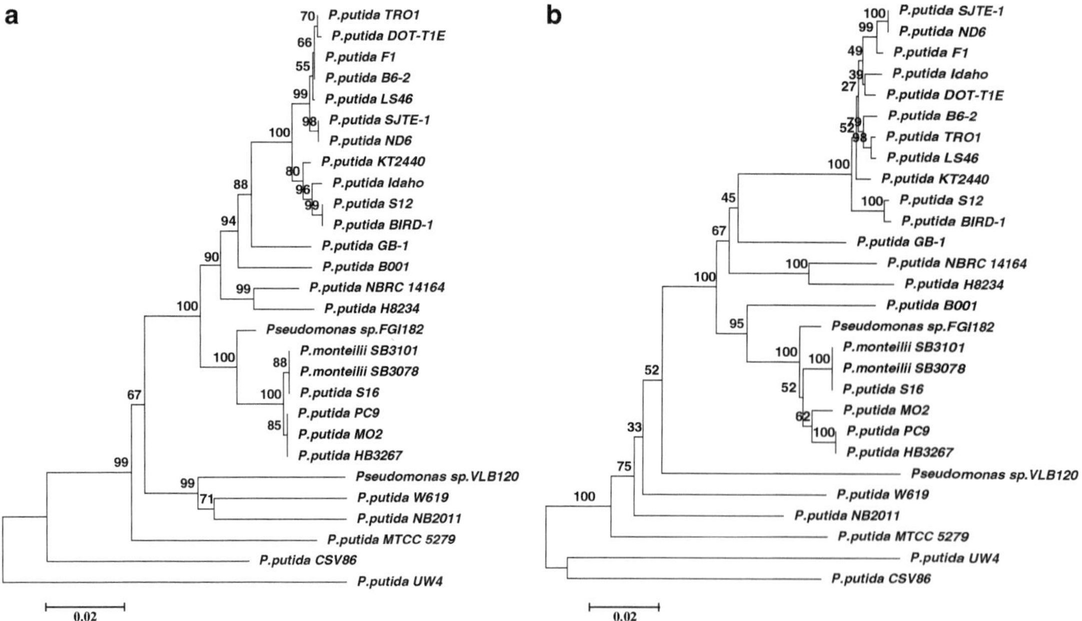

Fig. 4.2 Molecular phylogenetic analysis of *Pseudomonas* species by maximum likelihood method based on Juke and Cantor model. (**a**) *dnaJ* and (**b**) *gyrA*. The tree is drawn to scale, with branch lengths measured in the number of substitutions per site

similarity indices (0.99–1.00) among these species.

The analysis of 38 sequences of six genes (*cpn60*, *dnaJ*, *gyrA*, *gyrB*, *rpoA*, and *rpoD*) revealed wide variations in all these sequences. The *cpn60* (555), *dnaJ* (788), *gyrA* (2408), *gyrB* (1668), *rpoA* (1002), and *rpoD* (1220) genes contained 421, 331, 1006, 700, 262, and 457 polymorphic sites, respectively, and the nucleotide sequence identities ranged from 26.2 % to 75.8 % (Table 4.4). The average nucleotide differences calculated by Juke-Cantor (JC) were in the range of 54.1–287.8 in the six genes. The average pairwise distance in six genes ranged from 0.0 to 0.142. It was highest in *dnaJ* gene sequences and was least in *rpoA* gene sequences. The nucleotide diversity per site was highest in the *gyrB* gene, while minimum nucleotide diversity was detected in the *rpoA* gene. Likewise, a number of synonymous and non-synonymous sites and their diversity rates per site varied greatly in among the six genes. The phylogeny of 38 *Pseudomonas* strains based on individual housekeeping genes like *rpoA*, *rpoD*, *gyrA*, *gyrB*, and *dnaJ* was congruent with other gene phylogeny trees. The results were also in agreement with the phylogeny based on *cpn60* gene sequence analysis. Thus, phylogenetic analyses using both *cpn60* and five housekeeping genes indicated that our *Pseudomonas* sp. MO2 isolate clustered with *P. monteilii* SB3101 rather than *P. putida* KT2440.

Pseudomonas taxonomy has been revised from time to time using methods like DNA-rRNA reassociation data, as well as 16S rDNA sequence analysis (De Vos and De Ley 1983). Some *Pseudomonas* species were earlier transferred to new genera in the α-, β-, or γ-Proteobacteria. Even within the genus *Pseudomonas*, new species have been proposed based on inferred phylogenetic relationships using 16S rDNA. However, it has been observed that in the genus *Pseudomonas*, 16S rRNA-based phylogenies lack resolution at the intrageneric level due the slow rate of evolution of 16S rDNA sequences (Moore et al. 1996; Anzai et al. 2000). Ambiguities arising from the use of 16S rRNA-based phylogenetic analyses for *Pseudomonas* species have been overcome by using multilocus sequence analysis (MLSA) (Yamamoto and Harayama 1998; Yamamoto et al. 2000). Mulet

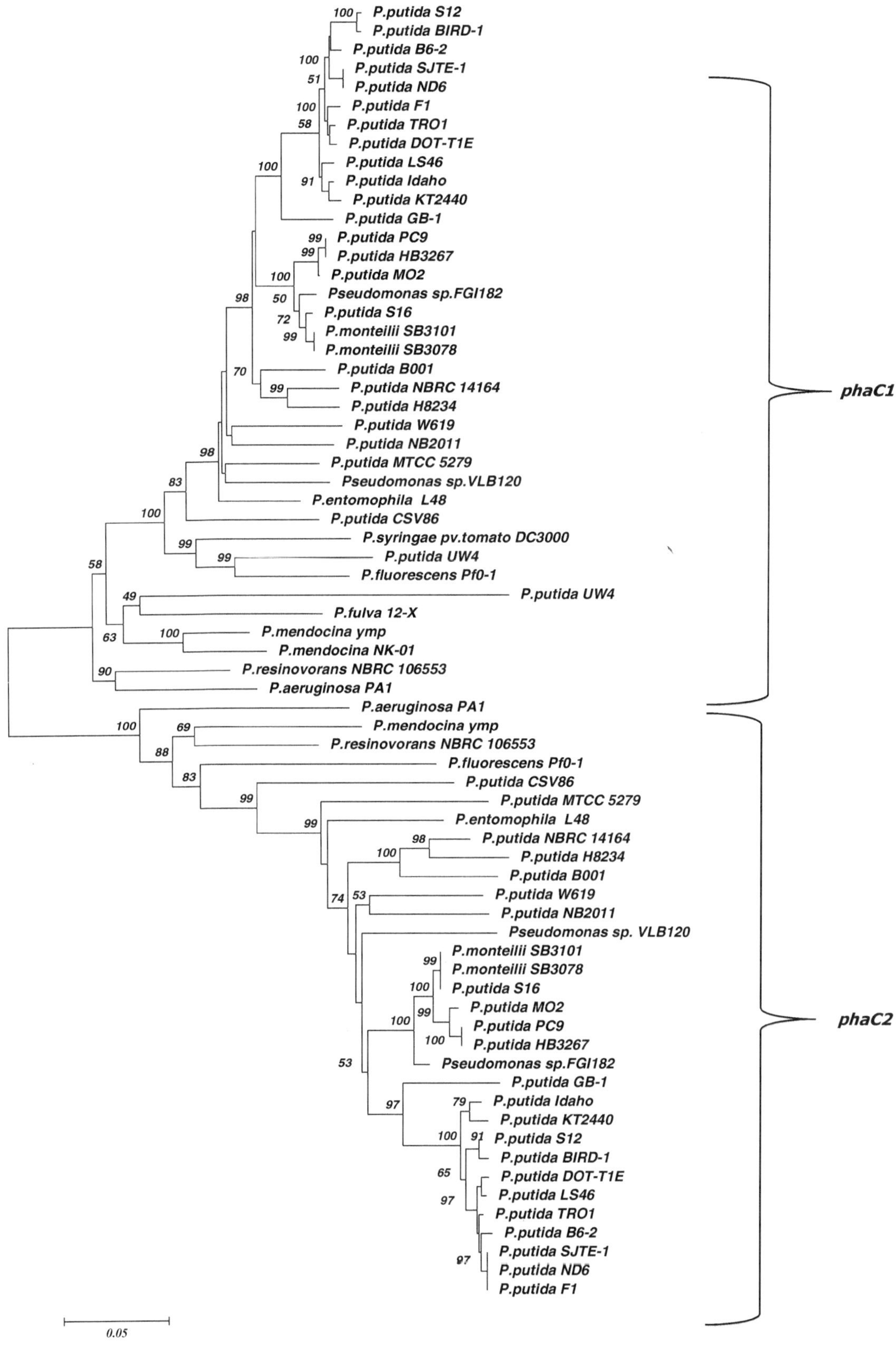

Fig. 4.3 Molecular phylogenetic analysis of *phaC* genes of *Pseudomonas* species by maximum likelihood method based on Juke and Cantor model. The tree is drawn to scale, with branch lengths measured in the number of substitutions per site

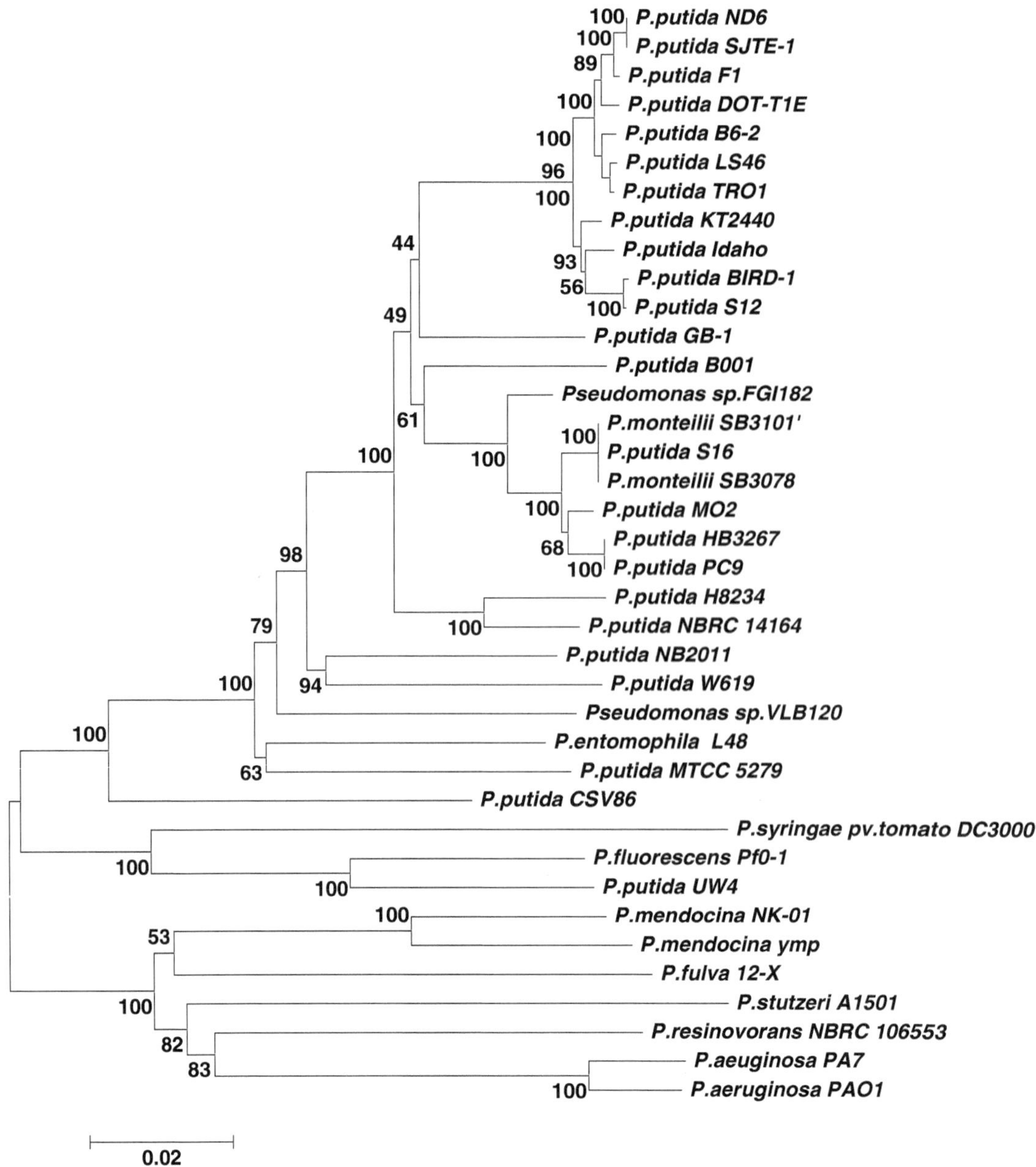

Fig. 4.4 Phylogenetic tree depicting the relationships among *Pseudomonas* species. The tree is based on concatenated sequences of five genes (*cpn60*, *rpoA*, *rpoD*, *gyrA*, and *dnaJ*), which were aligned by ClustalW and a neighbor-joining tree was generated using MEGA5 program. Bootstrap values are indicated at the nodes

et al. (2010) on the basis of the comparison of whole genome and phylogenetic similarity of *Pseudomonas* strains opined that DNA-DNA hybridization similarity of 70 % is cut off for species cutoffs, which was corresponding to 80 % genome similarity by blast analysis and to phylogenetic distance of a 97 % in multigene analysis. Other studies by Konstantinidis and Tiedje (2005) and Richter and Rosselló-Móra (2009), based on 94–95 % ANIb similarity corresponding to 97 % similarity in MLSA, placed *Pseudomonas* strains in separate species. They used a cutoff limit of 93 % (corresponding to 97 % similarity in MLSA) and this value separated *P. putida* from *P. aeruginosa*. Based on the *cpn60* gene similarity index, 13 *P. putida* strains,

Table 4.4 Analysis of *Pseudomonas* spp. sequences from *cpn60*, *dnaJ*, *grA*, *gyrB*, *rpoA*, and *rpoD genes*

Sequence information	Gene						
	16S rRNA	*cpn60*	*dnaJ*	*gyrA*	*gyrB*	*rpoA*	*rpoD*
Number of sequences	38	38	38	38	38	38	38
Number of sites	794	555	788	2408	1668	1002	1220
Number of polymorphic sites	668 (84.1)	421 (75.8)	331 (42.0)	1006 (41.8)	700 (41.9)	262 (26.2)	457 (37.4)
Average number of nucleotide differences	96.82 ± 5.83	54.1 ± 4.37	96.32 ± 5.81	287.8 ± 10.5	204.1 ± 8.46	63.05 ± 4.70	124.58 ± 6.61
Nucleotide diversity (per site)	0.487 ± 0.007	0.097 ± 0.009	0.136 ± 0.005	0.133 ± 0.003	0.136 ± 0.037	0.067 ± 0.003	0.112 ± 0.004
Synonymous sites PI (JC)	NA	144.36 (0.290)	203.54 (0.538)	576.20 (0.544)	401.91 (0.540)	251.88 (0.300)	326.9 (0.0499)
Non-synonymous sites PI (JC)	NA	410.64 (0.122)	582.46 (0.039)	1829.8 (0.044)	1266.09 (0.047)	747.12 (0.010)	891.1 (0.137)

including *P. putida* KT2440, had an SI of 99 %. *P. monteilii* SB3101 and *P. monteilii* SB3078 *cpn60* genes, however, had an SI of<97 % with *P. putida* KT2440. Therefore, the *P. monteilii* strains may be considered as species that are distinct from *P. putida* KT2440. Our *Pseudomonas* sp. MO2 showed>99 % sequence identity with the *cpn60* sequences of *P. monteilii* SB3101 and *P. monteilii* SB3078, but only 96 % sequence identity with *P. putida* KT2440 *cpn60* (Table 4.3). The comparison of similarity indices based on multilocus sequence analysis of five concatenated genes also confirmed the phylogenetic association of *Pseudomonas* sp. MO2 with the *P. monteilii* MO2 subgroup (Fig. 4.5). Thus, the *Pseudomonas* MO2 strain should be designated as *P. monteilii* MO2.

Pseudomonas monteilii has been identified as versatile microorganism associated with clinical specimens, bioremediation agent, and industrial production of indigo dye (Ma et al.2012). The species now designated *P. monteilii* was first isolated by Elomari et al. (1997). On the basis of a numerical analysis of the phenotypic characteristics of 39 strains, the Elomari et al. (1997) isolate was considered to be related to *P. putida*. However, on the basis of DNA-DNA hybridization, the isolate was reassigned as a new species, *P. monteilii*, with type strain designation CFML90-60 (= CIP104883). Bogaerts et al. (2011) isolated additional *P. monteilii* strains from clinical specimens. The clinical significance of *P. monteilii* is not known, but *P. monteilii* is assumed to be a rare opportunistic pathogen or colonizer. Recently another strain of *P. monteilii*, designated QM, was identified and shown to be capable of synthesizing indigoids from indoles (Qu et al. 2012; Ma et al.2012). Another *P. monteilii* strain has been identified as rhizosphere colonizer (Meyer et al. 2008). *P. monteilii* strains have also been associated with degradation of aromatic and heterocyclic compounds and dye compounds (Masuda et al.2007; Lie et al. 2009; Fulekar et al. 2013). Recently, the genomes of *P. monteilii* SB3078 and *P. monteilii* SB3101, which were identified as benzene-, toluene-, and ethylbenzene-degrading bacteria, were sequenced (Dueholm et al. 2014). *P. monteilii* strains are known to express endogenous extracellular lipases, which have been purified and characterized (Wang et al. 2009). For example, an extracellular lipase was isolated and purified from *P. monteilii* TKU009 (Wang et al. 2009).

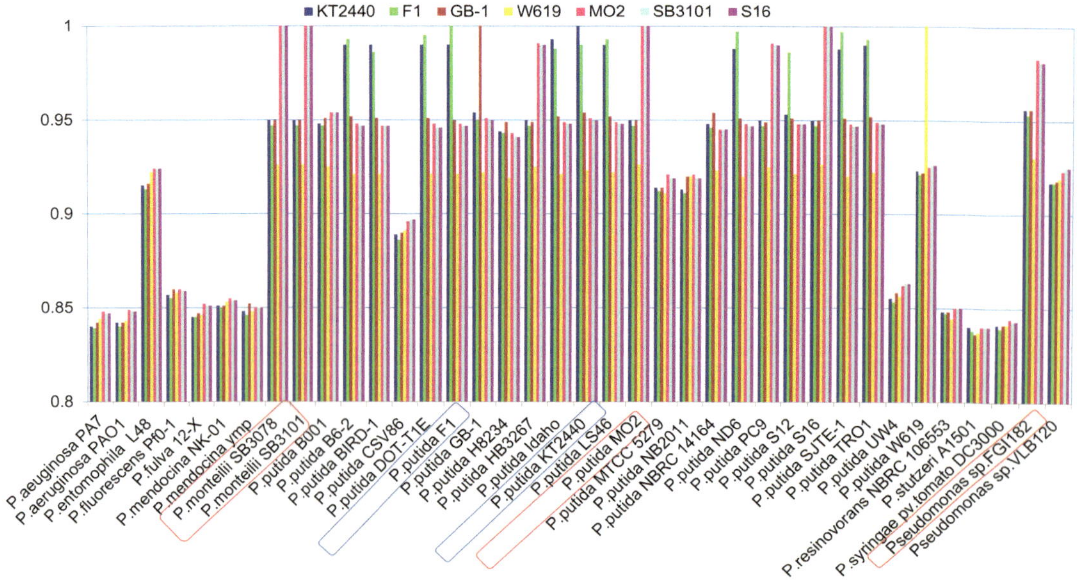

Fig. 4.5 Similarity indices of different *Pseudomonas* species based on concatenated sequence of the *cpn60*, *dnaJ*, *rpoA*, *rpoD*, and *gyrB* genes

These reports suggest that *P. monteilii* strains are versatile bacteria with multiple potential applications.

4.5 Conclusion

A novel *Pseudomonas* sp. (strain MO2) was identified as medium chain length polyhydroxyalkanoate producer from waste canola fryer oil. Based on phylogenetic analyses using 16S rRNA, *cpn60*, six individual housekeeping genes, the PHA synthase genes *phaC1* and *phaC2*, five concatenated gene sequences, and the sequence similarity index using the *cpn60* gene, we concluded that *Pseudomonas* sp. strain MO2 is different from *P. putida* KT2440 and most other *P. putida* strains and therefore designated the isolate as *P. monteilii* MO2. *Pseudomonas monteilii* MO2, unlike other *P. monteilii* strains isolated in 1997, was not a clinical isolate. It was more closely related to *P. monteilii* strains identified as bioremediation and rhizosphere isolates. In addition to this, *P. monteilii* MO2 can be developed as industrial strain for mcl-PHA production or for construction of recombinant strains to produce mcl-PHAs from oils.

Acknowledgments This work was supported by funds provided by the Natural Sciences and Engineering Research Council of Canada (NSERC), through a Strategic Programs grant (STPGP 306944-04), by Genome Canada, through the Applied Genomics Research in Bioproducts or Crops (ABC) program for the grant titled "Microbial Genomics for Biofuels and CoProducts from Biorefining Processes," and by the Province of Manitoba, through the Manitoba Research Innovation Fund (MRIF).

References

Anzai Y, Kim H, Park JY, Wakabayashi H, Oyaizu H (2000) Phylogenetic affiliation of the pseudomonads based on 16S rRNA sequence. Int J Syst Evol Microbiol 50:1563–1589. doi:10.1099/00207713-50-4-1563

Ashby RD, Foglia TA (1998) Poly(hydroxyalkanoate) biosynthesis from triglyceride substrates. Appl Microbiol Biotechnol 49:431–437. doi:10.1007/s002530051194

Barrett EL, Solanes RE, Tang JS, Palleroni NJ (1986) *Pseudomonas fluorescens* biovar V: its resolution into distinct component groups and the relationship of these groups to other *P. fluorescens* biovars, to *P. putida*, and to psychrotrophic pseudomonads associated with food spoilage. J Gen Microbiol 132:2709–2721. doi:10.1099/00221287-132-10-2709

Bhushan A, Joshi J, Shankar P, Kushwah J, Raju SC, Purohit HJ, Kalia VC (2013) Development of genomic tools for the identification of certain *Pseudomonas* up to species level. Indian J Microbiol 53:253–263. doi:10.1007/s12088-013-0412-1

Bogaerts P, Bouchahrouf W, Lissoir B, Denis O, Glupczynski Y (2011) IMP-13 producing *Pseudomonas monteilii* recovered in a hospital environment. J Antimicrob Chemother 66:2434–2435. doi:10.1093/jac/dkr294

Bossis E, Lemanceau P, Latour X, Gardan L (2000) The taxonomy of *Pseudomonas fluorescens* and *Pseudomonas putida*: current status and need for revision. Agronomie 20:51–63. doi:10.1051/agro:2000112

Braunegg G, Sonnleitner B, Lafferty RM (1978) A rapid gas chromatographic method for the determination of poly-β-hydroxybutyric acid in microbial biomass. Eur J Appl Microbiol Biotechnol 6:29–37. doi:10.1007/BF00500854

Champion AB, Barrett EL, Palleroni NJ (1980) Evolution in *Pseudomonas fluorescens*. J Gen Microbiol 120:485–511. doi:10.1099/00221287-120-2-485

Chan PL, Yu V, Wai L, Yu HF (2006) Production of medium-chain-length polyhydroxyalkanoates by *Pseudomonas aeruginosa* with fatty acids and alternative carbon sources. Appl Biochem Biotechnol 132:933–941. doi:10.1385/ABAB:132:1:933

Chen J, Liu T, Zheng Z, Chen J, Chen G (2004) Polyhydroxyalkanoate synthases PhaC1 and PhaC2 from Pseudomonas stutzeri 1317 had different substrate specificities. FEMS Microbiol Lett 234:231–237. doi:10.1016/j.femsle.2004.03.029

Chen YJ, Huang YC, Lee CY (2014) Production and characterization of medium-chain-length polyhydroxyalkanoates by *Pseudomonas mosselii* TO7. J Biosci Bioeng 118:145–152. doi:10.1016/j.jbiosc.2014.01.012

Chung AL, Jin HL, Huang LJ, Ye HM, Chen JC, Wu Q, Chen GQ (2011) Biosynthesis and characterization of poly (3-hydroxydodecanoate) by b-oxidation inhibited mutant of Pseudomonas entomophila L48. Biomacromolecules 12:3559–3566. doi:10.1021/bm200770m

Chung A, Zeng G, Jin H, Wu Q, Chen J, Chen G (2013) Production of medium-chain-length 3-hydroxyalkanoic acids by β-oxidation and phaC operon deleted Pseudomonas entomophila harboring thioesterase gene. Metab Eng 17:23–29. doi:10.1016/j.ymben.2013.02.001

Cromwick AM, Foglia T, Lenz RW (1996) The microbial production of poly(hydroxyalkanoates) from tallow. Appl Microbiol Biotechnol 46:464–469. doi:10.1007/s002530050845

Dabboussi F, Hamze M, Singer E, Geoffroy V, Meyer JM, Izard D (2002) *Pseudomonas mosselii* sp. nov., a novel species isolated from clinical specimens. Int J Syst Evol Microbiol 52:363–376. doi:10.1099/ijs.0.01541-0

Daird S, Carlier J-P, Ageron E, Patrick AD, Grimont PAD, Langlois V, Guérin P, Bouvet OMM (2002) Accumulation of poly(3-hydroxybutyrate) from octanoate in different *Pseudomonas* belonging to the rRNA homology group. Int Syst Appl Microbiol 25:183–188. doi:10.1078/0723-2020-00114

de Eugenio LI, Escapa IF, Morales V, Dinjaski N, Galán B, García JL, Prieto MA (2010) The turnover of medium chain-length polyhydroxyalkanoates in *Pseudomonas putida* KT2442 and the fundamental role of PhaZ depolymerase for the metabolic balance. Environ Microbiol 12:207–221. doi:10.1111/j.1462-2920.2009.02061.x

De Vos P, De Ley J (1983) Intra- and intergeneric similarities of Pseudomonas and Xanthomonas ribosomal ribonucleic acid cistrons. Int J Syst Bacteriol 33:487–509. doi:10.1099/00207713-33-3-487

Dueholm MS, Albertsen M, D'Imperio S, Tale VP, Lewis D, Nilsen PH, Nielsen JL (2014) Complete genomes of *Pseudomonas monteilii* SB3078 and SB3101, two benzene, toluene and ethylbenzene degrading bacteria used for bioaugmentation. Genome Announc 2(3):e00524-14. doi:10.1128/genomeA.00524-14

Elbahloul Y, Steinbüchel A (2009) Large-scale production of poly(3-hydroxyoctanoic acid) by *Pseudomonas putida* GPo1 and a simplified downstream process. Appl Environ Microbiol 75:643–651. doi:10.1128/AEM.01869-08

Elomari M, Izard D, Vincent D, Coroler L, Leclerc H (1994) Comparison of ribotyping analysis and numerical taxonomy studies of *Pseudomonas putida* biovar A. Syst Appl Microbiol 17:361–369. doi:10.1016/S0723-2020(11)80052-4

Elomari M, Coroler L, Verhille S, Izard D, Leclerc H (1997) *Pseudomonas monteilii* sp. nov., isolated from clinical specimens. Int J Syst Bacteriol 47:846–852. doi:10.1099/00207713-47-3-846

Fiedler S, Steinbuchel A, Rehm BHA (2002) The role of the fatty acid beta-oxidation multienzyme complex from *Pseudomonas oleovorans* in polyhydroxyalkanoate biosynthesis: molecular characterization of the *fadBA* operon from *P. oleovorans* and of the enoyl-CoA hydratase genes *phaJ* from *P. oleovorans* and *Pseudomonas putida*. Arch Microbiol 178:149–160. doi:10.1007/s00203-002-0444-0

Füchtenbusch B, Wullbrandt D, Steinbüchel A (2000) Production of polyhydroxyalkanoic acids by *Ralstonia eutropha* and *Pseudomonas oleovorans* from an oil remaining from biotechnological rhamnose production. Appl Microbiol Biotechnol 53:167–172

Fulekar MH, Wadgaonkar SL, Singh A (2013) Decolourization of dye compounds by selected bacterial strains isolated from dyestuff industrial area. Int J Adv Res Technol 2:182–192

Gamal RF, Abdelhady HM, Khodair TA, El-Tayeb TS, Hassan EA, Aboutaleb KA (2013) Semiscale production of PHAs from waste frying oil by Pseudomonas fluorescens S48. Braz J Microbiol 44(2):539–549. doi:10.1590/S1517-83822013000200034

Goris J, Konstantinidis KT, Klappenbach JA, Coenye T, Vandamme P, Tiedje JM (2007) DNA-DNA hybridization values and their relationship to whole genome sequence similarities. Int J Syst Evol Microbiol 57:81–91. doi:10.1099/ijs.0.64483-0

Guo WB, Song CJ, Kong MM, Geng WT, Wang YY, Wang SF (2011) Simultaneous production and characterization of medium-chain-length polyhydroxyalkanoates and alginate oligosaccharides by Pseudomonas mendocina NK-01. Appl Microbiol Biotechnol 92:791–901. doi:10.1016/j.micres.2012.11.003

Hilario E, Buckley TR, Young JM (2004) Improved resolution on the phylogenetic relationships among *Pseudomonas* by the combined analysis of *atpD, carA, recA* and 16S rDNA. Antonie Van Leeuwenhoek 86:51–64.doi:10.1023/B:ANTO.0000024910.57117.16

Hill JE, Penny SL, Crowell KG, Goh SH, Hemmingsen SM (2004) cpnDB: a chaperonin sequence database. Genome Res 14:1669–1675. doi:10.1101/gr.2649204

Itoh Y, Kawamura Y, Kasai H, Shah MM, Nhung PH, Yamada M, Sun X, Koyana T, Hayashi M (2006) *dnaJ* and *gyrB* gene sequence relationship among species and strains of genus *Streptococcus*. Syst Appl Microbiol 29:368–374. doi:10.1016/j.syapm.2005.12.003

Janse JD, Derks JHJ, Spit BE, Van Der Tuin WR (1992) Classification of fluorescent soft rot *Pseudomonas* bacteria, including *P. marginalis* strains, using whole cell fatty acid analysis. Syst Appl Microbiol 15:538–553. doi:10.1016/S0723-2020(11)80114-1

Jukes T, Cantor C (1969) Evolution of protein molecules. In: Munro HN (ed) Mammalian protein metabolism, vol 3. Academic, New York, pp 21–132. doi:10.1016/B978-1-4832-3211-9.50009-7

Konstantinidis K, Tiedje JM (2005) Genomic insights that advance the species definition for prokaryotes. Proc Natl Acad Sci U S A 102:2567–2592. doi:10.1073/pnas.0409727102

Konstantinidis KT, Ramette A, Tiedje JM (2006) The bacterial species definition in the genomic era. Philos Trans R Soc Lond B Biol Sci 361:1929–1940. doi:10.1098/rstb.2006.1920

Lazarovits G, Turnbull AL, Haug B, Links MG, Hill JE, Hemmingsen M (2013) Unraveling the rhizosphere using the *cpn*60 genomic marker and pyrosequencing. In: Bruijn FJ (ed) Molecular microbial ecology of the rhizosphere, vols 1 and 2. Wiley, London

Lee SY (1996) Bacterial polyhydroxyalkanoates. Biotechnol Bioeng 49:1–14. doi:10.1002/(SICI)1097-0290(19960105)49:1%3C1::AID-BIT1%3E3.0.CO;2-P

Lee SY, Lee Y, Wang FL (1999) Chiral compounds from bacterial polyesters: sugars to plastics to fine chemicals. Biotechnol Bioeng 65:363–368. doi:10.1002/(SICI)1097-0290(19991105)65:3%3C363::AID-BIT15%3E3.0.CO;2-1

Librado P, Rozas J (2009) DnaSP v5: a software for comprehensive analysis of DNA polymorphism data. Bioinformatics 25:1451–1452. doi:10.1093/bioinformatics/btp187

Lie F, Chen Y, Wang Z, Li Z (2009) Enantioselective benzylic hydroxylation of indan and tetralin with *Pseudomonas monteilii* TA-5. Tetrahedron Asymmetry 20:1206–1211. doi:10.1016/j.tetasy.2009.04.006

Ma Q, Qu Y, Tang H, Yu H, Ma F, Shi S, Zhang X, Zhou H, Zhou J, Xu P (2012) Genome sequence of a novel indigo-producing strain, *Pseudomonas monteilii* QM. J Bacteriol 194:4459–4460. doi:10.1128/JB.00867-12

Martınez-Murcia AJ, Monera A, Alperi A, Figueras MJ, Saavedra MJ (2009) Phylogenetic evidence suggests that strains of *Aeromonas hydrophila* subsp. *dhakensis* belong to the species *Aeromonas aquariorum* sp. nov. Curr Microbiol 58:76–80. doi:10.1007/s00284-008-9278-6

Masuda M, Yamasaki Y, Ueno S, Inoue A (2007) Isolation of bisphenol A-tolerant/degrading *Pseudomonas monteilii* strain N-502. Extremophiles 11:355–362. doi:10.1007/s00792-006-0047-9

Matsumoto K, Matsusaki H, Taguchi S, Seki M, Doi Y (2001) Biosynthesis of poly (3-hydroxybutyrate-co-3-hydroxyalkanoates) copolymer from sugars by recombinant *Ralstonia eutropha* harboring the phaC1Ps and the phaGPs genes of *Pseudomonas* sp. 61-3. Biomacromolecules 2:934–939. doi:10.1021/bm005604+

Meyer JM, Gruffaz C, Raharinosy V, Bezverbnaya I, Schäfer M, Budzikiewicz H (2008) Siderotyping of fluorescent *Pseudomonas*: molecular mass determination by mass spectrometry as a powerful pyoverdine siderotyping method. Biometals 21:259–271. doi:10.1007/s10534-007-9115-6

Molina L, Udaondo Z, Duque E, Fernández M, Molina-Santiago C, Roca A, Porcel M, de la Torre J, Segura A, Plesiat P, Jeannot K, Ramos JL (2014) Antibiotic resistance determinants in a *Pseudomonas putida* strain isolated from a hospital. PLoS One 9(1), e81604. doi:10.1371/journal.pone.0081604

Moore ERB, Mau M, Arnscheidt A, Böttger EC, Hutson RA, Collins MD, van De Peer Y, De Wachter R, Timmis KN (1996) The determination and comparison of the 16S rRNA gene sequences of species of the genus *Pseudomonas* (*sensu stricto*) and estimation of the natural intrageneric relationships. Syst Appl Microbiol 19:478–492. doi:10.1016/S0723-2020(96)80021-X

Morita Y, Maruyama S, Kabeya H, Nagai A, Kozawa K, Kato M, Nakajima T, Mikami T, Katsube Y, Kimura H (2004) Genetic diversity of the *dnaJ* gene in the *Mycobacterium avium* complex. J Med Microbiol 53:813–817. doi:10.1099/jmm.0.45601-0

Mulet M, Bennasar A, Lalucat J, Garcıa-Valdes E (2009) An *rpoD* based PCR procedure for the identification of *Pseudomonas* species and for their detection in environmental samples. Mol Cell Probes 23:140–147. doi:10.1016/j.mcp.2009.02.001

Mulet M, Lalucat J, Garcıa-Valdes E (2010) DNA sequence-based analysis of the *Pseudomonas species*. Environ Microbiol 12:1513–1530. doi:10.1111/j.1462-2920.2010.02181.x

Mulet M, Garcıa-Valdes E, Lalucat J (2013) Phylogenetic affiliation of *Pseudomonas putida* biovar A and B strains. Res Microbiol 164:351–359. doi:10.1016/j.resmic.2013.01.009

Nakazawa T (2003) Travels of a *Pseudomonas*, from Japan around the world. Environ Microbiol 4:782–786. doi:10.1046/j.1462-2920.2002.00310.x

Nelson KE, Weinel C, Paulsen IT, Dodson RJ, Hilbert H, Martins dos Santos VAP, Fouts DE, Gill SR, Pop M, Holmes M, Brinkac L, Beanan M, De Boy RT, Daugherty S, Kolonay J, Madupu R, Nelson W, White O, Peterson J, Khouri H, Hance I, Chris Lee P, Holtzapple E, Scanlan D, Tran K, Moazzez A, Utterback T, Rizzo M, Lee K, Kosack D, Moestl D, Wedler H, Lauber J, Stjepandic D, Hoheisel J, Straetz M, Heim S, Kiewitz C, Eisen J, Timmis KN, Düsterhoeft A, Tümmler B, Fraser CM (2002) Complete genome sequence and comparative analysis of the metabolically versatile *Pseudomonas putida* KT2440. Environ Microbiol 4:799–808. doi:10.1046/j.1462-2920.2002.00366.x

Otto K, Hofstetter K, Röthlisberger M, Witholt B, Schmidt A (2004) Biochemical characterization of *styAB* from *Pseudomonas* sp. strain VLB120 as a two-component flavin-diffusible monooxygenase. J Bacteriol 186:5292–5302. doi:10.1128/JB.186.16.5292-5302.2004

Palleroni NJ (1984) Genus I. *Pseudomonas* Migula 1894, 237AL. In: Krieg NJ, Holt JG (eds) Bergey's manual of systematic bacteriology, vol 1. Williams and Wilkins, Baltimore, pp 141–199

Palleroni NJ (2005) *Pseudomonas*. In: Garrity GM, Brenner DJ, Krieg NR, Staley JT (eds) Bergey's manual of systematic bacteriology, part B, the gammaproteobacteria, 2nd edn. Springer, New York, pp 323–379

Park SJ, Park JP, Lee Y (2002) Production of poly (3-hydroxybutyrate) from whey by fed-batch culture of recombinant Escherichia coli in a pilot-scale fermenter. Biotechnol Lett 24:185–189. doi:10.102 3/A:1014196906095

Pham TH, Webb JS, Rehm BHA (2004) The role of polyhydroxyalkanoate biosynthesis by Pseudomonas aeruginosa in rhamnolipid and alganiate production as well as stress tolerance and biofilm formation. Microbiology 150:3405–3413. doi:10.1099/mic.0.27357-0

Qu Y, Ma Y, Zhang X, Zhou H, Li X, Zhou J (2012) Optimization of indigo production by a newly isolated *Pseudomonas* sp. QM. J Basic Microbiol 52:1–8. doi:10.1002/jobm.201290001

Ramsay BA, Saracovan I, Ramsay JA, Marchessault RH (1991) Continuous production of long-sidechain poly-beta-hydroxyalkanoates by Pseudomonas oleovorans. Appl Environ Microbiol 3:625–629

Regenhardt D, Heuer H, Heim S, Fernandez DU, Strompl C, Moore ERB, Timmis KN (2002) Pedigree and taxonomic credentials of *Pseudomonas putida* strain KT2440. Environ Microbiol 4:912–915. doi:10.1046/j.1462-2920.2002.00368.x

Rehm BHA, Steinbüchel A (2001) PHA synthases – the key enzymes of PHA synthesis. In: Doi Y, Steinbüchel A (eds) Biopolymers, 1st edn. Wiley-VCH, Weinheim, pp 173–215

Rehm BHA, Mitsky TA, Steinbüchel A (2001) Role of fatty acid *de novo* biosynthesis in polyhydroxyalkanoic acid (PHA) and rhamnolipid synthesis by pseudomonads: establishment of the transacylase (PhaG)-mediated pathway for PHA biosynthesis in *Escherichia coli*. Appl Environ Microbiol 67:3102–3109. doi:10.1128/AEM.67.7.3102-3109.2001

Richter M, Rosselló-Móra R (2009) Shifting the genomic gold standard for the prokaryotic species definition. Proc Natl Acad Sci U S A 106:19126–19131. doi:10.1073/pnas.0906412106

Rozas J (2009) DNA sequence polymorphism analysis using DnaSP. In: Posada D (ed) Bioinformatics for DNA sequence analysis, vol 537, Methods in molecular biology series. Humana Press, New York, pp 337–350. doi:10.1007/978-1-59745-251-9_17

Saitou N, Nei M (1987) The neighbor-joining method: a new method for reconstructing phylogenetic trees. Mol Biol Evol 4:406–425. doi:0737-4038/87/0404-0007$02.00

Santos SR, Ochman H (2004) Identification and phylogenetic sorting of bacterial lineages with universally conserved genes and proteins. Environ Microbiol 6:754–759. doi:10.1111/j.1462-2920.2004.00617.x

Shah MM, Iihara H, Noda M, Song SX, Nhung PH, Ohkusu K, Kawamura Y, Ezaki T (2007) *dnaJ* gene sequence-based assay for species identification and phylogenetic grouping in the genus *Staphylococcus*. Int J Syst Evol Microbiol 57:25–30. doi:10.1099/ijs.0.64205-0

Sharma PK, Fu J, Cicek N, Sparling R, Levin DB (2012) Kinetics of medium-chain-length polyhydroxyalkanoates (mcl-PHAs) production by a novel isolate of *Pseudomonas putida* LS46. Can J Microbiol 58:982–989. doi:10.1139/w2012-074

Sharma PK, Fu J, Zhang XL, Fristensky BW, Davenport K, Chain PSG, Sparling R, Levin DB (2013) Draft genome sequence of medium-chain-length polyhydroxyalkanoate-producing Pseudomonas putida strain LS46. Genome Announc 1:e00151–13. doi:10.1128/genomeA.00151-13

Sharma PK, Fu J, Zhang XL, Fristensky BW, Sparling R, Levin DB (2014) Genome features of *Pseudomonas putida* LS46, a novel polyhydroxyalkanoate producer and its comparison with other *P. putida* strains. AMB Exp 4:37. doi:10.1186/s13568-014-0037-8

Solaiman DK, Ashby RD, Foglia TA (2000) Rapid and specific identification of medium chain-length polyhydroxyalkanoate synthase gene by polymerase chain reaction. Appl Microbiol Biotechnol 53:690–694. doi:10.1007/s002530000332

Smet MJ, Eggink G, Witholt B, Kingma J, Wynberg H (1983) Characterization of intracellular inclusions formed by *Pseudomonas oleovorans* during growth on octane. J Bacteriol 154:870–878

Stanier RY, Palleroni NJ, Doudoroff M (1966) The aerobic pseudomonads: a taxonomic study. J Gen Microbiol 43:159–271. doi:10.1099/00221287-43-2-159

Stead DE (1992) Grouping of plant-pathogenic and some other *Pseudomonas* spp. by using cellular fatty acid profiles. Int J Syst Bacteriol 42:281–295. doi:10.1099/00207713-42-2-281

Steinbüchel A (2001) Perspectives for biotechnological production and utilization of biopolymers: metabolic engineering of polyhydroxyalkanoate biosynthesis pathways as a successful example. Macromol Biosci 1:1–24. doi:10.1002/1616-5195(200101)1:1%3C1::AID-MABI1%3E3.3.CO;2-2

Steinbüchel A, Füchtenbusch B (1998) Bacterial and other biological systems for polyester production. Trends Biotechnol 16:419–427. doi:10.1016/S0167-7799(98)01194-9

Sun Z, Ramsay JA, Guay M, Ramsay B (2007) Increasing the yield of MCL-PHA from nonanoic acid by co-feeding glucose during the PHA accumulation stage in two-stage fed-batch fermentations of Pseudomonas putida KT2440. J Biotechnol 132:280–282. doi:10.1016/j.jbiotec.2007.02.023

Tamura K, Stecher G, Peterson D, Filipski A, Kumar S (2013) MEGA6: molecular evolutionary genetics analysis version 6.0. Mol Biol Evol 30:2725–2729. doi:10.1093/molbev/mst197

Tan IKP, Sudesh K, Theanmalar M, Gan SN, Gordon B (1997) Saponified palm kernel oil and its major free fatty acids as carbon substrates for the production of polyhydroxyalkanoates in *Pseudomonas putida* PGA1. Appl Microbiol Biotechnol 47:207–211. doi:10.1007/s002530050914

Tang H, Yao Y, Wang L, Yu H, Ren Y, Wu G, Xu P (2012) Genomic analysis of *Pseudomonas putida*: genes in a genome island are crucial for nicotine degradation. Sci Rep 2:377. doi:10.1038/srep00377

Tayeb LA, Ageron E, Grimont F, Grimont PAD (2005) Molecular phylogeny of the genus *Pseudomonas* based on *rpoB* sequences and application for the identification of isolates. Res Microbiol 156:763–773. doi:10.1016/j.resmic.2005.02.009

Thompson JD, Higgins DG, Gibson TJ (1994) CLUSTALW: improving the sensitivity of progressive multiple sequence alignment through sequence weighting, positions-specific gap penalties and weight matrix choice. Nucleic Acids Res 22:4673–4680. doi:10.1093/nar/22.22.4673

Timmis KN (2002) *Pseudomonas putida*: a cosmopolitan opportunist per excellence. Environ Microbiol 4:779–781. doi:10.1046/j.1462-2920.2002.00365.x

Timmis KN, Pieper DH (1999) Bacteria designed for bioremediation. Trends Biotechnol 17:200–204. doi:10.1016/S0167-7799(98)01295-5

Vancanneyt M, Torck U, Dewettinck D, Vaerewijck M, Kersters K (1996) Grouping of pseudomonads by SDS-PAGE of whole-cell proteins. Syst Appl Microbiol 19:556–568. doi:10.1016/S0723-2020(96)80027-0

Wang SN, Liu Z, Tang HZ, Meng J, Xu P (2007) Characterization of environmentally friendly nicotine degradation by *Pseudomonas putida* biotype A strain S16. Microbiology 153:1556–1565. doi:10.1099/mic.0.2006/005223-0

Wang SL, Lin YT, Liang TW, Chio SH, Ming LJ, Wu PC (2009) Purification and characterization of extracellular lipases from *Pseudomonas monteilii* TKU009 by the use of soybeans as the substrate. J Ind Microbiol Biotechnol 36:65–73. doi:10.1007/s10295-008-0473-z

Yamamoto S, Harayama S (1998) Phylogenetic relationships of *Pseudomonas putida* strains deduced from the nucleotide sequences of *gyrB, rpoD* and 16 rRNA genes. Int J Syst Bacteriol 48:813–819. doi:10.1099/00207713-48-3-813

Yamamoto S, Kasai H, Arnold DL, Jackson RW, Vivian A, Harayama S (2000) Phylogeny of the genus *Pseudomonas:* intrageneric structure reconstructed from the nucleotide sequences of *gyrB and rpoD* genes. Microbiology 146:2385–2394. doi:10.1099/mic.0.27096-0

Parveen Kumar Sharma received his M.Sc. and Ph.D. degree in Microbiology from Punjab Agricultural University, Ludhiana, India. Formerly a Professor of Microbiology at CCS Haryana Agricultural University, Hisar, India, he is now working as a Research Scientist at University of Manitoba, Winnipeg, Canada. His current interests are polyhydroxyalkanoate production using industrial and agricultural by-products and construction of recombinant bacteria for synthesis of novel PHA polymers.

Jilagamazhi FU, B.Sc., Inner Mongolia Agricultural University, Hohhot, Inner Mongolia, China, is a Ph.D. candidate at University of Manitoba, Winnipeg. He is presently in the last year in fulfillment of the Ph.D. degree at Department of Biosystems Engineering. His current research area is synthesis of medium chain length polyhydroxyalkanoates by *Pseudomonas putida.*

Xiang Zhang received the B.Eng. degree in Chemical Engineering from Tsinghua University, Beijing, China, and the B.Sc. degree in Bioinformatics from Simon Fraser University, Burnaby, BC Canada. He is currently working as a Bioinformatics Analyst at University of Manitoba. His current interests are data mining, systems biology, statistical modeling, genomics, and clinical trial.

Richard Sparling received his B.Sc. degree (Agr.) from McGill University (Canada) and Ph.D. degree in Microbiology from University of Iowa (USA). He is presently a Professor in the Microbiology Department at the University of Manitoba. His focus is to understand the metabolism and physiology of industrially relevant microorganisms using biochemical and genomic tools, with emphasis on bacteria producing bioplastics and on biofuel production from lignocellulosic materials using consolidated bioprocessing strategies.

David B. Levin received his B.Es. degree from University of Waterloo, M.Sc. degree from University of Guelph, and Ph.D. degree from McGill University. He is a Professor in Department of Biosystems Engineering at the University of Manitoba. His research focus is to understand the relationships between genome content, gene and gene product expression, metabolic pathway utilization, and end-product synthesis to increase the efficiencies of product synthesis by microorganisms.

Synthetic Biology Strategies for Polyhydroxyalkanoate Synthesis

5

Gunjan Arora, Andaleeb Sajid, Parijat Kundu, and Mritunjay Saxena

Abstract

Synthetic biologists are trying to apply engineering principles in biology to create an artificial world with unlimited possibilities. The guiding principle is to look beyond finding the solutions and to create new ones. Synthetic biologists are now routinely designing synthetic genetic circuits in bacteria and yeast. The major aim is to use microbes as cell factories for the production of bioactive compounds with a particular focus on drugs, biofuels, and biopolymers. In this chapter, we emphasize on understanding the synthetic biology approaches and their role in creating polyhydroxyalkanoate (PHA)-producing microbial factories.

5.1 Introduction

Why it is so difficult to apply engineering principles in biology? There are two likely explanations to this question. First, the complexity of biological systems is still not fully understood and the second is scientific community never tried to simplify it. The amalgamation of scientific disciplines in biology always fancied scientists. Historically, it led to creation of new branches like biochemistry, biophysics, recombinant DNA technology, fermentation technology, microbial technology, and systems biology. The recent progress in systems biology and rise of "omics" era intensified the application of engineering principles in biology to manipulate and concoct the cellular behavior (Cameron et al. 2014). The field of biology dedicated to such pursuits is prophesied as "synthetic biology."

5.2 Definition of Synthetic Biology

The broad literal definition of synthetic biology states that "it is an interdisciplinary branch of science that combines mathematical principles to diverse biological disciplines such as molecular and cellular biology, systems biology and biochemistry" (Gardner and Hawkins 2013; Cole 2014; Lienert et al. 2014). Synthetic biology merges biology and engineering to design and construct novel biological functions and proce-

G. Arora (✉) • A. Sajid • P. Kundu • M. Saxena
CSIR-Institute of Genomics and Integrative Biology (IGIB), Delhi University Campus, Mall Road, 110007 Delhi, India
e-mail: arorag1983@gmail.com; sajid.andaleeb@gmail.com; parijat@igib.in; mritunjaysaxena@gmail.com

V.C. Kalia (ed.), *Microbial Factories*, DOI 10.1007/978-81-322-2595-9_5

dures that help in delivering better technologies and products (Gardner and Hawkins 2013; Eils et al. 2015). This includes either creating new biological tools or artificial construction of existing products for novel purposes. The early synthetic biology manipulations were very much related to genetic engineering. The significant difference lies in the fact that genetic engineering has its roots in molecular biology, genetics, and biochemistry, whereas synthetic biology has foundations in systems biology and mathematics (Hu and Dhar 2015) (Fig. 5.1). The growth in synthetic biology can be attributed more to advancement in DNA sequencing and metabolomics.

Integration of different modules and standardization of the genetic circuits together at different levels is one big challenge in synthetic biology. Genetic circuit is composed of biological components that permit the flow of biological message by exerting regulatory, actuator, and/or signaling function(s) (Brophy and Voigt 2014). This will need an improvised genetic toolbox with fine precision. Further, there is need to pursue diverse biological applications to transform this new field into an independent engineering discipline. The focus would be to make engineering of biology quicker, reproducible, safer, and easy in practice.

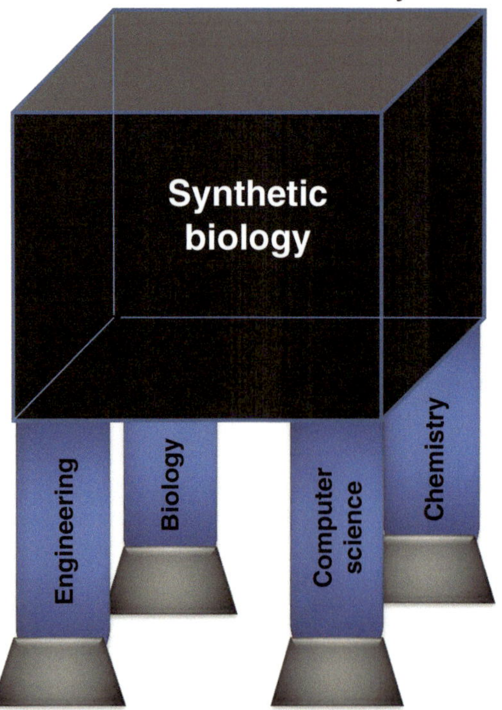

Fig. 5.1 Pillars of synthetic biology

5.3 History of Synthetic Biology

The classical study by Jacques and Monad incepted the very first idea in synthetic biology (Monod and Jacob 1961; Cameron et al. 2014). The discovery of regulatory circuits like *lac* operon in *Escherichia coli* indicated the cellular adaptability and genetic switches. However, it took almost four decades for biologists to understand the molecular details of these switches and how these switches can be modulated to generate specific responses spatiotemporally.

The word synthetic biology was coined in the distant past but was first described by Professor Waclaw Szybalski in 1974 (Benner et al. 2011). He suggested that use of advance molecular biology tools in manipulation of genetic systems will herald a new era of synthetic biology in the future. By the end of last century, advances in automated DNA sequencing, mass spectrometry, and computational biology enabled high-throughput analysis at the genome scale. These analyses reveal the molecular interaction networks in a cell involving different biomolecules: proteins, DNA, RNA, and metabolites. To elucidate these molecular networks, systems biology approaches were applied that revealed hierarchical functional modules, analogous to engineered modules (Andrianantoandro et al. 2006). This led to the renaissance of synthetic biology in the beginning of this century when the first major breakthrough was published (Elowitz and Leibler 2000; Gardner et al. 2000; Cameron et al. 2014).

In its journey of 15 years, synthetic biologists have tried to achieve ambitious goals such as designing novel therapeutic agents, large-scale production of low-cost drugs, and production of biofuels or even claimed creating artificial life (Cameron et al. 2014; Church et al. 2014). The advancement in genomics and metabolomics has tried to remove bottlenecks and to contribute toward the success of such pursuits. The particular focus on microorganisms, especially bacteria and yeast, to deliver engineered technologies and products has helped in this immense growth (Kalia et al. 2003, 2007; Anderson et al. 2006; Lu and Collins 2007; Atsumi and Liao 2008; Keasling 2008; Holtz and Keasling 2010; Paddon et al. 2013; Cameron et al. 2014). The field has provided some vital breakthroughs in such a short span that one expects synthetic biology principles will be extended to answer many challenging problems.

5.4 The Rationale of Synthetic Biology

For centuries, synthesis of diverse natural products and their derivatives such as metabolites, proteins, carbohydrates, and lipids in a test tube remained an ardent task. Enzymes that act in a concerted manner to produce desired outcome within a cell facilitate biological reactions. Production of the biological compounds requires a variety of intermediate components involved in the metabolic pathway. The traditional approach of organic synthesis of these intermediate components is very often difficult due to poor yields, instability, and hazardous nature of chemicals (Yeh and Lim 2007; Carothers et al. 2009; Wallace and Balskus 2014). A simple synthetic biology solution is to manipulate metabolic pathways and eliminate the need to purify or synthesize the intermediate chemicals. This will help in the production of complex natural products in a single-step reaction inside the cell. To achieve this, metabolic engineering and genetic switches are required that can respond to cellular demands. Additionally, when high concentrations of metabolic products become burden for the host cell, it may lead to cell death (Andersen and Krummen 2002; Rosano and Ceccarelli 2014). To overcome these limitations, it is necessary to keep the final product concentration below the toxic levels in the cell. Accurate operational management with engineering principles is needed to regulate native and heterogeneous reactions inside the cell. These measures will modify cellular metabolism to restore the product synthesis at amenable level. In addition to in vivo synthetic biology approaches, the field is also trying to come up with cell-free synthetic biology solutions (Smith et al. 2014).

In a cell, various genetic switches control the levels of diverse metabolites, and the first step is to understand these switches. In biology, we do not have arbitrary standard units to express the strength of biological signals, and thus, it is very difficult to apply engineering principles at the cellular level (Kelwick et al. 2014). Many metabolites, when accumulated, tend to be secreted outside the cell (Valle et al. 2008; Wintermute and Silver 2010). Mass production of such compounds requires an additional control on their secretion (Meyer and Schmidhalter 2012). Production of heterogenous compounds in large quantities creates burden for the biological system (Pickens et al. 2011; Immethun et al. 2013). In response to this metabolic burden, the biological systems tend to respond by generating spontaneous mutations. These mutations often switch off the metabolic pathways required for the production of compounds like polyhydroxyalkanoates (PHAs) (Zhang et al. 1994; Glick 1995; Kocharin et al. 2012). Such cells impede product synthesis and result in lower yields (Glick 1995). Therefore, it is necessary that we use such engineered cells with extreme caution. Further, many valuable biological components are protected by intellectual property rights, and application of such components will require cost-benefit analysis.

The basic difference between synthetic biology and other biological subfields is the emphasis on the "engineering" component that can be explained in mathematical terms, is amenable to manipulations, and can be designed to serve larger integrated systems (Heinemann and Panke 2006). Synthetic biologists are in the stage where early engineers would be, and they believe that in the near future, they will be as competent to design and accomplish large-scale biological engineering projects as engineers in designing the integrated circuits (Heinemann and Panke 2006; Schwille 2011). Research and development of new products in engineering fields is completely dependent on mathematical modeling, which is done by computer-aided simulation programs (Zeng and Yao 2009; Chandrasegaran et al. 2013). These simulations or computer-aided designs help in faster and improved analysis that aid in product testing in silico. Further, computer-aided simulations can even verify the performance of logical gates in integrated circuit design (Marchisio and Stelling 2009). This also decreases the cost and time in the engineering product development projects. However, as biological circuits are nonlinear, it has been very difficult to model such pathways by computational methods till now. Computational simulation programs that can help in protocol design and performance check will be needed to validate the synthetic biology components in the future (Chandran et al. 2009). One of the major considerations will be stochastic nature of biological components, which is different from reliable nature of engineering components. The stochastic behavior creates noise and ambiguity in the engineered biological component systems (Slusarczyk et al. 2012; Chen and Wu 2013; Wu et al. 2013). Therefore, the computer-aided program should be able to take such noise into account and define dynamic signal-to-noise ratio to explain the performance parameters on a real-time basis. The aim is to have computer-aided design software that not only can suggest the operational workflow but also can be integral component in designing the framework for synthetic biologists (Chandran et al. 2009; Wu and Rao 2012). Combinatorics of automatic lab assembly in biology is a pragmatic imagination that is still not conceptualized well and cannot be mathematically defined (Friedrich 2013). The gap between *in silico* and experimental biology will be negotiable, once matter can be expressed in pure digital terms that will bear ontological and functional ambivalences (Wu and Rao 2012). Till then lab automation tools may struggle to provide the right answers (Friedrich 2013).

Similar to electronics circuit design, modularization and standardization are the key components of synthetic biology. Synthetic biologists also came up with the "Registry of Standard Biological Parts" and "BioBricks," collection repository of genetic parts that can be mixed and matched to build new devices and systems (Endy 2005; Arkin 2008). Based on similar engineering design principles, these repositories are arguably the best IT infrastructures for the field at present. However, the computational setup needs further work. The first task is to integrate available chassis and design into powerful and intuitively usable workflows. The computational aspects in synthetic biology need more comprehensive and incorporated methods that can help even the non-specialists who want to apply synthetic biology in diverse applications. To achieve this, "Synthetic Biology Open Language" (SBOL; http://www.sbolstandard.org) has been created to advance the standards to expedite the addition of new software tools (Galdzicki et al. 2014).

5.5 Ethics and Synthetic Biology

Scientists of this era believe that the twenty-first century will be the century of biology (Ventor and Cohen 2004). In the first decade of this century, a prominent technology field "synthetic biology" was born. It aims to create artificial life or reengineer biological systems with innovative purposes to tackle numerous challenges that mankind is facing. However, due to sensational reporting, many people believe that the work of synthetic biologists will be threatening for biodiversity and speciation, therefore demeaning and disrespecting the meaning of life (Schmidt et al. 2009; Breitling et al. 2015). This may threaten the long-standing concepts of nature. Therefore, it is important to set the strategies for synthetic biology to contain social, environmental, and ethical risks (Schmidt et al. 2009; Dana et al. 2012). To ensure the progress of this field, US Presidential Committee has set up the guidelines for synthetic biology research. The report came up with 18 specific recommendations that can encourage synthetic biologist community to make specific guidelines (http://bioethics.gov/synthetic-biology-report). The report not only encourages the scientific pursuits in synthetic biology but also puts weight on the need of collaborative as well as coordinated efforts of scientists at the international level. It also discusses the importance of risk assessment review before releasing any synthetic organisms or products. This is to monitor and safeguard any conceivable threat through the inadvertent environmental release of organisms or bioactive compounds. It expects that synthetic biologists will find reliable containment and control mechanisms so that any synthetic product or organism designed will be unable to multiply and sustain in the natural environment (Kaebnick et al. 2014). The report also cautions synthetic biologists to refrain from claiming sensational terminology, for example, "creating life." It is also necessary that synthetic biologists should get ethical training and promote exchange of ideas with policy makers as well as religious and civil society groups about usefulness of synthetic biology and related emerging technologies.

5.6 Microbial Model Systems and Synthetic Biology

Single-celled bacteria have been utilized for many decades to perform genetic manipulations for the desired outcomes. Microbial factories are the key choice for most synthetic biology applications (Colin et al. 2011). The fact that our understanding of bacterial systems is far better than that of complex eukaryotic cells also indicates them being the system of choice. There are key considerations for synthetic biologists to use cellular systems (Keasling 2008). One of the main concerns is genetic stability and ability of the host to act on foreign elements (Keasling 2008). Additionally, for the mass production of biomolecules, cost estimation is a critical factor, and to reduce the cost, it is important to grow the cells in minimal media. Microbes can grow in minimal media, while eukaryotic cell culture needs additional growth supplements. Thus, it is one of the main considerations in mass production of large number of biomolecules as complex media requirements can be a bottleneck (Keasling 2008). A robust and well-characterized organism capable of growing on minimal, inexpensive carbon sources will be a perfect choice for many purposes including biopolymer and biofuel production (Keasling 2008; Pinto et al. 2012). The other considerations include quality control of production process, manipulations of regulatable genetic switches, and operational control to rectify problems. These considerations will be easier for engineered bacteria and other microorganisms (Keasling 2008). Therefore, most synthetic biologists will agree on the use of bacteria, and there are four reasons for this: (a) bacteria have simple genomes compared to eukaryotes, (b) genetic manipulations are simpler and faster in these organisms, (c) most genetic manipulations can be phenotypically calculated, and (d) lower costs of culturing systems.

It is considered that most of the biological hosts have the capability to inactivate or get rid themselves altogether of the foreign genes that create stress by metabolic burdens (Glass et al. 2006; Acevedo-Rocha et al. 2013). Thus, there is a need to design novel host strains that lack the capability to induce any mutation and can stably maintain and propagate foreign DNA (Glass et al. 2006; Acevedo-Rocha et al. 2013). The first such example is provided by two independent groups, Kolisnychenko et al. and Sharma et al., who designed "reduced-genome bacterial strains" for the stable expression of foreign elements (Kolisnychenko et al. 2002; Sharma et al. 2007). These strains indicate that it is much easier to manipulate microbial genomes for synthetic biology purpose that can be used for mass production of metabolites. The lower genome size of prokaryotic cell is easier to manipulate, and therefore, it is easier to create minimal genome by removing auxiliary genes. Besides genetic instability, the additional complexity is at transcriptional level where different promoters regulate the expression level of genes (Alper et al. 2005). Such transcriptional control is very complex, and thus, it is imperative to get tunable expression using a promoter system that can provide strong promoter strength (Keung et al. 2015).

To achieve the control on biological functions with predictability and reliability, synthetic biologists need simple biological models to work on (Keasling 2008). Microorganisms, specifically the prokaryotic bacteria *E. coli* and eukaryotic yeast *Saccharomyces cerevisiae*, were always the favorite ones due to our vast knowledge of these two organisms (Keasling 2008). In their native form these models are also very complex. Efforts were made to manipulate existing organisms at genetic level by removing additional and nonessential genetic components. One of such achievements was creation of "minimal *E. coli* strain" by different groups for efficient biotechnological production (Kolisnychenko et al. 2002; Posfai et al. 2006; Trinh et al. 2008). The next step is to generate amplified characteristics and functions within these systems by further replacing or adding specific genetic components.

5.7 The Early Advances in PHA Production by Synthetic Biology Approaches

Advances in genetic engineering, metabolic engineering, and systems biology led to the creation of novel microbes that can offer the product of choice. Pressure on environmental resources and associated economic problems in demand and supply are encouraging the research to develop economical microbial factories for mass production of a variety of substances (Kumar et al. 2009; Singh et al. 2009, 2013; Kumar et al. 2013, 2014; Liu et al. 2013). The use of microbial factories is far better than biomass extraction from other sources and chemical synthesis of many products. Using genetic and metabolic engineering approaches, microbial factories were reported for products when required genes were overexpressed in the surrogate host such as *E. coli*. Some successful examples of chemical production in microbial factories include succinic acid; fumaric acid; malic acid; adipic acid; propanol; butanol; isobutanol; diols such as 1,3-propanediol, 1,2-propanediol, 1,4-butanediol, and 2,3-butanediol; putrescine; cadaverine; polylactic acid; poly-gamma-glutamic acid; and biopolymers such as polyhydroxyalkanoates (PHAs), polyhydroxybutyrate (PHB), etc. (Liu et al. 2010; Lee et al. 2012a, b; Seo et al. 2013) (Fig. 5.2).

Polyhydroxyalkanoates (PHAs) comprise a class of biological polyesters accumulated by diverse bacteria when the carbon source is in excess and one essential growth nutrient is limited. PHAs have comparable material properties to plastics, specifically thermoplastic and elastomers (Madison and Huisman 1999; Reddy et al. 2003; Philip et al. 2007). The molecular masses of bacterially produced PHAs are generally of the order 50,000–1,000,000 Da that is similar to conventional plastics such as polypropylene (Madison and Huisman 1999). Natural PHA being biodegradable has the environmental advantage over conventional plastics. Using

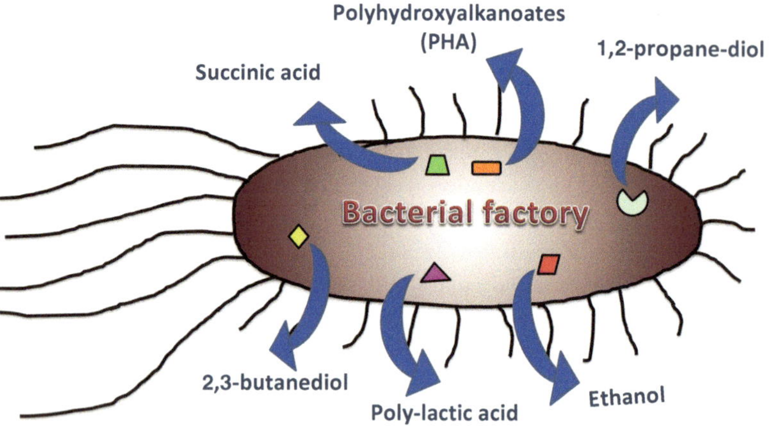

Fig. 5.2 Bacterial factory for synthetic biology

chemical hydrolysis of PHA, several commercially attractive molecules can be extracted that can be used as biodegradable solvents (Madison and Huisman 1999).

Despite having many advantages, PHAs, however, are not used commercially due to high costs (approximately $4.0–6.0 per kg) of mass production (Reddy et al. 2003; Rahman et al. 2013; Kumar et al. 2015a, b). The high production costs are due to lower yield, expensive feed materials, and extraction efficiency in natural environments (Linton 2010; Patel et al. 2012, 2015). The advent of recombinant DNA technology has made *E. coli* and *Bacillus subtilis* the preferred organisms for the production of recombinant proteins and metabolites (Verlinden et al. 2007; Singh et al. 2009; Zhou et al. 2012). The early efforts of recombinant PHA production in *E. coli* were based on expression of the *Ralstonia eutropha phaCAB* operon (Valentin and Dennis 1997). The *phaCAB* operon expresses three proteins by which acetyl-CoA is converted to polyhydroxybutyrate: *phaC* (PHA synthase), *phaA* (β-ketothiolase), and *phaB* (acetoacetyl-CoA reductase) (Rahman et al. 2013). PHA-associated proteins, called Phasins, adhere to the surface of PHA polymers and help in formation of spherical granules (Zhou et al. 2012). As the extraction of polymer from the cell is one of the most important steps in commercial processing of PHA polymer, development of its secretion mechanism is of potential interest (Linton 2010). Most of the cost of PHA processing is in extraction process. Synthetic modules designed to secrete polymers will eliminate the cell disruption step and decrease the production cost significantly. Application of such synthetic modules will not need bacterial cells lysis for the extraction of compounds, and therefore, it can help in developing continuous PHA production process. There is a lot of focus on the development of secretion mechanism by using "Phasins" as these proteins are suggested to increase the surface-to-volume ratio of granules for better accumulation (York et al. 2001; Rahman et al. 2013). Also, Phasin expression is shown to decrease the granule size. The translocation of smaller granules is much easier, and therefore, Phasin overexpression leads to more PHA production and reduced granular size (Maehara et al. 1999; Tian et al. 2005). Expression of recombinant secretory Phasin in *E. coli* paves the way for synthetic biological system to secrete PHAs (Linton 2010; Rahman et al. 2013). Secretion of PHA will be helpful in downstream processing whereby it can be easily separated from the cell mass, which can aid in better recovery and purification (Rahman et al. 2013).

Synthetic biology tools can also be applied to improve the PHA production by enhancing the activities of PHA-synthesizing enzymes. In the past, protein engineering approaches have been used for directed evolution to enhance the specific activity of PHA synthase from *Aeromonas punctata* and to synthesize higher amounts of PHA in recombinant *E. coli* (Amara et al. 2002; Foo et al. 2012). These studies indicate that synthetic biology in sync with metabolic engineering can help improving biocatalytic properties of PHA synthase to optimize engineered pathways in surrogate host like *B. subtilis* and *E. coli*.

PHA synthesis is an energy-consuming mechanism for the bacteria, and for commercial biopolymer production, it is imperative that we have fine control over the production process (Pham et al. 2004). Synthetic genetic switches as described earlier can help in the regulation of expression levels. However, traditional genetic switches can only be modified as "on" or "off," and during production they cannot sense when the cell is exhausted and stressed out. The inability of such genetic switches in responding to the dynamic needs of biological system and regulating the expression will result in process impediment. Therefore, the need for smart genetic switches arises that can sense, respond, and help the cell to adapt to every condition. These dynamic components that can cater to the energy demands of the cells and streamline the PHA production process will be required in a bioreactor. Synthetic biologists have come up with dynamic sensor-regulator system (DSRS) that can respond to dynamic changes in a cell (Zhang et al. 2012) and help in biofuel production. DSRS can be applied to controlled biopolymer production for better yields, and there is similar effort with technology named "CHIRP" (Circuit for Enhanced

In-vivo Regulated Bioplastics Production) (http://dev.nsta.org/evwebs/1282/index.html).

5.8 Perspective

The global market of synthetic biology is expected to reach 38.7 billion by 2020 (Singh 2014). This indicates synthetic biology is expected to deliver important products and technologies. Thus, synthetic biology can be applied to multiple aspects of PHA production to obtain better polymer quality, enhanced production, and ease in extraction/processing. These methods, if successful, can help decrease the production costs and large-scale commercial production of PHA derivatives.

Acknowledgments The authors wish to thank the CSIR-Institute of Genomics and Integrative Biology (IGIB), Delhi, Government of India for providing support.

References

Acevedo-Rocha CG, Fang G, Schmidt M, Ussery DW, Danchin A (2013) From essential to persistent genes: a functional approach to constructing synthetic life. Trends Genet 29:273–279

Alper H, Fischer C, Nevoigt E, Stephanopoulos G (2005) Tuning genetic control through promoter engineering. Proc Natl Acad Sci U S A 102:12678–12683

Amara AA, Steinbuchel A, Rehm BH (2002) In vivo evolution of the *Aeromonas punctata* polyhydroxyalkanoate (PHA) synthase: isolation and characterization of modified PHA synthases with enhanced activity. Appl Microbiol Biotechnol 59:477–482

Andersen DC, Krummen L (2002) Recombinant protein expression for therapeutic applications. Curr Opin Biotechnol 13:117–123

Anderson JC, Clarke EJ, Arkin AP, Voigt CA (2006) Environmentally controlled invasion of cancer cells by engineered bacteria. J Mol Biol 355:619–627

Andrianantoandro E, Basu S, Karig DK, Weiss R (2006) Synthetic biology: new engineering rules for an emerging discipline. Mol Syst Biol 2:2006.0028

Arkin A (2008) Setting the standard in synthetic biology. Nat Biotechnol 26:771–774

Atsumi S, Liao JC (2008) Metabolic engineering for advanced biofuels production from *Escherichia coli*. Curr Opin Biotechnol 19:414–419

Benner SA, Yang Z, Chen F (2011) Synthetic biology, tinkering biology, and artificial biology. What are we learning? C R Chim 14:372–387

Breitling R, Takano E, Gardner TS (2015) Judging synthetic biology risks. Science 347:107

Brophy JA, Voigt CA (2014) Principles of genetic circuit design. Nat Methods 11:508–520

Cameron DE, Bashor CJ, Collins JJ (2014) A brief history of synthetic biology. Nat Rev Microbiol 12:381–390

Carothers JM, Goler JA, Keasling JD (2009) Chemical synthesis using synthetic biology. Curr Opin Biotechnol 20:498–503

Chandran D, Bergmann FT, Sauro HM (2009) TinkerCell: modular CAD tool for synthetic biology. J Biol Eng 3:19

Chandrasegaran SK, Ramani K, Sriram RD, Horváth I, Bernard A, Harik RF, Gao W (2013) The evolution, challenges, and future of knowledge representation in product design systems. Comput Aided Des 45:204–228

Chen BS, Wu CC (2013) Systems biology as an integrated platform for bioinformatics, systems synthetic biology, and systems metabolic engineering. Cells 2:635–688

Church GM, Elowitz MB, Smolke CD, Voigt CA, Weiss R (2014) Realizing the potential of synthetic biology. Nat Rev Mol Cell Biol 15:289–294

Cole JA (2014) Synthetic biology: old wine in new bottles with an emerging language that ranges from the sublime to the ridiculous? FEMS Microbiol Lett 351:113–115

Colin VL, Rodriguez A, Cristobal HA (2011) The role of synthetic biology in the design of microbial cell factories for biofuel production. J Biomed Biotechnol 2011:601834

Dana GV, Kuiken T, Rejeski D, Snow AA (2012) Synthetic biology: Four steps to avoid a synthetic-biology disaster. Nature 483:29

Eils R, Ritzerfeld J, Wiechert W (2015) Editorial: Synthetic biology – ready for application. Biotechnol J 10:229–230

Elowitz MB, Leibler S (2000) A synthetic oscillatory network of transcriptional regulators. Nature 403:335–338

Endy D (2005) Foundations for engineering biology. Nature 438:449–453

Foo JL, Ching CB, Chang MW, Leong SS (2012) The imminent role of protein engineering in synthetic biology. Biotechnol Adv 30:541–549

Friedrich K (2013) Digital 'faces' of synthetic biology. Stud Hist Philos Biol Biomed Sci 44:217–224

Galdzicki M, Clancy KP, Oberortner E, Pocock M, Quinn JY, Rodriguez CA, Roehner N, Wilson ML, Adam L, Anderson JC, Bartley BA, Beal J, Chandran D, Chen J, Densmore D, Endy D, Grunberg R, Hallinan J, Hillson NJ, Johnson JD, Kuchinsky A, Lux M, Misirli G, Peccoud J, Plahar HA, Sirin E, Stan GB, Villalobos A, Wipat A, Gennari JH, Myers CJ, Sauro HM (2014) The Synthetic Biology Open Language (SBOL) provides a community standard for communicating designs in synthetic biology. Nat Biotechnol 32:545–550

Gardner TS, Hawkins K (2013) Synthetic biology: evolution or revolution? A co-founder's perspective. Curr Opin Chem Biol 17:871–877

Gardner TS, Cantor CR, Collins JJ (2000) Construction of a genetic toggle switch in *Escherichia coli*. Nature 403:339–342

Glass JI, Assad-Garcia N, Alperovich N, Yooseph S, Lewis MR, Maruf M, Hutchison CA, Smith HO, Venter JC (2006) Essential genes of a minimal bacterium. Proc Natl Acad Sci U S A 103:425–430

Glick BR (1995) Metabolic load and heterologous gene expression. Biotechnol Adv 13:247–261

Heinemann M, Panke S (2006) Synthetic biology-putting engineering into biology. Bioinformatics 22:2790–2799

Holtz WJ, Keasling JD (2010) Engineering static and dynamic control of synthetic pathways. Cell 140:19–23

Hu B, Dhar PK (2015) Introduction to systems biology. Systems and Synthetic Biology. Springer, Netherlands, pp 3–23

Immethun CM, Hoynes-O'Connor AG, Balassy A, Moon TS (2013) Microbial production of isoprenoids enabled by synthetic biology. Front Microbiol 4:75

Kaebnick GE, Gusmano MK, Murray TH (2014) The ethics of synthetic biology: next steps and prior questions. Hastings Cent Rep 44(S5):S4–S26

Kalia VC, Chauhan A, Bhattacharyya G (2003) Genomic databases yield novel bioplastic producers. Nat Biotechnol 21:845–846. doi:10.1038/nbt0803-845

Kalia VC, Lal S, Cheema S (2007) Insight in to the phylogeny of polyhydroxyalkanoate biosynthesis: horizontal gene transfer. Gene 389:19–26. doi:10.1016/j.gene.2006.09.010

Keasling JD (2008) Synthetic biology for synthetic chemistry. ACS Chem Biol 3:64–76

Kelwick R, MacDonald JT, Webb AJ, Freemont P (2014) Developments in the tools and methodologies of synthetic biology. Front Bioeng Biotechnol 2:60

Keung AJ, Joung JK, Khalil AS, Collins JJ (2015) Chromatin regulation at the frontier of synthetic biology. Nat Rev Genet 16:159–171

Kocharin K, Chen Y, Siewers V, Nielsen J (2012) Engineering of acetyl-CoA metabolism for the improved production of polyhydroxybutyrate in *Saccharomyces cerevisiae*. AMB Express 2:52

Kolisnychenko V, Plunkett G, Herring CD, Feher T, Posfai J, Blattner FR, Posfai G (2002) Engineering a reduced *Escherichia coli* genome. Genome Res 12:640–647

Kumar T, Singh M, Purohit HJ, Kalia VC (2009) Potential of *Bacillus* sp. to produce polyhydroxybutyrate from biowaste. J Appl Microbiol 106:2017–2023

Kumar P, Patel SK, Lee JK, Kalia VC (2013) Extending the limits of *Bacillus* for novel biotechnological applications. Biotechnol Adv 31:1543–1561

Kumar P, Singh M, Mehariya S, Patel SKS, Lee JK, Kalia VC (2014) Ecobiotechnological approach for exploiting the abilities of *Bacillus* to produce co-polymer of polyhydroxyalkanoate. Indian J Microbiol 54: 151–157. doi:10.1007/s12088-014-0457-9

Kumar P, Mehariya S, Ray S, Mishra A, Kalia VC (2015a) Biodiesel industry waste: a potential source of bioenergy and biopolymers. Indian J Microbiol 55:1–7. doi:10.1007/s12088-014-0509-1

Kumar P, Ray S, Patel SKS, Lee JK, Kalia VC (2015b) Bioconversion of crude glycerol to PHA by *Bacillus thuringiensis* under non-limiting nitrogen conditions. Int J Biol Macromol. doi:10.1016/j.ijbiomac.2015.03.046

Lee JW, Na D, Park JM, Lee J, Choi S, Lee SY (2012a) Systems metabolic engineering of microorganisms for natural and non-natural chemicals. Nat Chem Biol 8:536–546

Lee SY, Mattanovich D, Villaverde A (2012b) Systems metabolic engineering, industrial biotechnology and microbial cell factories. Microb Cell Fact 11:156

Lienert F, Lohmueller JJ, Garg A, Silver PA (2014) Synthetic biology in mammalian cells: next generation research tools and therapeutics. Nat Rev Mol Cell Biol 15:95–107

Linton E (2010) A synthetic biological engineering approach to secretion-based recovery of polyhydroxyalkanoates and other cellular products

Liu H, Xu Y, Zheng Z, Liu D (2010) 1,3-Propanediol and its copolymers: research, development and industrialization. Biotechnol J 5:1137–1148

Liu L, Liu Y, Shin HD, Chen RR, Wang NS, Li J, Du G, Chen J (2013) Developing *Bacillus* spp. as a cell factory for production of microbial enzymes and industrially important biochemicals in the context of systems and synthetic biology. Appl Microbiol Biotechnol 97:6113–6127

Lu TK, Collins JJ (2007) Dispersing biofilms with engineered enzymatic bacteriophage. Proc Natl Acad Sci U S A 104:11197–11202

Madison LL, Huisman GW (1999) Metabolic engineering of poly(3-hydroxyalkanoates): from DNA to plastic. Microbiol Mol Biol Rev 63:21–53

Maehara A, Ueda S, Nakano H, Yamane T (1999) Analyses of a polyhydroxyalkanoic acid granule-associated 16-kilodalton protein and its putative regulator in the *pha* locus of *Paracoccus denitrificans*. J Bacteriol 181:2914–2921

Marchisio MA, Stelling J (2009) Computational design tools for synthetic biology. Curr Opin Biotechnol 20:479–485

Meyer HP, Schmidhalter DR (2012) Microbial expression systems and manufacturing from a market and economic perspective, innovations in biotechnology, Dr. Eddy C. Agbo (Ed.), ISBN: 978-953-51-0096-6, InTech. doi:10.5772/29417

Monod J, Jacob F (1961) Teleonomic mechanisms in cellular metabolism, growth, and differentiation. Cold Spring Harb Symp Quant Biol 26:389–401

Paddon CJ, Westfall PJ, Pitera DJ, Benjamin K, Fisher K, McPhee D, Leavell MD, Tai A, Main A, Eng D,

Polichuk DR, Teoh KH, Reed DW, Treynor T, Lenihan J, Fleck M, Bajad S, Dang G, Dengrove D, Diola D, Dorin G, Ellens KW, Fickes S, Galazzo J, Gaucher SP, Geistlinger T, Henry R, Hepp M, Horning T, Iqbal T, Jiang H, Kizer L, Lieu B, Melis D, Moss N, Regentin R, Secrest S, Tsuruta H, Vazquez R, Westblade LF, Xu L, Yu M, Zhang Y, Zhao L, Lievense J, Covello PS, Keasling JD, Reiling KK, Renninger NS, Newman JD (2013) High-level semi-synthetic production of the potent antimalarial artemisinin. Nature 496:528–532

Patel SKS, Singh M, Kumar P, Purohit HJ, Kalia VC (2012) Exploitation of defined bacterial cultures for production of hydrogen and polyhydroxybutyrate from pea-shells. Bioma Bioenerg 36:218–225. doi:10.1016/j.biombioe.2011.10.027

Patel SKS, Kumar P, Singh M, Lee JK, Kalia VC (2015) Integrative approach to produce hydrogen and polyhydroxybutyrate from biowaste using defined bacterial cultures. Bioresour Technol 176:136–141. doi:10.1016/j.biortech.2014.11.029

Pham TH, Webb JS, Rehm BH (2004) The role of polyhydroxyalkanoate biosynthesis by *Pseudomonas aeruginosa* in rhamnolipid and alginate production as well as stress tolerance and biofilm formation. Microbiology 150:3405–3413

Philip S, Keshavarz T, Roy I (2007) Polyhydroxyalkanoates: biodegradable polymers with a range of applications. J Chem Technol Biotechnol 82:233–247

Pickens LB, Tang Y, Chooi YH (2011) Metabolic engineering for the production of natural products. Annu Rev Chem Biomol Eng 2:211–236

Pinto F, van Elburg KA, Pacheco CC, Lopo M, Noirel J, Montagud A, Urchueguia JF, Wright PC, Tamagnini P (2012) Construction of a chassis for hydrogen production: physiological and molecular characterization of a *Synechocystis* sp. PCC 6803 mutant lacking a functional bidirectional hydrogenase. Microbiol 158:448–464

Posfai G, Plunkett G, Feher T, Frisch D, Keil GM, Umenhoffer K, Kolisnychenko V, Stahl B, Sharma SS, de Arruda M, Burland V, Harcum SW, Blattner FR (2006) Emergent properties of reduced-genome *Escherichia coli*. Science 312:1044–1046

Rahman A, Linton E, Hatch AD, Sims RC, Miller CD (2013) Secretion of polyhydroxybutyrate in *Escherichia coli* using a synthetic biological engineering approach. J Biol Eng 7:24

Reddy CS, Ghai R, Rashmi, Kalia VC (2003) Polyhydroxyalkanoates: an overview. Bioresour Technol 87:137–146

Rosano GL, Ceccarelli EA (2014) Recombinant protein expression in *Escherichia coli*: advances and challenges. Front Microbiol 5:172

Schmidt M, Ganguli-Mitra A, Torgersen H, Kelle A, Deplazes A, Biller-Andorno N (2009) A priority paper for the societal and ethical aspects of synthetic biology. Syst Synth Biol 3:3–7

Schwille P (2011) Bottom-up synthetic biology: engineering in a tinkerer's world. Science 333:1252–1254

Seo SW, Yang J, Min BE, Jang S, Lim JH, Lim HG, Kim SC, Kim SY, Jeong JH, Jung GY (2013) Synthetic biology: tools to design microbes for the production of chemicals and fuels. Biotechnol Adv 31:811–817

Sharma SS, Blattner FR, Harcum SW (2007) Recombinant protein production in an *Escherichia coli* reduced genome strain. Metab Eng 9:133–141

Singh R (2014) Global synthetic biology market (products, technologies, applications and geography) – size, share, global trends, company profiles, demand, insights, analysis, research, report, opportunities, segmentation and forecast, 2013–2020. Report

Singh M, Patel SK, Kalia VC (2009) *Bacillus subtilis* as potential producer for polyhydroxyalkanoates. Microb Cell Fact 8:38

Singh M, Kumar P, Patel SKS, Kalia VC (2013) Production of polyhydroxyalkanoate co-polymer by *Bacillus thuringiensis*. Indian J Microbiol 53:77–83. doi:10.1007/s12088-012-0294-7

Slusarczyk AL, Lin A, Weiss R (2012) Foundations for the design and implementation of synthetic genetic circuits. Nat Rev Genet 13:406–420

Smith MT, Wilding KM, Hunt JM, Bennett AM, Bundy BC (2014) The emerging age of cell-free synthetic biology. FEBS Lett 588:2755–2761

Tian SJ, Lai WJ, Zheng Z, Wang HX, Chen GQ (2005) Effect of over-expression of phasin gene from *Aeromonas hydrophila* on biosynthesis of copolyesters of 3-hydroxybutyrate and 3-hydroxyhexanoate. FEMS Microbiol Lett 244:19–25

Trinh CT, Unrean P, Srienc F (2008) Minimal *Escherichia coli* cell for the most efficient production of ethanol from hexoses and pentoses. Appl Environ Microbiol 74:3634–3643

Valentin HE, Dennis D (1997) Production of poly(3-hydroxybutyrate-co-4-hydroxybutyrate) in recombinant *Escherichia coli* grown on glucose. J Biotechnol 58:33–38

Valle J, Da RS, Schmid S, Skurnik D, D'Ari R, Ghigo JM (2008) The amino acid valine is secreted in continuous-flow bacterial biofilms. J Bacteriol 190:264–274

Venter C, Cohen D (2004) The Century of biology. New Perspect Q 21:73–77

Verlinden RA, Hill DJ, Kenward MA, Williams CD, Radecka I (2007) Bacterial synthesis of biodegradable polyhydroxyalkanoates. J Appl Microbiol 102: 1437–1449

Wallace S, Balskus EP (2014) Opportunities for merging chemical and biological synthesis. Curr Opin Biotechnol 30:1–8

Wintermute EH, Silver PA (2010) Emergent cooperation in microbial metabolism. Mol Syst Biol 6:407

Wu K, Rao CV (2012) Computational methods in synthetic biology: towards computer-aided part design. Curr Opin Chem Biol 16:318–322

Wu M, Su RQ, Li X, Ellis T, Lai YC, Wang X (2013) Engineering of regulated stochastic cell fate determination. Proc Natl Acad Sci U S A 110:10610–10615

Yeh BJ, Lim WA (2007) Synthetic biology: lessons from the history of synthetic organic chemistry. Nat Chem Biol 3:521–525

York GM, Stubbe J, Sinskey AJ (2001) New insight into the role of the PhaP phasin of *Ralstonia eutropha* in promoting synthesis of polyhydroxybutyrate. J Bacteriol 183:2394–2397

Zeng Y, Yao S (2009) Understanding design activities through computer simulation. Adv Eng Inform 23:294–308

Zhang H, Obias V, Gonyer K, Dennis D (1994) Production of polyhydroxyalkanoates in sucrose-utilizing recombinant *Escherichia coli* and *Klebsiella* strains. Appl Environ Microbiol 60:1198–1205

Zhang F, Carothers JM, Keasling JD (2012) Design of a dynamic sensor-regulator system for production of chemicals and fuels derived from fatty acids. Nat Biotechnol 30:354–359

Zhou XY, Yuan XX, Shi ZY, Meng DC, Jiang WJ, Wu LP, Chen JC, Chen GQ (2012) Hyperproduction of poly(4-hydroxybutyrate) from glucose by recombinant *Escherichia coli*. Microb Cell Fact 11:54

Gunjan Arora received his Ph.D. degree in Zoology from University of Delhi and CSIR-Institute of Genomics and Integrative Biology (CSIR-IGIB, Delhi, India). He has also worked as vaccine research innovation awardee in Translational Health Science and Technology Institute and is currently working as postdoctoral fellow at the National Institute of Allergy and Infectious Diseases in the National Institutes of Health (NIH). His current research interests include systems and synthetic biology approaches.

Andaleeb Sajid received her Ph.D. degree in Biotechnology from the CSIR-Institute of Genomics and Integrative Biology (CSIR-IGIB, Delhi, India). Currently, she is working as postdoctoral fellow at Laboratory of Clinical Infectious Diseases at the National Institute of Allergy and Infectious Diseases in the National Institutes of Health (NIH). Her research interests include signal transduction events that regulate *Mycobacterium tuberculosis* physiology and drug discovery modules.

Parijat Kundu is a graduate in Biochemistry from Presidency College, Kolkata, India, and postgraduate in Biochemistry from University of Calcutta, Kolkata, India. He is currently working as Project Fellow at CSIR-Institute of Genomics and Integrative Biology, Delhi, India. His current research focuses on understanding posttranslational modifications network of *Mycobacterium tuberculosis*.

Mritunjay Saxena received his Ph.D. degree in Life Sciences from Devi Ahilya Vishwavidyalaya, Indore, India. He is currently working as a Senior Project Fellow at CSIR-Institute of Genomics and Integrative Biology, Delhi, India. His current interests are biosensors, *Plasmodium vivax* malaria, and malaria vaccine.

Frontiers in Biomedical Engineering: PHA-Fabricated Implants

6

Lalit K. Singh, Neha Dhasmana, Shashank S. Kamble, Aditya K. Sharma, and Yogendra Singh

Abstract

Polyhydroxyalkonoates (PHAs) are biological in origin, organic polyesters comprising the industrial and biomedical interest. This chapter summarizes the current advances, applications, limitations, and challenges of biopolymers in medicine. Biopolymers not only substitute the existing polymers, but novel combinations of diverse PHAs broaden the applicability and utility. The PHA-based implants are new dimensions of future in biomedical engineering.

6.1 Introduction

Polyhydroxyalkonoates (PHAs) have gained the significant attention, with diverse applications in medicine (Kalia et al. 2000). PHA is a microbial biopolymer having properties like biocompatibility, nontoxicity, biodegradability, and thermoplasticity. PHAs belong to the family of biopolymers containing polyesters of different hydroxyl-carboxylic acids. Bacteria synthesize polyhydroxyalkanoates as energy storage compound normally in excess with the carbon source and one component essential for growth like phosphorus, nitrogen, oxygen, and sulfur in limited concentration (Singh et al. 2015a). Novel blends are developed by mixing with variable amounts of polymers of 3-hydroxyoctanoate, 3-hydroxyvalerate, and 3-hydroxybutyrate (PHB) for their potential use in various biomedical applications. These blends showed higher Young's modulus and tensile strength compared to their parent molecules. PHAs are appealing implant material in biomaterial engineering and for mechanical supports because they are natural in origin, have enhanced biocompatibility, and lack cytotoxicity and capability to support cell adhesion and cell growth. PHAs can be categorized as short (*scl*), medium (*mcl*), and long (*lcl*), based on chain length (Fig. 6.1) (Misra et al. 2006; Singh et al. 2015b).

L.K. Singh (✉)
CSIR-Institute of Genomics and Integrative Biology, Allergy and Infectious Diseases, Delhi, India
e-mail: lalit.singh@igib.in

N. Dhasmana • S.S. Kamble • A.K. Sharma
Allergy and Infectious Diseases, CSIR-Institute of Genomics and Integrative Biology, 110007 Delhi, India
e-mail: neha.dhasmana@igib.in; shashank.kamble@igib.in; aditya.sharma@gmail.com

Y. Singh
Allergy and Infectious Diseases, CSIR-Institute of Genomics and Integrative Biology, Room No. 208, Mall Road, 110007 Delhi, India
e-mail: ysingh@igib.res.in

V.C. Kalia (ed.), *Microbial Factories*, DOI 10.1007/978-81-322-2595-9_6

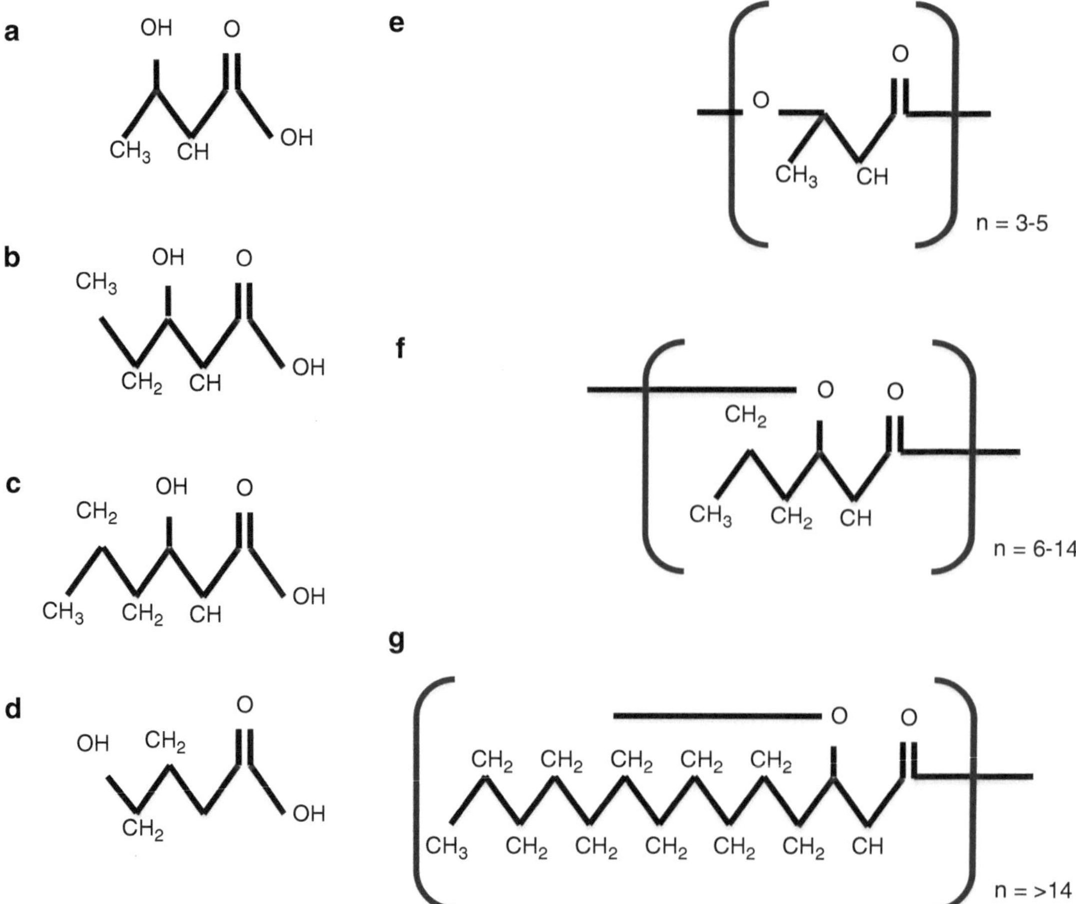

Fig. 6.1 Chemical representation of hydroxy-carboxylic acids, monomeric components of PHA* and category of PHA on the basis of carbon chain length present in the monomer. (**a**) 3HB (3-hydroxy butyric acid), (**b**) 3HV (3-hydroxy valeric acid), (**c**) 3HH (3-hydroxy hexanoic acid), (**d**) 4HB (4-hydroxy butyric acid), (**e**) *scl*-PHA (small chain length polyhydroxy butyric acid), (**f**) *mcl*-PHA (medium chain length polyhydroxy butyric acid), (**g**) *lcl*-PHA (longer chain length polyhydroxy butyric acid)

The *scl*-PHAs exhibit high melting temperature and are brittle in nature, while *mcl*-PHAs are lesser crystalline and having low melting temperature. The PHB is the most widely studied *scl*-PHA but is brittle in nature and has high crystallinity to make PHB industrially less valuable. On the other hand, PHO is the extensively studied *mcl*-PHA with higher tensile strength and low thermal stability and can be commercially developed.

PHAs have been discovered and investigated significantly for application in medicine and healthcare in the last two decades. PHAs have been incorporated to develop several medical implants like sutures, orthopedic anchors, and several repair devices to revolutionize the biomedical engineering. Hence, several PHA-based implants have been tested in animals, while some PHA-based implants are in practice for human applications recently. The purity of PHA always remains an issue since the pyrogenic contaminants gets co-purified (Singh et al. 2009). In the future, PHAs may serve as an attractive and promising biomaterial for diverse applications in biomedical engineering. Although in the industrial sector, PHAs are widely utilized as packaging material, biofuel, and drug delivery vehicle, in this chapter, we are mainly emphasizing and describing the use of PHA as biomedical implants. The potential implication of PHA appears excellent in the tissue engineering area near future.

6.2 PHA as In-Vivo Implants in Biomedical Engineering

PHAs have been developed over the years for various applications such as medical implant, mechanical supportive devices, drug delivery matrices, chips/patches to support the cell proliferation, and tissue regeneration by exploiting its multifacets properties like superior elasticity, adaptable mechanical strength, and biocompatibility (Chen 2009; Ihssen et al. 2009). The PHB and its copolymer production were demonstrated in *Bacillus* using different substrates to develop the PHA-based products with sustainable approach (Fig. 6.2) (Kumar et al. 2015a, b). Several investigators recommended that *Bacillus* could serve as a promising candidate for large-scale PHA production (Kumar et al. 2013, 2014; Patel et al. 2015; Singh et al. 2013). Numerous composites were formulated especially copolymers of HB with 3-hydroxyvalerate (P3HBV), 4-HB (P4HB), and 3-hydroxyoctanoate (P3HO). These different composites of PHAs were blended to design anchors, sutures, cardiovascular chips, repair patches, nerve guiding, slings, orthopedic anchors, guiding tissue, adhesion barriers, and devices for regeneration, cartilage and tendon repair, bone marrow scaffolds, etc. (Fig. 6.3) (Masood et al. 2014; Yang et al. 2014). We can achieve the desirable degradation time, biocompatibility, and favorable mechanical properties by manipulating the PHA compositions to meet the demands of specific physiological condition.

The PHA-derived implants/devices do not evoke the immune response inside the host and are thus considered as biologically safe materials (Chen and Wu 2005). Williams and colleagues first time demonstrated that PHB- and PHO-based microsphere and tubes were non-immunogenic in mice model. Lobler and coworkers investigated that the PHA-based gastrointestinal patches did not generate any inflammatory response. Moreover, PHA-based

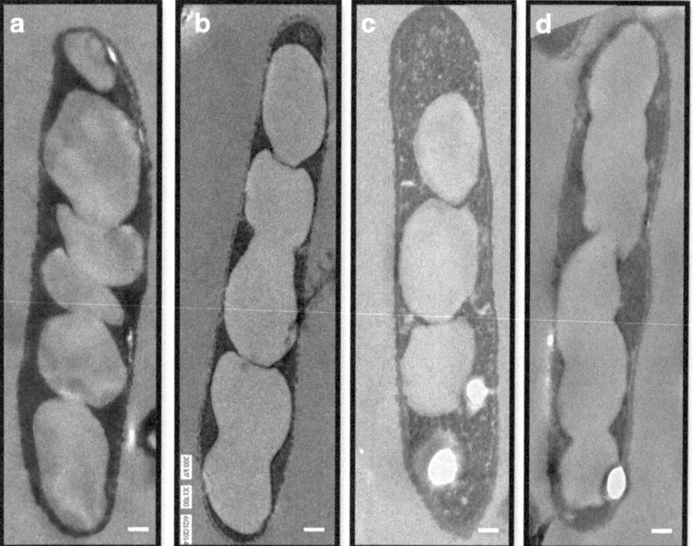

Fig. 6.2 Transmission electron microscope image of PHA granules accumulation in *Bacillus*. *Bacillus cereus* were grown on GM2 media supplemented with glycerol for 72 h. The cells were harvested at 8000 rpm and primarily fixed with Karnovasky's reagent for 16 h at 4 °C. Secondary fixation was performed with 1 % osmium tetroxide and then cells were gradually dehydrated followed by infiltration. The ultrathin sections were observed under Tecnai G2 Spirit Transmission Electron Microscope at 200 KV. Mainly three models – Micelle, Budding and Scaffold – are mentioned for PHA granules growth. Figure (**a**, **b**, **d**) shows the mixed Micelle and Budding model of PHA granule growth, while Fig. (**c**) represents Micelle model. Moreover interestingly granules occupy 70–80 % volume of the total cell (Scale bar 100 nm)

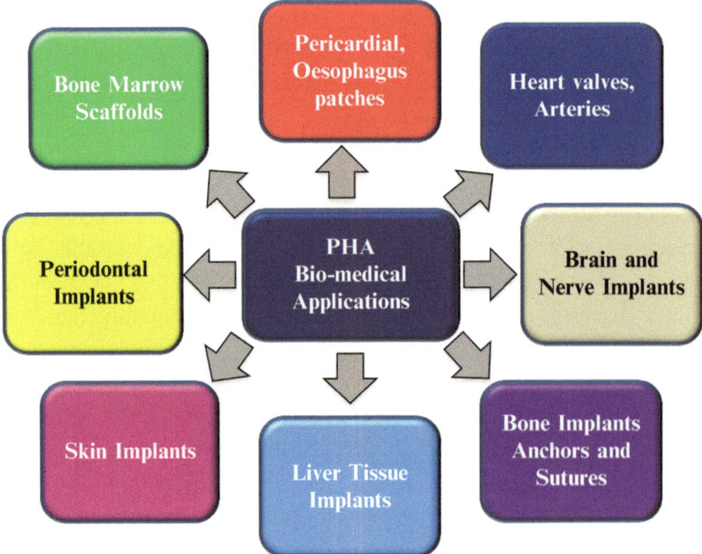

Fig. 6.3 Schematic representation of diverse applications of PHA in bio-medical engineering. PHAs classified into three major classes on the basis of carbon chain length present in the monomeric unit: short chain length (*scl*-PHA), medium chain length (mcl-PHA), and long-chain length (*lcl*-PHA). The various composites and blends were formulated, especially poly 3-hydroxybutyrate (P3HB), poly-3-hydroxyvalerate (P3HBV), poly-4-hydroxybutyrate (P4HB), copolymers of P3HB, and poly-3-hydroxyoctanoate (PHO). These variants of PHAs are blended to develop anchors, sutures, cardiovascular patches, heart valves, slings, skin implants, orthopedic implants/anchors, guiding tissue repair/regeneration devices like liver tissue implants, adhesion barriers, articular cartilage repair devices, nerve guiding, tendon repair devices, bone marrow scaffolds, periodontal implants, and wound dressings as described in the summary figure. We can achieve the desirable degradation time, biocompatibility, and favorable mechanical properties by manipulating the PHA compositions to meet the demands of specific physiological conditions. PHA has opened a new window of opportunity in the field of bio-medical engineering

implants are capable to support cell proliferation and adhesion and shown to be biodegradable in nature. The degraded hydroxybutyric acid from implant elevates the cytosolic calcium that enhances the attachment and cell proliferation (Xiao et al. 2007). An enhanced cell proliferation was observed in umbilical vein and smooth muscle cells on PHB copolymer like P3HB and poly-3-hydroxyhexanoate (P3HBHHx) coated with fibronectin (Qu et al. 2006). Poly-3-hydroxybutyrate-co-4-hydroxybutyrate-co-3-hydroxyhexanoate (P3HB4HB3HHx) is an improved implantation material for biomedical applications in comparison to P3HBHHx, poly-lactic acid (PLA) (Chen 2009). These PHB, P3HBHHx, and P3HB4HB3HHx composite exhibits increased mechanical properties when fabricated into nanostructure and behaves like natural extracellular matrix (Li et al. 2008). PHA-based implant materials were approved for human applications in 2007 by Food and Drug Administration USA.

Tissue engineering is multidisciplinary fields involving regeneration of the damaged cells or tissue using multiple stem cells, biomaterials, and signaling molecules (Bruder and Fox 1999). Tissue engineering can be categorized into two types: soft tissue engineering and hard tissue engineering. Highly porous, biodegradable scaffolds with specific cell types are utilized to promote ex-vivo growth in soft tissue engineering. The skin, heart valves, esophagus patches, vascular patches, liver, and nerve tissue guiding are the classical examples of soft tissue engineering (Du et al. 2009; Karageorgiou and Kaplan 2005). On the other hand, scaffolds fabricated with biological materials are implanted inside the body to provide the mechanical support and growth in the hard tissue engineering. The cartilage and bone tissue implants belong to hard tissue engineering (Yu and Fan 2008). PHA-based implants/devices contain all the characteristics that are mandatory to ideal biomaterial, i.e., capable to support the cell growth and maintain proper nutrients supply to cells and degradation in a short period after implantations.

6.2.1 Soft Tissue Implants: Esophagus, Pericardial Patches

Biocompatibility of the implants is a fundamental prerequisite for the biomaterial engineering applications. The other features including the material's shape, surface hydrophobicity, porosity, mechanical strength, biochemistry of the material, and degradation product also contribute to the success rate of the implant (Sun et al. 2005). The anchorage-dependent attachment and proliferation of mammalian cell fundamentally depend on the surface roughness and hydrophobicity of any scaffold. Myocardial infarction is the common cause of heart failure. However, heart transplantation is not always feasible due to limited number of organ donors. Tissue engineering serves as an alternative option to put the pericardial patch on affected area to repair the damaged heart. PHA products are fabricated into the pericardial patches, surgically placed between the sternum and heart to prevent the adhesion after the cardiac surgery. The immunogenicity of pericardial patches was first implanted and tested in the sheep. The safety was monitored in 19 human patients during bypass surgery for 6–24 months, which demonstrated that PHA-based implants can be used as pericardial patches (Malm et al. 1992; Duvernoy et al. 1995). The pericardial patch serves two major functions: first delivers healthy cardiomyocytes into the infarcted region and second provides the ventricular resistant (Fujimoto et al. 2007). Similarly, copolymer of poly-3-hydroxybutyric acid-co-3-hydroxyhexanoate (P3HB3HHx) can serve as a successful candidate to construct the artificial esophagus. Such composite P3HB3HHx blended artificial esophagus was implanted in the dog. The experiment stimulated regeneration of the tissue and negligible biodegradation of the composite (Chen and Wu 2005).

6.2.2 PHA-Based Heart Valves, Artery, and Vascular Grafting/Implants

The bicuspid/tricuspid valve malfunctioning also results in the cardiac failure. Sodian and colleagues demonstrated that *mcl*-PHA-based implants can serve an option in heart valve tissue engineering. The improved cell proliferation and cell adhesion were observed in poly-4-hydroxybutyric acid (P4HB) and polyglycolic acid (PGA) blended scaffolds (Fu et al. 2004). Generally, heart valves are implanted with supplementation of fibroblast growth factor and ascorbic acid. Wu and coworkers designed hybrid heart valve from decellularized porcine valve coated with P3HBHHx and implanted in the sheep for 16 weeks. These coated valves exhibited and promoted better proliferation of cardiomyocytes and endothelial cells and less calcification (Wu et al. 2007). Adamus and coworkers blended the P3HB and P3HO to design more elastic and tensile heart valves for the future biomedical engineering applications (Adamus et al. 2012).

The damaged or diseased blood vessels are repaired through the vascular grafting. The surgeons generally implant the vascular grafts composed of Dacron TM (polyethylene terephthalate) for larger diameter blood vessels. These grafts are not suitable for the narrow diameter blood vessels like in coronary artery bypass procedures. The synthetic grafts made with P3HB4HB were implanted in the dog, but degradation of the implant started within 2 weeks. The composite of P3HHx3HO was used as graft and was found stable till 6 months (Marois et al. 1999; Shum-Tim et al. 1999). Moreover, PHA blended vascular grafts are considered as valuable alternative for implication in vascular patches, atrial septal defect repair devices, and valves of the vein (Chen et al. 2001; Chen and Wu 2005; Dai et al. 2009; Wang et al. 2008). The P3HBHHx-based blended scaffolds coated with fibronectin increase the cell adhesion and cellular proliferation and were found suitable for vascular tissue engineering (Zhang et al. 2007; Wu et al. 2008). Gaudio and colleagues designed the composites of the polycaprolactone (PCL) and P3HBV in various proportions. These blended compositions of PHA could serve as future implants in vascular tissue engineering (Gaudio et al. 2012; Madhavan et al. 2013). These vascular grafts restore the malfunctioning of the blood vessels during injury and pathologic condition. However, the immune response, risk of infection during surgery, and stability of the scaffold/implant are some serious limiting factors.

6.2.3 Brain Nerve Guides and Implants

Peripheral nervous system injuries can lead to permanent disability during the surgical procedures. Nerve tissue engineering has made the neurological regeneration procedures feasible. The neurological recovery is considered less than 2 cm in humans. Apart from extremely low inflammatory, cytotoxic, and immunogenic response, the neurological implant material should mimic with the fibers scale of extracellular matrix and able to prevent the infiltration (Young et al. 2002; Mohanna et al. 2003). Although nonbiodegradable silicone-based biomaterial/implants are already in use to meet the demands in nerve tissue engineering, PHA-based nerve implants can serve as a good candidate for nerve regeneration. The PHB-based implants were introduced to regenerate radial nerves in the cats (Mohanna et al. 2003; Young et al. 2002). The axonal regeneration was demonstrated up to 2–3 mm distance with normal inflammatory response. Several studies indicate that PHB and P3HBHHx can be designed to fabricate nerve scaffolds for Schwann cell regeneration (Armstrong et al. 2007; Bian et al. 2009). In several studies, P3HBHHx was blended randomly and fabricated with poly-DL-lactide (PDLLA) in different studies. The fibronectin accumulation was higher on PDLLA biofilm sheet (Gao et al. 2006). The tubular PHBV was fabricated with the poly-L-lactide-co-D,L-lactide (PLDL) and PLGA poly-lactide-co-glycolide acid as an exterior component with electrospin technology. The PHBV–PLGA composites demonstrated the healing of nerve tissues in the injured area (Yucel et al. 2010). The composites/conduits made with the PHB and PHBHHx expand the limits of axonal regeneration up to 10 mm in the sciatic nerve of rat with good biocompatibility that is not feasible in the silicone-based implants (Armstrong et al. 2007; Bian et al. 2009). Mohanna and colleagues demonstrated artificial nerve implants supplemented with glial growth factor repair larger nerve gap up to 2–4 cm in rabbit (Mohanna et al. 2003). Chen and colleagues demonstrated that neural progenitor cells differentiate into neurons using the P3HBV microspheres (Chen and Tong 2012). Collagen is the structurally most abundant protein of the extracellular matrix of the nerves; Prabhakaran and colleagues designed the P3HBV-collagen composite nanowires to regenerate the nerve tissue (Prabhakaran et al. 2013). They demonstrated that collagen-associated P3HBV nanofibers provided proper orientations and bipolar extension during nerve cell proliferation. An enhanced cell proliferation was observed in the composite P3HBV/collagen (50:50) and P3HBV/collagen (75:25) nanofibers among the randomly fabricated nanofibers. Although P3HBV was blended with type I collagen in a random manner, still improved cell adhesion and cellular differentiation of Schwann cells were discovered in aligned PHB/P3HBV/collagen fibers. Masaeli and coworkers demonstrated the myelinated nerve cell regeneration and high expression of nerve growth factor on randomly fabricated PHB/PHBV/collagen nanofibers using electrospin technology (Masaeli et al. 2013; Merolli et al. 2014).

6.2.4 Bone Tissue Implants: Orthopedic Anchors, Sutures, and Stents

Bone is complex organic–inorganic hybrid tissue with characteristic mechanical properties such as high fracture strength and flexibility. Bone tissue engineering is an interdisciplinary field of kinesiology and material sciences and serves as an excellent tool for the treatment of damaged or lost bone due to aging or traumatic shock. The implant material must consist good mechanical strength and properties to regulate the cellular proliferations, differentiation of osteoblast, and development of bone extracellular matrix. Initially, *scl*-PHAs were considered suitable for the hard tissue engineering, but during the last decade, *mcl*-PHA-based blended materials were used with enhanced mechanical properties. Wang and colleagues evaluated PHB, P3HBHHx, and PLA-based 3D scaffolds during cellular differentiation and cell adhesion in rabbit model (Wang et al. 2004). The cell adhesion, proliferation of osteoblast, higher calcium deposition, and collagen synthesis in P3HBHHx-based implants were comparative to PHB and PLA (Li et al. 2007). Cool and coworkers demonstrated that PHBV/hydroxyapatite-based scaffolds exhibit low inflammatory response and high mineralization (Cool et al. 2007). Francis and colleagues designed novel multipurpose 45S5 bio-glass based on PHB and nano-sized hydroxyapatite (natural mineral of the bone matrix) used for gentamicin delivery during bone tissue engineering (Francis et al. 2011). Hyati and colleagues demonstrated that PHB/nanohydroxyapatite-based biomaterials are favorable in the bone tissue engineering. The PHB blended with 10–15 % of nanohydroxylapatite consequently yielded bone like highly porous material. All scaffolds contain the biocompatibility, mechanical properties, and porosity like cancellous bone (Hayati et al. 2012). Baek and colleagues demonstrated better adhesion, proliferation, and differentiation of osteoblast cells on PHBV/hydroxyapatite scaffolds immobilized with collagen I (Baek et al. 2012). Sultana and Wang designed the P3HBV blended with higher content of the hydroxyapatite; consequently, the scaffolds were found to have a more porous matrix and to be a favorable environment for osteoblast (Sultana and Wang 2012). Moreover, PHB blended with bio-glass 45S5 was shown to produce bone implant materials with 85 % porosity. Mirsa and colleagues demonstrated the scaffolds were non-immunogenic and provided better cellular attachment and proliferation of osteoblasts in rat (Misra et al. 2010). More recently, Wang and coworkers designed the PHBV/calcium silicate composites, which mimic natural extracellular matrix. The calcium silicate induces the expression of genes responsible for transforming growth factor-β1 and bone morphogenetic protein-7 that promotes early differentiation of human osteoblasts (Wang et al. 2013).

6.2.5 Cartilage Tissue Engineering

Cartilage is nonvascular tissue and hardly regenerates that supports the skeletal system. However, significant investigations have been carried out in PHA-based scaffolds during the last decades. The PHB, P3HBV, P3HBHHx, and various composites have been studied and were found suitable contenders for cartilage tissue engineering because of improved proliferation and enhanced differentiation of chondrocytes (Sun et al. 2005). Wang and colleagues developed the 3D-engineered P3HBHHx scaffold to repair articular cartilage in rabbit model. The engineered P3HBHHx cartilage scaffold was incubated with chondrocytes and showed effective cartilage repair with superior dissemination of extracellular matrix that was observed during 16 weeks in rabbit (Wang et al. 2008). Similarly, Liu and colleagues demonstrated the neocartilage formation by employing the chondrogenic differentiated human adipose stem cells on P3HBV scaffolds. The chondrogenic pre-differentiated human adipose stem cells constructed the neocartilage with good mechanical strength in nude mice after 16 weeks of implantation (Liu et al. 2010). Later, the mechanical strength was improved with the incorporation of poly-L-lactide-co-caprolactone to P3HBV microspheres for cartilage tissue engineering (Li et al. 2013).

6.2.6 Liver Tissue Implants

Although PHA-based liver implants have not been tested in human till date, some reports suggest that PHB-based 3D supports frameworks can be used in liver tissue regeneration. The investigators suggested that hepatocytes grow on the PHB beads in vitro. Zhu and coworkers demonstrated that composite of P3HBV and poly-3-hydroxybutyrate-co-3-hydroxyvalerate (P3HB3HV) microspheres allows the growth of hepatocytes in HepG2 and Hep3B cell lines. The microsphere allows proliferation of the HepG2 and Hep3B cells to construct the multilayer structure within 1 week. The HepG2 cells were unable to maintain the P450 activity. The cellular aggregation enhanced 2–4 folds upon bovine serum albumin secretion after 12 days (Zhu et al. 2007a). The research group identified the vital role of extracellular matrix during hepatocyte regeneration. This time PHBV microspheres were conjugated with three different extracellular matrix proteins: collagen, laminin, and fibronectin. The cellular hepatic function, P450 activity, and bovine serum albumin secretion were monitored for 2 weeks. The Hep3B cells demonstrated enhanced cellular proliferation and hepatic function on matrix-coated microspheres because coated microsphere mimics human physiological environment (Zhu et al. 2007b). Later, investigators identified the effect of copolymers on cellular aggregation in Hep3B cell lines. The P3HBV was blended with polymer PLGA serving as good framework for liver tissue engineering with the support of hepatocyte growth factor. The composite of P3HBV/PLGA microspheres demonstrated less degradation rate and maintained surface bioactivity indicating more suitability of the copolymer in liver tissue regeneration (Zhu et al. 2009).

6.2.7 Skin Tissue Implants

The human skin serves as the largest organ of the body that maintains temperature, regulates water loss, and works as an effective barrier against external environment. Skin tissue engineering has revolutionized the grafting procedures for the use of burn victims and non-healing lesions like diabetic and venous ulcers. Several investigators have improved the skin grafting technology for the betterment of human health by using various biomaterial sheets. Hyaluronic acid/gelatin/chitosan biofilms were used to propagate the human skin fibroblast cells and then transferred to the wound (Peschel et al. 2008; Tang et al. 2008). PHB copolymers such as P3HB4HB and P3HB3HHx with hyaluronic acid/chitosan have been tested in skin tissue engineering (Ji et al. 2008; Peschel et al. 2008). Ji and coworkers demonstrated PHB and P3HBV copolymers provide better mechanical properties and are excellent in supporting the growth of skin cells (Ji et al. 2008). Later, Kuppan and colleagues studied the human skin fibroblast cell proliferation, adhesion, and gene expression on the PHBV-based scaffolds and tissue culture polystyrene (Kuppan et al. 2011). P3HBV blended with chitosan scaffolds support better growth and enhanced cell proliferation and adhesion in wound healing model of rat in vivo (Veleirinho et al. 2012). The chitin composite P3HB3HV with hydrogel scaffold exhibited two-fold enhanced cell proliferation of human dermal fibroblast cells. These blends showed higher porosity and slow rate biodegradation (Sankar et al. 2012). The PHB-based hydrogels macroporous 3D scaffolds showed the promising application in skin tissue engineering and grafting.

6.3 Summary and Future Prospective

PHAs have opened new window of opportunity, with the diverse applications in biomedical engineering. The application of PHAs will be broader with the discovery of new bacterial strains capable to generate the homopolymers and copolymers, especially from biowastes as feed, in near future (Kalia et al. 2003; Porwal et al. 2008; Kumar et al. 2009; Patel et al. 2012; Kumar et al. 2015a, b). This is feasible to produce the copolymers through metabolic engineering with manifestation of diverse microbial metabolic background (Reddy et al. 2003; Kalia et al. 2007; Singh et al. 2009). Moreover, the host cell genome manipulations with engineered PHA synthase enzymatic machinery can design bacteria into robust microbial plastic factory. The government and institutions should take initiatives to promote the growth of biomaterial research in association with the medicine. The diverse groups of expertise can solve the problem of implant material limitations.

Acknowledgment The authors wish to thank the director of CSIR-Institute of Genomics and Integrative Biology (IGIB) and Defence Research and Development Establishment (DRDE), Jhansi Road, Gwalior (LSRB-268/BTB/2013 and BSC0123), Government of India, for providing the necessary funds and facilities. Authors are also thankful to the Academy of Scientific and Innovative Research (AcSIR), New Delhi. Lalit K Singh, Shashank S Kamble, and Aakriti Gangwal are thankful to UGC, and Neha Dhasmana is thankful to CSIR, for granting Senior Research Fellowships. We highly acknowledge Dr. V. C. Kalia and Mr. Prasun Kumar, from CSIR-IGIB, Delhi, and Ms. Aakriti Gangwal from UDSC New Delhi, India, for their critical comments in the manuscript.

References

Adamus G, Sikorska W, Janeczek H, Kwiecień M, Sobota M, Kowalczuk M (2012) Novel block copolymers of atactic PHB with natural PHA for cardiovascular engineering: synthesis and characterization. Eur Polym J 48:621–631. doi:10.1016/j.eurpolymj.2011.12.017

Armstrong SJ, Wiberg M, Terenghi G, Kingham PJ (2007) ECM molecules mediate both Schwann cell proliferation and activation to enhance neurite outgrowth. Tissue Eng 13:2863–2870. doi:10.1089/ten.2007.0055

Baek JY, Xing ZC, Kwak G, Yoon KB, Park SY, Park LS, Kang IK (2012) Fabrication and characterization of collagen-immobilized porous PHBV/HA nanocomposite scaffolds for bone tissue engineering. J Nanomater 171804:1–11. doi:10.1155/2012/171804

Bian YZ, Wang Y, Guli S, Chen GQ, Wu Q (2009) Evaluation of poly(3-hydroxybutyrate-co-3-hydroxyhexanoate) conduits for peripheral nerve regeneration. Biomaterials 30:217–225. doi:10.1016/j.biomaterials.2008.09.036

Bruder SP, Fox BS (1999) Tissue engineering of bone cell based strategies. Clin Orthop Relat Res 367:68–83, PMID:10546637

Chen GQ (2009) A microbial polyhydroxyalkanoates (PHA) based bio and materials industry. Chem Soc Rev 38:2434–2446. doi:10.1039/b812677c

Chen W, Tong YW (2012) PHBV microspheres as neural tissue engineering scaffold support neuronal cell growth and axon–dendrite polarization. Acta Biomater 8:540–548. doi:10.1016/j.actbio.2011.09.026

Chen GQ, Wu Q (2005) Polyhydroxyalkanoates as tissue engineering materials. Biomaterials 26:6565–6578. doi:10.1016/j.biomaterials.2005.04.036

Chen GQ, Zhang G, Park SJ, Lee SJ (2001) Industrial production of poly(hydroxybutyrate-co-hydroxyhexanoate). Appl Microbiol Biotechnol 57:50–55, PMID: 11693933

Cool SM, Kenny B, Wu A, Nurcombe V, Trau M, Cassady AI, Grøndahl L (2007) Poly(3-hydroxybutyrate-co-3-hydroxyvalerate) composite biomaterials for bone tissue regeneration, in vitro performance assessed by osteoblast proliferation, osteoclast adhesion and resorption, and macrophage pro-inflammatory response. J Biomed Mater Res A 82A:599–610. doi:10.1002/jbm.a.31174

Dai ZW, Zou XH, Chen GQ (2009) Poly(3-hydroxybutyrate-co-3- hydroxyhexanoate) as an injectable implant system for prevention of post-surgical tissue adhesion. Biomaterials 30:3075–3083. doi:10.1016/j.biomaterials.2009.02.015

Du DJ, Furukawa KS, Ushida T (2009) 3-D culture of osteoblast-like cells by unidirectional or oscillatory flow for bone tissue engineering. Biotechnol Bioeng 102:1670–1678. doi:10.1002/bit.22214

Duvernoy O, Malm T, Ramström J, Bowald S (1995) A biodegradable patch used as a pericardial substitute after cardiac surgery: 6- and 24-month evaluation with CT. Thorac Cardiovasc Surg 43:271–274. doi:10.1055/s-2007-1013226

Francis L, Meng D, Knowles J, Keshavarz T, Boccaccini AR, Roy I (2011) Controlled delivery of gentamicin using poly(3-hydroxybutyrate) microspheres. Int J Mol Sci 12:4294–4314. doi:10.3390/ijms12074294

Fu P, Sodian R, Lüders C, Lemke T, Kraemer L, Hübler M, Weng Y, Hoerstrup SP, Meyer R, Hetzer R (2004) Effects of basic fibroblast growth factor and transforming growth factor-beta on maturation of human pediatric aortic cell culture for tissue engineering of cardiovascular structures. ASAIO J 50:9–14, PMID: 14763486

Fujimoto KL, Guan J, Oshima H, Sakai T, Wagner WR (2007) In vivo evaluation of a porous, elastic, biodegradable patch for reconstructive cardiac procedures. Ann Thorac Surg 83:648–654. doi:10.1016/j.athoracsur.2006.06.085

Gao Y, Kong L, Zhang L, Gong Y, Chen G, Zhao N, Zhang X (2006) Improvement of mechanical properties of poly(DL-lactide) films by blending of poly(3-hydroxybutyrate-co-3-hydroxyhexanoate). Eur Polym J 42:764–775. doi:10.1016/j.eurpolymj.2005.09.028

Gaudio CD, Fioravanzo L, Folin M, Marchi F, Ercolani E, Bianco A (2012) Electrospun tubular scaffolds: on the effectiveness of blending poly(ε-caprolactone) with poly(3-hydroxybutyrate-co-3-hydroxyvalerate). J Biomed Mater Res B Appl Biomater 100:1883–1898. doi:10.1002/jbm.b.32756

Hayati AN, Hosseinalipour SM, Rezaie HR, Shokrgozar MA (2012) Characterization of poly(3-hydroxybutyrate)/nano-hydroxyapatite composite scaffolds fabricated without the use of organic solvents for bone tissue engineering application. Mater Sci Eng C 32:416–422. doi:10.1016/j.msec.2011.11.013

Ihssen J, Magnani D, Thony-Meyer L, Ren Q (2009) Use of extracellular medium chain length polyhydroxyalkanoate depolymerase for targeted binding of proteins to artificial poly[(3-hydroxyoctanoate)-co-(3-hydroxy-hexanoate)] granules. Biomacromol 10:1854–1864. doi:10.1021/bm9002859

Ji Y, Li XT, Chen GQ (2008) Interactions between a poly(3-hydroxybutyrate-co-3-hydroxyvalerate-co-3-hydroxyhexanoate) terpolyester and human keratinocytes. Biomaterials 29:3807–3814. doi:10.1016/j.biomaterials.2008.06.008

Kalia VC, Raizada N, Sonakya V (2000) Bioplastics. J Sci Ind Res 59:433–445

Kalia VC, Chauhan A, Bhattacharyya G (2003) Genomic databases yield novel bioplastic producers. Nat Biotechnol 21:845–846. doi:10.1038/nbt0803-845

Kalia VC, Lal S, Cheema S (2007) Insight in to the phylogeny of polyhydroxyalkanoate biosynthesis: horizontal gene transfer. Gene 389:19–26. doi:10.1016/j.gene.2006.09.010

Karageorgiou V, Kaplan D (2005) Porosity of 3-D biomaterial scaffolds and osteogenesis. Biomaterials 26:5474–5491. doi:10.1016/j.biomaterials.2005.02.002

Kumar T, Singh M, Purohit HJ, Kalia VC (2009) Potential of *Bacillus* sp. To produce polyhydroxybutyrate from biowaste. J Appl Microbiol 106:2017–2023. doi:10.1111/j.1365-2672.2009.04160.x

Kumar P, Patel SKS, Lee JK, Kalia VC (2013) Extending the limits of *Bacillus* for novel biotechnological applications. Biotechnol Adv 31:1543–1561. doi:10.1016/j.biotechadv.2013.08.007

Kumar P, Singh M, Mehariya S, Patel SKS, Lee JK, Kalia VC (2014) Ecobiotechnological approach for exploiting the abilities of *Bacillus* to produce co-polymer of polyhydroxyalkanoate. Indian J Microbiol 54:151–157. doi:10.1007/s12088-014-0457-9

Kumar P, Mehariya S, Ray S, Mishra A, Kalia VC (2015a) Biodiesel industry waste: a potential source of bioenergy and biopolymers. Indian J Microbiol 55:1–7. doi:10.1007/s12088-014-0509-1

Kumar P, Ray S, Patel SKS, Lee JK, Kalia VC (2015b) Bioconversion of crude glycerol to polyhydroxyalkanoate by *Bacillus thuringiensis* under non-limiting nitrogen conditions. Int J Biol Macromol. doi:10.1016/j.ijbiomac.2015.03.046

Kuppan P, Vasanthan KS, Sundaramurthi D, Krishnan UM, Sethuraman S (2011) Development of poly(3-hydroxybutyrate-co-3-hydroxyvalerate) fibers for skin tissue engineering: effects of topography, mechanical, and chemical stimuli. Biomacromolecules 12:3156–3165. doi:10.1021/bm200618w

Li Z, Lin H, Ishii N, Chen GQ, Inoue Y (2007) Study of enzymatic degradation of microbial copolyesters consisting of 3-hydroxybutyrate and medium-chain-length 3-hydroxyalkanoates. Polym Degrad Stab 92:1708–1714. doi:10.1016/j.polymdegradstab.2007.06.001

Li XT, Zhang Y, Chen GQ (2008) Nanofibrous polyhydroxyalkanoate matrices as cell growth supporting materials. Biomaterials 29:3720–3728. doi:10.1016/j.biomaterials.2008.06.004

Li C, Zhang J, Li Y, Moran S, Khang G, Ge Z (2013) Poly(L-lactide-co-caprolactone) scaffolds enhanced with poly(3-hydroxybutyrate-co-hydroxyvalerate) microspheres for cartilage regeneration. Biomed Mater 8025005:1–8. doi:10.1088/1748-6041/8/2/025005

Liu J, Zhao B, Zhang Y, Lin Y, Hu P, Ye C (2010) PHBV and pre-differentiated human adipose-derived stem cells for cartilage tissue engineering. J Biomed Mater Res Part A 94A:603–610. doi:10.1002/jbm.a.32730

Madhavan K, Elliott WH, Bonani W, Monnet E, Tan W (2013) Mechanical and biocompatible characterizations of a readily available multilayer vascular graft. J Biomed Mater Res B Appl Biomater 101:506–519. doi:10.1002/jbm.b.32851

Malm T, Bowald S, Bylock A, Busch C (1992) Prevention of postoperative pericardial adhesions by closure of the pericardium with absorbable polymer patches. An experimental study. J Thorac Cardiovasc Surg 104:600–607, PMID: 1513149

Marois Y, Zhang Z, Vert M, Beaulieu L, Lenz RW, Guidoin R (1999) In vivo biocompatibility and degradation studies of polyhydroxyoctanoate in the rat: a new sealant for the polyester arterial prosthesis. Tissue Eng 5:369–386, PMID: 10477858

Masaeli E, Morshed M, Nasr-Esfahani MH, Sadri S, Hilderink J, van Apeldoorn A, van Blitterswijk CA, Moroni L (2013) Fabrication, characterization and cellular compatibility of poly(hydroxy alkanoate) composite nanofibrous scaffolds for nerve tissue engineering. PLoS ONE 8(2), e57157. doi:10.1371/ journal.pone.0057157

Masood F, Yasin T, Hameed A (2014) Polyhydroxyalkanoates–what are the uses? Current challenges and perspectives. Crit Rev Biotechnol. doi: 10.3109/07388551.2014.913548

Merolli A, Rocchi L, Wang XM, Cui FZ (2014) Peripheral nerve regeneration inside collagen-based artificial nerve guides in humans. J Appl Biomater Funct Mater. doi:10.5301/jabfm.5000188

Misra SK, Valappil SP, Roy I, Boccaccini AR (2006) Polyhydroxyalkanoate (PHA)/inorganic phase composites for tissue engineering applications. Biomacromolecules 7:2249–2258. doi:10.1021/bm060317c

Misra SK, Ansari TI, Valappil SP, Mohn D, Philip SE, Stark WJ, Roy I, Knowles JC, Salih V, Boccaccini AR (2010) Poly(3-hydroxybutyrate) multifunctional composite scaffolds for tissue engineering applications. Biomaterials 31:2806–2815. doi:10.1016/j.biomaterials.2009.12.045

Mohanna PN, Young RC, Wiberg M, Terenghi G (2003) A composite poly-hydroxybutyrate–glial growth factor conduit for long nerve gap repairs. J Anat 203:553–565. doi:10.1046/j.1469-7580.2003.00243.x

Patel SKS, Singh M, Kumar P, Purohit HJ, Kalia VC (2012) Exploitation of defined bacterial cultures for production of hydrogen and polyhydroxybutyrate from pea-shells. Biomas Bioenerg 36:218–225. doi:10.1016/j.biombioe.2011.10.027

Patel SKS, Kumar P, Singh M, Lee JK, Kalia VC (2015) Integrative approach to produce hydrogen and polyhydroxybutyrate from biowaste using defined bacterial cultures. Bioresour Technol 176:136–141. doi:10.1016/j.biortech.2014.11.029

Peschel G, Dahse HM, Konrad A, Wieland GD, Mueller PJ, Martin DP, Roth M (2008) Growth of keratinocytes on porous films of poly(3-hydroxybutyrate) and poly(4-hydroxybutyrate) blended with hyaluronic acid and chitosan. J Biomed Mater Res A 85:1072–1081. doi:10.1002/jbm.a.31666

Porwal S, Kumar T, Lal S, Rani A, Kumar S, Cheema S, Purohit HJ, Sharma R, Patel SKS, Kalia VC (2008) Hydrogen and polyhydroxybutyrate producing abilities of microbes from diverse habitats by dark fermentative process. Bioresour Technol 99:5444–5451. doi:10.1016/j.biortech.2007.11.011

Prabhakaran MP, Vatankhah E, Ramakrishna S (2013) Electrospun aligned PHBV/collagen nanofibers as substrates for nerve tissue engineering. Biotechnol Bioeng 110:2775–2784. doi:10.1002/ bit.24937

Qu XH, Wu Q, Chen GQ (2006) In vitro study on hemocompatibility and cytocompatibility of poly(3-hydroxybutyrate-co-3-hydroxyhex-anoate). J Biomater Sci 17:1107–1121. doi:10.1163/156856206778530704

Reddy CSK, Ghai R, Kalia V (2003) Polyhydroxyalkanoates: an overview. Bioresour Technol 87:137–146. doi:10.1016/S0960-8524(02)00212-2

Sankar D, Chennazhi KP, Nair SV, Jayakumar R (2012) Fabrication of chitin/poly(3-hydroxybutyrate-co-3-hydroxyvalerate) hydrogel scaffold. Carbohyd Polym 90:725–729. doi:10.1016/j.carbpol.2012.05.041

Shum-Tim D, Stock U, Hrkach J, Shinoka T, Lien J, Moses MA, Stamp A, Taylor G, Moran AM, Landis W, Langer R, Vacanti JP, Mayer JE (1999) Tissue engineering of autologous aorta using a new biodegradable polymer. Ann Thorac Surg 68:2298–2304. doi:10.1016/S0003-4975(99)01055-3

Singh M, Patel SKS, Kalia VC (2009) *Bacillus subtilis* as potential producer for polyhydroxyalkanoates. Microb Cell Fact 8:38. doi:10.1186/1475-2859-8-38

Singh M, Kumar P, Patel SKS, Kalia VC (2013) Production of polyhydroxyalkanoate co-polymer by *Bacillus thuringiensis*. Indian J Microbiol 53:77–83. doi:10.1007/s12088-012-0294-7

Singh LK, Dhasmana N, Sajid A, Kumar P, Bhaduri A, Bharadwaj M, Gandotra S, Kalia VC, Das TK, Goel AK, Pomerantsev AP, Misra R, Gerth U, Leppla SH, Singh Y (2015a) *ClpC* operon regulates cell architecture and sporulation in *Bacillus anthracis*. Environ Microbiol 17:854–864. doi:10.1111/1462-2920.12548

Singh M, Kumar P, Ray S, Kalia VC (2015b) Challenges and opportunities for customizing polyhydroxyalkanoates. Indian J Microbiol 55:234. doi:10.1007/s12088-015-0528-6

Sultana N, Wang M (2012) PHBV/PLLA-based composite scaffolds fabricated using an emulsion freezing/freeze-drying technique for bone tissue engineering: surface modification and in vitro biological evaluation. Biofabrication 4:015003. doi:10.1088/1758-5082/4/1/015003

Sun JY, Wu J, Li HY, Chang J (2005) Macroporous poly(3-hydroxybutyrate-co-3-hydroxyvalerate) matrices for cartilage tissue engineering. Eur Polym J 41:2443–2449. doi:10.1016/j.eurpolymj.2005.04.039

Tang HY, Ishii D, Mahara A, Murakami S, Tetsuji Y, Kumar S, Razip S, Masahiro F, Mizuo M, Tadahisa I (2008) Scaffolds from electrospun polyhydroxyalkanoate copolymers: fabrication, characterization, bioabsorption and tissue response. Biomaterials 29:1307–1317. doi:10.1016/j.biomaterials.2007.11.031

Veleirinho B, Coelho DS, Dias PF, Maraschin M, Ribeiro-do-Valle RM, Lopes-da-Silva JA (2012) Nanofibrous poly(3-hydroxybutyrate-co-3-hydroxyvalerate)/chitosan scaffolds for skin regeneration. Int J Biol Macromol 51:343–350. doi:10.1016/j.ijbiomac.2012.05.023

Wang Y, Wu Q, Chen GQ (2004) Attachment, proliferation and differentiation of osteoblasts on random biopolyester poly(3-hydro-xybutyrate-co-3-hydroxyhexanoate) scaffolds. Biomaterials 25:669–675. doi:10.1016/S0142-9612(03)00561-1

Wang Y, Bian YZ, Wu Q, Chen GQ (2008) Evaluation of three-dimensional scaffolds prepared from poly(3-hydroxybutyrate-co-3-hydroxyhexanoate) for growth of allogeneic chondrocytes for cartilage repair in rabbits. Biomaterials 29:2858–2868. doi:10.1016/j.biomaterials.2008.03.021

Wang N, Zhou Z, Xia L, Dai Y, Liu H (2013) Fabrication and characterization of bioactive β-Ca_2SiO_4/PHBV composite scaffolds. Mater Sci Eng C 33:2294–2301. doi:10.1016/j.msec.2013.01.059

Wu S, Liu YL, Cui B, Qu XH, Chen GQ (2007) Study on decellularized porcine aortic valve/poly (3-hydroxybutyrate-co-3-hydroxyhexanoate) hybrid heart valve in sheep model. Artif Organs 31:689–697. doi:10.1111/j.1525-1594.2007.00442.x

Wu S, Liu YL, Cui B, Tang Y, Wang Q, Qu XH, Chen GQ (2008) Intravascular biocompatibility of poly (3-hydroxybutyrate-co-3-hydroxyhexanoate) (PHBHHx) for cardiovascular tissue engineering. Chinese J Biotechnol 24:610–616. doi:10.1016/S1872-2075(08)60030-8

Xiao XQ, Zhao Y, Chen GQ (2007) The effect of 3-hydroxybutyrate and its derivatives on the growth of glial cells. Biomaterials 28:3608–3616. doi:10.1016/j.biomaterials.2007.04.046

Yang Q, Wang J, Zhang S, Tang X, Shang G, Peng Q, Wang Rand Cai X (2014) The properties of poly (3-hydroxybutyrate-co-3-hydroxyhexanoate) and its applications in tissue engineering. Curr Stem Cell Res Ther 9:215–222

Young RC, Wiberg M, Terenghi G (2002) Poly-3-hydroxybutyrate (PHB): a resorbable conduit for long-gap repair in peripheral nerves. Br J Plast Surg 55:235–240. doi:10.1054/bjps.2002.3798

Yu GH, Fan YB (2008) Preparation of poly(D, L-lactic acid) scaffolds using alginate particles. J Biomater Sci-Polym Ed 19:87–98. doi:10.1163/156856208783227703

Yucel D, Kose GT, Hasirci V (2010) Polyester based nerve guidance conduit design. Biomaterials 31:1596–1603. doi:10.1016/j.biomaterials.2009.11.013

Zhang L, Zheng Z, Xi J, Gao Y, Aob Q, Gong Y, Zhao N, Zhang X (2007) Improved mechanical property and biocompatibility of poly(3-hydroxybutyrate-co-3-hydroxyhexanoate) for blood vessel tissue engineering by blending with poly(propylene carbonate). Europ Polym J 43:2975–2978. doi:10.1016/j.eurpolymj.2007.04.007

Zhu XH, Gan SK, Wang CH, Tong YW (2007a) Proteins combination on PHBV microsphere scaffold to regulate Hep3B cells activity and functionality: a model of liver tissue engineering system. J Biomed Mater Res 83A:606–616. doi:10.1002/jbm.a.31257

Zhu XH, Wang CH, Tong YW (2007b) Growing tissue-like constructs with Hep3B/HepG2 liver cells on PHBV microspheres of different sizes. J Biomed Mater Res B 82B:7–16. doi:10.1002/jbm.b.30698

Zhu XH, Wang CH, Tong YW (2009) In vitro characterization of hepatocyte growth factor release from PHBV/PLGA microsphere scaffold. Biomed Mater Res 89A:411–423. doi:10.1002/jbm.a.31978

Lalit K. Singh received the M.Sc. degree in Biotechnology from Rajiv Gandhi Biotechnology Centre, RTM Nagpur University, Nagpur, India. He is currently working as Senior Research Fellow, CSIR-IGIB, Delhi, India. His current research focuses on understanding the role of caseinolytic proteases and Ser/Thr kinases that mediated cellular architecture alterations, cell wall remodeling, and sporulation in *Bacillus*.

Neha Dhasmana graduated in Microbiology in 2008 from University of Delhi. She received her M. Sc. degree in Biochemistry in 2010 from Hamdard University. She is currently enrolled in Ph.D. with Academy of Innovative and Scientific Research (AcSIR) and works at CSIR-Institute of Genomics and Innovative Biology. She is studying the physiological significance of Ser/Thr kinases in *Bacillus anthracis*.

Shashank S. Kamble received his M.Sc. degree from University of Mumbai, India. Currently, he is pursuing Ph.D. under the supervision of Dr. Yogendra Singh at CSIR-IGIB, India. He is interested in the host pathogen cross talk during infection with *M. tuberculosis*. His work focuses on the modulation of host proteases by *M. tuberculosis*.

Aditya K. Sharma has done his graduation in Biomedical Sciences from University of Delhi and Master's in Biotechnology from Jamia Millia Islamia. Currently, he is pursuing his doctoral degree from Institute of Genomics and Integrative Biology under supervision of Dr. Yogendra Singh. For his Ph.D., he is working on regulation of kinase and its substrates via phosphorylation in *Mycobacterium tuberculosis*.

Yogendra Singh received his Ph.D. degree from Vallabhbhai Patel Chest Institute, Delhi University, and postdoctoral training at National Institute of Health (NIH), Bethesda. Currently, he is Chief Scientist at CSIR-IGIB and Professor of AcSIR. He is a Fellow of National Academy of Sciences, India, Fellow of Indian National Science Academy, and Fellow of Indian Academy of Sciences, India. He has been honored with Moselio Schaechter Distinguished Service Award in 2014 by American Society for Microbiology. He serves as an Editorial Board Member of *Journal of Biological Chemistry*. His current research interest is to understand the host response in pulmonary/extra-pulmonary tuberculosis and to decipher the role of eukaryotic like Ser/Thr kinases in *Mycobacterium tuberculosis* and *Bacillus anthracis*.

Sporulation, a Pitfall in the Path of PHB Production

7

Neha Dhasmana, Lalit K. Singh, Shashank S. Kamble, Nishant Kumar, and Yogendra Singh

Abstract

The concept of bioplastic is fascinating to our world, because of its potentiality to deal with one of the major global problems like plastic pollution (Kalia et al. J Sci Ind Res 59:433–445, 2000; Kalia et al. Nat Biotechnol 21:845–846, 2003). Polyhydroxybutyrates (PHB) are the best example for the polymers by plant or microorganisms from a wide range of habitats (Reddy et al. Bioresour Technol 87:137–146, 2003; Porwal et al. Bioresour Technol 99:5444–5451, 2008; Singh. Environ Microbiol 17:854–864, 2015). PHB refers to the polyesters of 3-hydroxybutyrate and can be extracted from various species like *Ralstonia*, *Bacillus*, *Streptomyces*, *Pseudomonas*, etc., which are extensively discussed in published reviews (Singh et al. Microb Cell Fact 8:38, 2009; Jendrossek and Pfeiffer. Environ Microbiol 16:2357–2373, 2014). Bioplastic has the distinct feature of being biodegradable. Further, the use of biowaste as substratum for bioplastic-producing organisms presents an interesting concept to deal with another global problem of waste management (Kumar et al. J Appl Microbiol 106:2017–2023, 2009; Kumar et al. Indian J Microbiol 55:1–7, 2015; Kumar et al. Int J Biol Macromol, 2015; Patel et al. Biomas Bioenerg 36:218–225, 2012; Patel et al. Bioresour Technol 176:136–141, 2015). Both Gram-positive and Gram-negative bacteria are reported to produce polyhydroxyalkanoates (PHA); among them Gram-negative bacteria, *Ralstonia eutropha* is the most extensively studied organism (Brigham

N. Dhasmana (✉) • S.S. Kamble • N. Kumar
Allergy and Infectious Diseases,
CSIR-Institute of Genomics and Integrative Biology,
110007 Delhi, India
e-mail: neha.dhasmana@igib.in;
shashank.kamble@igib.in; nishant.kumar@igib.in

L.K. Singh
CSIR-Institute of Genomics and Integrative Biology,
Allergy and Infectious Diseases, Delhi, India
e-mail: lalit.singh@igib.in

Y. Singh (✉)
Allergy and Infectious Diseases, CSIR-Institute
of Genomics and Integrative Biology,
Room No. 208, Mall Road, 110007 Delhi, India
e-mail: ysingh@igib.res.in

V.C. Kalia (ed.), *Microbial Factories*, DOI 10.1007/978-81-322-2595-9_7

et al. Appl Environ Microbiol 78:8033–8044, 2012). One major rationale to investigate the Gram-positive bacteria for their ability to produce PHB is the absence of immunogenic lipopolysaccharide which co-purifies with the PHB when Gram-negative organisms are employed, making PHB non-appealing for the use in medical purposes like various human tissue grafts (Valappil et al. Antonie Van Leeuwenhoek 91:1–17, 2007; Singh et al. Microb Cell Fact 8:38, 2009). Additional appeal for using Gram-positive bacteria, specifically *Bacillus* spp., is its ability to produce copolymers which are superior to their counterparts, that is, homopolymers given their enhanced characteristics like elasticity, etc. (Patel et al. Indian J Microbiol 51:418–423, 2011; Kumar et al. Indian J Microbiol 54:151–157, 2014; Kumar et al. Int J Biol Macromol, 2015). Gram-positive bacteria exist in two alternative phases in its life cycle, that is, vegetative cells and sporulation. The adverse environmental conditions drive the *Bacillus* vegetative cells into their transition to spores. Sporulation is an intrinsic characteristic of the *Bacillus* species and is mainly regulated by a master regulator of sporulation, namely, Spo0A (Slepecky and Law. J Bacteriol 82:37–42, 1961; Singh et al. Indian J Microbiol 55:234, 2015).

7.1 Introduction

The concept of bioplastic is fascinating to our world, because of its potentiality to deal with one of the major global problems like plastic pollution (Kalia et al. 2000, 2003). Polyhydroxybutyrates (PHB) are the best example for the polymers by plant or microorganisms from a wide range of habitats (Reddy et al. 2003; Porwal et al. 2008; Singh et al. 2015a). PHB refers to the polyesters of 3-hydroxybutyrate and can be extracted from various species like *Ralstonia*, *Bacillus*, *Streptomyces*, *Pseudomonas*, etc., which are extensively discussed in published reviews (Singh et al. 2009; Jendrossek and Pfeiffer 2014). Bioplastic has the distinct feature of being biodegradable. Further, the use of biowaste as substratum for bioplastic-producing organisms presents an interesting concept to deal with another global problem of waste management (Kumar et al. 2009, 2015a, b; Patel et al. 2012, 2015). Both Gram-positive and Gram-negative bacteria are reported to produce polyhydroxyalkanoates (PHA); among them Gram-negative bacteria, *Ralstonia eutropha* is the most extensively studied organism (Brigham et al. 2012). One major rationale to investigate the Gram-positive bacteria for their ability to produce PHB is the absence of immunogenic lipopolysaccharide which co-purifies with the PHB when Gram-negative organisms are employed, making PHB non-appealing for the use in medical purposes like various human tissue grafts (Valappil et al. 2007; Singh et al. 2009). Additional appeal for using Gram-positive bacteria, specifically *Bacillus* spp., is its ability to produce copolymers which are superior to their counterparts, that is, homopolymers given their enhanced characteristics like elasticity, etc. (Patel et al. 2011; Kumar et al. 2013, 2014, 2015b). Gram-positive bacteria exist in two alternative phases in its life cycle, that is, vegetative cells and spores. The adverse environmental conditions drive the *Bacillus* vegetative cells into their transition to spores. Sporulation is an intrinsic characteristic of the *Bacillus* species and is mainly regulated by a master regulator of sporulation, namely, Spo0A (Slepecky and Law 1961; Singh et al. 2015b).

7.2 Polyhydroxybutyrate in Sporulation

Life thrives on the energy acquired. In primitive forms of life, the energy requirement for the cellular processes is fulfilled by existing substrate availability or the intrinsic storage molecules.

Bacteria employ different strategies to store energy in forms that can be utilized in times of need (Singh et al. 2014). Under nutrient limiting conditions, excess carbon is stored in the form of PHB or glycogen. Under high C:N or C:P ratios or lesser dissolved oxygen, PHB biosynthesis is induced (Supono et al. 2013). For example, PHB synthesis in *Bacillus anthracis* is catalyzed by the three enzymes, β-ketothiolase, acetoacetyl-CoA reductase, and PHB polymerase encoded by the genes *phaA*, *phaB*, and *phaC*, respectively (Kalia et al. 2007; Jendrossek and Pfeiffer 2014). The two acetyl-CoA units are combined together to form acetoacetyl-CoA, which upon reduction gets transformed to 3-hydroxybutyrate by acetoacetyl-CoA reductase. The 3-hydroxybutyrate serves as a substrate for the polymerization reaction that causes the further extension of PHB template, catalyzed by PHB synthase PhaRC in *B. anthracis*. The size of PHB polymer ranges between 50,000 and 1,000,000 Da. Depending on their sizes, the PHB polymer can be largely categorized into short chain length (SCL) which has C_3–C_5 backbone and medium chain length (MCL) which has backbone of aliphatic or aromatic hydroxyalkanoate, specifically C_6–C_{14}. PHB exists in the cell as amorphous granules and these structures are stabilized as phasin proteins, PhaP (Mezzina et al. 2014).

A study by Emeruwa and Hawirko showed that synthesized PHB gets catabolized during the sporulation process in *Clostridium botulinum* sporogenic strains, while PHB content is unaffected in the asporogenic strains of *C. botulinum*. The results indicate that synthesized PHB serves as an endogenous source of energy for the spore component synthesis during the course of sporulation. Interestingly, asporogenic strains of *C. botulinum* were unable to utilize the accumulated PHB in response to nutritional stress resulting in 13 % of PHB (cell dry weight) as compared to sporogenic strains of *C. botulinum* (9 %) (Emeruwa and Hawirko 1973). Even in *B. anthracis*, the sporulation-defective strain (Δ*clpC*) has reported to accumulate higher PHB content as compared to the wild-type strain (Singh et al. 2015a). These results suggest that a major fraction of the energy acquired from PHB catabolism can be utilized to drive the sporulation, a high-energy-consuming process (Fig. 7.1; Wu et al. 2001). Moreover, the onset of PHB biosynthesis and sporulation is induced under the same environmental setup which renders us to intrigue their complex relationship. One major question still exists about the essentiality of PHB in the sporulation process of *Bacillus* species which has not been experimentally validated till now. Since both sporogenic and asporogenic strains of *Bacillus* and *Clostridium* are reported to synthesize the PHB under excess carbon in the surrounding. Therefore, the essentiality of PHB in the sporulation process remains to be elucidated. PHB has been found to be functionally relevant in maintaining the heat resistance of the *Bacillus cereus* spores. The study asserts the role of PHB in endotrophic sporulation process in *B. cereus* (Holmes and Christopher 1984). Weakly buffered medium supports the PHB accumulation and the endotrophic sporulation; further, it has been reported to yield heat-resistant spores as compared to the strongly buffered media, in which lesser PHB accumulates (Holmes and Christopher 1984). However, various reports state the insignificance of PHB in sporulation process of *B. cereus* and *Bacillus megaterium* (Slepecky and Law 1961). A report showed pH-dependent PHB accumulation in the *B. cereus* strain T with the highest accumulation at pH between 6.2 and 6.4, whereas the sporulation efficiency was similar in pH 6.4 or pH 7.4 (Nakata 1963).

7.3 Molecular Modulators of PHB Synthesis

7.3.1 Spo0A

At molecular level, the sporulation master regulator, Spo0A, has been shown to be crucial for the expression of PHB synthesis-related genes in *Bacillus thuringiensis*, hence causing PHB accumulation in the cells. A phosphorylated form of Spo0A positively regulates the cascade, which governs the transcription of sigma factor F, whose function is specifically in the forespore compartment of sporulation. The study has also revealed the non-importance of Spo0F in the context of the PHB synthesis in *B. thuringiensis* cells. The

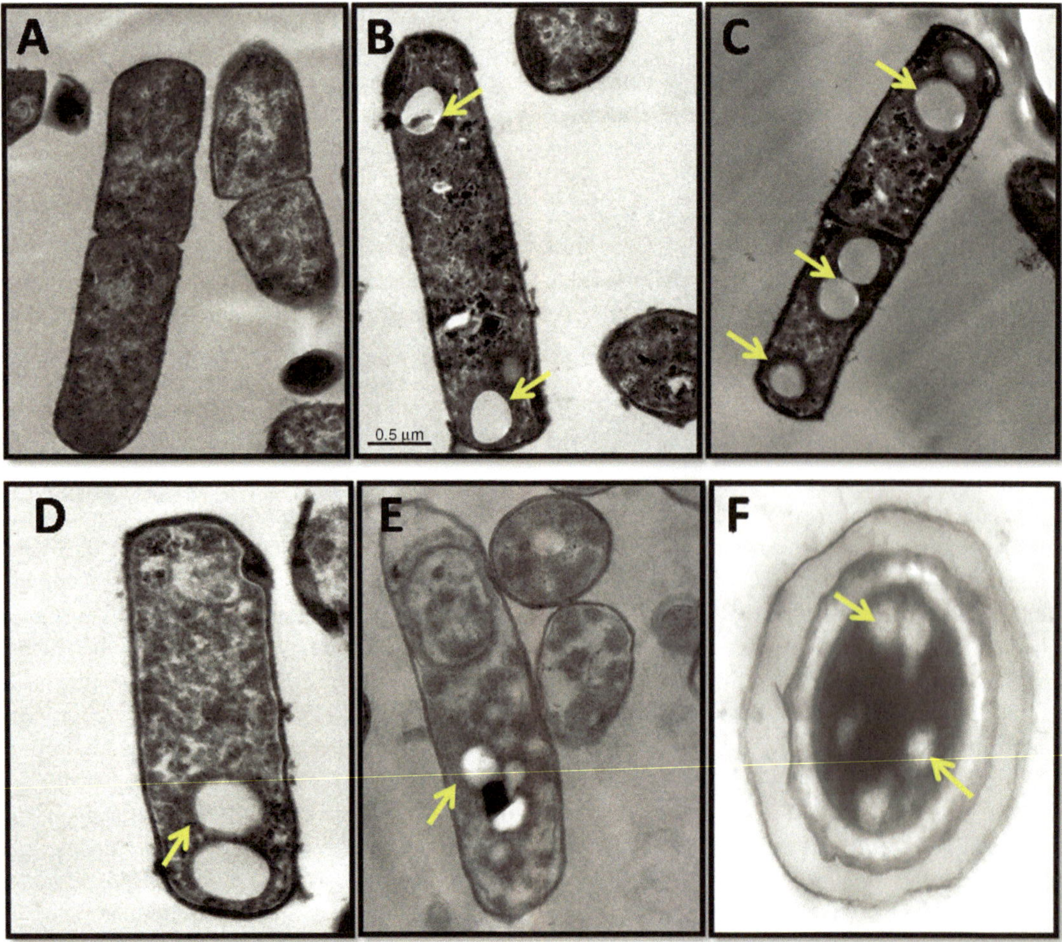

Fig. 7.1 *PHB accumulation profile in the Bacillus sp. cells during vegetative growth* (A–C) *or sporulation phases* (D–F). Sections of *Bacillus anthracis* Sterne observed under transmission electron microscopy. *Yellow arrows* indicate the PHB granules. Scale bar 0.5 μm

spo0F deletion mutant in *B. thuringiensis* was unaffected in PHB accumulation, suggesting that the molecule Spo0F and PHB synthesis are independent of each other at molecular levels. Broadly, Spo0A-phosphorylated form is the master regulator of many processes as it activates or represses the distinct set of genes. Among many, the AbrB protein, a global regulator of proteins, is responsible for the transition of cells from exponential to stationary phase of the growth cycle (Strauch et al. 2005). AbrB is repressed by the phosphorylated Spo0A form. The *abrB* and *spo0A* double mutant showed early PHB synthesis. Additionally, the *phaC* mutant in *B. thuringiensis* does not seem to affect the sporulation efficiency (Chen et al. 2010).

7.3.2 Phasin Proteins

Phasins are the amphiphilic proteins whose presence determines the number, size and surface to volume ratio of the PHB granules. Phasins perfectly surround the amorphous PHB granule (Pfeiffer and Jendrossek 2012; York et al. 2001). In *in vitro* conditions, phasin proteins are reported to activate the PhaCA synthase in *Aeromonas caviae*, while it masks the activities of PhaCR and PhaCD in *R. eutropha* and *Delftia acidovorans*, respectively. The activation of PhaCA through PhaP causes the change in substrate affinity toward 3-hydroxyhexanoyl-CoA. Additionally, the overexpression of phasin proteins resulted in increased PHA production by

2.3-fold (Ushimaru et al. 2014). The various knockout strains of phasin genes (*phaP1*, *phaP2*, *phaP3*, *phaP4*, or in combination) in *R. eutropha* were studied for their PHB accumulation ability and molecular weight. Phasins P2, P3, and P4 have been found to be crucial for the stability of PHB granules, while phasin P1 was reported to be important for the PHB degradation in *R. eutropha*. However, in this study, no change in the PHB molecular weight has been detected (Kuchta et al. 2007). In *Bradyrhizobium japonicum* USDA110, the PHB granules are stabilized by three PhaP paralogs, which also contribute to the organism's growth. Phasin protein in *B. japonicum* exhibited high affinity for PHB granules and can displace PhaR previously bound to PHB (Yoshida et al. 2013). On the other hand, phasins P2 and P4 inhibit the degradation mediated by PHB depolymerases (PhaZ2, Z3, and Z7) leading to higher PHB accumulations (Eggers and Steinbüchel 2014).

7.3.3 Catabolite Control Protein A

Catabolite control protein A has been known to play a role in various physiological processes, for example in biofilm formation and virulence of bacteria like *Clostridium difficile*, *Enterococcus faecalis*, *Staphylococcus epidermidis*, and *Streptococcus pneumoniae* (Sadykov et al. 2011; Antunes et al. 2012; Gao et al. 2013). CcpA inactivation in *Bacillus* sp. MA3.3 causes reduction in glucose catabolite repression and reduced PHB production on hydrolases of lignocellulose. CcpA activates the transcription of *gltAB* operon which is responsible for ammonium metabolism (Lopes et al. 2011).

7.3.4 Glucose and Peptones

Narayanan and Ramana have studied the effect of glucose and peptone on PHB production and sporulation of *Bacillus mycoides*. Since sporulation always raises an issue for maximum PHB production, the central composite rotatable design was used to maximize the PHB content with negligible sporulation (Narayanan and Ramana 2012).

7.3.5 Phosphotransacetylases and Phosphotransbutyrylase

Phosphotransacetylases (encoded by *pta*) cause the phosphorylation of acetyl-coenzyme A resulting in the formation of acetylphosphate. The metabolic intermediate, acetylphosphate, is shown to be an activator for the PHB synthase of *R. eutropha* in heterologous system, *Escherichia coli* (Miyake et al. 2000). On the other hand, *pta-deficient* mutant of *E. coli* results in increased acetyl-CoA concentration which acts as a substrate for the PHB synthesis, ultimately causing enhanced PHB accumulation in *pta* mutant as compared to the wild type (Miyake et al. 2000). Another study by Vazquez and colleagues revealed the positive correlation of PHB with phosphotransbutyrylase enzyme activity in *Bacillus megaterium*. Ptb activity and PHB accumulation are measured throughout the growth phases and correlation was plotted. Various activators (isoleucine and valine) and inhibitor (glucose) of this enzyme are shown to affect the PHB accumulation (Vazquez et al. 2003).

7.3.6 Endogenous Ethanol

Catabolism of ethanol yields acetyl-CoA and NADPH which inhibits the TCA cycle and thus drives the enhanced acetyl-CoA into the PHB biosynthetic pathway. Recently, it has been reported that exogenous as well as endogenous ethanol acts as an efficient chain transfer agent for PHB synthesis in *E. coli*. The endogenous ethanol exerts a positive effect on the molecular weight of PHB. The deletion of *adhE* gene (encoding alcohol dehydrogenase) in *E. coli* results in a drastic increase in PHB molecular weight that is 77 % (Hiroe et al. 2013). Even in *Cupriavidus necator*, ethanol accumulation is known to increase the total PHB content (Obruca et al. 2010).

7.3.7 Molecular Chaperones

Interestingly, apart from all the above mentioned factors, PHB synthesis is also governed by the expression of chaperones like GroEL and GroES. The heterologous expression of PHB synthase from *C. necator* in *E. coli* with chaperone co-expression results in higher fraction (six fold) of soluble PHB synthase. Reduced molecular weight of polymer was observed when PHB synthase is co-expressed with chaperones GroEL and GroES (Thomson et al. 2013).

7.4 Newly Identified *Bacillus* Species as Potent PHB Producers

7.4.1 *Bacillus licheniformis*

Recently, a *Bacillus licheniformis* MSBN12 has been isolated from marine sponge, specifically *Callyspongia diffusa*, and maximum PHB yield using this strain was standardized at 6.38 g/L. Comparison among different substrates reveals palm jiggery as the best carbon source for *B. licheniformis* MSBN12 (Sathiyanarayanan et al. 2013a).

7.4.2 *Bacillus subtilis*

Another strain of *Bacillus*, specifically *Bacillus subtilis* MSBN17, was isolated from marine sponge *Callyspongia diffusa*, and the PHB production efficiency was checked on the substrate like pulp industry waste, tamarind kernel powder, palm jiggery, and green gram flour. Among the various substrates, maximum PHB production has been standardized for pulp industry waste (PIW), which is 19.08 g/L (Sathiyanarayanan et al. 2013b).

7.4.3 *Bacillus shackletonii*

Another Gram-positive bacterium, *B. shackletonii* K5 has been isolated from biotrickling filter and has been proposed to accumulate PHB, recently. This strain could generate the maximum yield of 70 % cell dry weight on substratum of sodium succinate and glycerol as carbon sources. The optimum temperature for the growth of the strain was optimized as 45°C (Liu et al. 2014). Similar PHA yields were obtained with a unique strain of *B. thuringiensis*. It showed higher PHA production on crude glycerol to the tune of 74 % (w/w) as PHA even without the need of N-limitation.

7.5 Newly Investigated *Bacillus* Strains for Higher PHB Accumulation

7.5.1 *Bacillus megaterium* R11

The *Bacillus megaterium* strain R11 was isolated from Singapore and has the capacity to store PHB as 51 % of cell dry weight on a substratum of glucose and xylose. The maximum PHB yield was standardized up to 59 % on tryptone as a nitrogen source. Oil palm empty fruit bunch is a by-product of Malaysia palm oil refineries and is rich in cellulose and hemicellulose content (Zhang et al. 2013).

7.5.2 *Bacillus megaterium* BA-019

The PHB production in the *B. megaterium* BA-019 has been standardized to 45.84 % of cell dry weight, i.e., 1.73 g/L (Kanjanachumpol et al. 2013). The substrates used for the PHB production with BA-019 strain were sugarcane molasses (C source) and urea (N source); further, the C/N ratio was standardized to 25 for maximum cell growth (72 g/L) and PHB (1.27 g/L/h) production (Kulpreecha et al. 2009).

7.5.3 *Bacillus megaterium* H15, H16, and H26

Several halophilic bacterial strains were isolated from solar salterns from Indian western coast, specifically Goa. Three strains of *B.*

megaterium H15, H16, and H26 were identified and further characterized phenotypically and genotypically. The presence of NaCl in E2 mineral media induced extended lag phase (up to 24 h) but no significant difference in PHB production in the case of *B. megaterium* H16 strain. This strain is industrially more suitable for its higher tolerance to salt stress (Salgaonkar et al. 2013).

7.5.4 *Bacillus sphaericus* NCIM 5149

A strain of *Bacillus sphaericus*, specifically NCIM 5149, has been reported for the first time to produce PHB under submerged fermentation. PHB production was maximized to 49 % (2.2 g/L) of cell dry weight using a central composite design considering three parameters, i.e., pH (6), inoculum age (18 h), and jackfruit hydrolysate concentration (2.5 % reducing sugar) (Ramadas et al. 2010).

7.5.5 *Bacillus* spp. CFR-67 and CFR-256

Bacillus sp. CFR 67, a well-known species for the production of PHB as well as amylase, was further researched to maximize the PHB production through the addition of wheat bran hydrolysates, rice bran, or both. The most interesting fact is that the introduction of wheat bran or rice bran with ammonium acetate and corn starch results in significant increased polyhydroxybutyrate-co-hydroxyvalerate copolymer production from 5 to 10 %. The by-product amylase remains active for broad pH (4–9) and temperature ranges (40–60 °C) (Shamala et al. 2012). Another strain of *Bacillus* sp., namely, CFR 256, has been used to standardize the conditions for the fermentation setup using response surface methodology on a relatively cheap substrate, corn steep liquor (CSL). Various variables, including media composition (CSL, Na_2HPO_4, KH_2PO_4, sucrose) and inoculum concentration, were standardized to produce PHA which is 51 % of cell dry weight (Vijayendra et al. 2007).

7.6 Conclusion

The regulation of PHB synthesis converges from multifaceted pathways in different systems like *Bacillus*, *Ralstonia*, etc. Various endogenous as well as exogenous factors are involved in governing the final yield of PHB. The relevance of Gram-positive organisms in PHB production becomes more significant due to the absence of immunogenic LPS in purified PHB fraction. The PHB thus purified can be applied for a large number of medical uses like skin grafts, liver grafts, etc. The extensive research is going on for standardizing conditions for the maximum PHB production from *Bacillus* sp. Small changes in the substratum of specifically inexpensive by-products of several industries like corn steep liquor, sugarcane molasses, etc., cause drastic reduction in the PHB production cost. The above mentioned text specifies the newly discovered PHB producers from *Bacillus* species as well as different changes in the fed-batch fermentor conditions to maximize the PHB production.

Acknowledgment The authors wish to thank the Director of CSIR-Institute of Genomics and Integrative Biology (IGIB), Government of India, for providing the necessary funds and facilities (LSRB-268/BTB/2013 and BSC0123). Authors are also thankful to the Academy of Scientific and Innovative Research (AcSIR), New Delhi. ND is Shyama Prasad Mukherjee-Senior Research Fellow supported by CSIR, India. LKS and SSK are Senior Research Fellows supported by University Grant Commission, India. NK is Junior Research Fellow. We highly acknowledge Dr. V. C. Kalia from CSIR-IGIB, Delhi, India, for the inspiration and critical comments in the manuscript.

References

Antunes A, Camiade E, Monot M, Courtois E, Barbut F, Sernova NV, Rodionov DA, Martin-Verstraete I, Dupuy B (2012) Global transcriptional control by glucose and carbon regulator CcpA in *Clostridium difficile*. Nucleic Acid Res 40:10701–10718. doi:10.1093/nar/gks864

Brigham CJ, Speth DR, Rha C, Sinskey AJ (2012) Whole-genome microarray and gene deletion studies reveal regulation of the polyhydroxyalkanoate production cycle by the stringent response in *Ralstonia eutropha* H16. Appl Environ Microbiol 78:8033–8044. doi:10.1128/AEM.01693-12

Chen HJ, Tsai TK, Pan SC, Lin JS, Tseng CL, Shaw GC (2010) The master transcription factor Spo0A is required for poly(3-hydroxybutyrate) (PHB) accumulation and expression of genes involved in PHB biosynthesis in *Bacillus thuringiensis*. FEMS Microbiol Lett 304:74–81. doi:10.1111/j.1574-6968.2010.01888.x

Eggers J, Steinbüchel A (2014) Impact of *Ralstonia eutropha*'s poly(3-Hydroxybutyrate) (PHB) Depolymerases and Phasins on PHB storage in recombinant *Escherichia coli*. Appl Environ Microbiol 80:7702–7709. doi:10.1128/AEM.02666-14

Emeruwa AC, Hawirko RZ (1973) Poly-beta-hydroxybutyrate metabolism during growth and sporulation of *Clostridium botulinum*. J Bacteriol 116:989–993

Gao P, Pinkston KL, Bourgogne A, Cruz MR, Garsin DA, Murray BE, Harvey BR (2013) Library screen identifies *Enterococcus faecalis* CcpA, the catabolite control protein A, as an effector of Ace, a collagen adhesion protein linked to virulence. J Bacteriol 195:4761–4768. doi:10.1128/JB.00706-13

Hiroe A, Hyakutake M, Thomson NM, Sivaniah E, Tsuge T (2013) Endogenous ethanol affects biopolyester molecular weight in recombinant *Escherichia coli*. ACS Chem Biol 8:2568–2576. doi:10.1021/cb400465p

Holmes, Christopher A (1984) An intracellular polysaccharide that serves as a carbon and energy source for sporulation in "*Bacillus cereus*" Strain T. http://escholarshare.drake.edu/handle/2092/728

Jendrossek D, Pfeiffer D (2014) New insights in the formation of polyhydroxyalkanoate granules (carbonosomes) and novel functions of poly(3-hydroxybutyrate). Environ Microbiol 16:2357–2373. doi:10.1111/1462-2920.12356

Kalia VC, Raizada N, Sonakya V (2000) Bioplastics. J Sci Ind Res 59:433–445

Kalia VC, Chauhan A, Bhattacharyya G (2003) Genomic databases yield novel bioplastic producers. Nat Biotechnol 21:845–846. doi:10.1038/nbt0803-845

Kalia VC, Lal S, Cheema S (2007) Insight in to the phylogeny of polyhydroxyalkanoate biosynthesis: horizontal gene transfer. Gene 389:19–26. doi:10.1016/j.gene.2006.09.010

Kanjanachumpol P, Kulpreecha S, Tolieng V, Thongchul N (2013) Enhancing polyhydroxybutyrate production from high cell density fed-batch fermentation of *Bacillus megaterium* BA-019. Bioprocess Biosyst Eng 36:1463–1474. doi:10.1007/s00449-013-0885-7

Kuchta K, Chi L, Fuchs H, Pötter M, Steinbüchel A (2007) Studies on the influence of phasins on accumulation and degradation of PHB and nanostructure of PHB granules in *Ralstonia eutropha* H16. Biomacromolecules 8:657–662. doi:10.1021/bm060912e

Kulpreecha S, Boonruangthavorn A, Meksiriporn B, Thongchul N (2009) Inexpensive fed-batch cultivation for high poly(3-hydroxybutyrate) production by a new isolate of *Bacillus megaterium*. J Biosci Bioeng 107:240–245. doi:10.1016/j.jbiosc.2008.10.006

Kumar T, Singh M, Purohit HJ, Kalia VC (2009) Potential of *Bacillus* sp. to produce polyhydroxybutyrate from biowaste. J Appl Microbiol 106:2017–2023. doi:10.1111/j.1365-2672.2009.04160.x

Kumar P, Patel SKS, Lee JK, Kalia VC (2013) Extending the limits of *Bacillus* for novel biotechnological applications. Biotechnol Adv 31:1543–1561. doi:10.1016/j.biotech adv.2013.08.007

Kumar P, Singh M, Mehariya S, Patel SKS, Lee JK, Kalia VC (2014) Ecobiotechnological approach for exploiting the abilities of *Bacillus* to produce co-polymer of polyhydroxyalkanoate. Indian J Microbiol 54:151–157. doi:10.1007/s12088-014-0457-9

Kumar P, Mehariya S, Ray S, Mishra A, Kalia VC (2015a) Biodiesel industry waste: a potential source of bioenergy and biopolymers. Indian J Microbiol 55:1–7. doi:10.1007/s12088-014-0509-1

Kumar P, Ray S, Patel SKS, Lee JK, Kalia VC (2015b) Bioconversion of crude glycerol to polyhydroxyalkanoate by *Bacillus thuringiensis* under non-limiting nitrogen conditions. Int J Biol Macromol. doi:10.1016/j.ijbiomac.2015.03.046

Liu Y, Huang S, Zhang Y, Xu F (2014) Isolation and characterization of a thermophilic *Bacillus shackletonii* K5 from a biotrickling filter for the production of polyhydroxybutyrate. J Environ Sci (China) 26:1453–1462. doi:10.1016/j.jes.2014.05.011

Lopes MS, Steinert N, Rojas JD, Hillen W, Gomez JG, Silva LF (2011) Role of CcpA in polyhydroxybutyrate biosynthesis in a newly isolated *Bacillus* sp. MA3.3. J Mol Microbiol Biotechnol 20:63–69. doi:10.1159/000324502

Mezzina MP, Wetzler DE, de Almeida A, Dinjaski N, Prieto MA, Pettinari MJ (2014) A phasin with extra talents: a polyhydroxyalkanoate granule-associated protein has chaperone activity. Environ Microbiol. doi:10.1111/1462-2920.12636

Miyake M, Miyamoto C, Schnackenberg J, Kurane R, Asada Y (2000) Phosphotransacetylase as a key factor in biological production of polyhydroxybutyrate. Appl Biochem Biotechnol 84–86:1039–1044

Nakata HM (1963) Effect of pH on intermediates produced during growth and sporulation of *Bacillus cereus*. J Bacteriol 86:577–581

Narayanan A, Ramana KV (2012) Polyhydroxybutyrate production in *Bacillus mycoides* DFC1 using response surface optimization for physico-chemical process parameters. 3 Biotech 2:287–296. doi:10.1007/s13205-012-0054-8

Obruca S, Marova I, Stankova M, Mravcova L, Svoboda Z (2010) Effect of ethanol and hydrogen peroxide on poly(3-hydroxybutyrate) biosynthetic pathway in *Cupriavidus necator* H16. World J Microbiol Biotechnol 26:1261–1267. doi:10.1007/s11274-009-0296-8

Patel SKS, Singh M, Kalia VC (2011) Hydrogen and polyhydroxybutyrate producing abilities of *Bacillus* spp. from glucose in two stage system. Indian J Microbiol 51:418–423. doi:10.1007/s12088-011-0236-9

Patel SKS, Singh M, Kumar P, Purohit HJ, Kalia VC (2012) Exploitation of defined bacterial cultures for production of hydrogen and polyhydroxybutyrate from pea-shells. Biomas Bioenerg 36:218–225. doi:10.1016/j.biombioe.2011.10.027

Patel SKS, Kumar P, Singh M, Lee JK, Kalia VC (2015) Integrative approach to produce hydrogen and polyhydroxybutyrate from biowaste using defined bacterial cultures. Bioresour Technol 176:136–141. doi:10.1016/j.biortech.2014.11.029

Pfeiffer D, Jendrossek D (2012) Localization of poly(3-hydroxybutyrate) (PHB) granule-associated proteins during PHB granule formation and identification of two new phasins, PhaP6 and PhaP7, in *Ralstonia eutropha* H16. J Bacteriol 194:5909–5921. doi:10.1128/JB.00779-12

Porwal S, Kumar T, Lal S, Rani A, Kumar S, Cheema S, Purohit HJ, Sharma R, Patel SKS, Kalia VC (2008) Hydrogen and polyhydroxybutyrate producing abilities of microbes from diverse habitats by dark fermentative process. Bioresour Technol 99:5444–5451. doi:10.1016/j.biortech.2007.11.011

Ramadas NV, Soccol CR, Pandey A (2010) A statistical approach for optimization of polyhydroxybutyrate production by *Bacillus sphaericus* NCIM 5149 under submerged fermentation using central composite design. Appl Biochem Biotechnol 162:996–1007. doi:10.1007/s12010-009-8807-5

Reddy CSK, Ghai R, Rashmi, Kalia VC (2003) Polyhydroxyalkanoates: an overview. Bioresour Technol 87:137–146. doi:10.1016/S0960-8524(02)00212-2

Sadykov MR, Hartmann T, Mattes TA, Hiatt M, Jann NJ, Zhu Y, Ledala N, Landmann R, Herrmann M, Rohde H, Bischoff M, Somerville GA (2011) CcpA coordinates central metabolism and biofilm formation in *Staphylococcus epidermidis*. Microbiology 157:3458–3468. doi:10.1099/mic.0.051243-0

Salgaonkar BB, Mani K, Braganca JM (2013) Characterization of polyhydroxyalkanoates accumulated by a moderately halophilic salt pan isolate *Bacillus megaterium* strain H16. J Appl Microbiol 114:1347–1356. doi:10.1111/jam.12135

Sathiyanarayanan G, Saibaba G, Seghal Kiran G, Selvin J (2013a) Process optimization and production of polyhydroxybutyrate using palm jaggery as economical carbon source by marine sponge-associated *Bacillus licheniformis* MSBN12. Bioprocess Biosyst Eng 36:1817–1827. doi:10.1007/s00449-013-0956-9

Sathiyanarayanan G, Saibaba G, Seghal Kiran G, Selvin J (2013b) A statistical approach for optimization of polyhydroxybutyrate production by marine *Bacillus subtilis* MSBN17. Int J Biol Macromol 59:170–177. doi:10.1016/j.ijbiomac.2013.04.040

Shamala TR, Vijayendra SV, Joshi GJ (2012) Agro-industrial residues and starch for growth and co-production of polyhydroxyalkanoate copolymer and α-amylase by *Bacillus* sp. CFR-67. Braz J Microbiol 43:1094–1102. doi:10.1590/S1517-83822012000300036

Singh M, Patel SKS, Kalia VC (2009) *Bacillus subtilis* as a potential producer for polyhydroxyalkanoates. Microb Cell Fact 8:38. doi:10.1186/1475-2859-8-38

Singh LK, Dhasmana N, Singh Y (2014) Quorum sensing systems in *Bacillus*. Quorum sensing Vs Quorum Quenching: a battle with no end in sight. Springer, New Delhi. ISBN 978-81-322-1981-1

Singh LK, Dhasmana N, Sajid A, Kumar P, Bhaduri A, Bharadwaj M, Gandotra S, Kalia VC, Das TK, Goel AK, Pomerantsev AP, Misra R, Gerth U, Leppla SH, Singh Y (2015a) *clpC* operon regulates cell architecture and sporulation in *Bacillus anthracis*. Environ Microbiol 17:854–864. doi:10.1111/1462-2920.12548

Singh M, Kumar P, Ray S, Kalia VC (2015b) Challenges and opportunities for customizing polyhydroxyalkanoates. Indian J Microbiol 55:234. doi:10.1007/s12088-015-0528-6

Slepecky RA, Law JH (1961) Synthesis and degradation of poly-beta-hydroxybutyric acid in connection with sporulation of *Bacillus megaterium*. J Bacteriol 82:37–42

Strauch MA, Ballar P, Rowshan AJ, Zoller KL (2005) The DNA-binding specificity of the *Bacillus anthracis* AbrB protein. Microbiology 151:1751–1759. doi:10.1099/mic.0.27803-0

Supono, Hutabarat J, Prayitno SB, Darmanto YS (2013) The effect of different C:N and C:P ratio of media on the content of polyhydroxybutyrate in biofloc inoculated with bacterium *Bacillus cereus*. J Coast Dev 16:114–120

Thomson NM, Saika A, Ushimaru K, Sangiambut S, Tsuge T, Summers DK, Sivaniah E (2013) Efficient production of active polyhydroxyalkanoate synthase in *Escherichia coli* by coexpression of molecular chaperones. Appl Environ Microbiol 79:1948–1955. doi:10.1128/AEM.02881-12

Ushimaru K, Motoda Y, Numata K, Tsuge T (2014) *Aeromonas caviae* polyhydroxyalkanoate synthase is activated by phasin proteins: a comparative study with *Ralstonia eutropha* synthase. Appl Environ Microbiol 80:2867–2873. doi:10.1128/AEM.04179-13

Valappil SP, Boccaccini AR, Bucke C, Roy I (2007) Polyhydroxyalkanoates in Gram-positive bacteria: insights from the genera *Bacillus* and *Streptomyces*. Antonie Van Leeuwenhoek 91:1–17. doi:10.1007/s10482-006-9095-5

Vazquez GJ, Pettinari MJ, Méndez BS (2003) Evidence of an association between poly (3-hydroxybutyrate) accumulation and phosphotransbutyrylase expression in *Bacillus megaterium*. Int Microbiol 6:127–129

Vijayendra SV, Rastogi NK, Shamala TR, Anil Kumar PK, Kshama L, Joshi GJ (2007) Optimization of polyhydroxybutyrate production by *Bacillus sp.* CFR 256 with corn steep liquor as a nitrogen source. Indian J Microbiol 47:170–175. doi:10.1007/s12088-007-0033-7

Wu Q, Huang H, Hu G, Chen J, Ho KP, Chen GQ (2001) Production of poly-3-hydroxybutyrate by *Bacillus sp.* JMa5 cultivated in molasses media. Antonie Van Leeuwenhoek 80:111–118

York GM, Stubbe J, Sinskey AJ (2001) New insight into the role of the PhaP phasin of *Ralstonia eutropha* in promoting synthesis of polyhydroxybutyrate. J Bacteriol 183:2394–2397. doi:10.1128/JB.183.7.2394-2397.2001

Yoshida K, Takemoto Y, Sotsuka T, Tanaka K, Takenaka S (2013) PhaP phasins play a principal role in poly-β-hydroxybutyrate accumulation in free-living *Bradyrhizobium japonicum*. BMC Microbiol 13:290. doi:10.1186/1471-2180-13-290

Zhang Y, Sun W, Wang H, Geng A (2013) Polyhydroxybutyrate production from oil palm empty fruit bunch using *Bacillus megaterium* R11. Bioresour Technol 147:307–314. doi: 10.1016/j.biortech.2013.08.029

Neha Dhasmana graduated in Microbiology in 2008 from University of Delhi. She received her M. Sc. degree in Biochemistry in 2010 from Hamdard University. She is currently enrolled in Ph.D. with Academy of Innovative and Scientific Research (AcSIR) and works at CSIR-Institute of Genomics and Innovative Biology. She is studying the physiological significance of Ser/Thr kinases in *Bacillus anthracis*.

Lalit K. Singh received the M.Sc. degree in Biotechnology from Rajiv Gandhi Biotechnology Centre, RTM Nagpur University, Nagpur, India. He is currently working as Senior Research Fellow, CSIR-IGIB, Delhi, India. His current research focuses on understanding the role of caseinolytic proteases and Ser/Thr kinases that mediated cellular architecture alterations, cell wall remodeling, and sporulation in *Bacillus*.

Shashank S. Kamble received his M.Sc. degree from University of Mumbai, India. Currently, he is pursuing Ph.D. under the supervision of Dr. Yogendra Singh at CSIR-IGIB, India. He is interested in the host pathogen cross talk during infection with *M. tuberculosis*. His work focuses on the modulation of host proteases by *M. tuberculosis*.

Nishant Kumar graduated in Biotechnology in 2009 from HNB Garhwal University. He received his M.Sc. degree in Biosciences in 2012 from Jamia Millia Islamia University. He then completed M.Phil. from Zoology Department of University of Delhi in 2013. He is currently enrolled in Ph.D. with Academy of Innovative and Scientific Research (AcSIR) and works at CSIR-Institute of Genomics and Innovative Biology. He is studying the role of Ser/Thr kinases and phosphodiesterases in *Bacillus anthracis*.

Yogendra Singh received his Ph.D. degree from Vallabhbhai Patel Chest Institute, Delhi University, and postdoctoral training at National Institute of Health (NIH), Bethesda. Currently, he is Chief Scientist at CSIR-IGIB and Professor of AcSIR. He is a Fellow of National Academy of Sciences, India, Fellow of Indian National Science Academy, and Fellow of Indian Academy of Sciences, India. He has been honored with Moselio Schaechter Distinguished Service Award in 2014 by American Society for Microbiology. He serves as an Editorial Board Member of *Journal of Biological Chemistry*. His current research interest is to understand the host response in pulmonary/extra-pulmonary tuberculosis and to decipher the role of eukaryotic like Ser/Thr kinases in *Mycobacterium tuberculosis* and *Bacillus anthracis*.

Microbial Biopolymers: The Exopolysaccharides

8

Angelina and S.V.N. Vijayendra

Abstract

Microorganisms produce several biopolymers. Of these, intracellularly produced polyhydroxyalkanoates (PHAs) and extracellularly produced exopolysaccharides (EPS) are gaining importance over the other biopolymers. These naturally produced polymers can replace plant-based or petroleum-derived polymers. There are innumerable reports and reviews on the production of PHA and EPS by several bacteria, fungi, actinomycetes, and algae. This chapter briefly gives an introduction to PHA and provides recent developments in the genetic and metabolic pathways for the synthesis of microbial EPS. Different strategies used for fermentative production and various means of downstream processing are discussed. Possible ways to minimize the cost of production and downstream processing are covered in this chapter. Applications of these EPS in various fields such as agriculture, cosmetics, foods, medical and healthcare industry, mining, oil recovery, packaging, pharmaceuticals, printing and textile industry, wastewater treatment, etc., are presented. The potential of these polymers indicates that these microbial cell factories can be exploited for the better of mankind.

8.1 Introduction

Biopolymers are natural polymers derived from living organisms during their growth or from renewable resources in the form of polysaccharides and polyhydroxyalkanoates (PHAs) also known as bioplastics. Being biodegradable, these are studied for different properties from several microbial sources for its various applications that are of high value (Sutherland 1994). Microorganisms synthesize these biopolymers as intracellular, structural, and extracellular polymers for

Angelina • S.V.N. Vijayendra (✉)
Microbiology and Fermentation Technology,
CSIR-Central Food Technological Research Institute,
570020 Mysore, India
e-mail: ange.csk@gmail.com;
svnvijayendra@cftri.res.in,
svnvijayendra@yahoo.com

V.C. Kalia (ed.), *Microbial Factories*, DOI 10.1007/978-81-322-2595-9_8

their function and survival that play specific roles such as energy reserve materials, protective agents, cell functioning, symbiosis, and osmotic adaptation and support the microbes to function, adapt, multiply, and survive efficiently under changing environmental conditions (Vijayendra and Shamala 2014). This chapter gives an introduction to PHA and mainly focuses on recent developments in the area of microbial exopolysaccharides (EPS).

8.1.1 Polyhydroxyalkanoates

Polyhydroxyalkanoates (PHAs) are bioplastics produced from microorganisms that have gained attention as an alternative to petroleum-based plastics due to their similar properties and being biodegradable and biocompatible (environment friendly) are produced from renewable and waste resources. PHAs are accumulated as intracellular granules by a wide variety of microorganisms in the presence of an abundant carbon source and limited essential nutrients such as nitrogen, phosphorous, or oxygen and serve as reserve carbon and energy sources; also they do not substantially alter the cell's osmotic state (Laycock et al. 2013).

PHAs are polyoxoesters that are naturally synthesized as polymers of R-3-hydroxyalkanoic acids and constitutes of 150 different monomers (Steinbüchel 2005). PHB (polyhydroxybutyrate) is the most common of the PHAs produced, and it was the first PHA discovered in *Bacillus megaterium* by Lemoigne in 1927. Copolymers of PHB like poly(3-hydroxybutyrate-co-3-hydroxyvalerate), P(3HB-co-3HV), can be synthesized by substituting the side chain with functional groups, viz., hydroxyl, carboxylic, epoxy, phenoxy groups, halogens, etc., based on the carbon sources, thereby affecting its thermal and mechanical properties (Zinn and Hany 2005).

PHAs are classified on the basis of side chain length of monomers: (1) scl-PHAs (short chain length) containing monomers of 4–5 carbon atoms, such as P(3HB), are crystalline, brittle, and stiff, having high melting and low glass transition temperatures, whereas the copolymer P(3HB-co-3HV) is a strong and pliable thermoplastic; (2) mcl-PHAs (medium chain length) containing monomers of 6–14 carbon atoms, e.g., 3-hydroxydodecanoate (3HDD), 3-hydroxyoctanoate (3HO), 3-hydroxyhexanoate (3HHx), and 3-hydroxydecanoate (3HD), have low crystallinity and tensile strength and glass transition temperatures and lower melting points in comparison to scl-PHAs (Akaraonye et al. 2010); and (3) lcl-PHAs (long chain length) containing monomers of 15 and above carbon atoms (Luengo et al. 2003) are more ductile and easier to mold (Kabilan et al. 2012). The scl-PHA copolymer produced by *Rhizobium meliloti* was characterized (Shamala et al. 2009). The various monomer composition of PHAs results in a wide range of thermoplastic to elastomeric properties, viz., molecular weight, density, melting point, crystallinity, resistance to UV and solvents, water and O_2 permeability, glass transition temperature, and tensile strength and elongation to break (Du et al. 2012). Identification of PHA-producing *Bacillus* spp. by molecular methods was standardized (Shamala et al. 2003). Reports are also available for PHA production by yeasts (Kocharin and Nielsen 2013; Gumel et al. 2013), algae such as cyanobacteria (Balaji et al. 2013), and genetically modified diatoms (Hempel et al. 2011). The use of agro-industrial waste products such as corn steep liquor, mahua flower extracts, starch, wheat, or rice bran can reduce the cost of PHA polymer (Vijayendra et al. 2007; Anil kumar et al. 2007; Halami 2008; Saranya devi et al. 2012). Simultaneous production of either PHA and amylase or PHA and EPS by *Bacillus* sp. and *Sinorhizobium* sp., respectively (Saranya devi et al. 2012; Shamala et al. 2012; Sreekanth et al. 2013), or PHA and β-carotene by *Micrococcus* sp. was reported (Vijayendra et al. 2008a). Production of multiple biopolymers using fed batch reactor by employing a high cell density culture of *Sinorhizobium meliloti* was reported very recently (Shamala et al. 2014).

8.1.2 Microbial Exopolysaccharides

Microbial exopolysaccharides (EPS) are a kind of biopolymers synthesized by several microorganisms which include several genera of bacteria, molds, and yeasts. These are gumlike polymers synthesized by these organisms and released into the surrounding environment. EPS mainly protects the microorganisms from the surrounding environment and also acts as a reserve food material, and the producing organisms can use this as a main carbon source. Based on the sugar composition, EPS are classified as homopolymers (with a single type of sugar, glucose, xylose, etc.) and heteropolymers (with more than one type of sugar moieties, glucose, rhamnose, mannose, etc.). Based on the presence or absence of uronic acid, EPS are categorized as acidic or neutral EPS, respectively. EPS have several physical and chemical properties. Some of these EPS can form a film, some form gels, some more can increase the viscosity of solutions, etc. Besides, these EPS have several functional attributes; hence, these are being used in various foods and pharmaceutical, medical, and other industrial applications. Microbial EPS are an alternate to plant-based or seaweed-based water-soluble polymers, whose quality and production quantities depend on several environmental factors, whereas the production of microbial EPS even in large-scale fermentors can be controlled accurately and uniformity in the quality can also be achieved. However, even today, the cost of microbial EPS is higher than plant-based polymers. Hence, all over the world many researchers are exploring various ways to reduce the production cost by using various cheap and alternate substrates and optimizing its production and recovery and its applications, thus becoming a topic of many recent reviews (Nicolaus et al. 2010; Freitas et al. 2011; Palaniraj and Jayaraman 2011; Seviour et al. 2011; Donot et al. 2012; Patel et al. 2012; Singha 2012; Benny et al. 2014; Dhiya et al. 2014; Divyasri et al. 2014; Kaur et al. 2014; Prajapati et al. 2013; Finore et al. 2014).

8.1.2.1 EPS-Producing Microorganisms

Several microorganisms produce EPS. Table 8.1 depicts important EPS-producing microorganisms. *Agrobacterium* sp. produces curdlan, a neutral water-insoluble and alkali-soluble EPS. Curdlan has linear β-(1,3)-glycosidic linkages. It produces soft and hard gels when heated at 60 and 80 °C, respectively. Several lactic acid bacteria (LAB) are known to produce EPS, especially *Leuconostoc* spp. and *Lactobacillus* spp. (Vijayendra et al. 2008b, 2009; Badel et al. 2011; Yadav et al. 2011). *Aureobasidium pullulans*, a bimorphic fungi, produces an EPS known as pullulan. It is an extracellular, linear, unbranched, water-soluble EPS. It has maltotriose-repeating units linked through α-1,6-glucosidic bonds. Its molecular weight varies from 4.5×10^4 to 6×10^5 Da. *Sphingomonas paucimobilis*, formerly known as *Pseudomonas elodea*, is known to produce gellan, a gelling EPS. It can be used as a gelling and thickening agent or as a solidifying agent in tissue cultures. It is an acidic polysaccharide consisting of glucose, rhamnose, and glucuronic acid in 3:1:1 ratio. *Sinorhizobium*

Table 8.1 Important microorganisms producing exopolysaccharides

Microorganisms	Exopolysaccharides
Acetobacter xylinum	Acetan/cellulose
Agrobacterium tumefaciens	Succinoglucan
Alcaligenes faecalis	Curdlan
Alcaligenes faecalis/Sphingomonas sp.	Welan
Azotobacter vinelandii	Alginate
Aureobasidium pullulans	Pullulan
Epicoccum nigrum	Epiglucan
Grifola frondosa	Grifolan
Lactobacillus frumenti	Fructan
Lactobacillus reuteri	Mutan/reuteran
Lactobacillus sanfranciscensis	Levan
Lactic acid bacteria and yeasts	Kefiran
Pseudomonas aeruginosa/Lentinula edodes	Lentinan
Leuconostoc mesenteroides	Dextran
Pestalotia	Pestolotan
Saccharomyces sp.	Yeast glucan
Saccharomyces cerevisiae	Zymosan
Schizophyllum commune	Schizophyllan
Sclerotium glucanicum	Scleroglucan
Sphingomonas elodea	Gellan
Trametes versicolor	Krestin
Weissella confusa	Inulin and fructan
Xanthomonas campestris	Xanthan
Zymomonas mobilis	Levan

meliloti produces multiple biopolymers (EPS, cellular polysaccharides, and polyhydroxyalkanoates) simultaneously (Saranya Devi et al. 2012; Shamala et al. 2014). Several cyanobacteria like *Anabaena*, *Nostoc*, *Spirulina*, *Palmella mucosa*, etc., are also known to produce EPS (Chakraborty and Pal 2014), some of which are homopolysaccharides and some other are heteropolysaccharides. Several *Saccharomyces* spp. produce an EPS popularly known as yeast glucan. Finore et al. (2014) in their review on fermentative production of EPS by marine bacteria indicated several bacteria, such as *Aeribacillus pallidus*, *Alteromonas* sp., *Halomonas alkaliantarctica*, *Pseudoalteromonas* sp., *Salipiger mucosus*, *Zunongwangia profunda*, etc., as EPS producers. Very recently, it is reported that *Bacillus subtilis* produces an EPS that has antioxidant activity (Razack et al. 2014). An exhaustive review on EPS-producing fungi including lower filamentous fungi, higher basidiomycetes, and yeasts present in different ecological niches at laboratory scale has been made available (Mahapatra and Banerjee 2013). *Weissella cibaria* and *W. confusa* produce fructan and inulin (Malang et al. 2015). Antarctic bacterium *Pseudoalteromonas* sp. S-5 producing an EPS, having anticancer activity, was reported recently (Chen et al. 2015).

8.1.2.2 Genetics and Metabolic Pathway of EPS Synthesis

A majority of the microbial EPS are produced intracellularly and sent to the outside of the cell (Rehm 2009; Ullrich 2009). However, a few of the EPS like levan and dextran are synthesized and polymerized outside the producing bacterial cells through the action of secreted enzymes (Rehm 2009). The basics of the biosynthesis of microbial EPS have been reviewed by Freitas et al. (2011). Biosynthesis of EPS takes place in three steps. These are assimilation of carbon source, intracellular synthesis of polysaccharide, and release of the polysaccharide to the outside environment (Vandamme et al. 2002). The EPS production process is known to be costlier as it requires large volumes of the carbon source. However, keeping the cost of the EPS in today's market in view, it is considered to be economical if the rapid growth and higher survival of the producing organisms are addressed (Wolfaardt et al. 1999).

Biosynthesis of microbial EPS takes place using nucleotide-activated precursors. All the enzymes responsible for biosynthesis, assembly, and transport of EPS are derived from specific gene clusters. The mechanism of biosynthesis varies from EPS to EPS. Extracellular glycosyltransferases help in the polymerization of homopolysaccharides by increasing the chain length of the molecule, whereas in heteropolysaccharides polymerization takes place after the repeating units that are formed in the cytoplasm are transferred outside the cell through a lipophilic carrier (Finore et al. 2014); hence, molecular weight of the extruded polymer increases at the cell surface. Biosynthesis and characterization of EPS from lactic acid bacteria (LAB) had been reviewed in detail (Laws et al. 2001), and a comprehension of the glycosyltransferase gene map can help in predicting the structure of their EPS and its repeating unit (Remminghorst and Rehm 2009). For the optimization of the EPS biosynthesis by following a system biology approach, knowledge of the genome structure is essential as EPS production in microorganisms is controlled by a specific gene cluster (Ateş et al. 2013). The biosynthetic pathway and mode of export of EPS in alginate production (Rehm and Valla 1997) and cellulose production by *Acetobacter* (Chawla et al. 2009) and *Azotobacter* spp. (Gauri et al. 2012), *Rhizobium* spp. (Skorupska et al. 2006), EPS from lactobacilli spp. (Badel et al. 2011; Patten and Laws 2014), and in other microbes (Madhuri and Vidya Prabhakar 2014) were reviewed thoroughly. Metabolic pathway of curdlan synthesis is explained in detail earlier (Donot et al. 2012). The biosynthesis pathway of curdlan production by *Agrobacterium* was reviewed recently (Zhan et al. 2012). The EPS biosynthesis pathway exclusively in different species of bifidobacteria was reviewed recently (Hidalgo-Cantabrana et al. 2014).

Biosynthetic pathway of welan gum production was proposed to have two discrete systems for the metabolism of glucose (Li et al. 2010a).

In the first system, glucose is converted to glucose-1-phosphate and later to G-6-Pe and from there to fructose-6-phosphate and nucleotide sugar precursors. In the second system, the cell skeleton is synthesized from glucose using the pathways like the TCA cycle, Entner-Doudoroff pathway, and the PPP. A little later, Wang et al. (2012) analyzed 1.4 Mb genome sequence involved in welan production and annotated 55 coding sequences (CDSs) related to its monosaccharide metabolism and 10 CDSs responsible for its biosynthesis. Improvement in EPS production by overexpression of NADH oxidase gene in *L. casei* resulting in 46 % more EPS was attempted (Li et al. 2015). They have overexpressed a H2O-forming NADH oxidase gene in *L. casei* LC2W by cloning it from *Streptococcus mutans* under the control of constitutive promoter P_{23}; as a result, they could notice a 20-fold increase in the gene expression levels over the wild strain and a reduction of 22 % in lactate production.

8.1.2.3 Fermentative Production of EPS

Several physical and chemical factors affect the production of EPS by microorganisms. Aeration, agitation, fermentation period, rate of inoculum, and temperature are the important physical parameters, which can influence the production of EPS. Among the chemical factors, composition of the medium, source and concentration of carbon (Rajkumar et al. 2003; Vijayendra et al. 2003) and nitrogen, trace elements, pH, and dissolved oxygen are considered important that can affect the EPS production. All over the world, cheaper and alternate carbon and nitrogen sources are being used to reduce the production cost of the EPS. The use of high-yielding strains can increase the yield of EPS. Various strategies involved in large-scale production of microbial EPS such as type of fermentor, bioengineering and microbiological challenges in monitoring, and controlling conditions have been critically reviewed recently (Seviour et al. 2011). To increase the yield of EPS, various statistical methods are being used to optimize the nutritional and physical parameters. Besides these, batch, fed batch, and continuous fermentations are also in practice for the production of microbial EPS.

Various aspects of pullulan production are reviewed in the past (Seviour et al. 1992; Leathers 2003; Singh et al. 2008; Gaur et al. 2010; Cheng et al. 2011; Singh and Saini 2012; Yatmaz and Turhan 2012). Besides commonly used sugars like sucrose, maltose, glucose, and lactose, the use of the *jaggery*, a dry concentrate of sugarcane juice, as a cheap source of carbon, for pullulan production, is reported (Vijayendra et al. 2001). Using this carbon source at 5 % level, they could obtain 23 g/l of pullulan in 72 h. The other cheaper substrates used for pullulan production such as coconut water, cashew fruit juice, fuel ethanol by-products, grape skin and pulp extract, hydrolyzed potato starch, molasses, peat hydrolysate, olive oil/wastes, carob pod/extract, hydrolysates of inulin and cornmeal, corn syrup, fermentation stillage and sucrose, beet molasses, spent sulfite liquor, spent grain liquor, etc., are reviewed elsewhere (Vijayendra and Shamala 2014). Genome shuffling through protoplast fusion increased the productivity (179.7 %) of pullulan over the wild strain of *A. pullulans* N3387 (Kang et al. 2011). Shivakumar and Vijayendra (2006) for the first time used coconut water as a carbon source for producing curdlan.

Recent developments in the fermentative production of microbial EPS like fermentation conditions, mode of fermentation, and optimization methods of fermentative production of EPS like kefiran, bacterial cellulose, levan, and gellan and EPS of *Haloferax* have been reviewed recently (Vijayendra and Shamala 2014) and, hence, not mentioned here once again. In a fed batch fermentation by *Sphingomonas paucimobilis* ATCC 31461, 17.71 g/l higher gellan gum production and 57.12 % higher conversion efficiency were achieved using the Logistic and Luedeking-Piret models (Wang et al. 2006). Prior to this, the use of fed batch fermentation for xanthan production by *Xanthomonas campestris* was reported (Shamala and Prasad 2001).

Acetic acid stress after 24 and 26 h of fermentation increased the yield of xanthan production by *Xanthomonas campestris* (Shehni et al. 2011). An increase in the yield was dependent on increased concentration and the number of pulses of acetic acid addition. Later, to reduce the production cost, Costa et al. (2014) have used aqueous shrimp shell extract as a carbon and nitrogen source for xanthan production, and the yields were higher than that of control medium prepared with sucrose.

Using a statistical design (Plackett-Burman), higher yield of welan gum (22.85 g/l) was obtained with optimal medium combinations and fermentation conditions such as cornstarch (43.6 g/l), cottonseed cake flour (4.1 g/l), 3 % inoculum, initial pH 7.0, and temperature 30 °C (Li et al. 2010b). Similarly, very recently, Moshaf et al. (2014) have used a second-grade date palm for the production of xanthan and optimized its production by following statistical methods like response surface methodology (RSM) combined with a central composite design (CCD). Excepting dextran, which is being produced at the commercial level, other EPS of LAB could not be commercialized due to their low yields (<1 g/L) (Badel et al. 2011). However, production of 18 g/L of heteropolysaccharide by a *Leuconostoc* sp. CFR 2181 using a cheap semisynthetic medium in a short period of 4 h of fermentation at 25 °C was reported earlier (Vijayendra and Sharat Babu 2008). Galle and Arendt (2014) recently published an exhaustive review on fermentative production of EPS by LAB, exclusively isolated from sourdough, like *Leuconostoc mesenteroides*, *Lactobacillus sanfranciscensis*, *Lb. rossiae*, *Lb. brevis*, *Lb. spicheri*, *Lb. pontis*, *Lb. frumenti*, *Lb. reuteri*, *Weissella confusa*, etc. Method of levan production and various strategies being used for its production all over the world have been extensively reviewed recently (Srikanth et al. 2015).

8.1.2.4 Downstream Processing of EPS

Downstream processing consists of inactivation and removal of microbial cells from the fermented broth and recovery of the EPS from the broth by precipitation and subsequent drying using a drum drier. Inactivation of the cells is generally carried out by heating at pasteurization temperatures (García-Ochoa et al. 2000). This heating also inactivates the enzymes which may degrade the polymer at later stages of recovery. A number of chemical methods, physical methods, and their combinations are used for the recovery of EPS from either broth or sludges (Sheng et al. 2010). Comparatively, chemical methods are more efficient than physical methods. The physical methods of extraction include ultrasonic, centrifugation, microwave treatment, or heating (Donot et al. 2012). DomInguez et al. (2010) have indicated that the cationic exchange resin method resulted in 1.6 times more yield than thermal treatment. Extraction method for each EPS must be optimized individually as the properties of the EPS vary with each polymer. If EPS of higher purity is required, chemical deproteinization with trichloroacetic acid (Ayala-Hernández et al. 2008), treatment with enzymes like proteases (Wang et al. 2007), or membrane filtration either with ultrafiltration or diafiltration (Kumar et al. 2007; Bahl et al. 2010) can be used.

Optimization of downstream processing can also reduce the production cost of EPS. However, in the case of highly viscous EPS like xanthan, the recovery costs account for over 70 % (Torrestiana-Sanchez et al. 2007). Recent developments in the recovery of biopolymers including microbial EPS were reviewed (Kreyenschulte et al. 2014). Before solvent precipitation, heat treatment at 80 °C for 30 min can precipitate most of the thermosensitive proteins present in the pullulan fermentation broth without affecting the recovery of pullulan, and pullulan is recovered using cold isopropanol and subsequent drying at 60 °C for 40 min. (Mishra and Vuppu 2013). Using response surface methodology, enhanced recovery of pullulan was noticed with a combination of solvents of ethanol, acetone, isopropanol, and tetrahydrofuran as compared to ethanol alone (Choudhury et al. 2013).

Even for the recovery of xanthan, the most common strategy is precipitation using water-miscible non-solvents like acetone, ethanol, or isopropyl alcohol (García-Ochoa et al. 2000; Salah et al. 2011), or in some other cases the use of polyvalent cations such as aluminum, calcium, or quaternary ammonium salts is also reported (Pace and Righelato 1980; Palaniraj and Jayaraman 2011). However, the use of electrolytes such as potassium or sodium chloride can reduce the volume of the alcohol or isopropyl alcohol from 3 to 1.4 times (Galindo and Albiter 1996; García-Ochoa et al. 2000). Thermal treatment at 80 °C for 20 min at a pH of 6.3–6.9 enhances the recovery of xanthan, besides reducing the viscosity of the polymer, which eases the separation of insoluble material by filtration or by centrifugation (Smith and Pace 1982).

Agrobacterium sp. produced both water-soluble and water-insoluble EPS, simultaneously. However, the water-insoluble EPS is soluble in alkaline solution. Separation of both these EPS from the fermented broth is a tedious task. A detailed method of separating these EPS was reported elsewhere (Shivakumar and Vijayendra 2006). Recently, the downstream processing of curdlan was optimized by changing the quantities of NaOH and HCl (Kalyanasundaram et al. 2012). By optimization they could reduce the ratio of sample to alkali from 1:7.5 to 1:1, thus reducing the cost of the recovery process. Downstream processing of scleroglucan can be done in three different ways. After, initial heating of the broth at 80 °C for 30 min and subsequent centrifugation are common to all the three methods. Later, the polymer is precipitated either by using the solvent alone or addition of calcium chloride and by solvent extraction or addition of calcium chloride and then adjusting the pH of the broth to 10–12 using alkali solution (Survase et al. 2007).

8.1.2.5 Applications of EPS

The applications of the microbial EPS have been extensively reviewed recently (Rehm and Valla 1997; Giavasis et al. 2000; Bajaj et al. 2007; Survase et al. 2007; Chawla et al. 2009; Freitas et al. 2011; Poli et al. 2011; Freitas et al. 2014; Gauri et al. 2012; Singh and Saini 2012; Zhan et al. 2012; Kaur et al. 2014; Madhuri and Vidya Prabhakar 2014; Mahapatra and Banerjee 2013; Vijayendra and Shamala 2014). Table 8.2 sum-

Table 8.2 Application of some important microbial exopolysaccharides

Exopolysaccharide	Areas of application
Acetan/cellulose	Antimicrobial packaging films, wound dressings, as a parchment paper
Alginate	Ceramics, drug delivery, foods, pharmaceuticals, paper industry, textile printing, welding rods
Alternan	Bulking agent in foods, cosmetics, probiotics
Curdlan	Drug delivery, foods, as a prebiotic, water treatment
Dextran	Plasma extender, drug delivery, chromatography, foods, pharmaceuticals, paper industry
Fructan	Foods
Gellan	Capsules, controlled drug release, foods, gelling agent in dental and personal care items, microencapsulating agent, tissue culture media, wrapping of fruits
Inulin	Substitute for fat in foods, probiotics
Kefiran	Active food packaging film, gelatination
Levan	Biotech industry, beverages, food preservation, medicine, wound healing, nanoparticles of levan in protein/drug delivery
Pullulan	Oxygen barrier films for food and nonfood applications, cosmetics, capsules for drug delivery, foods, printing, textile, photography, plywood, dental care, pharmaceuticals, plasma extender
Reuteran	Foods (bakery)
Scleroglucan	Agriculture, ceramics, cosmetics, food, as a medicine, oil industry, ophthalmic solutions, tablet coating, paints, printing inks
Succinoglucan	Gelling agent and immobilizing agent
Welan	Cement manufacturing, cosmetics, foods, metal working compounds, oil recovery, pharmaceuticals, sealants
Xanthan	Agrichemical sprays, explosives, foods, oil drilling, ore extraction, pesticide and insecticide, printing, paints, water clarification
Yeast glucan	Foods, oil sparing agent

marizes major applications of some important EPS. Due to its varied functional properties, microbial EPS found applications in various fields such as agriculture, cosmetics, food, oil recovery, packaging, textile, wastewater treatment, pharmaceuticals, medicine, and in the form of membranes. Fungal EPS also have found to have many applications in food, cosmetics, pharmacy, medicine, feed, and other areas (Mahapatra and Banerjee 2013). Very recently, the application of microbial polymers exclusively for packaging food and nonfood items has been reviewed (Vijayendra and Shamala 2014). Various applications of several EPS of LAB such as dextran, alternan, reuteran, levan, kefiran, inulin, etc., have been reviewed (Patel et al. 2012). Several health benefits like heavy metal binding, antitumor activity, anti-atherosclerotic effect, immunomodulation activity, and prebiotic effect of LAB and other microbial EPS are reviewed recently (Patel and Prajapati 2013; Madhuri and Vidya Prabhakar 2014; Patten and Laws 2014). An exclusive review on the use of gellan in medicine in the form of oral, ophthalmic, and nasal formulations, tissue engineering, and dressing material is made available recently (Osmałek et al. 2014). Benny et al. (2014) have reviewed the applications of xanthan gum in drug delivery, due to its potential in retarding the drug release, in the form of liposomes, hydrogel, niosomes, nanoparticles, matrix system, or microspheres. In foods microbial EPS can be used as emulsifiers, gelling agents, thickeners, stabilizers, and viscosifying agents to improve the texture and stability, and their properties can be further strengthened by cross-linking, either by physical or chemical method, to prepare modified EPS (Ahmad et al. 2015).

Recently, phosphorylated curdlan microgels for in vitro drug release were prepared, and these microgels had excellent biocompatibility (Popescua et al. 2013). Prior to this, Chawla et al. (2009) have reviewed the production and applications of microbial celluloses in food, medical, pharmaceutical, mining, broadcasting, textile, refinery, waste treatment, etc. Bacterial cellulose is considered to be a best alternative to wound dressing material due to its water-holding capacity, porosity, and efficient barrier properties. Because of its stability at higher temperatures (150 °C) and varied pH (2–12), welan can be used in several areas like food, medicine, as an additive in concrete, as a coating material, and for enhancing the recovery of oil (Kang et al. 1983). The EPS production by *Azotobacter* in soil helps in the soil fertility by controlling the ecosystem through nutrient cycling and in maintaining the soil structure in different environments (Gauri et al. 2012).

Several applications of levan such as food, medicine, beverage, nanotechnology, biotechnology, purification of proteins, and other areas have been recently reviewed (Srikanth et al. 2015). Derivatization increased functionality of levan in terms of increased reducing power, antiproliferative activity, scavenging activity, antioxidant, and anticancer activity (Liu et al. 2012). At 0.1 to 1.0 % concentrations, levan is an excellent immunostimulant in fishes (Gupta et al. 2011). Fructooligosaccharides derived from acid hydrolysis of levan are considered as prebiotic agents (Huang et al. 2013). As reviewed by Srikanth et al. (2015), levan has several medical applications such as an anticlotting factor in heart surgery, healing wounds, after angioplasty anti-AIDS agent, and in the subcutaneous dental filling. Levan is an excellent stabilizer, emulsifier, and flavor enhancer and, hence, used in dairy beverages (Srikanth et al. 2015).

8.2 Future Perspectives

Microbial biopolymers are in great demand. However, the current production prices are inhibiting its practical use in many areas. Future research focus can be made on improving the fermentation strategies and use of alternative raw materials, mainly by-products of agro-industrial sources. Optimization of simultaneous synthesis of multiple polymers can also reduce the production cost, and more research could be focused on this aspect.

Acknowledgments The authors wish to thank the Director of CSIR-CFTRI, for providing necessary funds and facilities. Angelina is thankful to UGC for granting Maulana Azad National Fellowship.

References

Ahmad NH, Mustafa S, Man YBC (2015) Microbial polysaccharides and their modification approaches: a review. Int J Food Prop 18:332–347. doi:10.1080/10942912.2012.693561

Akaraonye E, Keshavarz T, Roy I (2010) Production of polyhydroxyalkanoates: the future green materials of choice. J Chem Technol Biotechnol 85:732–743. doi:10.1002/jctb.2392

Anil Kumar PK, Shamala TR, Lakshman K, Halami PM, Joshi GJ, Chandrashekar A, LathaKumari KS, Divyashree MS (2007) Bacterial synthesis of poly(hydroxybutyrate-co-hydroxyvalerate) using carbohydrate-rich mahua (*Madhuca* sp.) flowers. J Appl Microbiol 103:204–209. doi:10.1111/j.1365-2672.2006.03221.x

Ateş Ö, Arga KY, Öner EO (2013) The stimulatory effect of mannitol on levan biosynthesis: lessons from metabolic systems analysis of *Halomonas smyrnensis* AAD6T. Biotechnol Prog 29:1386–1397. doi:10.1002/btpr.1823

Ayala-Hernández I, Hassan A, Goff HD, Mira de Orduña R, Corredig M (2008) Production, isolation and characterization of exopolysaccharides produced by *Lactococcus lactis* subsp. *cremoris* JFR1 and their interaction with milk proteins: effect of pH and media composition. Int Dairy J 18:1109–1118. doi:10.1016/j.idairyj.2008.06.008

Badel S, Bernardi T, Michaud P (2011) New perspectives for lactobacilli exopolysaccharides. Biotechnol Adv 29:54–66. doi:10.1016/j.biotechadv.2010.08.011

Bahl MA, Schultheis E, Hempel DC, Nörtemann B, Franco-Lara E (2010) Recovery and purification of the exopolysaccharide PS-EDIV from *Sphingomonas pituitosa* DSM 13101. Carbohydr Polym 80:1037–1041. doi:10.1016/j.carbpol.2010.01.021

Bajaj IB, Survase SA, Saudagar PS, Singhal RS (2007) Gellan gum: fermentative production, downstream processing and applications. Food Technol Biotechnol 45:341–354

Balaji S, Gopi K, Bhaskaran M (2013) A review on production of poly β-hydroxybutyrates from cyanobacteria for the production of bio plastics. Algal Res 2:278–285. doi:10.1016/j.algal.2013.03.002

Benny IS, Gunasekar V, Ponnusami V (2014) Review on application of xanthan gum in drug delivery. Int J PharmTech Res 6:1322–1326

Chakraborty T, Pal R (2014) An overview of cyanobacterial exopolysaccharides: features, composition and effects of stress exposure. Int J Life Sci 8(4):1–9. doi:10.3126/ijls.v814.10891

Chawla PR, Bajaj IB, Survase SA, Singhal RS (2009) Microbial cellulose: fermentative production and applications. Food Technol Biotechnol 47:107–124

Chen G, Qiana W, Lia J, Xua Y, Chena K (2015) Exopolysaccharide of Antarctic bacterium *Pseudoaltermonas* sp. S-5 induces apoptosis in K562 cells. Carbohydr Polym 121:107–114. doi:10.1016/j.carbpol.2014.12.045

Cheng KC, Demirci A, Catchmark JM (2011) Pullulan: biosynthesis, production, and applications. Appl Microbiol Biotechnol 92:29–44. doi:10.1007/s00253-011-3477-y

Choudhury AR, Bhattacharjee P, Prasad GS (2013) Development of suitable solvent system for downstream processing of biopolymer pullulan using response surface methodology. PLoS One 8(10), e77071. doi:10.1371/journal.pone.0077071

Costa LAS, Campos MI, Druzian JI, de Oliveira AM, de Oliveira Jr EN (2014) Biosynthesis of xanthan gum from fermenting shrimp shell: yield and apparent viscosity. Int J Polymer Sci 2014, 273650. doi:10.1155/2014/273650

Dhiya C, Benny IS, Gunasekar V, Ponnusami V (2014) A review on development of fermentative production of curdlan. Int J ChemTech Res 6:2769–2773

Divyasri D, Gunasekar V, Benny IS, Ponnusami V (2014) A review on industrial applications of curdlan. Int J ChemTech Res 6:3000–3003

DomInguez L, RodrIguez M, Prats D (2010) Effect of different extraction methods on bound EPS from MBR sludges. Part I: influence of extraction methods over three-dimensional EEM fluorescence spectroscopy fingerprint. Desalination 261:19–26. doi:10.1016/j.desal.2010.05.054

Donot F, Fontana A, Baccou JC, Schorr-Galindo S (2012) Microbial exopolysaccharides: main examples of synthesis, excretion, genetics and extraction. Carbohydr Polym 87:951–962. doi:10.1016/j.carbpol.2011.08.083

Du C, Sabirova J, Soetaert W, Lin SKC (2012) Polyhydroxyalkanoates production from low-cost sustainable raw materials. Curr Chem Biol 6:14–25. doi:10.2174/2212796811206010014

Finore I, Di Donato P, Mastascusa V, Nicolaus B, Poli A (2014) Fermentation technologies for the optimization of marine microbial exopolysaccharide production. Mar Drugs 12:3005–3024. doi:10.3390/md12053005

Freitas F, Alves VD, Reis MAM (2011) Advances in bacterial exopolysaccharides: from production to biotechnological applications. Trends Biotechnol 29:388–398. doi:10.1016/j.tibtech.2011.03.008

Freitas F, Alves VD, Reis MA, Crespo JG, Coelhoso IM (2014) Microbial polysaccharide-based membranes: current and future applications. J Appl Polym Sci 40047. doi:10.1002/APP.40047

Galindo E, Albiter V (1996) High-yield recovery of xanthan by precipitation with isopropyl alcohol in a stirred tank. Biotechnol Prog 12:540–547. doi:10.1021/bp9600445

Galle S, Arendt EK (2014) Exopolysaccharides from sourdough lactic acid bacteria. Cri Rev Food Sci Nutr 54:891–901. doi:10.1080/10408398.2011.617474

García-Ochoa F, Santos VE, Casas JA, Gómez E (2000) Xanthan gum: production, recovery, and properties.

Biotechnol Adv 18:549–579. doi:10.1016/S0734-9750(00)00050-1

Gaur R, Singh R, Gupta M, Gaur MK (2010) *Aureobasidium pullulans*, an economically important polymorphic yeast with special reference to pullulan. Afr J Biotechnol 9:7989–7997. doi:10.5897/AJB10.948

Gauri SS, Mandal SM, Pati BR (2012) Impact of *Azotobacter* exopolysaccharides on sustainable agriculture. Appl Microbiol Biotechnol 95:331–338. doi:10.1007/s00253-012-4159-0

Giavasis I, Harvey LM, McNeil B (2000) Gellan gum. Cri Rev Biotechnol 20:177–211

Gumel AM, Annuar MSM, Chisti Y (2013) Recent advances in the production, recovery and applications of polyhydroxyalkanoates. J Polym Environ 21:580–605. doi:10.1007/s10924-012-0527-1

Gupta SK, Das P, Singh SK, Akhtar MS, Meena DK, Mandal SC (2011) Microbial levan, an ideal prebiotic and immunonutrient in aquaculture. World Aquacult 42(61):63, 66

Halami PM (2008) Production of polyhydroxyalkanoate from starch by the native isolate *Bacillus cereus* CFR06. World J Microbiol Biotechnol 24:805–812. doi:10.1007/s11274-007-9543-z

Hempel F, Bozarth AS, Lindenkamp N, Klingl A, Zauner S, Linne U, Steinbuchel A, Maier UG (2011) Microalgae as bioreactors for bioplastic production. Microb Cell Fact 10:81. doi:10.1186/1475-2859-10-81

Hidalgo-Cantabrana C, Sánchez B, Milani C, Ventura M, Margolles A, Ruas-Madiedoa P (2014) Genomic overview and biological functions of exopolysaccharide biosynthesis in *Bifidobacterium* spp. Appl Environ Microbiol 80:9–18. doi:10.1128/AEM.02977-13

Huang MY, Lee CF, Ho ST, Lin KJ, Pan CL (2013) High-yield levan produced by *Bacillus licheniformis* FRI MY-55 in high-sucrose medium and its prebiotic effect. J Pure Appl Microbiol 7:1585–1599

Kabilan S, Ayyasamy M, Jayavel S, Paramasamy G (2012) *Pseudomonas* sp. as a source of medium chain length polyhydroxyalkanoates for controlled drug delivery: perspective. Int J Microbiol 317828. doi:10.1155/2012/317828

Kalyanasundaram GT, Doble M, Gummadi SN (2012) Production and downstream processing of (1-3)-β-D-glucan from mutant strain of *Agrobacterium* sp. ATCC 31750. AMB Express 2:31. doi:10.1186/2191-0855-2-31

Kang KS, Veeder GT, Cottrell IW (1983) Some novel bacterial polysaccharides of recent development. Prog Ind Microbiol 18:231–253

Kang JX, Chen XJ, Chen WR, Li MS, Fang Y, Li DS, Ren YZ, Liu DQ (2011) Enhanced production of pullulan in *Aureobasidium pullulans* by a new process of genome shuffling. Process Biochem 46:792–795. doi:10.1016/j.procbio.2010.11.004

Kaur V, Bera MB, Panesar PS, Kumar H, Kennedy JF (2014) Welan gum: microbial production, characterization, and applications. Int J Biol Macromol 65:454–461. doi:10.1016/j.ijbiomac.2014.01.061

Kocharin K, Nielsen J (2013) Specific growth rate and substrate dependent polyhydroxybutyrate production in *Saccharomyces cerevisiae*. AMB Express 3:18. doi:10.1186/2191-0855-3-18

Kreyenschulte D, Krull R, Margaritis A (2014) Recent advances in microbial biopolymer production and purification. Crit Rev Biotechnol 34:1–15. doi:10.3109/07388551.2012.743501

Kumar AS, Mody K, Jha B (2007) Bacterial exopolysaccharides – a perception. J Basic Microbiol 47:103–117. doi:10.1002/jobm.200610203

Laws A, Gu Y, Marshall V (2001) Biosynthesis, characterisation, and design of bacterial exopolysaccharides from lactic acid bacteria. Biotechnol Adv 19:597–625. doi:10.1016/S0734-9750(01)00084-2

Laycock B, Halleya P, Pratt S, Werkerc A, Lanta P (2013) The chemomechanical properties of microbial polyhydroxyalkanoates. Prog Polym Sci 38:536–583. doi:10.1016/j.progpolymsci.2013.06.008

Leathers TD (2003) Biotechnological production and applications of pullulan. Appl Microbiol Biotechnol 62:468–473. doi:10.1007/s00253-003-1386-4

Li H, Xu H, Xu H, Li S, Ouyang PK (2010a) Biosynthetic pathway of sugar nucleotides essential for welan gum production in *Alcaligenes* sp. CGMCC2428. Appl Microbiol Biotechnol 86:295–303. doi:10.1007/s00253-009-2298-8

Li S, Xu H, Li H, Guo C (2010b) Optimizing the production of welan gum by *Alcaligenes faecalis* NX-3 using statistical experiment design. Afr J Biotechnol 9:1024–1030. doi:10.5897/AJB09.042

Li N, Wang Y, Zhu P, Liu Z, Guo B, Ren J (2015) Improvement of exopolysaccharide production in *Lactobacillus casei* LC2W by overexpression of NADH oxidase gene. Microbiol Res 171:73–77. doi:10.1016/j.micres.2014.12.006

Liu J, Luo J, Ye H, Zeng X (2012) Preparation, antioxidant and antitumor activities *in vitro* of different derivatives of Levan from endophytic bacterium *Paenibacillus polymyxa* EJS-3. Food Chem Toxicol 50:767–772. doi:10.1016/j.fct.2011.11.016

Luengo JM, Garcia B, Sandoval A, Naharro G, Olivera ER (2003) Bioplastics from microorganisms. Cur Opin Microbiol 6:251–260. doi:10.1016/S1369-5274(03)00040-7

Madhuri KV, Vidya Prabhakar K (2014) Microbial exopolysaccharides: biosynthesis and potential applications. Oriental J Chem 30:1401–1410, doi:10.13005/ojc/300362

Mahapatra S, Banerjee D (2013) Fungal exopolysaccharide: production, composition and applications. Microbiol Insights 6:1–16. doi:10.4137/MBI.S10957

Malang SK, Maina NH, Schwab C, Tenkanen M, Lacroix C (2015) Characterization of exopolysaccharide and ropy capsular polysaccharide formation by *Weissella*. Food Microbiol 46:418–427. doi:10.1016/j.fm.2014.08.022

Mishra B, Vuppu S (2013) A study on downstream processing for the production of pullulan by *Aureobasidium pullulans*-SB-01 from the fermentation broth. Res J Recent Sci 2:16–19

Moshaf S, Hamidi-Esfahani Z, Azizi MH (2014) Statistical optimization of xanthan gum production and influence of airflow rates in lab-scale fermentor. Appl Food Biotechnol 1:15–22

Nicolaus B, Kambourova M, Oner ET (2010) Exopolysaccharides from extremophiles: from fundamentals to biotechnology. Environ Technol 31:1145–1158. doi:10.1080/09593330903552094

Osmałek T, Froelich A, Tasarek S (2014) Application of gellan gum in pharmacy and medicine. Int J Pharm 466:328–340. doi:10.1016/j.ijpharm.2014.03.038

Pace GW, Righelato RC (1980) Production of extracellular microbial polysaccharides. Adv Biochem Eng 15:41–70. doi:10.1007/3540096868_2

Palaniraj A, Jayaraman V (2011) Production, recovery and applications of xanthan gum by *Xanthomonas campestris*. J Food Eng 106:1–12. doi:10.1016/j.jfoodeng.2011.03.035

Patel A, Prajapati JB (2013) Food and health applications of exopolysaccharides produced by lactic acid bacteria. Adv Dairy Res 1:2. doi:10.4172/2329-888X.1000107

Patel S, Majumder A, Goyal A (2012) Potentials of exopolysaccharides from lactic acid bacteria. Indian J Microbiol 52:3–12. doi:10.1007/s12088-011-0148-8

Patten DA, Laws AP (2014) *Lactobacillus*-produced exopolysaccharides and their health benefits: a review. Beneficial Microbes (in-press). doi:10.3920/BM2014.0117

Poli A, Di Donato P, Abbamondi GR, Nicolaus B (2011) Synthesis, production, and biotechnological applications of exopolysaccharides and polyhydroxyalkanoates by Archaea. Archaea 693253. doi:10.1155/2011/693253

Popescua I, Pelina IM, Butnarub M, Fundueanua G, Suflet DM (2013) Phosphorylated curdlan microgels. Preparation, characterization, and *in vitro* drug release studies. Carbohydr Polym 94:889–898. doi:10.1016/j.carbpol.2013.02.014

Prajapati VD, Jani GK, Khanda SM (2013) Pullulan: an exopolysaccharide and its various applications. Carbohydr Polym 95:540–549. doi:10.1016/j.carbpol.2013.02.082

Rajkumar R, Vijayendra SVN, Prasad MS (2003) Optimization of exopolysaccharides from *Alcaligenes eutrophus*. In: Soni PL (ed) Trends in carbohydrate chemistry. Surya International Publishers, Dehra Dun, pp 212–217

Razack SA, Velayutham V, Thangavelu V (2014) Medium optimization and *in vitro* antioxidant activity of exopolysaccharide produced by *Bacillus subtilis*. Korean J Chem Eng 31:296–303. doi:10.1007/s11814-013-0217-2

Rehm BHA (2009) Microbial production of biopolymers and polymer precursors: applications and perspectives. Caister Academic Press, Poole, p 293

Rehm BHA, Valla S (1997) Bacterial alginates: biosynthesis and applications. Appl Microbiol Biotechnol 48:281–288. doi:10.1007/s002530051051

Remminghorst U, Rehm BHA (2009) Microbial production of alginate: biosynthesis and applications. In: Microbial production of biopolymers and polymer precursor: applications and perspectives. Caister Academic Press, Palmerston North

Salah RB, Chaari K, Besbes S, Blecker C, Attia H (2011) Production of xanthan gum from *Xanthomonas campestris* NRRL B-1459 by fermentation of date juice palm by-products (*Phoenix dactylifera* L.). J Food Process Eng 34:457–474. doi:10.1111/j.1745-4530.2009.00369.x

Saranya Devi E, Vijayendra SVN, Shamala TR (2012) Exploration of rice bran, an agroindustry residue, for the production of intra and extra cellular polymers by *Sinorhizobium meliloti* MTCC 100. Biocatal Agric Biotechnol 1:80–84. doi:10.1016/j.bcab.2011.08.014

Seviour RJ, Stasinopoulos SJ, Auer DPF, Gibbs PA (1992) Production of pullulan and other exopolysaccharides by filamentous fungi. Crit Rev Biotechnol 12:279–298. doi:10.3109/07388559209069196

Seviour RJ, McNeil B, Fazenda ML, Harvey LM (2011) Operating bioreactors for microbial exopolysaccharide production. Crit Rev Biotechnol 31:170–185. doi:10.3109/07388551.2010.505909

Shamala TR, Prasad MS (2001) Fed-batch fermentation for rapid production of xanthan by *Xanthomonas campestris*. Food Biotechnol 15:169–177. doi:10.1081/FBT-100107628

Shamala TR, Chandrashekar A, Vijayendra SVN, Kshama L (2003) Identification of polyhydroxyalkanoate (PHA)-producing *Bacillus* spp. using the polymerase chain reaction (PCR). J Appl Microbiol 94:369–374

Shamala TR, Divyashree MS, Davis R, Latha Kumari KS, Vijayendra SVN, Raj B (2009) Production and characterization of bacterial polyhydroxyalkanoate copolymers and evaluation of their blends by fourier transform infrared spectroscopy and scanning electron microscopy. Indian J Microbiol 49:251–258. doi:10.1007/s12088-009-0031-z

Shamala TR, Vijayendra SVN, Joshi GJ (2012) Agro-industrial residues and starch for growth and co-production of polyhydroxyalkanoate copolymer and α-amylase by *Bacillus* SP. CFR-67. Braz J Microbiol 43:1094–1102. doi:10.1590/S1517-83822012000300

Shamala TR, Rohinishree YS, Vijayendra SVN (2014) Biosynthesis of multiple biopolymers by Sinorhizobium meliloti CFR 14 in high cell density cultures through fed batch fermentation. Biocatal Agric Biotechnol 3:316–322. doi:10.1016/j.bcab.2014.05.00

Shehni SA, Soudi MR, Hosseinkhani S, Behzadipour N (2011) Improvement of xanthan gum production in batch culture using stepwise acetic acid stress. Afr J Biotechnol 10:19425–19428. doi:10.5897/AJB11.1794

Sheng GP, Yu HQ, Li XY (2010) Extracellular polymeric substances (EPS) of microbial aggregates in biological wastewater treatment systems: a review. Biotechnol Adv 28:882–894. doi:10.5897/AJB11.1794

Shivakumar S, Vijayendra SVN (2006) Production of exopolysaccharides by *Agrobacterium* sp. CFR-24 using coconut water-a byproduct of food industry. Lett Appl Microbiol 42:477–482. doi:10.1111/j.1472-765X.2006.01881.x

Singh RS, Saini GK (2012) Biosynthesis of pullulan and its applications in food and pharmaceutical industry. In: Satyanarayana T, Jori BN, Prakash A (eds) Microorganisms in sustainable agricultural biotechnology, part 2. Springer, India, pp 509–553

Singh RS, Saini GK, Kennedy JF (2008) Pullulan: microbial sources, production and applications. Carbohydr Polym 73:515–531. doi:10.1016/j.carbpol.2008.01.003

Singha TK (2012) Microbial extracellular polymeric substances: production, isolation and applications. IOSR J Pharm 2:276–281. doi:10.9790/3013-0220276281

Skorupska A, Janczarek M, Marczak M, Mazur A, Król J (2006) Rhizobial exopolysaccharides: genetic control and symbiotic functions. Microb Cell Factor 5:7. doi:10.1186/1475-2859-5-7

Smith JH, Pace GW (1982) Recovery of microbial polysaccharides. J Chem Technol Biotechnol 32:119–129. doi:10.1002/jctb.5030320116

Sreekanth MS, Vijayendra SVN, Joshi GJ, Shamala TR (2013) Effect of carbon and nitrogen sources on simultaneous production of α-amylase and green food packaging polymer by *Bacillus* sp. CFR 67. J Food Sci Technol 50:404–408. doi:10.1007/s13197-012-0639-6

Srikanth R, Reddy CHSSS, Siddartha G, Ramaiah MJ, Uppuluri KB (2015) Review on production, characterization and applications of microbial levan. Carbohydr Polym 120:102–114. doi:10.1016/j.carbpol.2014.12.003

Steinbüchel A (2005) A commentary on "Biosynthesis of terpolyesters of 3-hydroxybutyrate, 3-hydroxyvalerate, and 5-hydroxyvalerate in *Alcaligenes eutrophus* from 5-chloropentanoic and pentanoic acids" by Doi Y, Tamaki A, Kunioka M, Soga K (Makromol Chem Rapid Commun 1987, 8:631–635). Macromol Rapid Commun 26:1025–1031. doi:10.1002/marc.200500262

Survase SA, Saudagar PS, Bajaj IB, Singhal RS (2007) Scleroglucan: fermentative production, downstream processing and applications. Food Technol Biotechnol 45:107–118

Sutherland IW (1994) Structure-function relationships in microbial exopolysaccharides. Biotechnol Adv 12:393–448. doi:10.1016/0734-9750(94)90018-3

Torrestiana-Sanchez B, Balderas-Luna L, Brito-De la Fuente E, Lencki RW (2007) The use of membrane-assisted precipitation for the concentration of xanthan gum. J Membr Sci 294:84–92. doi:10.1016/j.memsci.2007.02.014

Ullrich M (2009) Bacterial polysaccharides: current innovations and future trends. Caister Academic Press, Poole, p 358. ISBN 978-1-904455-45-5

Vandamme E, de Baets S, Steinbuchel A (2002) Biopolymers: polysaccharides I -polysaccharides from prokaryotes. Wiley-VCH Verlag GmbH, Weinheim

Vijayendra SVN, Shamala TR (2014) Film forming microbial biopolymers for commercial applications – a review. Crit Rev Biotechnol 34:338–357. doi:10.3109/07388551.2013.798254

Vijayendra SVN, Sharat Babu RS (2008) Optimization of a new hetero-polysaccharide production by a native isolate of *Leuconostoc* sp. CFR-2181. Lett Appl Microbiol 46:643–648. doi:10.1111/j.1472-765X.2008.02361.x

Vijayendra SVN, Bansal D, Prasad MS, Nand K (2001) *Jaggery*: a novel substrate for pullulan production by *Aureobasidium pullulans* CFR-77. Process Biochem 37:359–364. PII: S0032-9592(01)00214-X

Vijayendra SVN, Yamini D, Sudhamani SR, Prasad MS (2003) Effect of hexose sugars on exopolysaccharide production by selected bacterial cultures. J Food Sci Technol 40:611–614

Vijayendra SVN, Rastogi NK, Shamala TR, Anil Kumar PK, Kshama L, Joshi GJ (2007) optimization of polyhydroxybutyrate production by *Bacillus* sp. CFR 256 with corn steep liquor as a nitrogen source. Indian J Microbiol 47:170–175

Vijayendra SVN, Veeramani S, Shamala TR (2008a) Optimization of polyhydroxy-butyrate production by β-carotene producing strain of *Micrococcus* sp. J Food Sci Technol 45:506–509

Vijayendra SVN, Palanivel G, Mahadevamma S, Tharanathan RN (2008b) Physico-chemical characterization of an exopolysaccharide produced by a non-ropy strain of *Leuconostoc* sp. CFR 2181 isolated from *dahi*, an Indian traditional fermented milk product. Carbohydr Polym 72:300–307. doi:10.1016/j.carbpol.2007.08.016

Vijayendra SVN, Palanivel G, Mahadevamma S, Tharanathan RN (2009) Partial characterization of a new heteropolysaccharide produced by a native isolate of heterofermentative *Lactobacillus* sp. CFR-2182. Arch Microbiol 191:301–310. doi:10.1007/s00203-008-0453-8

Wang X, Xu P, Yuan Y, Liu C, Zhang D, Yang Z, Yang C, Ma C (2006) Modeling for gellan gum production by *Sphingomonas paucimobilis* ATCC 31461 in a simplified medium. Appl Environ Microbiol 72:3367–3374. doi:10.1128/AEM.72.5.3367-3374.2006

Wang X, Yuan Y, Wang K, Zhang D, Yang Z, Xu P (2007) Deproteinization of gellan gum produced by *Sphingomonas paucimobilis* ATCC 31461. J Biotechnol 128:403–407. doi:10.1016/j.jbiotec.2006.09.019

Wang X, Tao F, Gai Z, Tang H, Xu P (2012) Genome sequence of the welan gum producing strain

Sphingomonas sp. ATCC 31555. J Bacteriol 194:5989–5990. doi:10.1128/JB.01486-12
Wolfaardt GM, Lawrence JR, Korbe DR (1999) Function of EPS. In: Wingender J, Neu TR, Flemming HC (eds) Microbial extracellular polymeric substances: characterization, structure and function. Springer, New York, pp 171–200
Yadav V, Prapulla SG, Jha A, Poonia A (2011) A novel exopolysaccharide from probiotic *Lactobacillus fermentum* CFR 2195: production, purification and characterization. Biotechnol Bioinform Bioeng 1:415–421
Yatmaz E, Turhan I (2012) Pullulan production by fermentation and usage in food industry. GIDA J Food 37:95–102
Zhan XB, Lin CC, Zhang HT (2012) Recent advances in curdlan biosynthesis, biotechnological production, and applications. Appl Microbiol Biotechnol 93:525–531. doi:10.1007/s00253-011-3740-2
Zinn M, Hany R (2005) Tailored material properties of polyhydroxyalkanoates through biosynthesis and chemical modification. Adv Eng Mater 7:408–411. doi:10.1002/adem.200500053

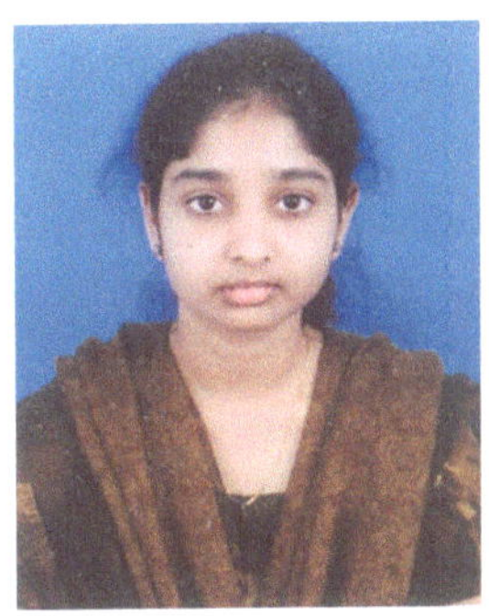

Angelina received her M.Sc. degree in Microbiology from University of Mysore, Mysore, India. She is currently enrolled for Ph.D. with Academy of Innovative and Scientific Research (AcSIR), New Delhi, India, and works as Senior Research Fellow at CSIR-Central Food Technological Research Institute, Mysore, India. Her current research work focuses on the production of polyhydroxyalkanoates (biopolymer) through high cell density fermentation.

S.V.N. Vijayendra received his M.Sc. degree (Microbiology) from SK University, Anantapur, and Ph.D. degree (Dairy Microbiology) from NDRI, Karnal, India. He is presently the Principal Scientist at CSIR-CFTRI and Associate Professor of AcSIR. He is a life member of several scientific associations. He is the former Editor of the Journal Food Science and Technology (2005–2013) and currently Editor of Indian Food Industry. His current interests are microbial biopolymers and fermented foods.

9 Innovations in Microalgal Harvesting Using Biopolymer-Based Approach

Chiranjib Banerjee, Rajib Bandopadhyay, Puneet Kumar Singh, Harsh Kumar Agrawal, and Pratyoosh Shukla

Abstract

Green unicellular microalgae increase their biomass content by the capability of entrapping CO_2 for photosynthesis and are crucial for important value product. Negative zeta value is imparted the presence of COOH and NH_2 groups. This review will give a detailing toward the forces that are responsible for making alga stable in a solution phase. Beside this, it also explains the various possibilities toward the recent advancement of bioharvesting in terms of technological aspects.

9.1 Stability of Microalgae Suspension

Van der Waals force exists in the case of colloidal particle in suspension and is characterized according to the nature of their water–solid interface. Electrostatic repulsion is a key factor for controlling hydrophobic and hydrophilic stability. In case of hydrophobicity, an excess of cations and anions may accumulate at the interface, resulting in electrical potential which exerts repulsion on the particulates of similar potential. However, for hydrophilic surfaces, typical electrical charges arise from dissociation of inorganic groups. Suspended particle in water usually carries a negative or a positive surface charge. The neutrality is maintained by repulsion between two precipitate particles when they approach each other, i.e., prevents the colloidal particles from sticking together and attracts opposite ions in the environs of the particle (Deryagin and Landau 1941) and thus creates a dispersed cloud

C. Banerjee
Department of Bio-Engineering, Birla Institute of Technology, 835215 Ranchi, Jharkhand, India
e-mail: cbanerjee310@gmail.com

R. Bandopadhyay
Department of Botany, The University of Burdwan, 713109 Golapbag, Bardhaman, West Bengal, India
e-mail: rajibindia@gmail.com; rajib_bandopadhyay@yahoo.com

P.K. Singh
Department of Microbiology, Enzyme Technology and Protein Bioinformatics Laboratory, Maharshi Dayanand University, 124001 Rohtak, Haryana, India
e-mail: puneet.micro.rs@mdurohtak.ac.in

P. Shukla (✉)
Professor and Head, Department of Microbiology, Maharshi Dayanand University, 124001 Rohtak, Haryana, India
e-mail: pratyoosh.shukla@gmail.com

H.K. Agrawal
Microbiology Lab, Department of Bio-Engineering, Birla Institute of Technology, Mesra, 835215 Ranchi, Jharkhand, India
e-mail: harsh1681.11@bitmesra.ac.in

V.C. Kalia (ed.), *Microbial Factories*, DOI 10.1007/978-81-322-2595-9_9

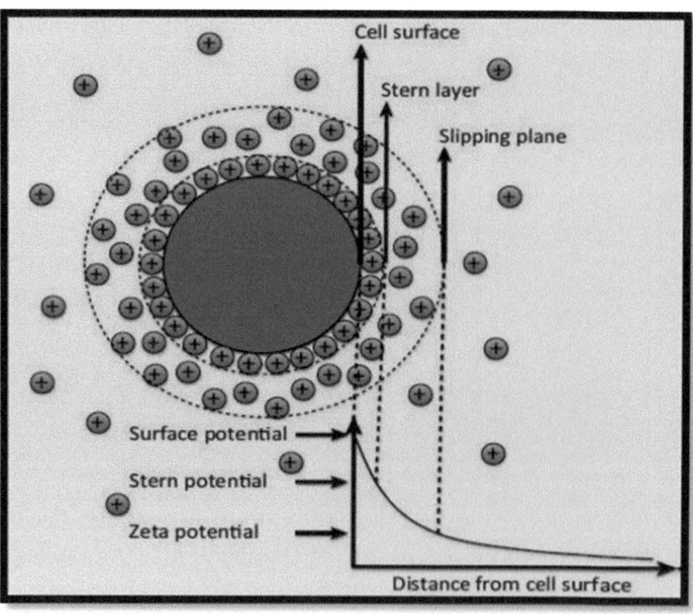

Fig. 9.1 Algal cells showing the nature of electrical forces (Source: Vandamme et al. 2013)

of ions around the particulate. Thus, a double layer is created between charged surfaces and coupled counter ions in a surrounding solution (Fig. 9.1).

A diffusion layer of counterions is formed far from the particle surface due to the stability of electrostatic pull and thermal diffusion. Due to this, the potential difference between the bulk solution and particle surface declines exponentially. When a particle is moving, it withdraws its counterions toward itself and leaves behind the ions which are far away from its surface. Thus, a plane is set which consists of shear and the potential difference, defined as the zeta potential (ζ).

Thus, the zeta value "ζ" can be measured by the mobility of the charged particle under electric field and thereby act as an indicator for the degree of repulsion in suspension.

Zeta potential gives a clue headed for the stability of the system. If the particles in the suspension have a large positive or negative zeta potential, they will repel and there is dispersion stability. Particles having low zeta values could not be prevented from coming together, resulting in dispersion instability. Particles which have zeta potentials more positive than (+) 30 mV are normally considered stable. On other hand (−) 30 mV are also considered to be stable. The surface charges of the microalgal cells are responsible for their stability in the suspension. Carboxylic (-COOH) and amine (-NH_2) are responsible for the charge on the algal cell surface. Above pH 4–5 carboxylic groups dissociate and are negatively charged, whereas there is no charge on amine groups above this pH. A negative charged is observed above 4–5 (Vandamme et al. 2013).

9.2 Flocculation Mechanism

How flocculants can act on the small particle (microalgae and cyanobacteria) which have a negative surface charge that repels among each other. Basically, flocs are generated by adhering to each other; at high pH flocculants block this surface charge by binding (Henderson et al. 2008a, b) and thus settle down.

The proposed mechanisms are *charged neutralization*, *electrostatic patch mechanism*, *bridging*, *sweeping flocculation*, and *Singh's Easy Approachability Model*.

In *charge neutralization*, charges on the colloidal particle surface are neutralized by adsorption of oppositely charged ion thus leading to canceling the colloid particle charge. This ultimately leads to flocculation where the electrostatic repulsion between the colloidal particles disappears. In *electrostatic patch mechanism*, charged polymer binds to oppositely charged particle. "Island" patches are enclosed by the opposite charge on polymer and are created prior to the absorption of polyelectrolytes, in which oppositely charged particles come in and get in touch with other particles giving rise to the strong attraction that leads to flocculation of particles. Mabire et al. in 1984 described the sedimentation rate, the clarification and the height of the sediment bed with stirring, adsorption isotherms, and electrokinetic potential (Mabire et al. 1984).

Bridging phenomenon was first proposed by Ruehrwein and Ward in 1952. In this mechanism, a small dosage of long chain polymer when added to colloidal suspension results to form a bridge with the adsorption of two or more particles (Ruehrwein and Ward 1952). Here it should be considered that adequate unoccupied space also must be there to form the polymer clustering. The presently described process is being achieved equal to a specific dose of polymeric material and this process is termed as steric stabilization. The dose of polymer should be meticulous to achieve this process. Therefore, in low polymer dosage, flocculation is low because no significant bridging occurs and in higher dosage, there will be insufficient surface space for connection of polymer due to which flocs destabilize. The literature suggests that flocculants, their geometrical parameters, and the Kozeny–Carman permeability equation explain the disparities of rate of filtration with concentration of flocculants (Smellie and La Mer 1958). In flocculation kinetics of nanosized particle of silica, cationic amylopectin also supports the bridging flocculation phenomenon (Larsson and Wall 1998).

According to Singh's Easy Approachability Model, flocculation is characterized into hydrolyzed and unhydrolyzed polyacrylamides and grafted and cationic polysaccharides. As per this model the hanging branches of polyacrylamide/cationic moiety will be easily accessible if attached to the backbone. This leads to the formation of aggregates, thus providing the finest flocculation character (Singh et al. 2000; Brostow et al. 2007). Flocculants are divided into two categories: natural or synthetic. Natural flocculants are low in molecular weights; they are to be added in larger dose, do not last long, but are completely safe (nontoxic). Synthetic flocculants in contrast are effective even in small doses (<1 ppm or mgL-1) and comprise of long shelf life.

9.3 Microalgae Harvesting Technologies

Biomass could be used as feedstock for fuel, but it has its own disadvantages. It will be overburdened on the agriculture which also feeds human and livestock and triggers food shortage consequently. Microalgae could be the solution of this problem. The major challenges lie in the harvesting process. The recovery process has been estimated to account for 20–30 % of the total cost of biomass produced (Grima et al. 2003). A description of different microalgae-harvesting technologies is summarized in Table 9.1.

The different processes of harvesting are as follows:

- Electrical
- Biological
- Autoflocculation
- Physical
- Chemical
- Polymer based

9.3.1 Electrical

Electrical based flocculation comprises of coagulant generation by oxidation of electrode, destabilization of particulate suspension followed by aggregation of the destabilized particles which will form flocs. The sequential events are, first, the coagulant generation by oxidation of electrode; second, destabilization of particulate suspension; and third, aggregation of the destabilized particles which will form flocs. It needs electricity for flocculation instead of any flocculants and to flocculate the microalgae from solution and subsequently float the algal flocs with 95 % removing efficiency (Poleman et al. 1997). Continuous harvesting of microalgae has also been reported via electrolysis with polarity exchange (Kim et al. 2012).

Table 9.1 Different algae-harvesting techniques and their efficiency in separating algal biomass

Method	Material used	Yield (%)	References
Flotation	Saponin and chitosan with CTAB	93.7	Kurniawati et al. 2014
	Tetradecyl trimethylammonium bromide	–	Garg et al. 2012
Foam flotation	CTAB	–	Jonathan et al. 2014
Bacterial biocoagulation	PEI-coated *E. coli*	83	Agbakpe et al. 2014
Electrocoagulation	Aluminum electrodes	100	Gao et al. 2010
Flocculation			
Physical flocculation	Ultrasound chitosan	97–99	Fast and Gude 2014
Auto flocculation	*Phaeodactylum tricornutum*	85	Spilling et al. 2010
Chemical flocculation	Aluminum sulfate and poly-aluminum-silicate-sulfate	–	Boisvert et al. 1998
Polymer-based flocculation	Cationic tamarind kernel polysaccharide	–	Pal et al. 2009

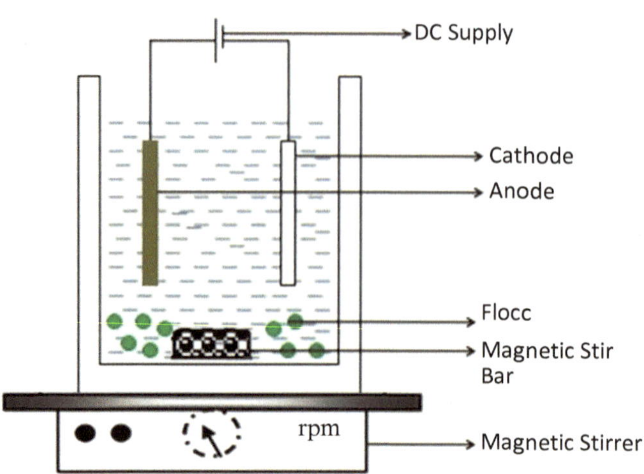

Fig. 9.2 Electrical-based flocculation technique

Electrocoagulation–flocculation method has also been applied for harvesting a *Chlorella vulgaris* (freshwater) and a *Phaeodactylum tricornutum* (marine) microalgal species (Fig. 9.2). Electro-based method with aluminum anode was shown to be more efficient than using an iron anode. Efficiency can be considerably improved by dropping the initial pH and by escalating the turbulence in the microalgal suspension (Vandamme et al. 2011). Electro-flocculation method produces 93.6 % of algae from *Botryococcus braunii* after 30 min of experimentation. Around 98.9 % of the harvest was observed in 14 min when electroflocculation was combined with dispersed-air flotation (Xu et al. 2010).

9.3.2 Biological

Biological-based flocculation or *bioflocculation* is thought to be caused by extracellular polymeric substance in the suspension. Bioflocculation is used to harvest algae where microalgae are used for wastewater treatment; additionally algal–bacterial biomass could be an imperative energy source for the production of biofuel (Craggs et al. 2012). Five different infochemicals were isolated from a flocculating and senescent culture of *Skeletonema marinoi* and *Dunaliella salina*. Water-soluble extracts of *S. marinoi* induce flocculation in the *Nannochloropsis oculata* (Taylor et al. 2012). A novel bioflocculation technology involving fungi (*Aspergillus* sp.)

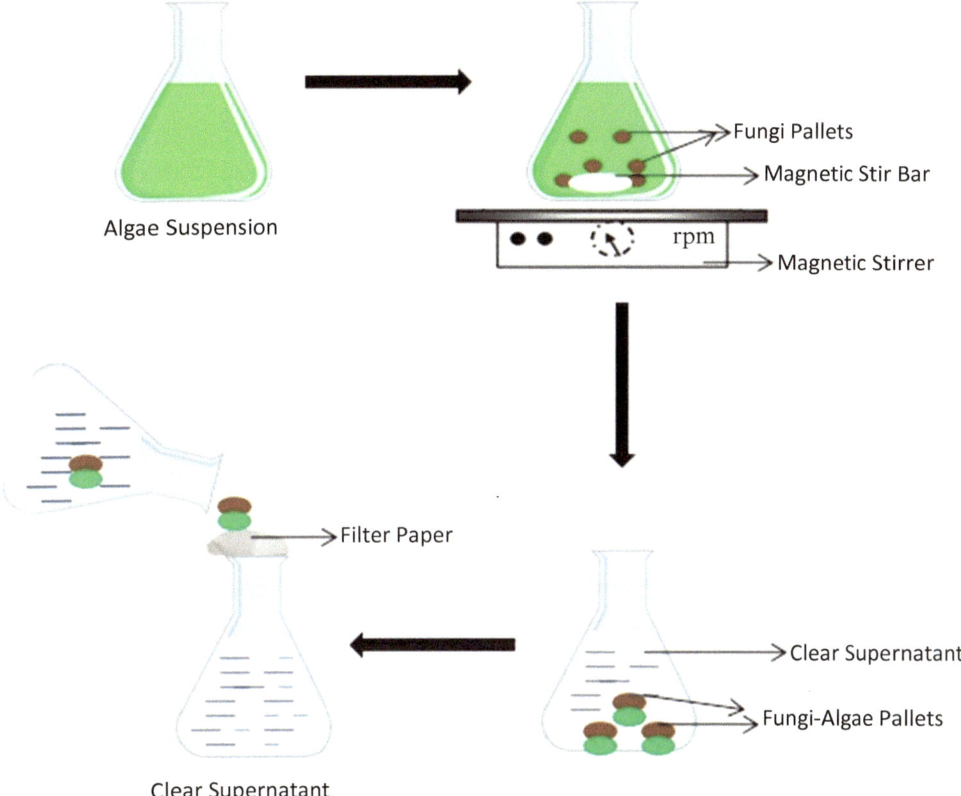

Fig. 9.3 Biological-based flocculation technique

leads to bioflocculation of *Chlorella vulgaris* (Fig. 9.3). The novel technology developed is wastewater treatment technologies (Zhou et al. 2012). Microalgae cells can be co-palletized into fungal pellets for easier harvest; this process can be described as co-palletized cell cultivation. Due to the large size of fungi–algae pellet, it can be easily harvested through sieve (Zhang and Hu 2012). For harvesting marine microalgae, microbial flocculation is one of the efficient separation techniques for the biodiesel production (Lee et al. 2009). Simple separation of the algal biomass by gravity sedimentation through algae–bacteria aggregates is also described (Gutzeit et al. 2005). Cell wall-associated polysaccharides were responsible for self-flocculating microalga *Scenedesmus obliquus* (Guo et al. 2013).

9.3.3 Autoflocculation

Autoflocculation occurs due to precipitation of carbonate salts by algae at high pH values. This is primarily because of photosynthetic CO_2 consumption by algae (Sukenik and Shelef 1984). Microalgae autoflocculation was studied, and about 80–90 % removal of algal cells was observed; it was in combination with chemical flocculation with the help of alum or C-31 polymer (Koopman and Lincoln 1983). The process is shown in Fig. 9.4.

9.3.4 Physical

The removal of algal biomass with the implementation of physical forces is referred to as

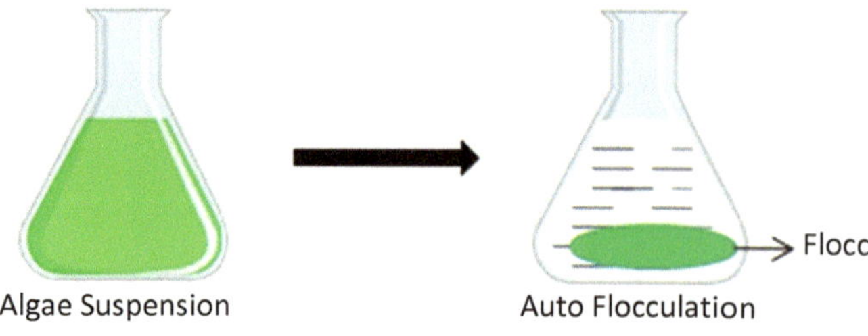

Fig. 9.4 Auto flocculation

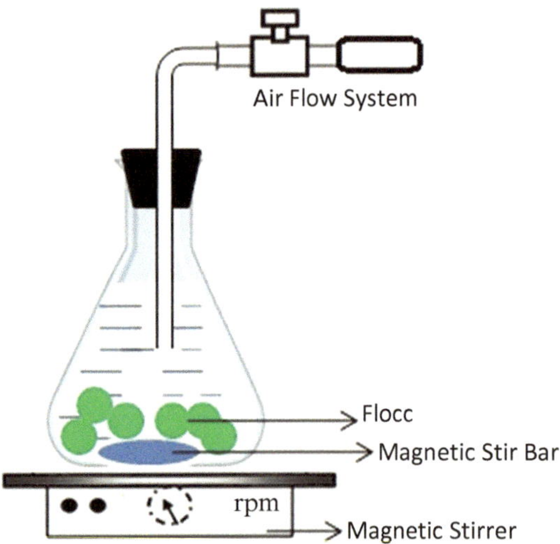

Fig. 9.5 Physical-based flocculation

physical-based flocculation. Magnetic nanoparticle (Fe_3O_4/magnetite) based separation method is reliable and thus can be used in algal biomass recovery. It depends upon high dose of nanoparticles and a pH suitable for recovery of microalgae (Fig. 9.5). From time and energy saving, this process of flocculation is inexpensive and nontoxic as well. It is also engaged in the elimination of detrimental algae from freshwater. Microalgae revival is performed by the phenomenon of electrostatic attraction among nanoparticle and microalgal cells (Xu et al. 2011). Freshwater and marine algae are removed by silica-coated magnetic particles with the help of high-gradient magnetic filtration and have also been studied (Cerff et al. 2012).

In flotation gaseous bubbles are attached to the solid particles, which are thus carried to the liquid surface. Here, gravity plays an important role in the separation process. Flotation is having more advantages than sedimentation especially for removing microalgae (Chen et al. 1998). Expressions for the probability of collision and adhesion have been derived from fine particle on flotation leading to their adhesion to each other (Yoon and Luttrell 1989). Dispersed-air flotation entails 700–1500 μm bubbles, which are formed by a high-speed agitator (Rubio et al. 2002). This process was utilized to remove algae (*Scenedesmus quadricauda*) from water. CTAB was found to be effective to dewater *S. quadricauda* among three types of collectors, i.e., cationic (CTAB), anionic (SDS), and the nonionic (Triton X-100) (Chen et al. 1998). *Scenedesmus obliquus* has great capacity for capturing CO_2. The lipid so produced was harvested using dispersed O_3 flotation (Cheng et al. 2011).

Cultivation of selected thermotolerant microalgae in the heat discharge water and subsequent harvesting of the algal biomass with microstainer (Wilde et al. 1991) have been described. Tangential flow filtration (TFF) system involves 0.45 μm membrane to concentrate phytoplankton for water treatment (Petrusevski et al. 1995). Submerged microfiltration was also carried out and energy consumption to dewater C. *vulgaris* and *P. tricornutum* was well described (Bilad et al. 2012). The economic assessment revealed that dynamic filtration is economically more efficient than tangential cross-flow filtration (Rios et al. 2012). Flotation under vacuum for harvesting microalgae (Barrut et al. 2013) and combination of chemical and biological flocculation have also been used (Zamalloa et al. 2013).

9.3.5 Chemical

Chemical-based flocculation is mainly used in the pretreatment of algal cell to increase its size. It is used before using another method of floccu-

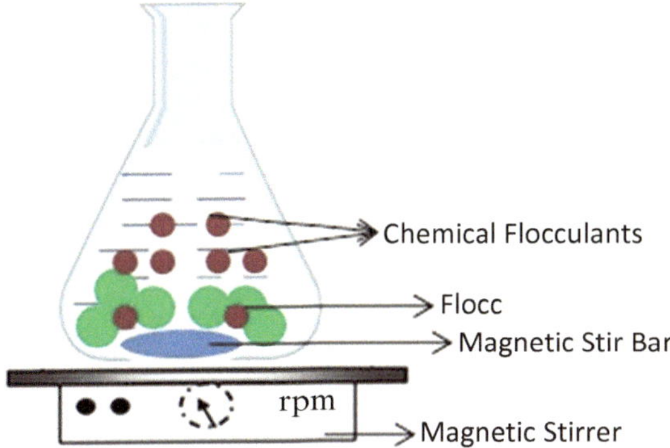

Fig. 9.6 Chemical-based flocculation

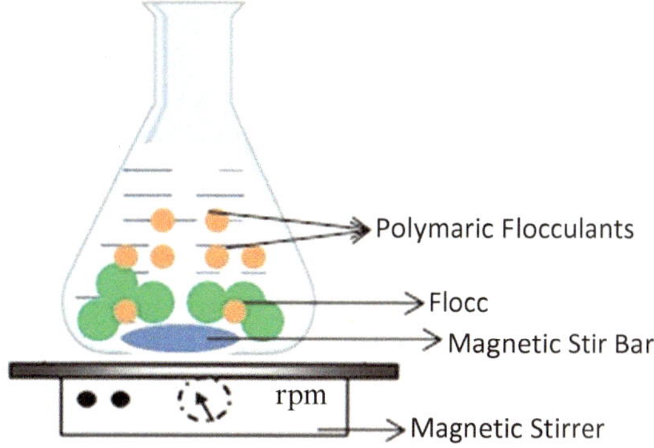

Fig. 9.7 Polymer-based flocculation

lation such as flotation. It is mainly performed from the salts like alum and ferric chloride (Fig. 9.6). The testing of different harvesting methods for *B braunii* was performed once a week. The following conditions were applied: pH adjustment, treatment with aluminum sulfate and Pestan, and biopolymer treatment. The efficient harvesting method is to be adjusted at pH 11 (Lee et al. 1998). Likewise, *Phaeodactylum* sp. diatom flocculation was done with 0.4–1.3 mM calcium hydroxide (Veloso et al. 1991) and *Tetraselmis* sp. (Millamena et al. 1990). A high pH was needed to flocculate with the use of calcium and magnesium: where pH 10–12 (Ayoub et al. 1986; Yahi et al. 1994; Blanchemain and Grizeau 1999; McCausland et al. 1999) initiated flocculation of algae. A nonionic polymer Magnafloc LT-25 was employed for flocculation at pH 10–10.6 (Knuckey et al. 2006). Here, a large amount of blend of base and calcium was also used (Davis and Foust 1969). *Scenedesmus* and *Chlorella* were harvested by alum by means of charge neutralization (Grima et al. 2003). With the use of inorganic flocculants, microalgae can also be flocculated by satisfactorily low pH (Uduman et al. 2010).

Flocculation can be induced by increasing the pH of the medium followed by reusing the flocculated medium for reculturing. About 90 % of the microalgae was harvested by increasing the pH (*Scenedesmus* sp., *Chlorella vulgaris* and *Chlorococcum* sp., *Phaeodactylum tricornutum, Nannochloropsis oculata*) (Wu et al. 2012). The flocculation–sedimentation is the costliest steps of production of microalgae. For that reason *Scenedesmus obliquus* and *Chlorella vulgaris* were studied at high pH induced to improve the efficiency of harvesting. Here, the required pH values were achieved by using sodium hydroxide and calcium hydroxide (Castrillo et al. 2013).

Poly γ-glutamic acid, a microbial flocculant, was used to harvest oleaginous microalgae. Response Surface Methodology (RSM) was used to optimize the flocculation condition for *Chlorella vulgaris* (marine) *and Chlorella protothecoides* (freshwater) (Zheng et al. 2012). The highest flocculating activity of 95 % was achieved for *Scenedesmus* cultures by employing coagulants (8.5 mM $CaCl_2$ and 0.2 mM $FeCl_3$) along with 1 % bioflocculant of *P. polymyxa* (Kim et al. 2011). Chemical coagulants [$FeCl_3$ and $Fe_2(SO_4)_3$] along with commercial polymeric flocculant like Drewfloc 447, Flocudex CS/5000, Chemifloc CV/300 and chitosan are used for flocculation. Their ability was compared for the removal of algal–bacterial biomass during piggery wastewater treatment (De Godos et al. 2011).

9.3.6 Polymer Based

Polymer-based flocculation includes the water-soluble high-molecular weight linear or modified polymers that can be used as flocculants. Flocculation of textile industry wastewater, coal suspension, kaolin, and iron ore was achieved by using synthetic cationic tamarind kernel polysaccharide (Pal et al. 2009), cationic glycogen (Pal et al. 2008), and amphoteric amylopectin (Singh et al. 2013).

Flocculation of freshwater algae, *Spirulina*, *Oscillatoria Chlorella*, and one brackish alga, *Synechocystis*, by means of chitosan was studied in the pH range of 4–9. The settled algal cells are intact and live but will not be redispersed by mechanical agitation (Fig. 9.7). The agitated water may be reused to produce fresh cultures of algae (Divakaran and Pillai 2002). The efficiency of cationic starch as flocculant is a not very proficient for marine microalgae (*Phaeodactylum*, *Nannochloropsis*), but for freshwater (*Parachlorella*, *Scenedesmus*) it is very effective. There was destabilization at high cationic starch concentration. The requirement of the cationic starch dose for the stimulation of flocculation increased linearly with algal biomass quantity. Among the flocculants, namely, Greenfloc 120 and cationic starch, cationic starch was effective (Vandamme et al. 2010).

The required dose of chitosan is lesser than that of inorganic metal salts due to the presence of higher number of functional groups in chitosan (Renault et al. 2009). Recovered algal biomass can be used directly in industrial production of food and fuels for using chitosan as a natural flocculant (Ahmad et al. 2011). Chitosan was found to be the most effective flocculant when it was used with dose of 100 mg/L algae broth when it was compared with ferric sulfate and alum (Beach et al. 2012). Chitosan was used as natural flocculant to explore the green microalga *Chlorella sorokiniana*. Results showed that the clarification efficiency of the process could reach above 99 % below pH 7 (Xu et al. 2013). Microalga *Nannochloropsis* sp. was harvested by flocculant chitosan which was modified to nanochitosan by cross-linking with sodium tripolyphosphate. After the recovery of biomass, the effects of flocculants and type, dosage, and pH of the culture were examined (Farid et al. 2013).

Microalgae are stable in suspension culture, but they are difficult to flocculate due to their small size and negative charge on their surface. Cationic guar gum with the introduction of quaternary amine groups could be used for flocculation. For the flocculation of green algae, viz., *Chlorella* sp. CB4 and *Chlamydomonas* sp. CRP7, different doses of the modified synthesized biopolymer were used (Banerjee et al. 2013). Recently, synthesis of aminoclays with Mg^{2+} or Fe^{3+}, by sol–gel reaction-based methodology with 3-aminopropyltriethoxysilane as a precursor molecule, is exhibited, which produces $(CH_2)3NH_2$ showing covalent bonding. In aqueous solution the proton-rich amine groups validate the competent harvesting of microalgal biomass for various marine algae (Farooq et al. 2013). We understand that genetic modification includes insertion of flocculin protein in the cell wall of *Saccharomyces cerevisiae,* which leads to efficient settling to enhance product clarification and recovery (Govender et al. 2008).

9.4 Concluding Remarks

Production of microalgae is quite expensive in today's scenario. Harvesting microalgae through polymer is still lower compared to centrifugation. Flocculation using ultrasound is still more costly. It is evident that cationic starch and biopolymers are cheaper to some extent, but probably these are too costly for their application in biofuel production (Vandamme et al. 2013). Relatively, autoflocculation (alkalinity induced) is cost-effective than other methods (Vandamme et al. 2012). On other hand, bioflocuulation by microbes (bacteria and fungus) is not feasible owing to their culturing property.

Chemical flocculation has a disadvantage of contaminating biomass; however, a physical separation process does not have this problem. Primary research into flocculation that is stimulated by infochemicals in microalgae is extremely needed, which would lead to an extremely controlled method for suggesting contamination-free flocculation. Genetic modification based flocculation is not a cost effective process and for assessing a innovation flocculation technique economy feasibility should be an important factor for consideration (Vandamme et al. 2013). Nevertheless, we must also ensure that flocculation step is to be taken into account for cost evaluation and extra efforts are also to be made to understand its effect on the entire biofuel production process using flocculation-based technologies.

Acknowledgements Dr. Chiranjib Banerjee highly acknowledges Department of Science and Technology (DST), Government of India for providing financial support as well as project grant from INSPIRE Faculty award scheme.

References

Agbakpe M, Ge S, Zhang W, Zhang X, Kobylarz P (2014) Algae harvesting for biofuel production: influences of UV irradiation and polyethylenimine (PEI) coating on bacterial biocoagulation. Bioresour Technol 166:266–272

Ahmad AL, Mat Yasin NH, Derek CJC, Lim JK (2011) Optimization of microalgae coagulation process using Chitosan. Chem Eng J 173:879–882

Ayoub GM, Lee SL, Koopman B (1986) Seawater induced algal flocculation. Water Res 20:1265–1271

Banerjee C, Ghosh S, Sen G, Mishra S, Shukla P et al (2013) Study of algal biomass harvesting using cationic guar gum from the natural plant source as flocculant. Carbohydr Polym 92:675–681

Barrut B, Blancheton JP, Muller Feuga A, Rene F, Narvaez C et al (2013) Separation efficiency of a vacuum gas lift for microalgae harvesting. Bioresour Technol 128:235–240

Beach ES, Eckelman MJ, Cui Z, Brentner L, Zimmerman JB (2012) Preferential technological and life cycle environmental performance of chitosan flocculation for harvesting of the green algae *Neochloris oleoabundans*. Bioresour Technol 121:445–449

Bilad MR, Vandamme D, Foubert I, Muylaert K, Vankelecom IF (2012) Harvesting microalgal biomass using submerged microfiltration membranes. Bioresour Technol 111:343–352

Blanchemain A, Grizeau D (1999) Increased production of eicosapentaenoic acid by Skeletonema costatum cells after decantation at low temperature. Biotechnol Tech 13:497–501

Boisvert JP, To TC, Berrak A, Jolicoeur C (1998) Phosphate adsorption in flocculation processes of aluminium sulphate and poly-aluminium-silicate-sulphate. Water Res 31:1939–1946

Brostow W, Pal S, Singh RP (2007) A model of flocculation. Mater Lett 61:4381–4384

Castrillo M, Lucas-Salas LM, Rodríguez-Gil C, Martínez D (2013) High pH-induced flocculation-sedimentation and effect of supernatant reuse on growth rate and lipid productivity of *Scenedesmus obliquus* and *Chlorella vulgaris*. Bioresour Technol 128:324–329

Cerff M, Morweiser M, Dillschneider R, Michel A, Menzel K et al (2012) Harvesting fresh water and marine algae by magnetic separation: screening of separation parameters and high gradient magnetic filtration. Bioresour Technol 118:289–295

Chen YM, Liu JC, Ju YH (1998) Flotation removal of algae from water. Colloids Surf B Biointerfaces 12:49–55

Cheng YL, Juang YC, Liao GY, Tsai PW, Ho SH et al (2011) Harvesting of *Scenedesmus obliquus* FSP-3 using dispersed ozone flotation. Bioresour Technol 102:82–87

Craggs R, Sutherland D, Campbell H (2012) Hectare-scale demonstration of high rate algal ponds for enhanced wastewater treatment and biofuel production. J Appl Phycol 24:329–337

Davis NS, Foust OJ (1969) Flocculation of suspensions. US Patent 3,431,200, 1969

De Godos I, Guzman HO, Soto R, García-Encina PA, Becares E et al (2011) Coagulation/flocculation-based removal of algal-bacterial biomass from piggery wastewater treatment. Bioresour Technol 102:923–927

Deryagin BV, Landau LD (1941) Theory of the stability of strongly charged lyophobic sols and the adhesion of strongly charged particles in solutions of electrolytes. Acta Phys Chim URSS 14:633–662

Divakaran R, Pillai VNS (2002) Flocculation of algae using chitosan. J Appl Phycol 14:419–422

Farid MS, Shariati A, Badakhshan A, Anvaripour B (2013) Using nano-chitosan for harvesting microalga *Nannochloropsis* sp. Bioresour Technol 131:555–559

Farooq W, Lee YC, Han JI, Darpito CH, Choi M et al (2013) Efficient microalgae harvesting by organo-building blocks of nanoclays. Green Chem 15:749–755

Fast SA, Gude VG (2014) Ultrasound-chitosan enhanced flocculation of low algal turbid waters. J Ind Eng Chem. doi:10.1016/j.jiec.2014.09.023

Gao S, Yang J, Tian J, Ma F, Tu G, Du M (2010) Electro-coagulation–flotation process for algae removal. J Hazard Mater 177:336–343

Garg S, Li Y, Wang L, Schenk PM (2012) Flotation of marine microalgae: effect of algal hydrophobicity. Bioresour Technol 121:471–474

Govender P, Domingo JL, Bester MC, Pretorius IS, Bauer FF (2008) Controlled expression of the dominant flocculation genes FLO1, FLO5, and FLO11 in *Saccharomyces cerevisiae*. Appl Environ Microbiol 74:6041–6052

Grima ME, Belarbi EH, Fernandez FGA, Medina AR, Chisti Y (2003) Recovery of microalgal biomass and metabolites: process options and economics. Biotechnol Adv 20:491–515

Guo SL, Zhao XQ, Wan C, Huang ZY, Yang YL et al (2013) Characterization of flocculating agent from the self-flocculating microalga *Scenedesmus obliquus* AS-6-1 for efficient biomass harvest. Bioresour Technol 145:285–289

Gutzeit G, Lorch D, Weber A, Engels M, Neis U (2005) Bioflocculent algal-bacterial biomass improves low-cost wastewater treatment. Water Sci Technol 52:9–18

Henderson RK, Parsons SA, Jefferson B (2008a) Successful removal of algae through control of the zeta potential. Sep Sci Technol 43:1653–1666

Henderson RK, Parsons SA, Jefferson B (2008b) The impact of algal properties and pre-oxidation on solid–liquid separation of algae. Water Res 42:1827–1845

Jonathan TC, Lee GM, Caldwell GS (2014) Harvesting microalgae by CTAB-aided foam flotation increases lipid recovery and improves fatty acid methyl ester characteristics. Biomass Bioenergy 67:354–362

Kim DG, La HJ, Ahn CY, Park YH, Oh HM (2011) Harvest of *Scenedesmus* sp. with bioflocculant and reuse of culture medium for subsequent high-density cultures. Bioresour Technol 102:3163–3168

Kim J, Ryu BG, Kim BK, Han JI, Yang JW (2012) Continuous microalgae recovery using electrolysis with polarity exchange. Bioresour Technol 111:268–275

Knuckey RM, Brown MR, Robert R, Frampton DMF (2006) Production of microalgal concentrates by flocculation and their assessment as aquaculture feeds. Aquac Eng 35:300–313

Koopman B, Lincoln EP (1983) Autoflotation harvesting of algae from high-rate pond effluents. Agric Wastes 5:231–246

Kurniawati HA, Ismadji S, Liu JC (2014) Microalgae harvesting by flotation using natural saponin and chitosan. Bioresour Technol 166:429–434

Larsson A, Wall S (1998) Flocculation of cationic amylopectin starch and colloidal silicic acid. The effect of various kinds of salt. Colloids Surf A Physicochem Eng Asp 139:259–270

Lee SJ, Kim SB, Kim JE, Kwon GS, Yoon BD et al (1998) Effects of harvesting method and growth stage on the flocculation of the green alga *Botryococcus braunii*. Lett Appl Microbiol 27:14–18

Lee AK, Lewis DM, Ashman PJ (2009) Microbial flocculation, a potentially low-cost harvesting technique for marine microalgae for the production of biodiesel. J Appl Phycol 21:559–567

Mabire F, Audebert R, Quivoron C (1984) Flocculation properties of some water-soluble cationic copolymers toward silica suspensions: a semiquantitative interpretation of the role of molecular weight and cationicity through a "patchwork" model. J Colloid Interface Sci 97:120–136

McCausland MA, Braun MR, Barrett SM, Diemar JA, Heasman MP (1999) Evaluation of live microalgae and microbial pastes as supplementary food for Pacific oysters. Aquaculture 174:323–342

Millamena OM, Aujero EJ, Borlongan IG (1990) Techniques on algae harvesting and reservation for use in culture and a larval food. Aquac Eng 9:295–304

Pal S, Sen G, Karmakar NC, Mal D, Singh RP (2008) High performance flocculating agents based on cationic polysaccharides in relation to coal fine suspension. Carbohydr Polym 74:590–596

Pal S, Ghosh S, Sen G, Jha U, Singh RP (2009) Cationic tamarind kernel polysaccharide (Cat TKP): a novel polymeric flocculant for the treatment of textile industry wastewater. Int J Biol Macromol 45:518–523

Petrusevski B, Bolier G, Van Breemen AN, Alaerts GJ (1995) Tangential flow filtration: a method to concentrate freshwater algae. Water Res 29:1419–1424

Poleman E, DePauw N, Jeurissen B (1997) Potential of electrolytic flocculation for recovery of micro-algae. Resour Conserv Recycl 19:1–10

Renault F, Sancey B, Badot PM, Crini G (2009) Chitosan for coagulation/flocculation processes – an eco-friendly approach. Eur Polym J 45:1337–1348

Rios SD, Salvado J, Farriol X, Torras C (2012) Antifouling microfiltration strategies to harvest microalgae for biofuel. Bioresour Technol 119:406–418

Rubio J, Souza ML, Smith RW (2002) Overview of flotation as a wastewater treatment technique. Miner Eng 15:139–155

Ruehrwein RA, Ward DW (1952) Mechanism of clay aggregation by polyelectrolytes. Soil Sci 73:485–492

Singh RP, Karmakar GP, Rath SK, Karmakar NC, Tripathy T et al (2000) Biodegradable drag reducing agents and flocculants based on polysaccharides: materials and applications. Polym Eng Sci 40:46–60

Singh RP, Pal S, Rana VK, Ghorai S (2013) Amphoteric amylopectin: a novel polymeric flocculant. Carbohydr Polym 91:294–299

Smellie RH Jr, La Mer VK (1958) Flocculation, subsidence and filtration of phosphate slimes: VI. A quantitative theory of filtration of flocculated suspensions. J Colloid Sci 13:589–599

Spilling K, Seppälä J, Tamminen T (2010) Inducing autoflocculation in the diatom *Phaeodactylum tricornutum* through CO2 regulation. J Appl Phycol 23:959–966. doi:10.1007/s10811-010-9616-5

Sukenik A, Shelef G (1984) Algal autoflocculation – verification and proposed mechanism. Biotechnol Bioeng 26:142–147

Taylor RL, Rand JD, Caldwell GS (2012) Treatment with algae extracts promotes flocculation, and enhances growth and neutral lipid content in *Nannochloropsis oculata* – a candidate for biofuel production. Marine Biotechnol 14:774–781

Uduman N, Qi Y, Danquah MK, Forde GM, Hoadley A (2010) Dewatering of microalgal cultures: a major bottleneck to algae-based fuels. J Renewable Sustainable Energy 2:012701

Vandamme D, Foubert I, Meesschaert B, Muylaert K (2010) Flocculation of microalgae using cationic starch. J Appl Phycol 22:525–530

Vandamme D, Pontes SC, Goiris K, Foubert I, Pinoy LJ et al (2011) Evaluation of electro-coagulation-flocculation for harvesting marine and freshwater microalgae. Biotechnol Bioeng 108:2320–2329

Vandamme D, Foubert I, Fraeye I, Meesschaert B, Muylaert K (2012) Flocculation of *Chlorella vulgaris* induced by high pH: role of magnesium and calcium and practical implications. Bioresour Technol 105:114–119

Vandamme D, Foubert I, Koenraad M (2013) Flocculation as a low-cost method for harvesting microalgae for bulk biomass production. Trends Biotechnol 31:233–239

Veloso V, Reis A, Gouveia L, Fernandes HL, Empis JA et al (1991) Lipid production in *Phaeodactylum tricornutum*. Bioresour Technol 38:115–119

Wilde EW, Benemann JR, Weissman JC, Tillett DM (1991) Cultivation of algae and nutrient removal in a waste heat utilization process. J Appl Phycol 3:159–167

Wu Z, Zhu Y, Huang W, Zhang C, Li T et al (2012) Evaluation of flocculation induced by pH increase for harvesting microalgae and reuse of flocculated medium. Bioresour Technol 110:496–502

Xu L, Wang F, Li HZ, Hu ZM, Guo C et al (2010) Development of an efficient electroflocculation technology integrated with dispersed-air flotation for harvesting microalgae. J Chem Technol Biotechnol 85:1504–1507

Xu L, Guo C, Wang F, Zheng S, Liu CZ (2011) A simple and rapid harvesting method for microalgae by in situ magnetic separation. Bioresour Technol 102:10047–10051

Xu Y, Purton S, Baganz F (2013) Chitosan flocculation to aid the harvesting of the microalga *Chlorella sorokiniana*. Bioresour Technol 129:296–301

Yahi H, Elmaleh S, Coma J (1994) Algal flocculation–sedimentation by pH increase in a continuous reactor. Water Sci Technol 30:259–267

Yoon RH, Luttrell GH (1989) The effect of bubble size on fine particle flotation. Miner Process Extr Metall Rev 5:101–122

Zamalloa C, Boon N, Verstraete W (2013) Decentralized two-stage sewage treatment by chemical-biological flocculation combined with microalgae biofilm for nutrient immobilization in a roof installed parallel plate reactor. Bioresour Technol 130:152–160

Zhang J, Hu B (2012) A novel method to harvest microalgae via co-culture of filamentous fungi to form cell pellets. Bioresour Technol 114:529–535

Zheng H, Gao Z, Yin J, Tang X, Ji X et al (2012) Harvesting of microalgae by flocculation with poly (γ-glutamic acid). Bioresour Technol 112:212–220

Zhou WG, Cheng YL, Li Y, Wan YQ, Liu YH et al (2012) Novel fungal pelletization-assisted technology for algae harvesting and wastewater treatment. Appl Biochem Biotechnol 167:214–228

Chiranjib Banerjee received the M.Sc. degree in Biotechnology from Vidyasagar University, India, and Ph.D. from Department of Bioengineering, BIT Mesra, Ranchi, India. He has been awarded with the DST Inspire Faculty Scheme. His current interests are algal biofuel, algal harvesting, and molecular taxonomy. He is a life member of the Association of Microbiologists of India (AMI).

Rajib Bandopadhyay received his M.Sc. and Ph.D. degree (Molecular Biology) from University of Calcutta. He participated for Southern Ocean Expedition in 2011. He is presently serving as Associate Professor in the University of Burdwan, WB, India. He is recipient of BOYSCAST Fellowship of DST and Korean Research Fund Foreign Scientist Exchange Programme. His current interests are algal biotechnology and plant molecular biology.

Puneet Kumar Singh graduated in Bioinformatics in 2009 from Devi Ahilya University, Indore. He received his M.Sc. degree in Bioinformatics in 2011 from BIT Mesra, Ranchi, India. He is currently pursuing Ph.D. in Microbiology, Maharshi Dayanand University, Rohtak, India. His current research interests are in systems biology, protein engineering, and algal biofuel production.

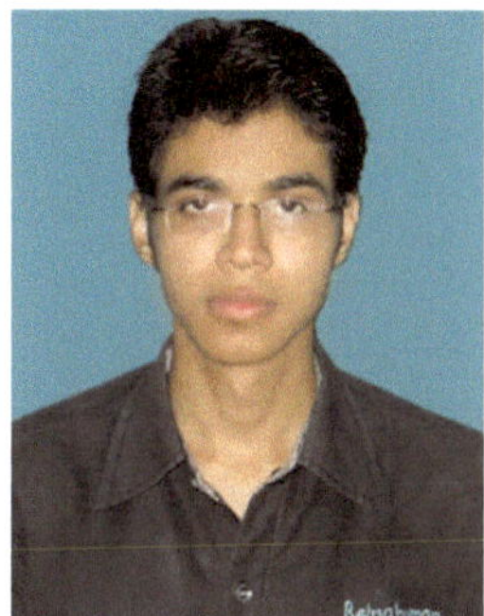

Harsh Kumar Agrawal is a graduate student B.E. (Biotechnology), from Birla Institute of Technology, Mesra, Ranchi, India. His research interests are algal harvesting and bioinformatics. He is currently engaged in final semester project work related to algal harvesting.

Pratyoosh Shukla received his M.Sc. degree (Applied Microbiology Biotechnology) from Dr. H.S. Gour University, Sagar, India, and Ph.D. degree in Microbiology and presently working as Professor and Head at Department of Microbiology, Maharshi Dayanand University, Rohtak, India. He was awarded with Indo-US Visiting Research Professorship by American Society of Microbiology (ASM) in 2014. His current interests are bioenergy, enzyme technology, and protein bioinformatics.

From Microbial Biopolymers to Bioplastics: Sustainable Additives for PHB Processing and Stabilization

10

Stefania Angelini, Pierfrancesco Cerruti, Barbara Immirzi, Merima Poskovic, Gabriella Santagata, Gennaro Scarinzi, and Mario Malinconico

Abstract

The term biopolymers refers to a broad class of materials that derive from naturally occurring resources. Biopolymers can be obtained through extraction from biomasses, but also through chemical or biotechnological methods from raw natural substrates. They are used to produce bioplastics, which could substitute fossil fuel-derived commodities. Among them, polyhydroxyalkanoates (PHAs) are polyesters synthesized by microorganisms as energy reserve. The most important member of PHA family is poly(3-hydroxybutyrate) (PHB). PHB is mechanically similar to polypropylene, even though its thermal instability, brittleness, and stiffness hinder its applicability. Improving PHB physical properties can be achieved by blending it with natural additives or by-products of industrial processes. This work takes the form of a case study about the effects of three natural, phenol-based, and polysaccharidic compounds on PHB properties. In particular, data on blending of two PHB matrices with a grape pomace extract (EP), a lignocellulosic biomass (LC), and tannic acid (TA) are reported. The preparation and characterization of PHB compounds and the effects of the additives on processing, thermal and photooxidative stability, crys-

S. Angelini (✉) • P. Cerruti • B. Immirzi
G. Santagata • G. Scarinzi • M. Malinconico
Polymer Science, IPCB-Institute for Polymers, Composites and Biomaterials, 80078 Pozzuoli (NA), Italy
e-mail: stefania.angelini@ictp.cnr.it; cerruti@ictp.cnr.it; barbara.immirzi@ictp.cnr.it; gabriella.santagata@ictp.cnr.it; gennaro.scarinzi@ictp.cnr.it; mario.malinconico@ictp.cnr.it

M. Poskovic
Polymer Science, Novamont, 28100 Novara, Italy
e-mail: merima.poskovic@novamont.com

V.C. Kalia (ed.), *Microbial Factories*, DOI 10.1007/978-81-322-2595-9_10

tallization rate, and microbial digestion of PHB are also shown. An overall improvement of polymer processability and photostability, along with changes in crystallization rates, was observed. The study provides evidence that natural additives have the potential for promoting the transition from biopolymers to bioplastics in a sustainable way, both from an environmental and economical point of view.

10.1 Introduction

The use of synthetic polymers, derived from oil-based resources and aimed to the production of plastics, is widely spread in the fields of medicine, food packaging, agriculture, cosmetics, furniture, composites, etc. (Lee et al. 2011; Kasirajan and Ngouajio 2012). Features as durability, toughness, and elasticity have provided great benefits to modern society but have also raised significant economical and ecological issues. Indeed, most plastics exhibit a remarkable resistance to biodegradation once landfill is disposed and, in some cases, their recycling is not economically sustainable (Dintcheva et al. 1997; Kim et al. 2001; Guo and Xu 2009). Hence, due to the rising environmental awareness, a large and growing body of studies has been devoted to the development of novel and sustainable polymers based on renewable resources (Scott 2000). Biopolymers are a class of materials that derive from naturally occurring supplies (Niaounakis 2013). This definition refers to a heterogeneous group of bio-based, mostly biodegradable polymers; a first classification is reported in Fig. 10.1.

Currently, a univocal method of classification for this category of macromolecules does not exist. In this chapter the biopolymers will be classified on the basis of the approach used to produce them. The main methods for the production of these materials can be divided into extraction and synthetic routes. The first approach regards the extraction of the product from biomasses and dedicated crops, agro-industrial and municipal wastes, and, last but not least, microorganisms. The second approach involves the chemical/biotechnological polymerization of bio-derived monomers. Among the more common biopolymers, we can list polysaccharides (alginate, starch, and starch blends such as Mater-Bi® or Solanyl®, chitin), vegetable and animal proteins, nucleic acids, lipids, polylactic acids (PLAs), polyhydroxyalkanoates (PHAs), and bio-based polyethylene terephthalate (PET) or polyethylene (PE), even if the latter two are not biodegradable. Biopolymers, with respect to their petrochemical-based counterparts, are generally more sustainable (Mac Gregor 2001). The use of these materials is aimed to the production of bioplastics. Their main advantage, besides their natural origin, is the possibility to be disposed as compost. This point is particularly attractive under the environmental perspective as it contributes to solve one of the main problems of traditional plastics, that is, their lack of biodegradability.

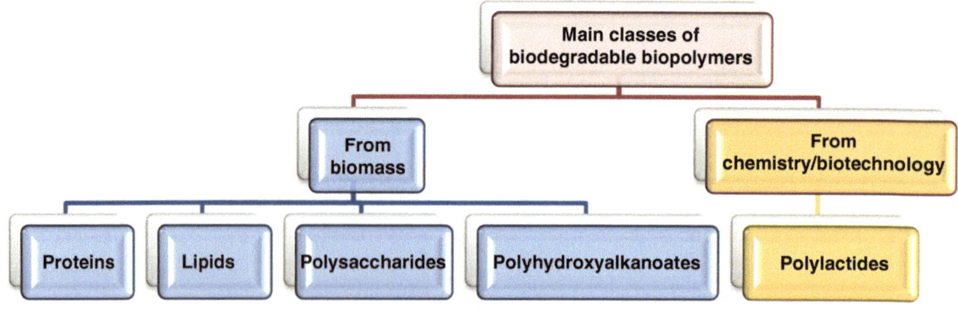

Fig. 10.1 Different classes of biodegradable biopolymers

10.2 The Blending Technique: A Strategy to Obtain Bioplastics from Biopolymers

In many cases, biopolymers are not able to fully replace synthetic polymers. This is due to their poor chemical–physical and mechanical properties, along with difficulties in processability (Van de Velde and Kiekens 2002; Mitra et al. 2014). These shortcomings strongly restrict the potential technological applications of such materials. A typical way developed to modify biopolymer physical properties consists in blending (Mohanty et al. 2002). The blending method is a cheap and simple technique, based on mixing polymers with additives that can act as processing aids, plasticizers, and UV or thermo-oxidative stabilizers. The blending technique is often used to modify the properties of biopolymers with the aim of widening their field of applicability (Marsano et al. 2004; Sionkowska 2011; Wang et al. 2011). Starch-based blends represent a typical example of this approach. Natural starch is one of the most studied storage polysaccharides (Ellis et al. 1998). It is present in the form of crystalline beads in plants as corn, maize, potato, and other crops. Structurally, it consists of amylose and amylopectin, a linear and a branched polymer, both constituted by D-glucose repeating units linked together by α-1,4 and α-1,6 linkages (Buléon et al. 1998; Miao et al. 2015).

Several studies have been devoted to the conversion of starch into a thermoplastic material (Kaseem et al. 2012; Rosa et al. 2009). Most of the proposed methods are based on the application of heat and shear forces or the addition of plasticizers such as glycerol. The purpose is to demolish the crystalline structure and weaken the intermolecular interactions existing between starch chains (Bastioli 1998; Stepto 2006). After this treatment, starch can be processed like other thermoplastics in typical injection-molding machines and extruders (Mahieu et al. 2013). The aim of this physical manipulation of the polymer is to produce processable materials suitable for rigid/flexible packaging. Nevertheless, the applications of thermoplastic starch (TPS) are limited due to humidity sensitivity and retrogradation, which cause surface stickiness as well as poor mechanical and barrier properties (Thunwall et al. 2008). Therefore, plenty of studies have been dealing with blending TPS (usually from 20 % to 80 %) with biopolymers as chitosan (Pelissari et al. 2012), cellulose and proteins (Gáspár et al. 2005), PHAs (Yu 2009), and PLAs (Ferreira et al. 2015) with the aim of widening the application window of TPS. In particular, much effort has been made to develop TPS/PLA blends (Wang et al. 2007). The low interfacial adhesion between hydrophilic starch and hydrophobic PLA has been recognized as the major problem to be addressed. To promote compatibility of these two materials, compatibilizers such as maleic anhydride are often used (Huneault and Li 2007). Another example of blending is reported in the recent study carried out by Ortega-Toro et al. (2014), who examined the effect of hydroxypropyl methylcellulose on the microstructure, thermal, and physical properties of glycerol-plasticized starch films in the presence of citric acid. The authors recorded a slowdown of starch retrogradation, due to the addition of hydroxypropyl methylcellulose and citric acid.

Blending TPS with fossil fuel-based polymers as PVOH or PCL has also been tested (Kalambar and Rizvi 2006). This successful approach has been adopted by several companies as Rodenburg (Netherlands), Biotec (Germany), Limagrain (France), Livan (Canada), and Novamont (Italy), resulting in a number of commercially available products. As an example, the European leading company Novamont offers four classes of biodegradable materials under the Mater-Bi® trademark (Shen et al. 2009); all of them were based on starch and synthetic components. The four classes are listed below (Bastioli 1998):

- Class Z: biodegradable and compostable blends made of TPS and PCL for films and sheets
- Class A: biodegradable and not compostable blends made of starch and ethylene vinyl alcohol copolymers for molded items,
- Class Y: biodegradable and compostable blends made of TPS and cellulose derivatives for rigid injection-molded items

- Class V: biodegradable, compostable, and soluble materials for rigid and expanded articles (TPS > 85 %)

Although Mater-Bi® is used in packaging, agriculture, or consumer goods, its properties are often unsuitable for commercial purposes (Briassoulis 2007). In order to improve Mater-Bi® properties, several approaches have been pursued. In particular several studies have attempted to produce Mater-Bi®-based composites using cellulose fibers as fillers in order to improve processability and chemical–physical performances of the material (Alvarez et al. 2006). It is worth noting that the addition of the filler produces an enhancement of mechanical properties and water barrier capability. In recent years Cerruti et al. (2011) have used a polyphenolic extract (EP), the biowaste of winery production, as an additive for Mater-Bi. The authors found that EP behaved as a plasticizing agent and as a modifier of Mater-Bi processing, mechanical and thermal properties. Moreover, it also produced a decrease in the microbial digestion rate.

Another class of promising biopolymers is represented by PLAs, a family of biodegradable, biocompatible, and nontoxic polyesters based on lactic acid as monomer. Lactic acid can be obtained by the fermentation of various hexoses. These sugars are produced by the hydrolysis of several polysaccharides, including starch, cellulose, and lignocellulosic materials (Coelho et al. 2010). Although PLAs have a wide range of applications, they biodegrade slowly in soil. This feature limits their use in agriculture. Therefore, PLA blending represents a useful method that allows to overcome this shortcoming. It was shown that the addition of useful fillers can enhance the biodegradability of the matrix and improve its mechanical properties as well. Model examples are those reported by some authors who studied blends of PLAs with starch (Lv et al. 2015; Guzman-Sielicka et al. 2013) and chitosan (Nugraha et al. 2004).

Another commercial member of the polyester class is Ecoflex®. This is the registered trademark of a family of aliphatic–aromatic biodegradable polyesters, which are produced by BASF. The original Ecoflex® grade is a fossil-based copolyester that originates from the reaction between 1,4-butanediol, adipic acid, and terephthalic acid. Blending Ecoflex® with biological polymers such as starch or poly(lactic acid) (PLA) provides biodegradable bioplastics with interesting technological properties (Pack et al. 2012). Compounds of Ecoflex® and PLA are produced by BASF under the registered trademark Ecovio®. Ibrahim (2009) described the preparation and characterization of new biodegradable composites made by Ecoflex with kenaf fibers as filler. The effects of the filler loading on the mechanical and thermal performances were examined, and an enhancement of tensile/flexural properties was observed.

10.3 The PHA Family: A Particular Focus on Poly(3-hydroxybutyrate) (PHB)

PHAs are a family of microbial polyesters with a wide range of properties that find application as biodegradable and biocompatible thermoplastic polymers (Riggi et al. 2011; Zheng et al. 2015). They are produced by microorganisms and accumulated in microbial cytoplasm as energy storage materials. An alternative route for the production of PHAs is based on the use of algae. This approach is still at an experimental level but can contribute to solve some of PHA restrictions, such as the high cost. Generally, these production processes are carried out in tanks, and they consist in two stages, the first one in which the algae growth is promoted by optimal conditions and the second step where the algae begin to accumulate PHAs as secondary metabolites. A deeply studied member of PHAs is poly(3-hydroxybutyrate) (PHB). PHB is a linear polyester, homopolymer of 3-hydroxybutyrate, which is accumulated in the cytoplasm of a wide variety of gram-positive and gram-negative microorganisms under physiological stress (Dawes and Senior 1973). PHB is a thermoplastic polymer that shows biodegradability in compost and in different environments, such as marine water (Volova et al. 2010). Due to its mechanical and barrier characteristics, which are similar to those of oil-based polymers such as polyethylene, polypropylene, and polyethylene terephthalate, this material can be proposed as a substitute for petroleum-derived plastics. Nevertheless, the diffusion of PHB as a commodity polymer is hampered by its brittleness (with elongation at break below 15 %) and excessive stiffness (Mekonnen et al. 2013). This last shortcoming is the result of the recrystallization phenomenon and physical aging that the material undergoes at room temperature (de Koning and Lemstra 1993). Moreover, another serious drawback of PHB is represented by its thermal decomposition at temperatures just above the melting point (~175 °C) (Chodak 2002). This phenomenon strongly restricts its processing window and limits its workability. Both the recrystallization phenomenon and polymer degradation during processing can be prevented and reduced by the addition of nucleating agents or lubricants (Bugnicourt et al. 2014). In addition to the problems described above, another strong limitation to the commercial diffusion of PHB is represented by its high production costs. One of the main cost factors, related to PHB production, is due to high price of the carbon substrates. Generally, the substrates are pure carbohydrates as glucose or sucrose with competing food value. Currently, in order to reduce PHB costs, the use of less expensive raw materials has been considered. These new substrates include biomasses from the maintenance of green spaces, wastes, and coproducts of industrial processes such as glycerol, sugarcane bagasse, and lignocellulosic feedstocks from agricultural or forestry residues. The major advantage of this approach is the conversion of these wastes into value-added products. Another route to reduce PHB costs and to widen its applicability field is represented by the blending of this polyester with suitable fillers and additives.

In this regard, blends of PHB with synthetic polymers have been reported since the early industrial availability of PHB from ICI Chemical Industries. Since then, a considerable amount of literature has been published on the formulation and characterization of PHB/polymer blends. The effect of the polymeric second component on blend miscibility, mechanical behavior, morphology, and crystallization rate of the PHB matrix has been largely investigated. From around the 1990s, a great deal of scientific papers concerning polymer blends based on PHB and poly(3-hydroxybutyrate-hydroxyvalerate) (PHBV) has appeared. The interest was mainly focused on blending these natural polymers with polyolefins such as poly(vinyl alcohol) (PVA) (Greco and Martuscelli 1989), polyethers as poly(oxyethylene) (PEO) or poly(methyleneoxide) (POM) (Avella and Martuscelli 1988; Avella et al. 1997), polyesters as polycaprolactone (PCL) (Kumagai and Doi 1992; Immirzi et al. 1994), and polyacrylates as poly(methyl methacrylate) (PMMA) (Lotti et al. 1993; Yoon et al. 1993) or poly(butyl acrylate) (Immirzi et al. 1994). Another field that attracted a great deal of interest was the blending of PHBV with polysaccharides as cellulose and starch derivatives (Lotti and Scandola 1992; Buchanan et al. 1992; Shogren 2009) or natural fibers as wheat straw or hemp (Paramasivan and Abdul Kalam 1974; Felix and Gatenholm 1991). In the last few years, the scientific interest has been devoted to blending PHB with natural additives or industrial biowastes. The latter perspective deserves particular attention, because the disposal of industrial by-products represents a significant environmental issue (Rossini et al. 2013). The incorporation of these biowastes into a biopolymer offers the possibility to modify the matrix properties and, at the same time, to address an environmental concern. In this frame, several authors investigated the effects of the addition of natural fibers such as coconut, hemp, jute, and flax into PHB matrix (Gunning et al. 2013). In particular, Macedo et al. (2010) used coir dust, an abundant and cheap biowaste that derives from the fiber separation process of the coconut, as filler material for PHB matrix. The addition of the bio-charge produced an improvement of stress–strain properties and thermal stability. Other examples of blending PHB with natural additives are also reported in the literature, and some authors investigated the effects of chitin and chitosan (Ikejima and Inoue 2000); starch and its derivatives, in the form of starch–adipate or grafted starch–urethane (Innocentini-Mei et al. 2003); and soda lignin extracted from bagasse (Mousavioun et al. 2010). Biodegradability of PHB/chitin and PHB/chitosan films was found to be enhanced. PHB-based blends with starch and starch derivatives were found to be more brittle than neat PHB, but also characterized by an improved processability due to the reduction of the polymer melting temperature. Soda lignin improved the polymer thermal stability, owing to the intermolecular interactions established between PHB and lignin.

In the next paragraph, the effects of blending PHB with organic renewable fillers, some of which are biowastes of industrial productions, will be discussed in depth.

10.4 Case Study: Use of Grape Pomace Extract, Lignocellulosic Biomass, and Tannic Acid as Additives in PHB

In the subsequent sections, the outcomes acquired on the processing, thermal stability, photodegradation, and microbial digestion of two grades of PHB filled with a pomace extract (EP), a lignocellulosic biomass (LC), and the natural polyphenolic tannic acid (TA) are reviewed in the perspective of tailoring several PHB properties relevant for its technological application. Before discussing blend preparation and characterization, a focus on the properties of the additives is given.

10.4.1 Chemical and Physical Properties of the Natural Additives

In this paragraph the characterization of three materials deriving from renewable resources, used as additives in two PHB matrices, will be discussed. These materials are:

- A dried hydroalcoholic extract from grape pomace extract (EP)
- A lignocellulosic biomass (LC)
- Tannic acid (TA)

Their visual appearance is showed in Fig. 10.2. EP is a dark purple and granular soft paste, while LC and TA are two fine powders, brownish and off-white, respectively.

EP is obtained from a dried Cabernet pomace through a high-pressure hydroalcoholic extraction (Persico et al. 2012). Briefly, a mixture of 70:30 (v/v) ethanol–water was used, adjusting the solution pH at 3 and using a volume ratio between pomace and extracting solvent of 1:3. After solvent evaporation, a soft paste (water amount of 20 wt%) is recovered. From the analysis of the dried extract, a content in simple carbohydrates and polysaccharides equal to 79.0 ± 4.2 wt% (Dubois 1956) and a phenol amount of 4.3 ± 1.2 wt%, mostly composed by catechin derivatives, was found (Garaguso and Nardini 2015). The remaining fraction was mainly composed of organic acids derived from tartrates and malates (Kammerer et al. 2004). Some of the

Fig. 10.2 An optical image of EP (*left*), LC (*middle*), and TA (*right*)

Fig. 10.3 Typical components found in the grape pomace extract: (**a**) tartaric acid, (**b**) pectins, (**c**) catechins, and (**d**) malic acid

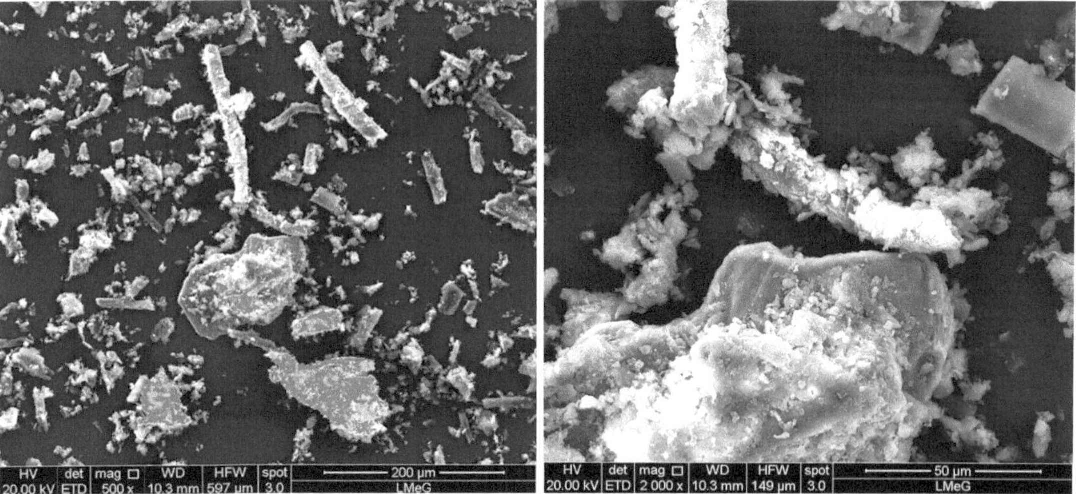

Fig. 10.4 SEM micrographs of LC at two different magnifications

typical components present in EP are illustrated in Fig. 10.3.

LC is a lignocellulosic biomass obtained as a biowaste from the second-generation bioethanol production process. The definition "second generation" refers to the use of fermentable substrates not interfering with food industry, such as agricultural residues and by-products (Wyman 1996). LC is produced using Arundo donax stems as substrates, and it is recovered as a solid waste after the fermentation process. Due to mild processing conditions upon fermentation, the highly crystalline cellulose fraction is not completely converted, so that LC still contains an amount of polysaccharide component. In fact, the scanning electron microscopy (SEM) observation of LC (Fig. 10.4) reveals the heterogeneous morphology of the biomass characterized by the presence of irregularly shaped lignin particles along with some cellulose/hemicellulose fibrous material. The lignin content of LC was investigated by means of the Klason method (Angelini et al. 2014), and an acid-insoluble lignin content of 56 wt% was found.

TA is the commercial form of tannin, one of the most important polyphenolic ingredients in the bark of trees, where it provides thermal and antimicrobial protection (Larif et al. 2013). TA and other polyphenols have also antimutagenic, anticarcinogenic, and antioxidant activities (Lopes et al. 1999). Tannins can be classified into two categories: hydrolysable and nonhydrolysable or condensed tannins. TA consists of a central carbohydrate (glucose) surrounded by a number of galloyl groups that range from 2 up to 10 according to the plant source (Fig. 10.5) (Sahiner 2015).

As underlined before, the aim of the work is to evaluate the effect of these natural substances as additives in PHB. This polymer is processed at a temperature, around 200 °C, which can trigger thermal degradation phenomena of the organic molecules. Therefore, the thermal behavior of the additives is an important parameter to be considered when PHB-based blends are prepared. The thermal behavior of EP, LC, and TA was investigated through two thermal characterization techniques: thermogravimetric analysis (TGA) and differential scanning calorimetry (DSC) under nitrogen. The TGA thermograms and the DSC second heating scan are shown in Fig. 10.6a, b, respectively.

Table 10.1 lists temperatures of onset (T_{onset}), degradation peak in the DTG curves (T_{DTG}), glass transition (T_g), and residue at 700 °C. EP weight loss started right above 100 °C and a steep decrease was noticed until 500 °C. 70 % of initial weight was lost in the temperature between 150 °C and 500 °C, and a char value of 23 wt% was recorded at 700 °C. It is likely that the fast degradation of EP is due to the large amount of carbohydrate structures, which undergo massive dehydration processes at temperatures lower than 200 °C. LC appeared to be more stable than EP, as demonstrated by the weight loss onset shifted toward significantly higher temperatures. LC degradation started at around 250 °C, and at 700 °C, a char of 36 wt% was calculated. The higher onset temperature of this material was related to the presence of a remarkable aromatic fraction, which degraded more slowly due to a low content of free hydroxyls. The major degradation step of tannic acid occurred from 230 °C to 350 °C and a char yield of 27 % was recorded. At the end of the thermal analysis experiment, a sort of carbonaceous aerogel structure formed due to the dehydration of hydroxyl groups of tannic acid. As shown by Xia et al. (2015) under air, tannic acid burns almost completely leaving only 1.4 % of char residue.

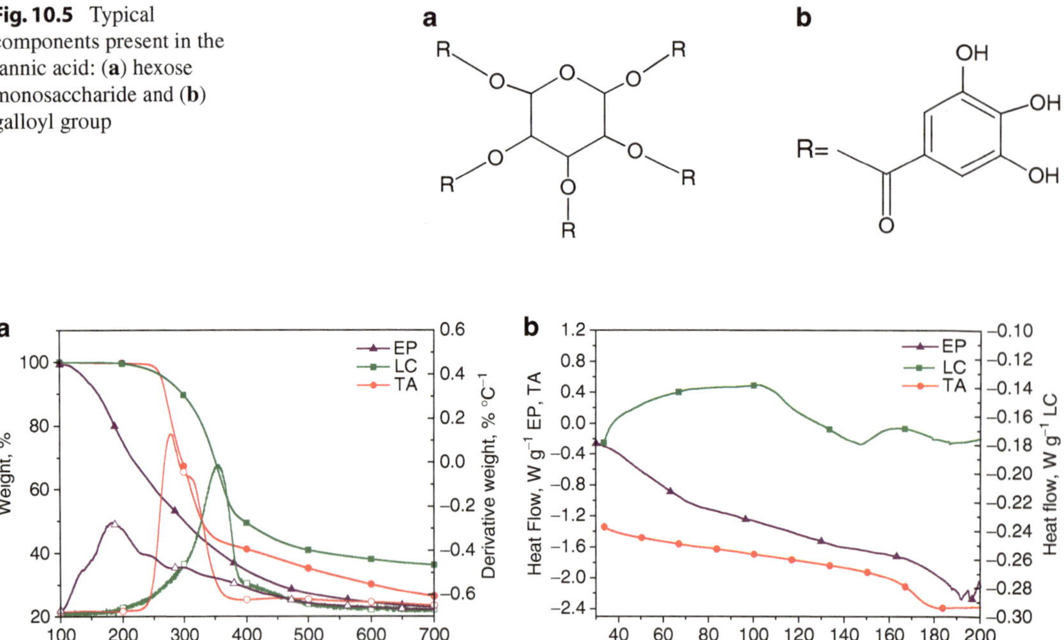

Fig. 10.5 Typical components present in the tannic acid: (**a**) hexose monosaccharide and (**b**) galloyl group

Fig. 10.6 (**a**) TGA (*full symbols*) and DTG (*empty symbols*) curves and (**b**) DSC second heating curve of EP, LC, and TA

DSC curves of EP, LC, and TA were characterized by an irregular baseline and did not display any signals due to melting. A very broad endotherm step ascribed to glass transition of LC was visible at temperatures higher than 100 °C. Since the glass transition temperatures were not clearly appreciable, their values were assessed using the first derivative of the DSC traces (Gordobil et al. 2014). By means of this procedure, a T_g value of 113 °C was obtained. EP displayed a softening temperature at about 50 °C and started decomposing at 180 °C as also indicated by TGA, while the thermogram of TA was more regular, evidencing a relaxation phenomenon at 171 °C.

10.4.2 Preparation and Characterization of PHB/EP and PHB/TA Blends and PHB/LC Biocomposite

Two bacterial poly(3-hydroxybutyrate) (PHB) grades, coded T3 and T19 (supplied by Biomer, Germany), were used after drying under vacuum in oven. The polymers' average molecular weights (M_w), as determined by gel permeation chromatography (GPC), were 840 and 890 Kg mol^{-1}, respectively. EP was added in PHB-T3, while LC and TA were added in PHB-T19, at 15 wt% for each additive. The vacuum oven-dried additives were dissolved (EP and TA) or suspended (LC) in methanol. A proper amount of PHB was added to the resulting mixtures, and then the formulations were held at 50 °C under stirring for solvent evaporation. They were then dried under vacuum until a constant weight was achieved. The samples were then milled in a mortar to a fine powder. The three obtained formulations will be referred to as PHB/15EP, PHB/15LC, and PHB/15TA in the following. In the case of PHB/15EP, dumbbell-shaped specimens were obtained by means of an injection-molding equipment, in which the material was processed at 200 °C for 5 min. Conversely, PHB blended with LC was compression molded at 180 °C in order to obtain films 150 μm thick. Both EP and TA were homogeneously and finely blended in PHB, while LC particles were dispersed in the matrix, acting as a filler fraction. A proof of this statement is given by SEM micrographs of the fracture surfaces of neat PHB, PHB/15LC, and PHB/15EP, shown in Fig. 10.7. It is remarkable to note that the surface of neat polymer is smooth due to the brittle behavior of PHB, while it appears to be rough and more ductile in the presence of LC and EP. In particular, a biphasic system consisting of a filler fraction dispersed in a polymeric matrix was evident for the PHB/15LC biocomposite. Lignocellulosic fibers and particles as large as 50 μm were embedded in the PHB matrix, suggesting the presence of weak interactions at the interface between the filler and the polymer. Conversely, PHB/15EP appeared characterized by several globular particles of size

Table 10.1 Thermal parameters of EP, LC, and TA under nitrogen

Sample	T_{onset} (°C)	T_{DTG} (°C)	T_g (°C)	$Char_{700\,°C}$ (%)
EP	148	182	50	23
TA	239	277	171	27
LC	289	354	113	36

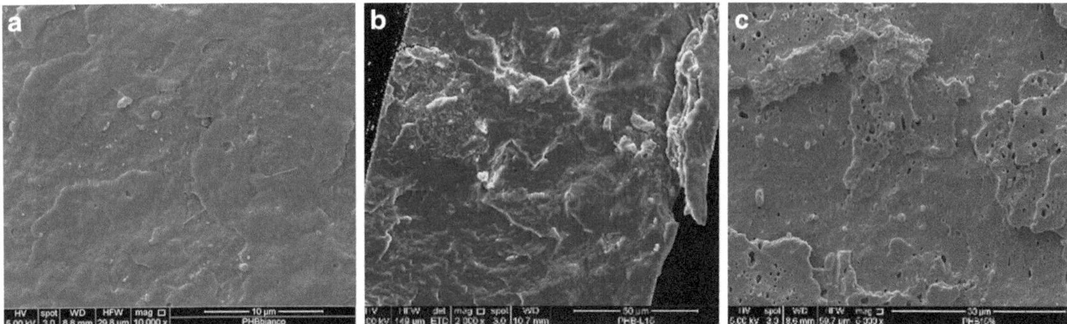

Fig. 10.7 Fracture surfaces micrographs of (**a**) neat PHB, (**b**) PHB/15LC, and (**c**) PHB/15EP

Table 10.2 M_w of unprocessed PHB-T19 compared with that of processed PHB-T19, PHB/15TA, and PHB/15LC held at 190 °C for 4000 s

	Unprocessed	Processed		
	PHB-T19	PHB-T19	PHB/15TA	PHB/15LC
M_w (Kg mol^{-1})	890	47	50	34

lower than 5 μm well dispersed in the polymeric matrix. Moreover, the observation of several voids on the surface was attributed to the EP granules pullout caused by the cryogenic fracture.

10.4.3 Effect of the Additives on Processing and Thermal Stability of PHB

It has been demonstrated that high temperatures adopted upon PHB processing are responsible for a dramatic change in molecular weight (Vogel et al. 2007), due to the thermally triggered polymer chain scission. In this case, the effect of processing was evaluated by holding PHB-T19, PHB/15TA, and PHB/15LC samples for a time period of 4000 s in the chamber of a rotational rheometer at 190 °C. GPC measurements were performed dissolving PHB-based samples in chloroform, and molecular weights were measured before and after the thermal treatment (Table 10.2).

Neat PHB-T19, PHB/15LC, and PHB/15TA underwent massive thermal degradation when held at 190 °C for long times. Therefore, the addition of such fillers did not improve the polymer thermal stability. GPC analysis was also performed on unprocessed and injection-molded PHB-T3 blends with EP. M_w of unprocessed PHB-T3 (840 Kg mol^{-1}) was significantly higher than that of the injection-molded counterpart (134 Kg mol^{-1}), indicating that even processing times as long as 5 min bring about thermal degradation (Chen et al. 2013), causing the molecular weight to decrease significantly. In this regard, it is worth noticing that the addition of 15 wt% EP was able to efficiently hinder polymer chain scission during processing, since a M_w value of 648 Kg mol^{-1} was recorded for PHB/15EP. This finding reveals the efficiency of EP as processing stabilizer. Such

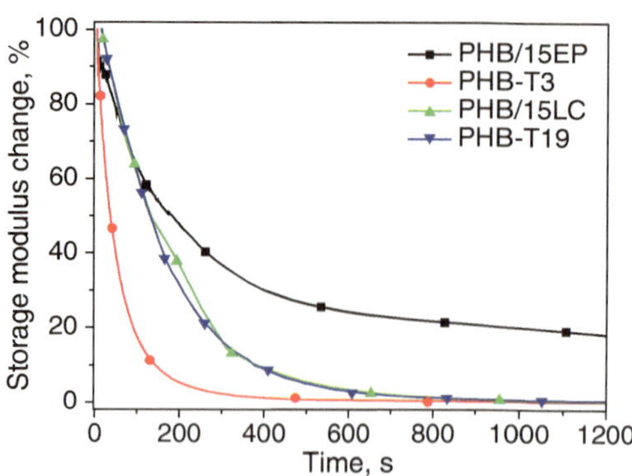

Fig. 10.8 Percent change of storage modulus versus time at 10 rad s^{-1} for the PHB-based samples at 190 °C

aspect was investigated in detail by measuring the rheological properties of PHB/15EP, PHB/15LC, and PHB/15TA melts. Actually, the rheological properties are strongly related to changes in molecular weight, entanglement density, effect of additives, and fillers. The rheological behavior of PHB was measured by means of parallel-plate rotational rheometry under oscillatory conditions at 190 °C. As the linear viscoelastic properties in dynamic tests are sensitive to polymer chain scission, valuable insight can be gained about polymer stability under these severe thermal conditions. Figure 10.8 displays the percent change of storage modulus G' of the PHB-based samples as a function of time.

From this figure, it can be noticed that, due to thermal degradation, the G' values relative to the two PHB grades dropped down dramatically. However, T3 resulted to be more sensitive to high temperature than T19. It can also be observed that the presence of the lignocellulosic filler had no significant impact on the course of the process. A different finding was obtained for PHB/15EP, as the decrease of G' was significantly slowed down, confirming that EP improved the polymer thermal stability.

According to these data, it can be supposed that EP can stabilize PHB thermally; thereby, the polymer can withstand the residence times usually adopted in common processing machines. A similar result was observed with the addition of tannic acid, which determined higher viscoelastic properties and clearly delayed the thermal degradation process to longer times (Auriemma et al. 2015). From the literature data, the mechanism for PHB decomposition consists in a statistical chain scission occurring by means of a β-elimination reaction, which generates crotonic acid and oligomers with a crotonate end group as thermal degradation products (Ariffin et al. 2008). This process is characterized by a stage in which a pseudo-six-membered structure is formed. In this structure, the carbonyl oxygen abstracts hydrogen ion from the γ-methylene. Therefore, any species bearing acidic hydrogen ions may interfere with this autocatalytic cycle, thus affecting the polymer thermal stability. EP and TA possess a number of phenolic OH groups able to act as proton donors: it can be claimed that PHB C=O oxygen atoms interact with the hydrogen atoms of OH (Chen et al. 2002). These hydrogen bonds are likely to interfere with the synthesis of the transition structure. Such interaction validates EP and TA capability of delaying the polymer degradation process. The physical bondings between TA hydroxyls and PHB carbonyls are confirmed by the results of FTIR analysis. Figure 10.9 shows the result of the spectral subtraction between PHB/15TA and PHB-T19 spectra acquired at 190 °C.

Herein, in addition to the obvious increases due to the TA absorption peaks at about 3400 cm^{-1} (phenolic O–H), 1610 cm^{-1} (resonance of the aromatic C=C of TA), and 1520 cm^{-1} (in-plane bending of phenyl C–H bonds), a decrease of the 1740 cm^{-1} ester band of PHB is evident, accompanied by the appearance of a very strong increase of the signal at 1705 cm^{-1}. This finding is related to the formation of intermolecular hydrogen bonds between the carbonyl of PHB and the hydroxyl of TA, which is responsible for the band shift toward lower frequencies, analogous to what is observed for catechin-PHB blends (Li et al. 2003).

In Fig. 10.10, the TGA thermogram of PHB-based specimens under nitrogen is reported. The main thermal parameters of the studied samples are reported in Table 10.3. Neat PHB-T3 and PHB-T19 degraded through a single weight loss step; however, the T19 grade was endowed with a higher thermal stability with respect to T3. In fact, a shift of about 15 °C exists between the two thermogravimetric curves. This is particularly remarkable, since the two polymers had comparable molecular weight and crystallinity. It has been recently reported that acidic treatments can increase thermal stability of PHB (Arza et al. 2014). In this regard, it is then likely that the extraction protocol and work-up during manufac-

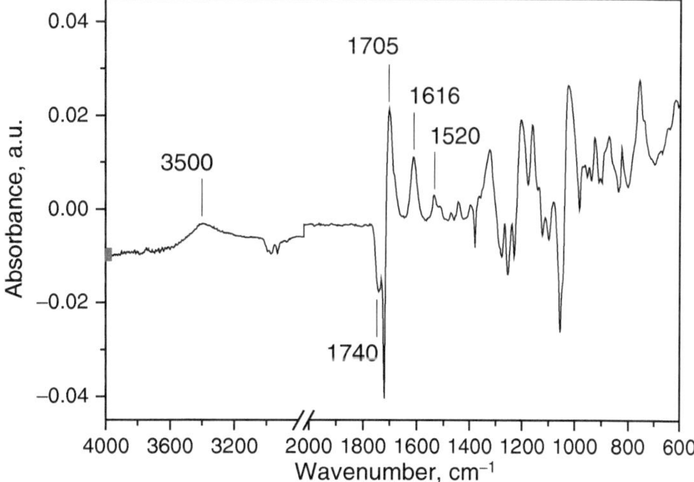

Fig. 10.9 Spectral subtraction between PHB/15TA and PHB-T19 FTIR spectra acquired at 190 °C

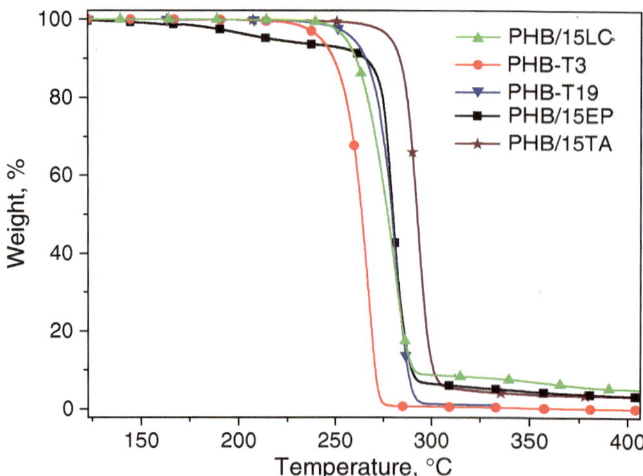

Fig. 10.10 TGA of neat PHB-T3 and T19 and of their compounds with EP, LC, and TA

Table 10.3 Thermal parameters of PHB and PHB compounds measured by TGA

	T_{onset} (°C)	T_{DTG} (°C)	Char $_{400\,°C}$ (%)
PHB-T19	264	282	0
PHB-T3	252	266	0
PHB/15EP	272	280	3
PHB/15LC	259	280	5
PHB/15TA	283	293	3

turing operations play a role. From the figure, it is also clear that both the phenol-containing additives own the potential to increase the thermal stability of the polymer. An increase in degradation temperature by about 15 °C was observed for PHB/15EP and PHB/15TA, with respect to the parent PHB-T3 and T19, respectively, indicating a relevant stabilization effect ascribed to the additives rich in phenols.

On the other hand, from the figure, the effect of LC on PHB-T19 reveals that this additive has no influence on the thermal stability of the microbial polyester. This finding accounts for the importance of the physical state and dispersion of the additive. In fact, SEM observations (Fig. 10.7) showed that LC was not actually blended with PHB, since large insoluble particles of LC were visible in the PHB matrix. The absence of filler–matrix interactions at a molecular level was also confirmed by FTIR spectroscopy. In fact, notwithstanding the presence of phenolic hydroxyls in the lignin fraction of LC, no hydrogen bonding was detected by comparing the FTIR spectrum of PHB/15LC with that of PHB-T19 (Angelini et al. 2014).

10.4.4 Effects of EP, LC, and TA on the Crystallization of PHB

Both the processing parameters and mechanical properties of polymeric materials depend on the crystallization behavior. Moreover, the knowledge of crystallization mechanisms is important for the design of materials with tailored properties. This is particularly relevant in applications in the field of medicine and packaging. In our case, PHB shows a low nucleation density. This results in the growth of very large spherulites that are responsible for the brittleness of this polymer (El-Hadi et al. 2002a).

The crystallization kinetics of PHB and its blends was investigated by differential scanning calorimetry (DSC). Figure 10.11 reports the DSC curves of the cooling step (rate −50 °C min^{-1}) after fusion and the second heating scan (rate 10 °C min^{-1}) of PHB-T3 and PHB-T19 and their compounds with EP and LC, respectively. The thermal data recorded through DSC measurements are listed in Table 10.4.

The experimental results show that the process of crystallization depends upon the presence and type of additive. Neat PHB-T3 (Fig. 10.11a) had a melt crystallization at 59 °C; on the contrary, PHB/15EP did not show any specific melt-crystallization signal. This indicates that EP is responsible for delaying the process during rapid cooling, restricting the mobility of polymer chain (Mousavioun et al. 2010). This result confirms that EP enhances macromolecule entanglements,

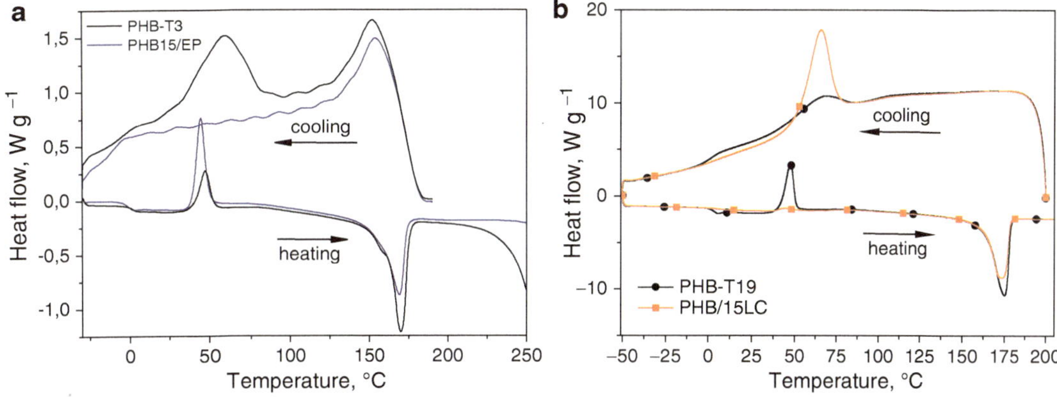

Fig. 10.11 DSC of (**a**) neat PHB-T3 and PHB/15EP and (**b**) neat PHB-T19 and PHB/15LC. Only the cooling and the second heating ramps are reported

Table 10.4 Thermal properties of PHB-based compounds measured by DSC

	$T_{c, melt}$ (°C)	$\Delta H_{c, melt}$ (J g^{-1})	T_g (°C)	$T_{c, cold}$ (°C)	$\Delta H_{c, cold}$ (J g^{-1})	T_m (°C)	ΔH_m (J g^{-1})
PHB-T19	67	12.0	4	47	68.7	175	87.4
PHB-T3	59	27.2	2	47	17.6	170	74.0
PHB/15EP	n.d.	n.d.	−1	45	40.7	169	66.6
PHB/15LC	65	42.0	5	44	26.1	173	84.3
PHB/15TA	n.d.	n.d.	3	75	–	172	–

which allow to get enhanced viscous melts. This results in the delay in the chain rearrangements during the melt crystallization. The DSC second heating ramp showed for both samples a notable glass transition and a cold-crystallization exotherm, which were followed by a single melting endotherm. The introduction of additive, which acted as a mild plasticizer, affected the glass transition temperature of the polymer. PHB/15EP exhibited a cold-crystallization peak in the temperature range, which varied from 50 to 45 °C. The slow heating process promotes crystallization of the amorphous part of PHB/15EP, which was not able to crystallize while the rapid cooling process was taking place. A very similar result was recorded in the presence of TA (Auriemma et al. 2015), which also restricted the crystallization of PHB from the melt, giving rise to a less-ordered crystalline form. Moreover, during subsequent heating, the cold-crystallization peak was detected at a temperature higher by about 30 °C when compared to the neat polymer. Angelini et al. (2014) found a similar behavior, adding a commercial alkali lignin (AL) in PHB. They recorded that AL did not promote melt crystallization of PHB, while in the second heating scan, the glass transition was found to be easily detectable. This outcome was ascribed to the pro-degrading effect of AL on PHB, which provokes depolymerization and reduction of the molecular entanglements. In DSC traces relative to PHB and PHB/15EP, at temperatures above 200 °C, a remarkable effect was observed. For neat PHB, a drop in the thermogram suggested that a thermal degradation is going on. On the contrary, in the blend a steady signal was recorded. These features confirm that the additives help in enhancing the thermal stability of PHB.

Figure 10.11b reports the DSC response of PHB-T19 and PHB/15LC. The cooling ramp subsequent to the first heating run and the second heating scan is displayed. It is worth highlighting that neat PHB-T19 displayed a small, broad peak of melt crystallization, since the rapid cooling inhibited chain mobility and hindered the crystallization process of the polymer. El-Hadi et al. (2002b) showed that the broadened signal of melt crystallization during a fast dynamic cooling scan could be ascribed to the formation of spherulites of large size and low density, representative of bacterially synthesized PHB (El-Hadi et al. 2002b). In the second heating step, the PHB-T19 glass transition was clearly evident, thus confirming that most of the amorphous phase did not crystallize upon cooling. Nevertheless, the slow heating of the second scan induced the formation of crystallites, as shown by the presence of a sharp cold-crystallization peak, followed by a melting endothermic signal. It is interesting to emphasize that the introduction of LC strongly influenced the crystallization behavior of PHB. In particular, it was evidenced that during cooling, the melt crystallization of PHB was markedly enhanced by the presence of the filler to an extent proportional to its amount (Angelini et al. 2014). Literature data suggest that melt- and cold-crystallization temperatures are indirect parameters of the crystallization rate. Generally, a decrease of $T_{c, cold}$ indicates faster crystallization, whereas lower $T_{c,melt}$ indicates slower crystallization (Weihua et al. 2004). From the data recorded for PHB/15LC, a drop of cold-crystallization enthalpy ($\Delta H_{c, cold}$) values in PHB/LC-based biocomposites confirms that the crystallization process mainly occurred during the cooling scan. These results indicate that LC was a heterogeneous nucleating agent, potentially able to control the physical aging of the polymer (Weihua et al. 2004). Finally, it should be observed that T_m values of PHB-T19 and PHB/15LC were not significantly different, suggesting that the introduction of the filler as a nucleating agent did not influence the overall crystallinity of PHB.

10.4.5 Photodegradation and Microbial Digestion of PHB in the Presence of EP and LC

Polymeric materials may degrade during processing at high temperatures as well as during use in the presence of oxygen, light, and heat. Therefore, stabilization against the oxidative degradation phenomena is important to preserve PHB physical and mechanical properties. In a bioactive environment, degradation of polymer is initiated through material fragmentation, which is then followed by mineralization. The enzymatic activity of microorganisms shortens and weakens the polymer chains (Mohee et al. 2008). In particular, for biodegradable polyesters, molecular chain breakage proceeds by enzyme-catalyzed hydrolytic chain scission of ester bonds (Sarasa et al. 2009). In the present section, the effect of the natural additives on the photooxidative stability of PHB will be discussed. Moreover, the behavior of PHB-T19 and PHB/15LC compounds from the microbial degradation point of view was qualitatively assessed through SEM observation of garden soil-buried specimens.

First, the effect of UV light irradiation on the properties of plain PHB-T19 and the same polymer modified with 2.5 wt% and 5 wt% of EP was studied. Dumbbell-shaped films were exposed to UV radiation in a weathering chamber containing a mercury vapor discharged lamp kept at 40 °C. The exposure times used were 0, 1, 2, 3, 6, and 12 days. After a visual inspection of the samples exposed to UV light, the dumbbell-shaped films were subjected to tensile tests, and Young modulus (E), strength (σ_b), and strain (ε_b) at break were recorded and reported in Fig. 10.12a–c.

PHB exposure to UV radiation caused several changes in the mechanical response of the material. Long residence times in the chamber (3, 6, and 12 days) damaged the polymer to such an extent that it broke down. Conversely, when EP was added, it was observed that tensile tests could be also performed on the samples subjected to longer irradiation times. It is worth to highlight that, with increasing EP amount, both strength and strain at break decreased (Fig. 10.12b, c). These results are attributed to scission reactions that become predominant upon UV exposure (Sadi et al. 2010), which cause a reduction on the chain entanglements. The drop of molecular weight, followed by the chain rearrangement, leads also to an increase of crystallinity,

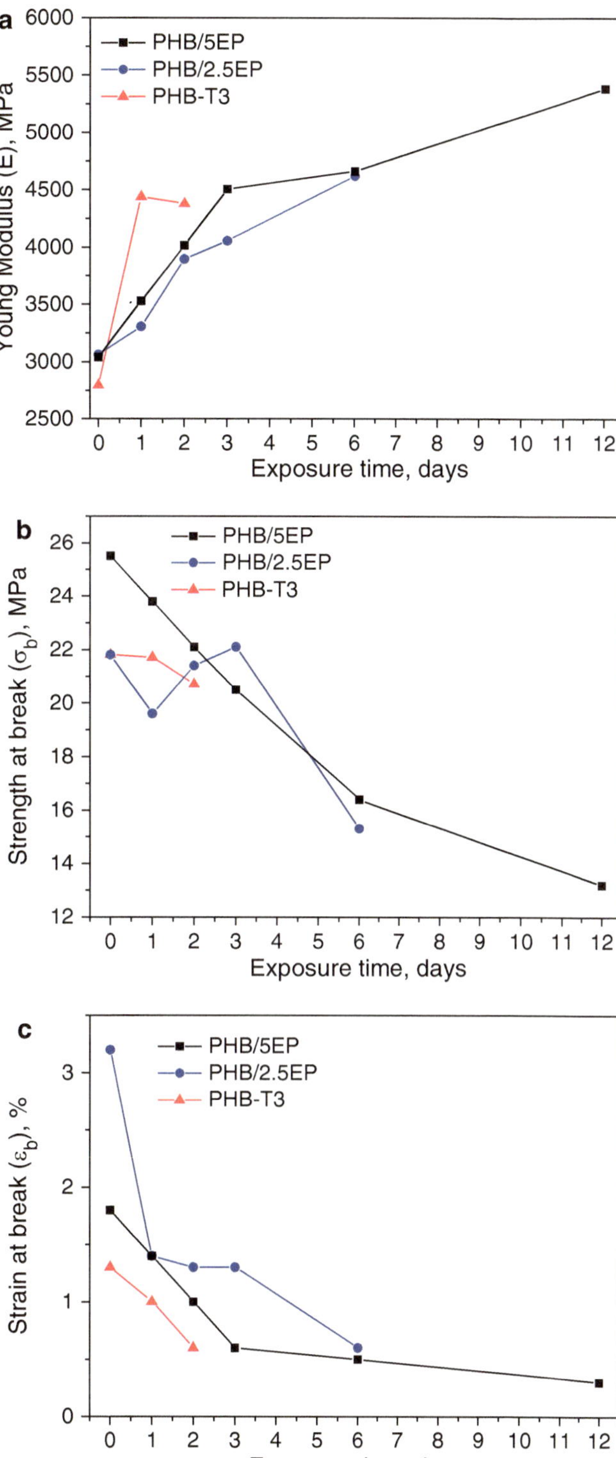

Fig. 10.12 Young modulus (**a**) stress at break (**b**) and strain at break (**c**) versus time of exposure to UV radiation of PHB-T3, PHB/2.5EP, and PHB/5EP

responsible for a rise of the elastic modulus value of more than 40 % (Fig. 10.12a) (Law et al. 2008).

Photooxidation of polymers occurs through a mechanism involving several steps. First, peroxy radicals and hydroperoxides form. These species are unstable to near ultraviolet (UV) exposure and cleave to give radicals capable of initiating further oxidative reactions. As a result, new oxidized products such as alcohols and primary carbonyl compounds (predominantly aldehydes and ketones) are produced. The primary carbonyl-based oxidation products can undergo photolysis according to the so-called Norrish pathways, giving rise to carboxylic acids as ultimate products. The observed evidence suggests that EP primarily acts by scavenging highly reactive oxygen radicals and preventing free radical damage to PHB by effective H-atom transfer, as already reported in the case of Mater-Bi (Cerruti et al. 2011). According to these results, EP is proposed as an efficient, renewable, and biocompatible additive for a sustainable approach to photooxidative stabilization of PHB.

The microbial biodegradation behavior of PHB-T19 and PHB/15LC compounds was qualitatively assessed through SEM observation of garden soil-buried specimens, in order to evaluate the effect of the lignocellulosic filler on the biodegradation rate of PHB. Film specimens were buried in soil and taken out at time intervals, and their surface was observed through electron microscopy. Figure 10.13 displays SEM pictures of PHB-T19 and PHB/15LC after 0, 20, and 80 days of burial.

From the micrographs, significant degradation due to the microbial attack caused by the microorganisms present in the soil was evidenced by the appearance of a rough surface of the film. Many small cavities were formed in the early burial period, which merged together over the course of the degradation process, eventually giving rise to large holes. The SEM analysis showed that degradation process started in the amorphous

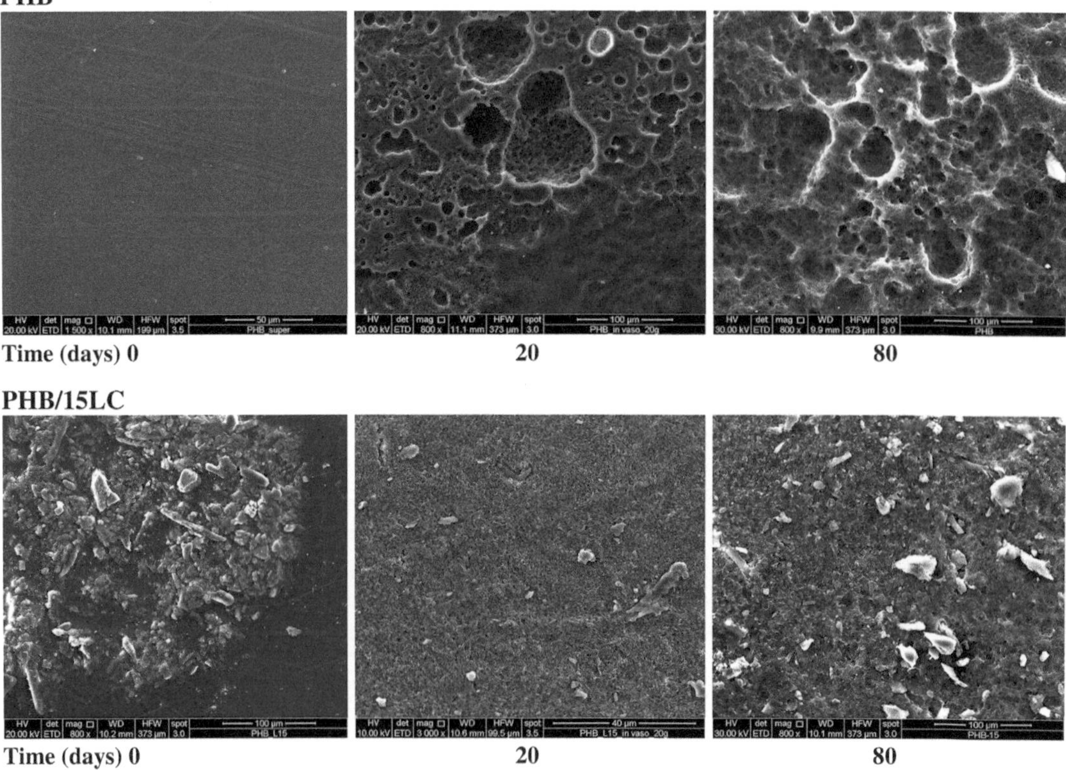

Fig. 10.13 SEM micrographs of PHB and PHB/15LC after 0, 20, and 80 days in soil

regions of the polymer. This was in accordance with data on the degradation of PHB and poly(hydroxybutyrate-*co*-hydroxyvalerate) reported by Boyandin and Prudnikova (2012) that observed a faster degradation of the less crystalline copolymer specimens. The soil burial tests also showed that the presence of LC affected the surface degradation rate of PHB. In comparison with neat PHB, after 20 days, the surface of the biocomposite was still rather smooth and compact. Degradation was more evident with increasing burial times, and bare lignin particles distinctly emerged on the sample surface; however, no large holes were visible. The observed retarding effect of LC can be explained taking into account different aspects. First, as shown before (Sect. 10.4.4), LC was observed to act as a nucleating agent that promoted PHB crystallization and reduced the sensitivity of PHB to hydrolysis and microbial attack. Moreover, it should be reminded that a slight antimicrobial activity can be ascribed to LC (Dong et al. 2011). In fact, after 80 days, PHB surface was cracked, while lignin and cellulose fibers were apparently less damaged. It is likely that during the degradation process, the developing filler surface layer acted as a barrier and delayed water diffusion and microbial digestion of the polymer film. These data were in agreement with those reported in the study conducted by Mousavioun et al. (2012), based on the burial of PHB/lignin films in the soil of a plant pot for up to 12 months. Their work revealed that lignin inhibited the colonization of the microorganisms and made the blends more resistant to microbial and fungal attack. This was attributed to the oxygen-containing lignin functionalities (hydroxyl and carboxylic acid) which affect the antimicrobial activity.

10.5 Conclusions

Poly(3-hydroxybutyrate) (PHB) is a biodegradable polymer, whose marketing potential is limited by some flaws such as brittle nature and melting temperature close to degradation stage, which restricts the processing window. In this study, several additives from renewable sources, some of which are currently regarded as biowaste, were used to improve thermal, processing, and crystallization properties of this polymer, which are targeted to engineer completely bio-based plastic materials. It has been shown experimentally that the phenol-based additives, EP and TA, enhanced the thermal stability of biopolymer, PHB. As a consequence, the polymer could maintain high molecular weights after processing. This finding was in accordance with the slower decay in storage modulus, which was observed through rheological tests. Physical interactions between the polymer and the additive were considered as a key factor to interpret the experimental data. LC, a lignocellulosic biomass, influenced the melt-crystallization kinetics and the crystallinity as a whole, acting as a heterogeneous nucleating agent, with a positive effect on the physical aging of PHB. It was also demonstrated that EP improved the polymer photooxidative stability. Finally, with increasing contents of LC, the biocomposites were more resistant to the biotic degradation due to the antimicrobial activity of the lignin, suggesting the possibility to modulate the rate of degradation of the products made with these materials.

10.6 Opinion

The use of natural additives is an attracting approach to modulate the properties of biopolymers such as PHB. The described results are remarkable with a view to developing sustainable alternatives to synthetic polymer additives. Overall, these outcomes demonstrated the feasibility of using agro-food by-products and natural biomasses as bio-resources, aimed at improving natural biopolymer features in the transition from biopolymers to bioplastics. Future efforts should address environmentally friendly approaches to chemical modification of such additives aimed to get better compatibility between matrix and additives/fillers. The improvement of the interactions between matrix and additives/fillers can lead to a significant enhancement of properties of the resulting blends and biocomposites.

Acknowledgment The authors wish to thank the Italian Minister of Research for the financial support (Enerbiochem PON01_01966).

References

Alvarez VA, Ruseckaite R, Vazquez A (2006) Degradation of sisal fibre/Mater Bi-biocomposites buried in soil. Polym Degrad Stab 91:3156–3162. doi:10.1016/j.polymdegradstab.2006.07.011

Angelini S, Cerruti P, Immirzi B, Santagata G, Scarinzi G, Malinconico M (2014) From biowaste to bioresource: effect of a lignocellulosic filler on the properties of poly(3-hydroxybutyrate). Int J Biol Macromol 71:163–173. doi:10.1016/j.ijbiomac.2014.07.038

Ariffin H, Nishida H, Shirai Y, Hassan MA (2008) Determination of multiple thermal degradation mechanisms of poly(3-hydroxybutyrate). Polym Degrad Stab 93:1433–1439. doi:10.1016/j.polymdegradstab.2008.05.020

Arza CR, Jannasch P, Johansson P, Magnusson P, Werker A, Maurer FHJ (2014) Effect of additives on the melt rheology and thermal degradation of poly[(R)-3-hydroxybutyric acid]. J Appl Polym Sci 132:41836–41844. doi:10.1002/app.41836

Auriemma M, Piscitelli A, Pasquino R, Cerruti P, Malinconico M, Grizzuti N (2015) Blending poly(3-hydroxybutyrate) with tannic acid: influence of a polyphenolic natural additive on the rheological and thermal behaviour. Eur Polym J 63:123–131. doi:10.1016/j.eurpolymj.2014.12.021

Avella M, Martuscelli E (1988) Poly-D(-)(3-hydroxybutyrate)/poly(ethylene oxide) blends: phase diagram, thermal and crystallization behaviour. Polymer 29:1731–1737. doi:10.1016/0032-3861(88)90384-9

Avella M, Martuscelli E, Orsello G, Raimo M (1997) Poly(3-hydroxybutyrate)/poly(methyleneoxide) blends: thermal, crystallization and thermal behaviour. Polymer 38:6135–6143. doi:10.1016/S0032-3861(97)00166-3

Bastioli C (1998) Properties and application of Mater-Bi starch-based material. Polym Degrad Stab 59:263–272. doi:10.1016/S0141-3910(97)00156-0

Boyandin A, Prudnikova S (2012) Biodegradation of polyhydroxyalkanoates by soil microbial communities of different structures and detection of PHA degrading microorganisms. Appl Biochem Microbiol 48:28–36. doi:10.1134/S0003683812010024

Briassoulis D (2007) Analysis of the mechanical and degradation performances of optimised agricultural biodegradable films. Polym Degrad Stab 92:1115–1132. doi:10.1016/j.polymdegradstab.2007.01.024

Buchanan CM, Gedon SC, White AW, Wood MD (1992) Cellulose acetate butyrate and poly(hydroxybutyrate-co-valerate) copolymer blends. Macromolecules 25:7373–7381. doi:10.1021/ma00052a046

Bugnicourt E, Cinelli P, Lazzeri A, Alvarez V (2014) Polyhydroxyalkanoate (PHA): review of synthesis, characteristics, processing and potential applications in packaging. Express Polym Lett 8:791–808. doi:10.3144/expresspolymlett.2014.82

Buléon A, Colonna P, Planchot V, Ball S (1998) Starch granules: structure and biosynthesis. Int J Biol Macromol 23:85–112. doi:10.1016/S0141-8130(98)00040-3

Cerruti P, Santagata G, Gomez d'Ayala G, Ambrogi V, Carafagna C, Malinconico M, Persico P (2011) Effects of a natural polyphenolic extract on the properties of a biodegradable starch-based polymer. Polym Degrad Stab 96:839–846. doi:10.1016/j.polymdegradstab.2011.02.003

Chen C, Fei B, Peng S, Zhuang Y, Dong L, Feng Z (2002) The kinetics of the thermal decomposition of poly(3-hydroxybutyrate) and maleated poly(3-hydroxybutyrate). J Appl Polym Sci 84:1789–1796. doi:10.1002/app.10463

Chen Y, Chou IN, Tsai YH, Wu HS (2013) Thermal degradation of poly(3-hydroxybutyrate) and poly(3-hydroxybutyrate-co-3-hydroxyvalerate) in drying treatment. J Appl Polym Sci 130:3659–3667. doi:10.1002/app.39616

Chodak I (2002) Polyhydroxyalkanoates: properties and modifications for high volume applications, biodegradable polymers (book). Kluwer, Dordrecht

Coelho LF, De Lima CJB, Bernardo MP, Alvarez GM, Contiero J (2010) Improvement of L(+) –lactic acid production from cassava wastewater by *Lactobacillus rhamnosus* B 103. J Sci Food Agric 90:1944–1950. doi:10.1002/jsfa.4039

Dawes EA, Senior PJ (1973) The role and regulation of energy reserve polymers in micro-organisms. Adv Microb Physiol 10:135–266. doi:4594739

de Koning GJM, Lemstra PJ (1993) Crystallization phenomena in bacterial poly[3-hydroxybutyrate]: 2. Embrittlement and rejuvenation. Polymer 34:4089–4094. doi:10.1016/0032-3861(93)90671-V

Dintcheva NT, Jilov N, La Mantia FP (1997) Recycling of plastics from packaging. Polym Degrad Stab 57:191–203. doi:10.1016/S0141-3910(96)00232-7

Dong X, Dong M, Lu Y, Turley A, Jin T, Wu C (2011) Antimicrobial and antioxidant activities of lignin from residue of corn stover to ethanol production. Ind Crop Prod 34:1629–1634. doi:10.1016/j.indcrop.2011.06.002

Dubois M, Gilles DA, Hamilton JK, Rebers PA, Smith F (1956) Colorimetric method for the determination of sugars and related substances. Anal Chem 28:350–356. doi:10.1021/ac60111a017

El-Hadi A, Schnabel R, Straube E, Muller G (2002a) Correlation between degree of crystallinity, morphology, glass temperature, mechanical properties and biodegradation of poly (3-hydroxyalkanoate) PHAs and their blends. Polym Test 21:665–674. doi:10.1016/S0142-9418(01)00142-8

El-Hadi A, Schnabel R, Straube E, Müller G, Riemschneider M (2002b) Effect of melt processing on crystallization behavior and rheology of poly(3-hydroxybutyrate) (PHB) and its blends. Macromol Mater Eng 287:363–372. doi:10.1002/1439-2054(20020501)287:5<363::AID-MAME363>3.0.CO;2-D

Ellis RP, Cochrane MP, Dale MFB, Duþus CM, Lynn A, Morrison IM, Prentice RDM, Swanston JS, Tiller SA (1998) Starch production and industrial use. J Sci Food Agric 77:289–311. doi:10.1002/(SICI)1097-0010(199807)77:3

Felix JM, Gatenholm P (1991) The nature of adhesion in composites of modified cellulose fibers and polypropylene. J Appl Polym Sci 42:609–620. doi:10.1002/app.1991.070420307

Ferreira WH, Carmo MMIB, Silva ALN, Andrade CT (2015) Effect of structure and viscosity of the components on some properties of starch-rich hybrid blends. Carbohydr Polym 117:988–995. doi:10.1016/j.carbpol.2014.10.018

Garaguso I, Nardini M (2015) Polyphenols content, phenolics profile and antioxidant activity of organic red wines produced without sulfur dioxide/sulfites addition in comparison to conventional red wines. Food Chem 179:336–342. doi:10.1016/j.foodchem.2015.01.144

Gáspár M, Benkő Z, Dogossy G, Réczey K, Czigány T (2005) Reducing water absorption in compostable starch-based plastics. Polym Degrad Stab 90:563–569. doi:10.1016/j.polymdegradstab.2005.03.012

Gordobil O, Egüès I, Llano-Ponte R, Labidi J (2014) Physicochemical properties of PLA lignin blends. Polym Degrad Stab 108:330–338. doi:10.1016/j.polymdegradstab.2014.01.002

Greco P, Martuscelli E (1989) Crystallization behaviour of poly (ethylene oxide) from poly (3-hydroxybutyrate)/poly (ethylene oxide) blends: phase structuring, morphology and thermal behaviour. Polymer 30:1475–1483. doi:10.1016/0032-3861(91)90401-4

Gunning MA, Geever LM, Killion JA, Lyons JG, Higginbotha CL (2013) Mechanical and biodegradation performance of short natural fibre polyhydroxybutyrate composites. Polym Test 32:1603–1611. doi:10.1016/j.polymertesting.2013.10.011

Guo J, Xu Z (2009) Recycling of non-metallic fractions from waste printed circuit boards: a review. J Hazard Mater 168:567–590. doi:10.1186/2193-1801-2-521

Guzman-Sielicka A, Janik H, Sielicki P (2013) Proposal of new starch-blends composition quickly degradable in marine environment. J Polym Environ 21:802–806. doi:10.1007/s10924-012-0558-7

Huneault MA, Li H (2007) Morphology and properties of compatibilized polylactide/thermoplastic starch blends. Polymer 48:270–280. doi:10.1016/j.polymer.2006.11.023

Ibrahim NA (2009) Effect of fiber treatment on mechanical properties of kenaf fiber-Ecoflex composites. J Reinf Plast Compos 29:2192–2198. doi:10.1177/0731684409347592

Ikejima T, Inoue Y (2000) Crystallization behaviour and environmental biodegradability of the blend films of poly(3-hydroxybutyric acid) with chitin and chitosan. Carbohydr Polym 41:351–356. doi:10.1016/S0144-8617(99)00105-8

Immirzi B, Malinconico M, Martuscelli E, Volpe MG (1994) Reactive blending of bioaffine polyesters through free-radical processes initiated by organic peroxides. Macromol Symp 78:243–258. doi:10.1002/masy.19940780121

Innocentini-Mei LH, Bartoli JR, Baltieri RC (2003) Mechanical and thermal properties of poly(3-hydroxybutyrate) blends with starch and starch derivatives. Macromol Symp 197:77–87. doi:10.1002/masy.200350708

Kalambar S, Rizvi SSH (2006) An overview of starch-based plastic blends from reactive extrusion. J Plast Film Sheet 22:39–58. doi:10.1177/8756087906062729

Kammerer D, Claus A, Carle R, Schieber A (2004) Polyphenol screening of pomace from red and white grape varieties (*Vitis vinifera* L.) by HPLC-DAD-MS/MS. J Agric Food Chem 52:4360–4367. doi:10.1021/jf049613b

Kaseem M, Hamad K, Deri F (2012) Thermoplastic starch blends: a review of recent works. Polym Sci Ser A 54(2):165–176. doi:10.1134/S0965545X1202006X

Kasirajan S, Ngouajio M (2012) Polyethylene and biodegradable mulches for agricultural applications: a review. Agron Sustain Dev 32:501–529. doi:10.1007/s13593

Kim BK, Hwang GC, Bae SY, Yi SC, Kumazawa H (2001) Depolymerization of polyethyleneterephthalate in supercritical methanol. J Appl Polym Sci 81:2102–2108. doi:10.1002/app.1645

Kumagai Y, Doi Y (1992) Enzymatic degradation and morphologies of binary blends of microbial poly(3-hydroxybutyrate) with poly(ε-caprolactone), poly(1,4-butylene adipate) and poly(vinyl acetate). Polym Degrad Stab 36:241–248. doi:10.1016/0141-3910(92)90062-A

Larif M, Ouhssine M, Soulaymani A, Elmidaoui A (2013) Potential effluent oil mills and antibacterial activity polyphenols against some pathogenic strains. Res Chem Intermed 41:1213–1225. doi:10.1007/s11164

Law A, Simon L, Lee-Sullivan P (2008) Effects of thermal aging on isotactic polypropylene crystallinity. Polym Eng Sci 48:627–633. doi:10.1002/pen.20987

Lee HW, Cho HJ, Yim JH, Kim JM, Sohn JM, Yoo KS, Kim SS, Jeon JK, Park YK (2011) Removal of Cu(II)-ion over amine-functionalized mesoporous silica materials. J Ind Eng Chem 17:504–509. doi:10.1016/j.jiec.2010.09.022

Li J, Zhu B, He Y, Inoue Y (2003) Thermal and infrared spectroscopic studies on hydrogen-bonding interaction between poly(3-hydroxybutyrate) and catechin. Polym J 35:384–392. doi:10.1295/polymj.35.384

Lopes GKB, Shulman HM, Hermes-Lima M (1999) Polyphenol tannic acid inhibits hydroxyl radical formation from Fenton reaction by complexing ferrous ions. Biochim Biophys Acta 1472:142–152. doi:10.1016/S0304-4165(99)00117-8

Lotti N, Scandola M (1992) Miscibility of bacterial poly(3-hydroxybutyrate-co-3-hydroxyvalerate) with ester substituted celluloses. Polym Bull 29:407–413. doi:10.1007/BF00944838

Lotti N, Pizzoli M, Ceccorulli G, Scandola M (1993) Binary blends of microbial poly(3-hydroxybutyrate) with polymethacrylates. Polymer 34:4935–4940. doi:10.1016/0032-3861(93)90022-3

Lv S, Cao J, Tan H, Zhang Y (2015) Effect of annealing on the thermal properties of poly(lactic acid)/starch blends. Int J Biol Macromol 74:297–303. doi:10.1016/j.ijbiomac.2014.12.022

Macedo J de S, Costa MF, Travares MIB, Thirè RMSM (2010) Preparation and characterization of composites based on polyhydroxybutyrate and waste powder from coconut fibers processing. Polym Eng Sci 50:1466–1475. doi:10.1002/pen.21669

MacGregor EA (2001) Biopolymers. In: Meyers RA (ed) Encyclopedia of physical science and technology, polymers. Academic, New York

Mahieu A, Terrie C, Agoulon A, Leblanc N, Youssef B (2013) Thermoplastic starch and poly(epsilon-caprolactone) blends: morphology and mechanical properties as a function of relative humidity. J Polym Res 20:229. doi:10.1007/s10965-013-0229-y

Marsano E, Vicini S, Skopinska J, Sionkowska A (2004) Chitosan and poly(vinyl pyrrolidone): compatibility and miscibility of blends. Macromol Symp 218:251–260. doi:10.1002/masy.200451426

Mekonnen T, Mussone P, Khalil H, Bressler DJ (2013) Progress in bio-based plastics and plasticizing modifications. J Mater Chem 1:13379–13398. doi:10.1039/C3TA12555F

Miao M, Li R, Huang C, Jiang B, Zhang T (2015) Impact of β-amylase degradation on properties of sugary maize soluble starch particles. Food Chem 177:1–7. doi:10.1016/j.foodchem.2014.12.101

Mitra T, Sailakshmi G, Gnanamani A (2014) Could glutaric acid (GA) replace glutaraldehyde in the preparation of biocompatible biopolymers with high mechanical and thermal properties? J Chem Sci 126:127–140. doi:10.1007/s12039-013-0543-2

Mohanty AK, Misra M, Drzal LT (2002) Sustainable biocomposites from renewable resources: opportunities and challenges in the green materials world. J Polym Environ 10:19–26. doi:10.1023/A:1021013921916

Mohee R, Unmar GD, Mudhoo A, Khadoo P (2008) Biodegradability of biodegradable/degradable plastic materials under aerobic and anaerobic conditions. Waste Manag 28:1624–1629. doi:10.1016/j.wasman.2007.07.003

Mousavioun P, Doherty WOS, George G (2010) Thermal stability and miscibility of poly(hydroxybutyrate) and soda lignin blends. Ind Crop Prod 34:656–661. doi:10.1016/j.indcrop.2010.08.001

Mousavioun P, George G, Doherty WOS (2012) Environmental degradation of lignin/poly(hydroxybutyrate) blends. Polym Degrad Stab 97:1114–1122. doi:10.1016/j.polymdegradstab.2012.04.004

Niaounakis M (2013) Biopolymers: reuse, recycling, and disposal, PDL handbook series. Willliam Andrew, Norwich

Nugraha E, Suyatma AC, Tighzert L, Coma V (2004) Mechanical and barrier properties of biodegradable films made from chitosan and poly (lactic acid) blends. J Polym Environ 12:1–6. doi:10.1023/B:JOOE.0000003121.12800.4e

Ortega-Toro R, Jimenez A, Talens P, Chiralt A (2014) Properties of starch-hydroxypropyl methylcellulose based films obtained by compression molding. Carbohydr Polym 109:155–165. doi:10.1016/j.carbpol.2014.03.059

Pack S, Lewin M, Rafailovich MH (2012) A review of engineering biodegradable polymer blends: morphology, mechanical property, and flame retardancy. ACS Symp Ser 1118:427–443. doi:10.1021/bk-2012-1118.ch027

Paramasivan T, Abdul Kalam APJ (1974) On the study of indigenous natural-fibre composites. Fibre Sci Tech 7:85–88. doi:10.1016/0015-0568(74)90020-7

Pelissari FM, Yamashita F, Garcia MA, Martino MN, Zaritzky NE, Gross-mann MVE (2012) Constrained mixture design applied to the development of cassava starch–chitosan blown films. J Food Eng 108:262–267. doi:10.1016/j.jfoodeng.2011.09.004

Persico P, Ambrogi V, Baroni A, Santagata G, Carfagna C, Malinconico M (2012) Enhancement of poly(3-hydroxybutyrate) thermal and processing stability using a bio-waste derived additive. Int J Biol Macromol 51:1151–1158. doi:10.1016/j.ijbiomac.2012.08.036

Riggi E, Santagata G, Malinconico M (2011) Bio-based and biodegradable plastics for use in crop production. Recent Pat Food Nutr Agric 3:49–63. doi:10.2174/2212798411103010049

Rosa DS, Bardi MAG, Machado LDB, Dias DB, Silva LGA, Kodama Y (2009) Starch plasticized with glycerol from biodiesel and polypropylene blends. J Therm Anal Calorim 102:181–186. doi:10.1007/s10973

Rossini G, Toscano G, Duca D, Corinaldesi F, Foppa Pedretti E, Riva G (2013) Analysis of the characteristics of the tomato manufacturing residues finalized to the energy recovery. Biomass Bioenergy 51:177–182. doi:10.1016/j.biombioe.2013.01.018

Sadi RK, Fechine GJM, Demarquette NR (2010) Photodegradation of poly(3-hydroxybutyrate). Polym Degrad Stab 95:2318–2327. doi:10.1016/j.polymdegradstab.2010.09.003

Sahiner N, Sagbas S, Aktas N (2015) Single step natural poly(tannic acid) particle preparation as multitalented biomaterial. Mater Sci Eng C 49:824–834. doi:10.1016/j.msec.2015.01.076

Sarasa J, Gracia JM, Javierre C (2009) Study of the bio disintegration of a bioplastic material waste. Bioresour Technol 100:3764–3768. doi:10.1016/j.biortech.2008.11.049

Scott G (2000) Green Polymers. Polym Degrad Stab 68:1–7. doi:10.1016/S0141-3910(99)00182-2

Shen L, Haufe J, Patel MK (2009) Product overview and market projection of emerging bio-based plastics. PRO-BIP European Polysaccharide network of Excellence (EPNOE) and European Bioplastics, Utrecht

Shogren NR (2009) Starch–poly(hydroxyalkanoate) composites and blends. In: Biodegradable polymer blends and composites from renewable resources. Wiley, Hoboken. doi:10.1002/9780470391501.ch9

Sionkowska A (2011) Current research on the blends of natural and synthetic polymers as new biomaterials: review. Prog Polym Sci 36:1254–1276. doi:10.1016/j.progpolymsci.2011.05.003

Stepto RFT (2006) Understanding the processes of thermoplastic starch. Macromol Symp 246–245:571–577. doi:10.1002/masy.200651382

Thunwall M, Kuthanova V, Boldizar A, Rigdahl M (2008) Film blowing of thermoplastic starch. Carbohydr Polym 71:583–590. doi:10.1016/j.carbpol.2007.07.001

Van de Velde K, Kiekens P (2002) Biopolymers: overview of several properties and consequences on their applications. Polym Testing 21:433–442. doi:10.1016/S0142-9418(01)00107-6

Vogel C, Morita S, Sato H, Noda I, Ozaki Y, Siesler HW (2007) Thermal degradation of poly(3-hydroxybutyrate) and poly(3-hydroxybutyrate-co-3-hydroxyhexanoate) in nitrogen and oxygen studied by thermogravimetric-Fourier transform infrared spectroscopy. Appl Spectrosc 61:755–764. doi:10.1366/000370207781393370

Volova TG, Boyandin AN, Vasiliev AD, Karpov VA, Prudnikova SV, Mishukova OV, Boyarskikh UA, Filipenko ML, Rudnev VP, Xuan BB, Dung VV, Gitelson II (2010) Biodegradation of polyhydroxyalkanoates (PHAs) in tropical coastal waters and identification of PHA-degrading bacteria. Polym Degrad Stab 95:2350–2359. doi:10.1016/j.polymdegradstab.2010.08.023

Wang N, Yu J, Ma X (2007) Preparation and characterization of thermoplastic starch/PLA blends by one-step reactive extrusion. Polym Int 56:1440–1447. doi:10.1002/pi.2302

Wang X, Sang L, Luo D, Li X (2011) From collagen–chitosan blends to three-dimensional scaffolds: the influences of chitosan on collagen nanofibrillar structure and mechanical property. Colloids Surf B 82:233–240. doi:10.1016/j.colsurfb.2010.08.047

Weihua K, He Y, Asakawa N, Inoue Y (2004) Effect of lignin particles as a nucleating agent on crystallization of poly(3-hydroxybutyrate). J Appl Polym Sci 94:2466–2474. doi:10.1002/app.21204

Wyman CE (1996) Ethanol production from lignocellulosic biomass: overview. In: Handbook on bioethanol: production and utilization. Taylor & Francis, Washington, DC

Xia Z, Singh A, Kiratitanavit W, Mosurkal R, Kumarc J, Nagarajan R (2015) Unraveling the mechanism of thermal and thermo-oxidative degradation of tannic acid. Thermochim Acta 605:78–85. doi:10.1016/j.tca.2015.02.016

Yoon JS, Choi CS, Maing SJ, Choi HJ, Lee H, Choi SJ (1993) Miscibility of poly-d(-)(3-hydroxybutyrate) in poly(ethylene oxide) and poly(methyl methacrylate). Eur Polym J 29:1359–1364. doi:10.1016/0014-3057(93)90194-K

Yu L (2009) Starch-poly(hydroxyalkanoate) composite and blends. In: Biodegradable polymer blends and composites from renewable resources. doi:10.1002/9780470391501.ch9

Zheng B, Lu J, Tong Y, Li H, Chen Q (2015) Isolation and characterization of poly(3-hydroxybutyrate)-producing bacteria from aerobic sludge. Appl Biochem Biotech 175:421–427. doi:10.1007/s12010-014-1271-x

Stefania Angelini received her M.Sc. degree in Molecular–Industrial Biotechnology from Perugia University, Italy. She is currently enrolled in a Ph.D. in Agricultural and Food Science with Naples University, Italy, and she works at IPCB-Institute for Polymers, Composites and Biomaterials in Pozzuoli, Italy. She is working on lignocellulosic biomasses and ligninsulfonates and bioresidue streams of biorefineries and paper industries. Her current interests are microbial polymers, polysaccharide hydrogels, and solar cells.

Pierfrancesco Cerruti received his M.Sc. degree in Chemistry and his Ph.D. degree in Materials Engineering from University of Naples, Italy. He is presently full-time Researcher at IPCB-Institute for Polymers, Composites and Biomaterials in Pozzuoli, Italy. His current interests are natural and synthetic polymers, composites, and stimuli-responsive materials.

Barbara Immirzi received her M.Sc. degree in Chemistry from University of Naples, Italy. She is presently a Researcher at IPCB-Institute for Polymers, Composites and Biomaterials in Pozzuoli, Italy. Her expertise is in gel permeation chromatography (GPC). Her current interests are biodegradable polymers applied in agriculture and in nursery.

Merima Poskovic received her bachelor's degree and Master I level in Engineering in Plastic Material from Polytechnic of Turin, Italy. She is currently student grant full time at Novamont S.p.A. Her expertise is in plastic materials, innovative technologies, and laboratory characterization: analytical and mechanical.

Gabriella Santagata received her M.Sc. degree in Chemistry from University of Naples, Italy. She is presently a Researcher at IPCB-Institute for Polymers, Composites and Biomaterials in Pozzuoli, Italy. Her current interests are biodegradable polymers applied in agriculture, food packaging, and biomedical field.

Gennaro Scarinzi received his M.Sc. degree in Chemistry from University of Naples, Italy. He is presently a Researcher at IPCB-Institute for Polymers, Composites and Biomaterials in Pozzuoli, Italy. His current interests are thermoset polymers and lignin-based materials.

Mario Malinconico received his M.Sc. degree in Industrial Chemistry from University of Naples, Italy. He is presently the Chief Scientist at IPCB-Institute for Polymers, Composites and Biomaterials in Pozzuoli, Italy. He has been appointed as Italian Representative at IUPAC by the President of the Italian CNR for 2011. His current interests are biodegradable polymers in packaging, aeronautics, and biomedical and agricultural fields.

The Survivors of the Extreme: Bacterial Biofilms

11

Neha Dubey, Raja Singh, Aditya K. Sharma, Sharmila Basu-Modak, and Yogendra Singh

Abstract

Biofilms are bacteria's way of behaving like a multi-cellular organism. Bacteria have been constantly evolving in the face of myriad natural challenges since the first life form appeared. Over the centuries, they have survived under extreme conditions by virtue of their ability to form biofilms. Biofilms have proved to be an immensely strong collaborative effort of bacteria, and this interaction is administered via their quorum-sensing mechanism. A biofilm plays a crucial role in survival, dispersal, transfer of resistance genes, and generation of diversity among bacteria. It can also act as a pathogenic factor for virulent bacteria and biofilms are often listed as the major cause of many diseases, such as endocarditis, cystic fibrosis, etc. Bacterial biofilms can be formed on almost every surface and, thus, have deleterious effects on many indwelling medical devices and industrial equipment. Many methods—from antibiotics to ultraviolet (UV) radiation—are currently being used to eliminate or reduce biofilm. However, the most effective and eco-friendly measure involves the targeting of the root phenomenon, quorum sensing. Biofilm formation and dispersal mechanisms are being studied to increase the efficiency of biofilm elimination. Despite the many harms they can pose, these biofilms have been efficiently manipulated to be used for various purposes such as wastewater treatment, microbial fuel cells, drug delivery, nanobiotechnol-

N. Dubey (✉) • S. Basu-Modak
Department of Zoology, University of Delhi, 110007 Delhi, India
e-mail: nnehadubey18@gmail.com; sbasumodak@zoology.du.ac.in

R. Singh
Special center for molecular Medicine (SCMM), Jawaharlal Nehru University, 110067 New Delhi, India
e-mail: t.rajasingh1786@gmail.com

A.K. Sharma
Lab 208, Allergy and Infectious Diseases, CSIR-Institute of Genomics and Integrative Biology, Delhi University Campus, Mall Road, New Delhi 110007, India

Academy of Scientific & Innovative Research (AcSIR), 2- Rafi Marg, Anusandhan Bhawan, New Delhi 110001, India

Y. Singh
Allergy and Infectious Diseases, CSIR-Institute of Genomics and Integrative Biology, Room No. 208, Mall Road, 110007 Delhi, India
e-mail: ysingh@igib.res.in

V.C. Kalia (ed.), *Microbial Factories*, DOI 10.1007/978-81-322-2595-9_11

ogy, etc. Biofilms bring about a very detailed level of complexity that helps for better persistence of the bacterial population and at the same time, provides us a valuable tool to address several important environmental issues. Thus, it will be appropriate to term bacterial biofilms as remarkably proficient assemblies of the life forms.

11.1 Introduction

Microorganisms are an important part of our ecosystem that have evolved over the centuries. During evolution, microbes must undergo many stresses to successfully survive and divide in extreme conditions. The organisms that counteract these external stresses eventually direct the path of their own evolution. One of the major mechanisms behind their survival is the tendency to 'stay together' (Kalia 2014a; Kalia and Kumar 2015a). Bacteria tend to reside in the form of adherent communities to enable them to cope with environmental pressure. It is widely established that bacteria in communities survive much better than their planktonic versions. These adherent, self-supporting communities are termed 'biofilms'. Biofilms are the most successful bacterial environment, evolved to sustain microorganisms from stressed environments and fulfill the nutrient requirements of the bacteria. The first report defining the existence of bacterial communities was by Zobell (1943). Later on, Costerton devoted a majority of his work to understanding the mechanisms and benefits of biofilm formation (Costerton et al. 1978). Within biofilms, the bacterial cells secrete a gluey material called exopolysaccharide (EPS) that helps them adhere to any living or non-living surface, such as medical implant materials, soil particles, metals, plastics, and, most significantly, human or animal tissue. A vast number of pathogens often stay as biofilms to combat the hosts' immune system. The biofilm bacterial population shows a reduced growth rate and an altered transcriptional profile. Additionally, the bacterial community in biofilms are up to 1000 times more resistant to anti-microbial stresses (Kalia et al. 2014; Kalia and Kumar 2015b).

11.2 Biofilm Structure and Functions

Biofilm is mainly composed of the same or other species of bacteria comprising around 15 % population and 85 % exo-polymer material, mainly polysaccharides. Biofilms can vary in thickness from a monolayer to a thickness of 6–8 cm but, on average, are about 100 μm thick. Biofilm formation is a complex process involving a number of steps starting from initial adhesion to the final community establishment (Donlan 2002). The first bacterial counterparts that adhere to the surface generally form a weak van der Waals force-like interaction; this step is called the initial attachment. Later, upon non-dispersal of the first commensals, the attachment is strengthened using cell adhesion molecules such as bacterial pili, fimbriae, and cell surface proteins leading to the irreversible attachment (Hinsa et al. 2003). These bacterial pioneers facilitate the arrival and adherence of the bacterial population; some attach directly to the adhesion sites while others attach to the colonizing bacteria and exopolysaccharide. One interesting aspect of biofilm biology is the communication aid. The colonizing bacteria in the biofilm tend to communicate via quorum sensing, which is both an intra-species as well as an inter-species phenomenon that serves as a simple communication network. It regulates a number of different processes involving various chemical messengers that bring together bacteria under a highly organized societal frame. This cellular communication marks the maturity of the biofilm (De Kievit et al. 2001). The most important aspect is the dispersal of the biofilm. Dispersal is necessary to recruit the new microbial population and for proliferation of the biofilm entity. This periodic dispersal of planktonic species is mediated by some detachment signals.

Biofilms are coordinated structures that accomplish their own metabolic requirements; hence, they are often compared to the higher organism tissues that orchestrate their own development and survival (Hall-Stoodley et al. 2004). (Figs. 11.1, 11.2, 11.3, 11.4, and 11.5)

The substrate is an important part of the biofilm assembly. Biofilms can be formed on almost every living or non-living surface, for example, water bodies, seashore, rocks, pipelines, stagnant pools, hot and cold springs, sewage pipes, prosthetics, body tissue, enamel, gut, nails, the inner lining of lungs, and almost every body tissue. These communities harbor interstitial water networks that channel the diffusion of nutrients across the biofilm. Broadly speaking, it has been shown that biofilms often form comfortably on a solid rough surface by virtue of fewer non-tripping forces and large surface areas. The moistness of a surface enhances the likelihood of biofilm formation by providing a conditional matrix. The moistness immediately conditions the surface with different polymers and organic matters that facilitate the biofilm (Love 2010). Conditional films can range from organic polymers to human secretions like mucous, blood, tears, gingival fluid, etc., as in the case of dental plaques, where the bacteria rapidly form biofilms on the conditioned enamel of the tooth (Rosen et al. 2005). Furthermore, the surface energy and hydrophobicity of the suspended matter influence the bond strength between the microbes and the surface by altering the substrate characteristics. Other characteristics of the substrate, like temperature, pH, metal ion composition, etc., add to the complexity of biofilm dynamics. The seasonal behavior of biofilms along water bodies has been studied in detail by researchers and has been attributed to local temperature fluctuations (Else et al. 2003). Tidal and wind velocity also affects the establishment of primary pioneers of the biofilm organization (Soini et al. 2002). Fletcher and coworkers found an increase in cationic metal concentrations minimized the repulsive forces between the bacterial cell and glass material, thereby affecting the attachment of *Pseudomonas fluorescens* to the surface (Fletcher 1988). An increase in the nutrient levels enhances the survival of the microbial community (Bowden and Li 1997). Cellular appendages such as pili and fimbriae are found to play an important role in minimizing electrostatic repulsion by virtue of their hydrophobicity caused by the presence of large numbers of non-polar amino acids. Several studies have provided evidence of the relation between the surface protein hydrophobicity and substrate adherence (Rodrigues and Elimelech 2009). Gram-negative bacteria consist of a hydrophilic O layer (LPS); this is why it is unable to create a functional biofilm. Perhaps this is

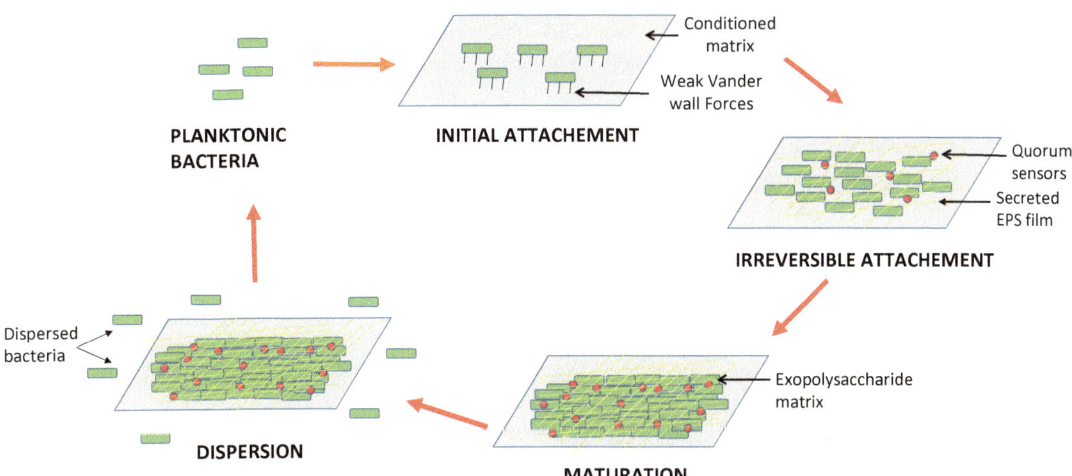

Fig. 11.1 Biofilm formation. Steps of biofilm formation: 1. Arrival of planktonic bacteria on a conditioned substrate. 2. Establishment of pioneer community. 3. Growth of the bacterial communities within the biofilm. 4. Maturation of a biofilm. 5. Dispersal

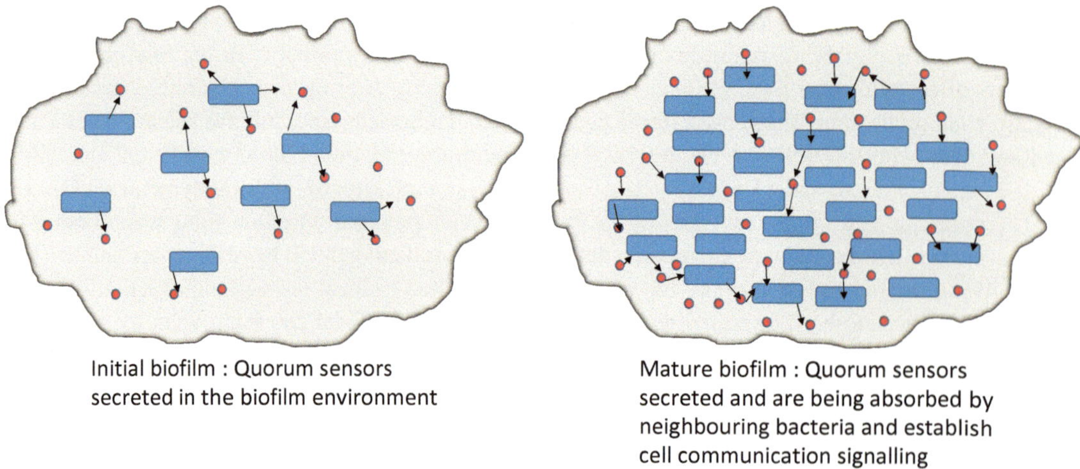

Fig. 11.2 Population dependent initiation and stimulation of biofilm development of 'quorum sensor'

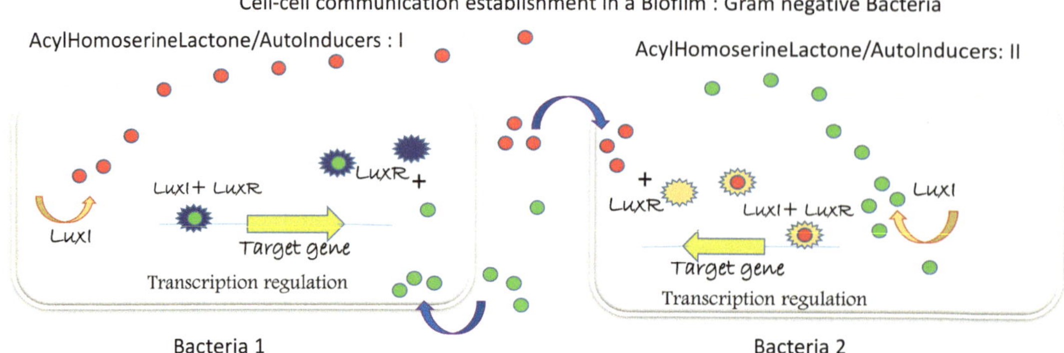

Fig. 11.3 Most Gram-negative bacteria have a LuxI/LuxR system. LuxI encodes the quorum-sensing molecules called autoinducers, namely AI-1, AI-2, and Acyl homoserine lactones. These quorum-sensing molecules are secreted in the environment constitutively. As soon as the population density crosses a threshold level, these secreted molecules are taken up by neighboring cells. Inside cells, these molecules bind to a protein sensor receptor encoded by LuxR. This complex then binds to the promoter regions of the target genes and hence control the transcription of the quorum-sensing-related processes

responsible for the differential biofilm formation capacity between Gram-positive and Gram-negative bacteria.

The key organic element of the biofilm are the EPSs, which are mainly high-molecular-weight polymers secreted by the bacteria to maintain the structural and functional integrity of the biofilms (Flemming and Wingender 2010). It mainly comprises sugar moiety along with non-sugar elements like protein and nucleic acids other than organics like acetate, pyruvate, succinate, and phosphate. It is anionic in nature, conferred by mannuronic acids, D-glucuronic, D-galacturonic, and other ketal-linked pyruvates (Nwodo et al. 2012). The EPS matrix depends on two important properties governing the architecture and the strength of the biofilms. (1) The composition and structure of the polysaccharide components. Most of the Gram-positive bacterial EPSs are composed of hexose residues with either a 1, 3- or 1, 4-β-linked backbone that confers the strength and non-deformity to the biofilm; and (2) The shape and age of the matrix. Biofilms are structurally non-uniform, and the amount of EPS varies hugely between young and old biofilms. Additionally, over time, the EPS accumulates

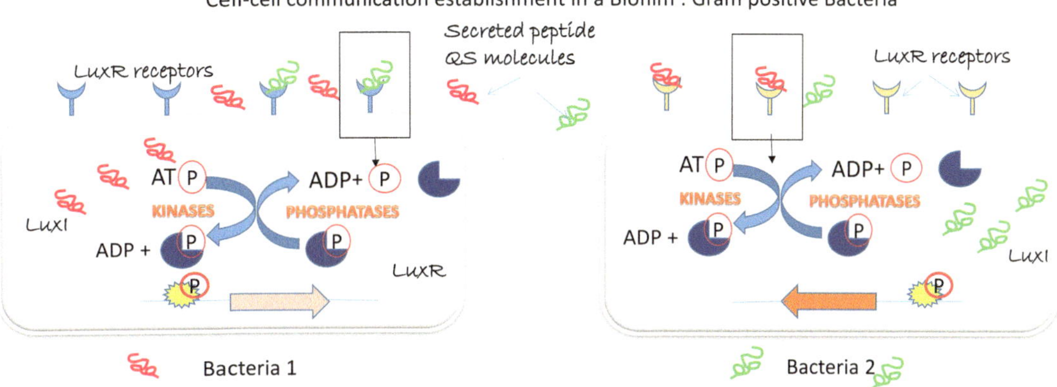

Fig. 11.4 Most Gram-positive bacteria have a LuxI/LuxR system. In this system, LuxI encodes the quorum-sensing oligopeptide moieties. These oligopeptides are secreted in the environment and, upon proper signal stimulation, binds to the surface-located LuxR receptors on the other cells. This binding facilitates a series of phosphorylation and dephosphorylation events, affecting downstream transcription factors. These specifically activated transcription factors bind to the promoter regions of specific target genes and hence regulate quorum-sensing-specific transcriptome

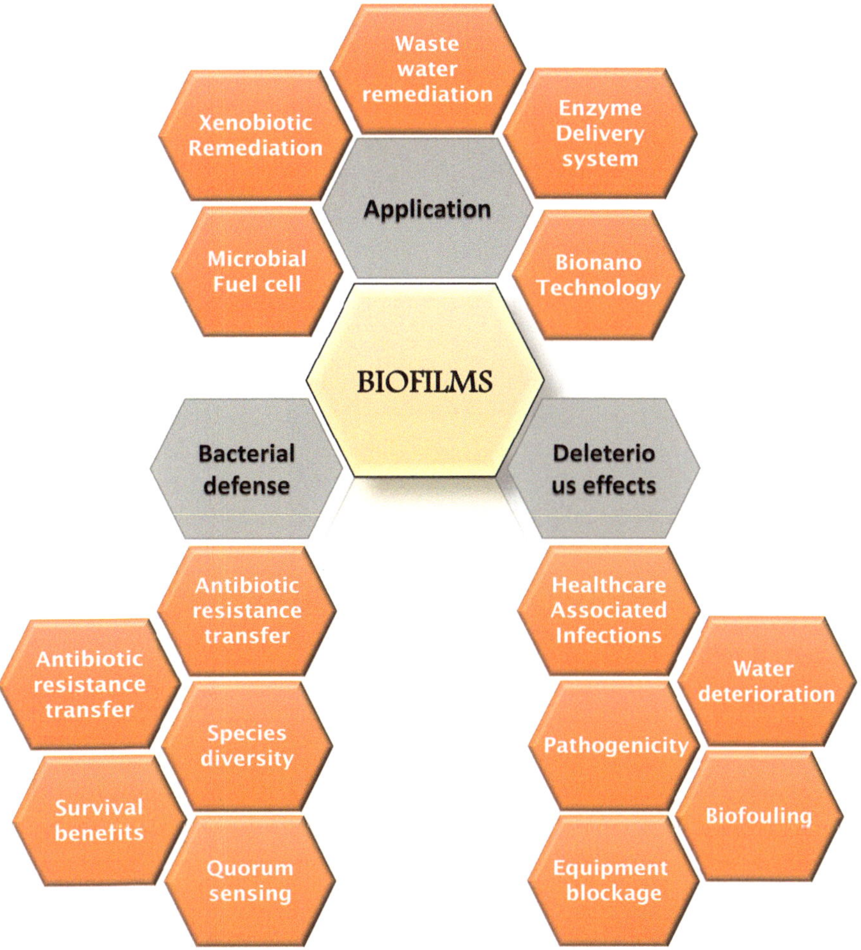

Fig. 11.5 An overview of biofilm: characteristics, functions and applications

components such as substratum particles, metal ions, other cell macromolecules (such as proteins, DNA, lipids, etc.) (Sutherland 2001). EPS secretion is affected by the nutrient status of the adherence medium. Hydration of EPS prevents the desiccation of the biota. Extracellular DNA is part of the outer layer of the biofilm and, in some cases, has been seen to provide strength to the biofilm structure. A recent report proves that physical interaction of the extracellular DNA and EPS component *Psl* is crucial for the formation and stabilization of the biofilm skeleton in *Pseudomonas aeruginosa* (Wang et al. 2015). Among the membrane components, glucosidases and N-acetylglucosaminidase are shown to reduce the attachment of bacteria with substrates in bacteria, e.g., *P. fluorescence* and *Desulfovibrio desulfuricans.* Additionally, LapA, a secretory protein, is well elucidated to act as an anchorage between bacterial cells. In *P. aeruginosa*, SadB is known to mediate adhesion and establish a more permanent stage of biofilm formation (Caiazza and O'Toole 2004).

Bacteria benefits from biofilm assembly in many ways. The mode of communication established between the communities acts as a method to channel nutrients, water, and cellular signals across the biofilm (Karatan and Watnick 2009). It is also an established mode of antimicrobial resistance transfer (Hannan et al. 2010). The biofilm population is more resistant to many antimicrobial stresses than its planktonic counterparts. A biofilm can be composed of single as well as multiple species. Many natural biofilms formed across the global atmosphere usually comprise unions of multiple species. The biofilms that are composed of different bacterial communities are often stronger and thicker as different species synchronize together to stabilize the biofilm. They has also been shown to enable the transfer of nucleic acid across strains in the form of conjugation (Molin and Tolker-Nielsen 2003). The act of conjugation is synergistically coordinated amongst bacterial communities in the system. The DNA released in the environment for the process is autocatalytic and helps in biofilm community stabilization. These inter- and intraspecies DNA transfers not only generate the diversity and distributes antibiotic resistance but also synchronizes the biofilm cycle and population (Nguyen et al. 2010). Thus, these biofilms serve as a diverse source of species (Peyyala and Ebersole 2013). The ability to secrete EPSs is higher in the concerted population spheres of the biofilm, while the outer edges with a looser population helps in dispersion and migration of the population. Alginate is the major component that regulates the formation and the dissemination of bacterial biofilms (Hay et al. 2013). The dissemination of biofilm is regulated by the expression of the gene alginate lyase. Researchers have proved the important role of *algL* and *algX* genes in alginate biosynthesis and have also showed an increase in the expression of algL substantially increased the shedding of biofilm, while down-regulation of the gene resulted in a firm biofilm production (Robles-Price et al. 2004; Albrecht and Schiller 2005). However, another consecutive report claimed the necessity of these genes for *Pseudomonas* biofilm architecture and tenacity yet denies its importance in the formation or seeding of the biofilm (Stapper et al. 2004). The basis behind this fascinating collaboration between bacteria in the biofilm lies in the more dense and compact assembly, which leads to a more precise and coordinated exchange of signals and thus enhances the efficiency compared with planktonic forms. Cyclic diguanosine-5'monophosphate (cyclic di-GMP) is the foremost intracellular secondary messenger that positively regulates initiation and formation of biofilm (Hickman et al. 2005; Borlee et al. 2010).

Biofilm formation relies on cell-to-cell communication systems that are coordinated by quorum sensing. Planktonic bacterial populations, upon reaching a threshold number, secrete chemical messengers called quorum-sensing molecules into the environment and thus initiate and execute a strictly population-dependent quorum-sensing signaling system. These chemical autoinducers are transcription factors that are activated only when the population reaches a certain number. These inducers bind to specific receptors and hence regulate the up-regulation or down-regulation of genes in a manner that can benefit the bacterial community and prevent assets from the competing planktonic or non-planktonic bacteria (Waters and Bassler 2005). In Gram-negative bacteria, the chemical inducer is acyl homoserine lactones (AHLs) – auto inducer-1 (AI-1) – which is secreted by the bacterial cells and, under proper threshold stimulus, enters the neighboring cells and binds to the protein receptor LuxR (Whitehead et al. 2001; Laverty et al. 2014). Gram-positive bacteria, on the other hand, secrete peptides that bind to the cell surface receptor on other bacterial cells (Fleuchot et al. 2011; Monnet et al. 2014). These signals are detected by two component sensors, leading to the phosphorylation of the rear end proteins, which eventually bind to the DNA to impact the transcription of the target genes (Lyon et al. 2002; Jimenez and Federle 2014). The common quorum-sensing mediator of both Gram-positive and Gram-negative is AI-2. AI-2 consists of boron in a protein assembly that is produced by catalysis of LuxI enzyme. In most organisms, the LuXI/LuxR system composes the quorum-sensing machinery, LuxR being the receptor. This decentralization of the quorum sensing brings together the bacterial micro-colonies as a well-synchronized system. Quorum sensing is a particularly crucial phenomenon during processes like biofilm formation, conjugation, antibiotic resistance, and pathogenicity. Some enteric pathogens, like *Escherichia coli, Shigella, Salmonella, Yersinia*, and other Gram-negative species have auto inducer type 3, which is a epinephrine/norepinephrine-regulated system (Walters and Sperandio 2006a, b). Other known quorum-sensing molecules are indole; 3-hydroxy palmitic acid methyl ester (3OH PAME); 3,4-dihydroxy-2-heptylquinolone (PQS); butyrolactones; and cyclic dipeptides (Kalia and Purohit 2011). *Campylobacter jejuni* is a food-borne pathogen capable of forming biofilms in food as well as in related industrial equipment. It has been shown that the LuxS gene mutant of *C. jejuni* showed a significant decrease in biofilm-forming capability compared with wild-type (Reeser et al. 2007). The same effect is seen in *Aeromonas hydrophila*, a facultative aerobe in which the mutation in Acyl homoserine lactone (AHL) synthesis gene resulted in a desolated biofilm (Lynch et al. 2002). In organisms like *Bacillus cereus* and *Helicobacter pylori*, AI-2 is shown to negatively regulate the biofilm formation on glass and polystyrene surfaces, respectively. *Listeria monocytogenes* has a regulon that has been proved to play an important role in biofilm formation. This regulon has four genes, *agrB, agrD, agrC,* and *agrA*. In *L. monocytogenes*, the deletion mutants of agrA and agrD displayed a defect in biofilm formation (Vivant et al. 2014). *E. coli* has a LuxS/AI-2 system, but its role in biofilm formation is unclear. Rather, a LuxR homolog/ sdiA (suppressor of cell division inhibition) deletion mutant in *E. coli* showed an enhanced biofilm contributed to its function to suppress the motility and biofilm formation in wild type (Sharma et al. 2010). In *P. aeruginosa*, disruption of the AHL-producing gene *lasI* displayed a major defect in biofilm formation (Heydorn et al. 2002). Such wide examples clearly establish the inevitable role of quorum sensing in biofilm formation. Researchers are trying to find ways to break this quorum-sensing-mediated rhythm to prevent the formation of biofilms. This inhibition can be accomplished in many ways, mostly targeting the auto inducer and receptor genes along with interfering with the secretion of quorum-sensing molecules (Kjelleberg et al. 2008; Kalia 2013, 2014b; Kalia et al. 2011; Kumar et al. 2015). Recently, one of the most robust *P. aeruginosa* biofilms has been shown to be inhibited by a synthetic compound meta-bromo-thiolactone, which is targeted against the quorum-sensing receptors – LasR and RhlR (O'Loughlin et al. 2013).

11.3 Epidemiology Caused by Biofilms

Globally, much focus has been placed on epidemic diseases. These epidemics are often caused by certain planktonic bacteria that are able to aggravate our immune system all at once and give rise to severe to chronic infections. In the past, a large number of these epidemics has been controlled by targeting the specific structure of the bacteria, such as proteins, DNA, membranes, etc. Over time, we have been minimally successful in getting rid of a few epidemics, but it would not be wrong to say that we still have organisms around us that can prevail and cause many life-threatening diseases. One major problem that scientists face during their research is the non-reproducibility of the data when dealing with bacteria in vitro and in vivo. The root of this difference lies with the adaptation and survival tactics of the bacteria in the form of biofilms. Biofilms are often seen succeeding in the cases of most relapsing diseases that are acute but consistent. These diseases progress through seasonal inertness and subsequent exacerbations, for example, diarrhea (*E. coli*), dental plaque (*H. pylori*), *Staphylococcus aureus* infection, middle ear infections (*Haemophilus influenza*), and many more. Time and again, such incidences have been seen to be frequent in the case of biomedical engineering, be it life related or equipment related. The implantation of grafts and prosthetics in advanced medicine unveiled several such cases of biofilm dispersal that eventually give rise to secondary chronic infections. Therefore, it would not be wrong to consider biofilms as a successful factor for pathogens (Donlan and Costerton 2002).

Among the pathogenic bacterial population, different bacteria are known to form persistent biofilms, e.g., *L. monocytogenes*, *Legionella pneumophila*, *S. aureus*, *Campylobacter* spp., *E. coli*, *Salmonella typhimurium*, *Vibrio cholera*, *Pseudomonas* spp., *Rhizobium* spp., *Enterococcus* spp., *Candida* spp., etc. These, along with many other opportunistic pathogens, are the main causes of many diseases in humans, namely cystic fibrosis, dental plaques, native valve endocarditis, periodontitis, prostatitis, etc. The disease is spread through various methods, mostly upon establishment of the biofilm in a vulnerable area; the cells detach in a timely manner from the site and are circulated across the body. The susceptible areas catch infections and therefore increase the complexity of the disease. The other method of progression is the secretion of toxins by the Gram-negative population of the biofilm.

Native valve endocarditis is a disease of the vascular connective and endothelial tissue. The common bacterial population involved belongs to the coccus group like *Streptococci, Enterococci*, and *Staphylococci*, further inhabited by fungal groups such as *Candida* and *Aspergillus* (Tunkel and Mandell 1992; Verma et al. 2006). Any wound or injury at the endothelial point results in the formation of thrombotic lesions primarily made up of platelets, fibronectin, and red blood cells (RBC). Under these conditions, the persisting blood bacteria adhere and colonize the site of infection. Fibronectin acts as a major point of adhesion for these bacteria. Several bacteria possess fibronectin receptors that strengthen the binding. *S. aureus* executes this adhesion via many receptors such as surface receptors SasC and SasG along with biofilm-associated protein Bap, serine aspartate repeat protein SdrC, and fibronectin/fibrinogen-binding proteins – FnBPA and FnBPB (Vazquez et al. 2011). In *Staphylococcus epidermidis*, accumulation-associated protein (Aap) and extracellular matrix-binding protein (Embp) are known to mediate cellular adhesion and biofilm formation. In *Staphylococcus lugdunensis*, surface-localized AtlL is shown to affect biofilm formation as well as its internalization in eukaryotic cells (Hussain et al. 2015). Upon embedment and stabilization in the endothelial fibrin matrix, the bacteria soon start to multiply and coat themselves in fibrin capsules, which protects it from white blood cells. These biofilms are known to shed cell clumps, which eventually cause emboli. These bacterial- and fungal-loaded infective emboli then spread the infection and can cause severe complications throughout the body. Depending on the type of bacterial endocarditis, various antibiotic doses are ascertained. Penicillin was used against streptococcal endocarditis, while vancomycin and rifampicin have been used as antibacterials against staphylococcal endocarditis (Riedel et al. 2008). Candidal endocarditis has been treated successfully with fluconazole (Lye et al. 2005). Although the antibacterial dosage proved sufficient to control the endocarditis complication, the very basis of biofilm establishment is poorly understood.

Other common forms of severity caused by opportunistic bacteria are periodontitis, which can be caused by different bacterial species such as *Fusobacterium* spp., *Bacteroides* spp., and *Porphyromonas gingivalis* (Darveau 2010). These bacteria utilize the conditioning film of pellicle over the enamel, which comprises phosphoproteins, gingival crevice fluid, albumin, lysozymes, glycoproteins, and lipids. The bacteria reside in subgingival crevice and attach with the pellicle via extracellular glucan, further followed by characteristic polysaccharide secretion. This congregation eventually turns into dental plaques of 50–100um thickness. The administration of tetracycline and metronidazole, along with N-acyl homo serine lactone analog, is suggested to be preventive for periodontitis (Gamboa et al. 2014).

In cystic fibrosis, *P. aeruginosa* usually benefits from transmembrane conductance defects. Disease is caused by the genetic mutagenesis of the cystic fibrosis transmembrane conductance regulator (CFTR), responsible for mucus, sweat, and digestive juice production. The disturbance in the ionic conductance within the body cavities results in uncontrolled mucous production, which eventually thickens the cough and thereby acts as a suitable habitat for the bacteria (Høiby et al. 2010). The bacteria are highly antibiotic tolerant because of the hypoxic, mucus-rich collaborative biofilm environment; thus, treating the *Pseudomonas* biofilm in cystic fibrosis patients is extremely difficult. The infection caused by *P. aeruginosa* is resistant to immune system clearance methods such as opsonization because the thick mucoid alginate layer prevents the antibody coating necessary for opsonization. However, a few reports have shown an effective concentration of colistin and tobramycin in combination with meropenem and ciprofloxacin to kill the bacterium when tested on over 15 clinical isolates (Herrmann et al. 2010). As the pathology of *P. aeruginosa* largely depends on mucus production, the use of hypertonic solution has been demonstrated to indirectly reduce the bacterial persistence as it prompts the clearance of mucus obstruction when carried out on a regular basis, although the long-term effect is yet to be examined (Elkins et al. 2006). Species other than *P. aeruginosa* that aid the disease severity are *H. influenza* (Starner et al. 2006; Cardines et al. 2012) and *S. aureus* (Hirschhausen et al. 2013). The focus has long since shifted towards infringement of the signaling mechanism responsible for the tenacity of biofilm (Davies et al. 1998; Bjarnsholt et al. 2010). A recent report has shown that the mucoid phenotype and biofilm formation by *Pseudomonas* is ion-regulated, favoring the concept of targeting the iron homeostasis of the bacterium in the lung airway (Wiens et al. 2014).

Biofilms are a major impediment to the use of indwelling medical devices. The incidence of biofilm-dependent complications is usually seen in cases of biomedical devices such as urinary catheters, prosthetic valves, and central venous catheters (Trautner and Darouiche 2004; Donlan 2011). The source of initial contamination could be many, per se, blood- and tissue-borne microbes, ill health, administration practices, and external factors, etc. When a catheter is inserted into the body, it stays in direct contact with blood and tissues and eventually is covered by cells, blood platelets, tissue secretions, mucous, urine, and fibrin. These accumulations act as a conditioning film for the circulating bacteria. The most common pathogenic colonized cases are *S. aureus*, *Klebsiella pneumonia*, *P. aeruginosa*, *Enterococcus faecalis*, and *Candida albicans*. Intrauterine devices are a major source of intrauterine infections. Researchers have shown heavy depositions of biofilm in the distal end of such devices, which are heavily loaded with bacteria such as *E. coli, S. epidermidis*, *Enterococcus* spp., and anaerobic *Lactobacilli* spp., *Corynebacterium* spp., *Micrococcus spp*., *C. albicans*, and *S. aureus*, etc. (Wang et al. 2010).

Many methods are being tried to eliminate or reduce infections originating from biofilms. Antibiotics such as tobramycin, chlorhexidine, and ciprofloxacin have been conservatively used to treat catheter-based biofilm infections to prevent biofilm accumulation on medical devices. It has been shown that the ultrasound treatment of indwelling medical devices prior to administration reduces the risk of bacterial biofilms such as *Pseudomonas diminuta*, *P. aeruginosa*, and *L. monocytogenes* (Holá et al. 2010; Torlak and Sert 2013). An anti-parasitic drug, nitazoxanide, has been reported as inhibiting staphylococcal biofilms, probably by blocking biofilm accumulation rather than affecting the dispersal (Tchouaffi-Nana et al. 2010). Pentasilver hexaoxoiodate (Ag_5IO_6) coating on medical devices has been shown to have strong anti-biofilm activity (Incani et al. 2014). Another well-known approach targeting the EPS component of the biofilm via antagonistic and hydrolyzing enzymes could be effective for the known lab-grown organisms. An earlier study demonstrated the up-regulation of alginate lyase with the diffusion of the antibiotics gentamicin and tobramycin across *P. aeruginosa* biofilms because of high EPS degradation, resulting in a loosened matrix (Hatch and Schiller 1998). However, a major setback of these studies was the variability of the EPS being secreted by different biofilm-composing bacteria. Though EPSs include many polymorphisms, the genetic variation of the EPS-producing genes is comparatively conserved and this could possibly be used to identify certain drugs or compounds that can block or inhibit the transcription and translation of the participating proteins. One interesting method for disrupting the biofilm is targeting the quorum-sensing mechanism. Researchers are putting effort into discovering the genetic basis of quorum sensing. Davis et al. showed the significance of acyl-homoserine lactones (signaling molecules) in biofilm segregation (Davies et al. 1998). Thus, continuing on the same line, if we were able to explore the crucial signaling paradigm behind it, we might be able to intervene in biofilm formation.

11.4 Applications

Microbes constitute the largest biomass of the earth. We have been living amongst all sorts of microbes since the rise of human beings. Not all microorganism are harmful; some actually contribute to maintenance of the well-being of the human system, e.g., gut colonizers. Microorganisms actually tend to recycle some of the vital elements of life, e.g., minerals, water, and atmospheric gases, etc. Similarly, not all the biofilms around us are harmful. Researchers have been trying to devise techniques and methods that can utilize the beneficial part of the biofilm (Kalia and Kumar 2015a; Clark et al. 2012) .

11.4.1 Waste Water Treatment

Water is the most essential element of nature. Water scarcity is now a major issue being faced by the human race. Many national government agencies have been trying to find an effective method to overcome this problem, via water conservation and waste water treatment. Waste water has many contaminants, including particulate matter, microorganisms, organic matter, heavy metals, and excess nutrients such as nitrogen, phosphorus, etc. Current industrial wastewater treatment subunits are high-maintenance and high-cost consuming processes. Thus, microorganisms have been utilized to produce biofilm-based bioreactors (Wagner and Loy 2002). Biofilm process treatment has several advantages over conventional waste water methods: convenient operation, environmentally friendly, high biomass activity, higher toxin resistance, and low cost (Tao et al. 2010). Biofilm-based bioreactors are established either on a static substrate or on a conditioned synthetic substrate system. In fixed-type bioreactors, matter such as rock, sand, or plastic is used to condition biofilm, e.g., trickling filters, rotating biological contactors, and sand filters. Second are the suspension type in which microorganisms are suspended in the waste water where they multiply and eventually settle at the bottom of the reactor as activated sludge. This bio-sludge filters the water, and the sludge can be changed as necessary, e.g., activated sludge, extended aeration, and sequential batch reactor systems. Third are the lagoon systems in which waste water is treated in water bodies such as lakes by the persisting micro flora of the water bodies (Mancl 2002). The organisms in a biofilm-based bioreactor are selectively used to remove the organic matter and to cause denitrification and dephosphorylation of the waste water (Yang et al. 2010). This process is implemented in either fixed bed or moving reactors. Advantages of the moving reactors include better usage of reactor volume. In 1860, the first sand filter treatment methods were used for water purification where sand acts as a support for biofilm establishment. In industrial systems, rotating biological contactor, bioWev, linpor sponge, capture sponge, and ring lace are the preferred media for the support of biofilms. The bacteria that grow on biofilms start with the selective development of autotrophs. It is very important that easier and effective methods for waste water treatment are introduced to obtain high-quality recycled water; thus, application of biofilm technology along with industrial expertise will be a constructive approach.

11.4.2 Biofouling

Many industries across the globe rely on high-throughput machinery with an intricate chain of supply pipelines and filters. These pipelines and membranes in a bioreactor system are often choked to insufficiency by the accumulation of certain growing entities such as plants, algae, and microbial biofilms, etc. This phenomenon is called 'biofouling' (Flemming 2002). This problem, apart from creating unnecessary financial and technical burdens, also adds to probable contamination of the produce; hence, it turns out to be a huge nightmare for the business globally. The agents used to prevent biofouling are called 'antifouling agents' (Yebra et al. 2004; Gupta et al. 2014; Kalia and Kumar 2015a; Kalia et al. 2015) and include chemical, physical, mechanical, thermal, electrical, and biological methods. Chemical methods utilize many artificially synthesized 'biocides' that are designed to kill any organism present. The most commonly used biocide is tributyltin (TBT). Although this method is effective to a certain extent, it has some serious implications on existing environmental and marine fauna. Nowadays, copper-based antifouling coating or paints are used as a self-polishing copolymers that react with ion in sea water and gradually leaches off. Yet again the safety of released copper in sea water is debatable (Guardiola et al. 2012; UK Marines). The use of oligo and poly ethylene glycols, fluoropolymers, and silicone coatings as inert surface coating material is also favorable, attributed to the material's non-reactivity, a disadvantage being their short-term stability (Rosenhahn et al. 2010). Thermal methods imply the treatment of equipment with hot water for a certain time period, causing detachment of the microbial communities (Piola and Hopkins 2012). Electrical methods involve the administration of short-term pulses to remove diatoms and algal biomass. Another recent method uses thin nanofiltration (NF) film membranes of electrically conductive polymer-nanocomposite (ECPNC), a strong and consistent electric flux to prevent biofilm deposition (De Lannoy et al. 2013). The most useful eco-friendly method for antifouling is the use of quorum-quenching systems. Quorum quenchers are the enzymes or inhibitors produced by bacteria, including acylases, lactonases, paraoxonases, and oxidoreductases. These molecules suppress the communication-related genes and signaling cascades (Mitchell 2001; Huma et al. 2011; Kalia and Purohit 2011; Kumar et al. 2013). This phenomenon was first reported in *Bacillus* spp., which secretes autoinducer inactivation gene *aiiA* lactonases, disrupting the lactone moieties of the acyl homoserine lactones and rendering the signaling inactive (Huma et al. 2011; Kalia and Purohit 2011; Kumar et al. 2013). Apart from the naturally occurring quorum-sensing inhibitors, many synthetic compounds have been used and frequently researched to be used against biofilm-based biofouling (Kalia 2013; Brackman and Coenye 2015). Biofouling can be prevented first by using quorum quenchers to specifically inhibit different crucial building steps such as blocking the transcription of quorum-sensing molecules like AHL, AI-2, etc.; second, by disrupting the efflux pumps of quorum-sensing molecules; third, by directly blocking the quorum-sensing molecules and transcription activators at the protein level using analogs; and, last and most effective, by using hydrolyzing enzymes such as lactones, acylases, etc. (Sio et al. 2006; Kalia et al. 2015). All these methods have been thoroughly researched and are constantly being modified to become more eco-friendly, either by immobilizing the quorum-quenching enzymes on the matrix or magnetic beads or by embedding the enzymes or inhibitors on the cell-entrapping beads (Yeon et al. 2009; Kim et al. 2013). Bacteriophages have also recently been constructively modified for use as quorum-quencher delivery agents (Pei and Lamas-Samanamud 2014). Bacteria have developed a fascinating level of evolution over the years; therefore, we cannot deny the ever-increasing emergence of bacterial resistance towards quorum-sensing inhibitors (García-Contreras et al. 2013; Kalia et al. 2014; Kalia and Kumar 2015b). This resistance will probably be our next step in the chase.

11.4.3 Bioremediation

Bioremediation deals with the removal of metallic waste from contaminated water, industrial effluent sites, etc. The environmental and health hazards caused by heavy metal toxicity is well known. Therefore, it is very important to dislodge heavy metals and other hazardous hydrocarbon impurities from waste water, which is often loaded with heavy metals such as iron, copper, lead, etc. Teitzel et al. showed a significant reduction of heavy metal contamination when passed through a *P. aeruginosa*-based rotating disk bioreactor because of the entrapment or chelation of metals by the EPS matrix (Teitzel and Parsek 2003). Scott and coworkers showed the successful entrapment of metal and other contaminants using species NCIMB11592 in an EPS matrix bound to granular-activated carbon. This setup provided a high surface area matrix with a very good absorption rate of metals (Scott et al. 1995). *Bacillus subtilis,* along with *Bacillus cereus* has been known to remove chromium ions (Sundar et al. 2011). Similarly *E. coli* biofilm on NaY zeolite and kaolin removed the Cr(VI), Cd(II), Fe(III), and Ni(II) from wastewater (Cristina et al. 2009). Another report demonstrated the removal of hexavalent chromium by *E. coli* biofilm supported on a granulated activated carbon from waste water bodies (Rabei et al. 2009). Brandy and coworkers reported the absorption of zinc by *Pseudomonas putida* by virtue of the functional metal chelators present on the bacterial cells (Toner et al. 2005). Diels used a moving-bed biofilm sand filter to bio precipitate heavy metals such as cadmium, copper, lead, etc. from the waste water. They used a cocktail of heavy metal-resistant bacteria *Alcaligenes eutrophus* CH34, *Pseudomonas mendocina* AS 302 and *Arthrobacter* spp. BP7/26 (Diels et al. 2001). Later on, Costley and coworkers used the rotating biological contactor for the removal of metals like zinc, cadmium, and copper using activated sludge enriched with the nutrient broth in a multiple absorption desorption cycle (Costley and Wallis 2001).

One of the most sought after applications of biofilms is the severance of oil spillage and other toxic compounds from water bodies. The common pollutants of this category are hydrocarbons, petrol, diesel, and xenobiotic waste such as synthetic chemicals. These threatening pollutants not only render unusable the available water on earth, but also strongly affect aqueous flora and fauna. Fortunately, some microorganisms in nature are equipped to eliminate these deleterious contents. The source of such pollutants is often industrial waste, accidental or negligent. In the past, bacteria such as *Planococcus alkanoclasticus, Marinobacter hydrocarbonoclasticus, Alcanivorax borkumensis* have been known to degrade hydrocarbon (Yakimov et al. 1998; Engelhardt et al. 2001; Mounier et al. 2014). Recently, by virtue of biotechnological development, we now know of a number of microorganisms that can prove beneficial for remediation. Marine cyanobacteria have long been known for their oil-degrading functions. Different marine bacterial species like *Plectonema terebrans, Aphanocapsa Alcanivorax*, and *Cycloclasticus* have been established for their exceptional role in the degradation of crude oil from water bodies (Raghukumar et al. 2001; Harayama et al. 2004). Another bacterial strain, *Pseudomonas stutzeri* T102, biofilm is described for its naphthalene-degrading characteristics as compared with its planktonic forms (Shimada et al. 2012). Marine bacteria *A. borkumensis* is reported for the degradation of linear and branched hydrocarbon chains. *Candida tropicans* was confirmed to display a capability of degrading almost 90 % of the diesel oil on a gravel bed (Chandran and Das 2011). This was the first report to promote the role of yeast flora in biodegradation. In 2013, new stains of *Pseudomonas* spp. were identified (KPW.1-S1, HRW.1-S3, and DSW.1-S4) for crude oil degradation (Dasgupta et al. 2013). These above reports and many more are suggestive of the new methods and different bacterial biofilms that can be utilized in a wise way, with much simpler operational costs to counteract hydrocarbon pollutants.

Chlorinated aromatic compounds such as dichlorophenols are major contaminants of the soil and groundwater. It is crucial that these contaminants be removed from the environment. So far, *P. putida* remains the most efficient microorganism for biodegradation of these xenobiotic compounds (Park and Kim 2000; Fernández et al. 2009). It was shown to remove almost 100 % of the dichlorophenols and around 70 % of the tri and tetrachlorophenols when used in combination with *Rhodococcus* species on a fluidized bed biofilm (Park et al. 1999). Polychlorinated biphenyls were also degraded using the granular activated carbon bed when kept in incubation with a mixture of microorganisms for a period of 2 months, which is suggestive of the acclimatization and altered metabolic profile of the bacterial strains. *Burkholderia vietnamiensis* G4 is used for elimination of toluene-based compounds (Amit et al. 2009). The removal of alkyl benzene sulphonates from water bodies by *Stenotrophomonas maltophilia* biofilm on silanized glass beads was shown by Hosseini and co-workers (Farzaneh et al. 2010). Along with this organic waste, one of the most fearsome wastes from human populations is the polyethene waste depicted with a common formula $(C_2H_4)_nH_2$. Polyethene is not readily biodegradable and, therefore, disposal is a very important issue worldwide. Apart from the role of different algal biofilms that are known to media plastic degradation, per se, *Phormidium Coleochaete Scutata, Coleochaete Soluta, Chroococcus, Aphanochaete, Aphanothece, Navicula, Gloeotaenium, Oedogonium, Chaetophora, Oocystis, Oscillatoria, Fragillaria, Cocconeis*, and *Cymbellan* (Suseela and Toppo 2007). Many new bacteria species have been identified for the purpose. A biofilm-producing actinomycete bacteria *Rhodococcus rubber* C208 has been discovered that displays a 0.86 % per week degradation rate of plastics (Orr et al. 2004; Mor and Sivan 2008). Previously, the same group reported the bio-degrading ability of a thermophillic bacteria *Brevibacillus borstelensis* (Hadad et al. 2005). The next step should be the efficient replication of these laboratory-validated strains on a commercial level in order to successfully remediate and recycle a major human resource.

11.4.4 Microbial Fuel Cells

Microorganisms have the capability to convert the metabolic energy of chemical bonds to generate free electrons and photons in the system. The oxidation of organic compounds carried out by every microbial cell results in the release of a large number of electrons. These charged particles are used to create an electrochemical gradient across the cell. The utilization of the electric potential-generating properties of bacteria is one of the most constructive forms of renewable energy, using the largest biomass population on earth. Microbial fuel cells (MFCs) are the devices that work on the bio-electrochemical principle, where microbes are grown as a biofilm in a conductivity device (Lovley 2006). An MFC assembly includes two chambers, aerobic and anaerobic, separated by a semi-permeable membrane. The aerobic chamber has a positively charged cathode electrode that is bubbled with oxygen gas, while the anaerobic chamber has an immersed anode electrode in the organic matter enriched with the bacterial population. The bacteria oxidize the organic matter and transfer electrons to the anode, which is connected to the cathode through an outer circuit (Logan et al. 2006). Microbes were first utilized for this purpose in 1911 by Potter, where he used *E. coli* as a source of electron beneficiary at anode (Potter 1911). Later on, in 1931, another scientist, Barnet Cohen, created several microbial half cells, and produced 35 V of energy on being arranged in series (Cohen 1931). Suzuki et al., in 1977, established the current working model of MFCs using *Clostridium butyricum* as the first commercialized organism (Karube et al. 1977). The initial working model used a chemical mediator like thionine, methyl viologen, and neutral red to transfer electrons from microbes to anode. Later on, the technology advanced to the mediator-less MFCs. The microorganism transfer electrons either by the surface localized cytochrome C or by their cellular appendages, pili. These MFC used highly electrochemically active microorganisms like *Shewanella putrefacians* and *A. hydrophila* (Kim et al. 1999; Pham et al. 2003). MFCs are the most promising application of bacterial biofilms in the area of energy production and waste water treatment. MFCs are equipped with a wide range of highly productive substrates to be catalyzed by the bacteria (Pant et al. 2010). As waste water is an extremely rich source of organic matter, many researchers have invested in methods that demonstrate the resumption of electricity from waste water using MFCs (Capodaglio et al. 2013). This approach practically prevents two major global problems. Many reports show the prevailing enthusiasm for the technology. Other than being used in recycling and regenerating electricity from waste water, MFCs have great potential in electrifying wireless remote sensors, underwater monitoring devices, and determining biological oxygen demands, denitrification of groundwater, etc. (Chang et al. 2004; Shantaram et al. 2005; Donovan et al. 2008; Pous et al. 2013). The potentials for this application are still unfolding and will probably be lifesaving in the coming years for the human population.

11.4.5 Bacterial Nanotechnology

Biofilm constituting bacteria produce adhesive nanofibers to attach themselves to surfaces and to each other. Under repulsive and distressed conditions, these bacterial nanofibrils provide flexibility and strength to the matrix. These nanofibers confer upon the biofilm matrix high adhesive quality and strength (Zhong et al. 2014). The bacterial community of the biofilm can be utilized to create synthetic matrices by genetically altering the matrix composition. A very recent technology has been introduced by scientists from Harvard University. This technology exploits the self-assembly and rhythmicity of the biofilm-forming bacteria. As we know, bacterial biofilms are highly organized structures that have self-healing properties. This feature has been used as a synthetic biological platform to self-assemble the recombinant proteins introduced in the matrix. This technology is called 'biofilm integrated nanofiber display' (BIND). Scientists have developed a strategy to reprogram the matrix material by genetically modulating the peptide domains of an *E. coli* secretory amyloid protein CsgA (Uhlich et al. 2009). These genetically engineered CsgA fusion proteins are secreted and undergo self-assembly into amyloid nanofibers that are capable of adhering to stainless steel and contribute to the robustness, substrate adhesiveness, and functionality of biofilms (Nguyen et al. 2014). BIND can be a very promising technology that uses these microbial biofilms as factories to create programmable artificial nano assemblies and nanomaterial in the biofilm matrix and thereby can be utilized to create robust biomaterials relevant to biomedical technology and nanotechnology.

11.4.6 Drug/Enzyme Delivery

To achieve efficient and biocompatible targeted drug delivery, various technologies have been developed to deliver chemotherapeutic agents, antibiotics, therapeutic proteins, and biomolecules. One novel aspect could be the application of biofilm formation by drug-carrying bacteria to deliver drugs/antibiotics to the site of infection. Biofilm formation in the human body via various commensals is very common. These commensal biofilms can be productively used to deliver various compounds such as drugs, enzymes, etc. (Claesen and Fischbach 2014). A nonpathogenic commensal surviving in the human system as a biofilm can be further augmented with the population enriched with specific enzyme-secreting engineered bacteria. The same principle can be applied to eliminate biofilm via targeted delivery of alginate lyase or mucinase-overproducing bacterial strains. Otsuka and co-workers recently designed a chimeric *E. coli* strain expressing the glucan-digesting enzymes mutanase and dextranase, which decompose the *Streptococcus mutans* biofilm in vitro (Otsuka et al. 2015). Previous reports claim the potential treatment of murine colitis by implanting the engineered *Lactococcus lactis*-secreting interleukin-10 in mice colon (Steidler et al. 2000). Naznein and co-wokers recently demonstrated the inhibition of *P. aeruginosa* biofilms by developing a synthetic system comprising the genetically engineered *E. coli* secreting the toxin pyocin. The synthetic bacteria targets and kill the planktonic as well as the biofilm of *P. aeruginosa* with almost 90 % efficiency (Saeidi et al. 2011). Other *E. coli* engineered systems were used for gut inflammation (Archer et al. 2012). In an identical scenario, the attenuated strains of *Salmonella typhimurium* are genetically modified to trigger the production of the tumor necrosis factor-related apoptosis-inducing ligand (TRAIL) under a hypoxia-induced targeted promoter system *nirB* (Chen et al. 2012). When administered, this strain specifically targeted the malignant melanoma of mice. These reports, along with many others, provide a strong platform for antimicrobial technology. This technology has been approved by extensive literature and various scientific communities working in the area of targeted drug delivery. Pitfalls in this technology can include specificity, biocompatibility, and toxicity.

11.5 Summary

Bacterial biofilms are nature's work of art. The bacteria stays and functions as an efficient cellular society capable of sustaining its own needs. The bacteria, by means of community establishment, not only excel in survival but also gain many important characteristics. This report describes in general many aspects of bacterial biofilms, such as structure, bacterial benefits, infections caused by biofilms, and, finally, the exploitation of bacterial biofilms for human welfare. The havoc caused by bacterial biofilms in biomedical industries, food industries, and water pipelines is a major concern. Today, with the rise of technological development and advances, it is possible to quarantine or even destroy biofilms. The scientific explorations utilizing bacteria as a source of drug/enzyme delivery is one of the foremost applications that can be further elaborated to cure many anomalies given the basic behavior of organism and disease. The site-specific delivery of lacking enzymes can be achieved by introducing genetically engineered bacteria to establish in a commensal biofilm. *E. coli* is a very common organism for this purpose because of its universal locality. It is equally important to ensure the reproducibility of the in vitro parameters in the in vivo systems. The technology has its own disadvantages. Since bacteria are an ever-evolving population, it becomes very important to prevent bacterial evolution, which may allow bacteria to undergo genetic transformations to gain antibiotic resistance and surpass effective measures. Survival in harsh and stressed environments is the forte of bacterial biofilms. It is very important to understand the proper physiological and mechanistic aspects behind biofilm formation and bacterial defenses. A lot of research work is aimed at better understanding the steps of biofilm formation. Rather than dislodging the membrane merely by administration of antibiotics, it is important to find the right cord that disables bacterial biofilm formation or irreversibly disperses it. Biofilm is an immensely successful survival tactic of the world's largest biomass, and we should focus our research towards creating methods that can utilize it in the best possible way.

Acknowledgments The authors are thankful to Dr. V.C. Kalia and Dr. Yogendra Singh for providing us the opportunity to contribute a chapter to this book. ND, RS, SK and AKS are thankful to ICMR, CSIR, and UGC for providing fellowships. We are also thankful to the Dean, Department of Zoology, Delhi University, and the Director, CSIR-IGIB for providing necessary facilities and support for this work.

References

Albrecht MT, Schiller NL (2005) Alginate lyase (AlgL) activity is required for alginate biosynthesis in *pseudomonas aeruginosa*. J Bacteriol 187:3869–3872. doi:10.1128/JB.187.11.3869-3872.2005

Amit K, Dewulf J, Wiele TV, Langenhove HV (2009) Bacterial dynamics of biofilm development during toluene degradation by *Burkholderia vietnamiensis* G4 in a gas phase membrane bioreactor. J Microbiol Biotechnol 19:1028–1033

Archer EJ, Robinson AB, Süel GM (2012) Engineered *E. coli* that detect and respond to gut inflammation through nitric oxide sensing. ACS Synth Biol 1:451–457. doi:10.1021/sb3000595

Bjarnsholt T, Tolker-Nielsen T, Høiby N, Givskov M (2010) Interference of *Pseudomonas aeruginosa* signalling and biofilm formation for infection control. Expert Rev Mol Med 12:e11. doi:10.1017/S1462399410001420

Borlee BR, Goldman AD, Murakami K, Samudrala R, Wozniak DJ, Parsek MR (2010) *Pseudomonas aeruginosa* uses a cyclic-di-GMP-regulated adhesin to reinforce the biofilm extracellular matrix. Mol Microbiol 75:827–842. doi:10.1111/j.1365-2958.2009.06991.x

Bowden GH, Li YH (1997) Nutritional influences on biofilm development. Adv Dent Res 11:81–99

Brackman G, Coenye T (2015) Quorum sensing inhibitors as anti-biofilm agents. Curr Pharm Des 21:5–11. doi:10.2174/1381612820666140905114627#sthash.Z926xlPS.dpuf

Caiazza NC, O'Toole GA (2004) SadB is required for the transition from reversible to irreversible attachment during biofilm formation by *Pseudomonas aeruginosa* PA14. J Bacteriol 186:4476–4485. doi:10.1128/JB.186.14.4476-4485.2004

Capodaglio AG, Molognoni D, Dallago E, Liberale A, Cella R, Longoni P, Pantaleoni L (2013) Microbial fuel cells for direct electrical energy recovery from urban wastewaters. Sci World J 2013:634738. doi:10.1155/2013/634738

Cardines R, Giufrè M, Pompilio A, Fiscarelli E, Ricciotti G, Di Bonaventura G, Cerquetti M (2012) *Haemophilus influenzae* in children with cystic fibrosis: antimicrobial susceptibility, molecular epidemiology, distribution of adhesins and biofilm formation. Int J Med Microbiol 302:45–52. doi:10.1016/j.ijmm.2011.08.003e

Chandran P, Das N (2011) Degradation of diesel oil by immobilized *Candida tropicalis* and biofilm formed on gravels. Biodegradation 22:1181–1189. doi:10.1007/s10532-011-9473-1

Chang IS, Jang JK, Gil GC, Kim M, Kim HJ, Cho BW, Kim BH (2004) Continuous determination of biochemical oxygen demand using microbial fuel cell type biosensor. Biosens Bioelectron 19:607–613. doi:10.1016/S0956-5663(03)00272-0

Chen J, Yang B, Cheng X, QiaoY TB, Chen G, Wei J, Liu X, Cheng W, Du P, Huang X, JiangW HQ, Hu Y, Li J, Hua ZC (2012) *Salmonella*-mediated tumor-targeting TRAIL gene therapy significantly suppresses melanoma growth in mouse model. Cancer Sci 103:325–333. doi:10.1111/j.1349-7006.2011.02147.x

Claesen J, Fischbach MA (2014) Synthetic microbes as drug delivery systems. ACS Synth Biol. doi:10.1021/sb500258b

Clark Ehlers GA, Turner SJ (2012) "Chapter 6: Abstract." Microbial biofilms current research and applications. Caister Academic Pr,. Web. 21 Jan 2012

Cohen B (1931) The bacterial culture as an electrical half-cell. J Bacteriol 21:18–19

Costerton JW, Geesey GG, Cheng KJ (1978) How bacteria stick? Sci Am 238:86–95

Costley SC, Wallis FM (2001) Bioremediation of heavy metals in a synthetic wastewater using a rotating biological contactor. Water Res 35:3715–3723. doi:10.1016/S0043-1354(01)00072-0

Cristina Q, Zelia R, Bruna F, Hugo F, Teresa T (2009) Biosorptive performance of an *Escherichia coli* biofilm supported on zeolite NaY for the removal of Cr(VI), Cd(II), Fe(III) and Ni(II). Chem Eng J 152:110–115. doi:10.1016/j.cej.2009.03.039

Darveau RP (2010) Periodontitis: a polymicrobial disruption of host homeostasis. Nat Rev Microbiol 8:481–490. doi:10.1038/nrmicro2337

Dasgupta D, Ghosh R, Sengupta TK (2013) Biofilm-mediated enhanced crude oil degradation by newly isolated *Pseudomonas* species. ISRN Biotechnol 2013:250749. doi:10.5402/2013/250749

Davies DG, Parsek MR, Pearson JP, Iglewski BH, Costerton JW, Greenberg EP (1998) The involvement of cell-to-cell signals in the development of a bacterial biofilm. Science 280:295–298. doi:10.1126/science.280.5361.295

De Kievit TR, Gillis R, Marx S, Brown C, Iglewski BH (2001) Quorum-sensing genes in *Pseudomonas aeruginosa* biofilms: their role and expression patterns. Appl Environ Microbiol 67:1865–1873. doi:10.1128/AEM.67.4.1865-1873.2001

de Lannoy CF, Jassby D, Gloe K, Gordon AD, Wiesner MR (2013) Aquatic biofouling prevention by electrically charged nanocomposite polymer thin film membranes. Environ Sci Technol 47:2760–2768. doi:10.1021/es3045168

Diels L, Spaans PH, Van Roy S, Hooyberghs L, Wouters H, Walter E, Winters J, Macaskie L, Pernfuss B, Woebking H, Pümpel T (2001) Heavy metals removal by sand filters inoculated with metal sorbing and precipitating bacteria. Biohydro metall Fundam Technol and Sustain Dev 11:317–326. doi:10.1016/S0304-386X(03)00161-0, part B

Donlan RM (2002) Biofilms: microbial life on surfaces. Emerg Infect Dis 8:881–890

Donlan RM (2011) Biofilm elimination on intravascular catheters: important considerations for the infectious disease practitioner. Clin Infect Dis 52:1038–1045. doi:10.1093/cid/cir077

Donlan RM, Costerton JW (2002) Biofilms: survival mechanisms of clinically relevant microorganisms.

Clin Microbiol Rev 15:167–193. doi:10.1128/CMR.15.2.167-193.2002

Donovan C, Dewan A, Heo D, Beyenal H (2008) Batteryless, wireless sensor powered by a sediment microbial fuel cell. Environ Sci Technol 42:8591–8596. doi:10.1021/es801763g

Elkins MR, Robinson M, Rose BR, Harbour C, Moriarty CP, Marks GB, Belousova EG, Xuan W, Bye PT, National Hypertonic Saline in Cystic Fibrosis (NHSCF) Study Group (2006) A controlled trial of long-term inhaled hypertonic saline in patients with cystic fibrosis. N Engl J Med 354:229–240. doi:10.1056/NEJMoa043900

Else TA, Pantle CR, Amy PS (2003) Boundaries for biofilm formation: humidity and temperature. Appl Environ Microbiol 69:5006–5010. doi:10.1128/AEM.69.8.5006-5010.2003

Engelhardt MA, Daly K, Swannell RPJ, Head IM (2001) Isolation and characterization of a novel hydrocarbon-degrading, gram-positive bacterium, isolated from intertidal beach sediment, and description of *Planococcus alkanoclasticus* sp. nov. J Appl Microbiol 90:237–247. doi:10.1046/j.1365-2672.2001.01241.x

Farzaneh H, Fereidon M, Noor A, Naser G (2010) Biodegradation of dodecyl benzenesulfonate sodium by *Stenotrophomonas maltophilia* biofilm. Afr J Biotechnol 9:55–62

Fernández M, Duque E, Pizarro-Tobías P, Van Dillewijn P, Wittich RM, Ramos JL (2009) Microbial responses to xenobiotic compounds. Identification of genes that allow *Pseudomonas putida* KT2440 to cope with 2,4,6-trinitrotoluene. Microb Biotechnol 2:287–294. doi:10.1111/j.1751-7915.2009.00085.x

Flemming HC (2002) Biofouling in water systems – cases, causes and countermeasures. Appl Microbiol Biotechnol 59:629–640. doi:10.1007/s00253-002-1066-9

Flemming HC, Wingender J (2010) The biofilm matrix. Nat Rev Microbiol 8:623–633. doi:10.1038/nrmicro2415

Fletcher M (1988) Attachment of *Pseudomonas fluorescens* to glass and influence of electrolytes on bacterium-substratum separation distance. J Bacteriol 170:2027–2030

Fleuchot B, Gitton C, Guillot A, Vidic J, Nicolas P, Besset C, Fontaine L, Hols P, Leblond-Bourget N, Monnet V, Gardan R (2011) Rgg proteins associated with internalized small hydrophobic peptides: a new quorum-sensing mechanism in *Streptococci*. Mol Microbiol 80:1102–1119.doi:10.1111/j.1365-2958.2011.07633.x

Gamboa F, Acosta A, García DA, Velosa J, Araya N, Ledergerber R (2014) Occurrence of *Porphyromonas gingivalis* and its antibacterial susceptibility to metronidazole and tetracycline in patients with chronic periodontitis. Acta Odontol Latinoam 27:137–144. doi:10.1590/S1852-48342014000300007

García-Contreras R, Martínez-Vázquez M, Velázquez Guadarrama N, Villegas Pañeda AG, Hashimoto T, Maeda T, Quezada H, Wood TK (2013) Resistance to the quorum-quenching compounds brominated furanone C-30 and 5-fluorouracil in *Pseudomonas aeruginosa* clinical isolates. Pathog Dis 68:8–11. doi:10.1111/2049-632X.12039

Guardiola FA, Cuesta A, Meseguer J, Esteban MA (2012) Risks of using antifouling biocides in aquaculture. Int J Mol Sci 13:1541–1560. doi:10.3390/ijms13021541

Gupta K, Barua S, Hazarika SN, Manhar AK, Nath D, Karak N, Namsa ND, Mukhopadhyay R, Kalia VC, Mandal M (2014) Green silver nanoparticles: enhanced antimicrobial and antibiofilm activity with effects on DNA replication and cell cytotoxicity. RSC Adv 4:52845–52855. doi:10.1039/C4RA08791G

Hadad D, Geresh S, Sivan A (2005) Biodegradation of polyethylene by the thermophilic bacterium *Brevibacillusborstelensis*. J Appl Microbiol 98:1093–1100. doi:10.1111/j.1365-2672.2005.02553.x

Hall-Stoodley L, Costerton JW, Stoodley P (2004) Bacterial biofilms: from the natural environment to infectious diseases. Nat Rev Microbiol 2:95–108. doi:10.1038/nrmicro821

Hannan S, Ready D, Jasni AS, Rogers M, Pratten J, Roberts AP (2010) Transfer of antibiotic resistance by transformation with eDNA within oral biofilms. FEMS Immunol Med Microbiol 59:345–349. doi:10.1111/j.1574-695X.2010.00661.x

Harayama S, Kasai Y, Hara A (2004) Microbial communities in oil-contaminated seawater. Curr Opin Biotechnol 15:205–214. doi:10.1016/j.copbio.2004.04.002

Hatch RA, Schiller NL (1998) Alginate lyase promotes diffusion of aminoglycosides through the extracellular polysaccharide of mucoid *Pseudomonas aeruginosa*. Antimicrob Agents Chemother 42:974–977

Hay ID, Rehman ZU, Moradali MF, Wang Y, Rehm BHA (2013) Microbial alginate production, modification and its applications. Microb Biotechnol 6:637–650. doi:10.1111/1751-7915.12076

Herrmann G, Yang L, Wu H, Song Z, Wang H, Høiby N, Ulrich M, Molin S, Riethmüller J, Döring G (2010) Colistin-tobramycin combinations are superior to monotherapy concerning the killing of biofilm *Pseudomonas aeruginosa*. J Infect Dis 202:1585–1592. doi:10.1086/656788

Heydorn A, Ersbøll B, Kato J, Hentzer M, Parsek MR, Tolker-Nielsen T, Givskov M, Molin S (2002) Statistical analysis of *Pseudomonas aeruginosa* biofilm development: impact of mutations in genes involved in twitching motility, cell-to-cell signaling, and stationary-phase sigma factor expression. Appl Environ Microbiol 68:2008–2017. doi:10.1128/AEM.68.4.2008-2017.2002

Hickman JW, Tifrea DF, Harwood CS (2005) A chemosensory system that regulates biofilm formation through modulation of cyclic diguanylate levels. Proc Natl Acad Sci 102:14422–14427. doi:10.1073/pnas.0507170102

Hinsa SM, Espinosa-Urgel M, Ramos JL, O'Toole GA (2003) Transition from reversible to irreversible attachment during biofilm formation by *Pseudomonas fluorescens* WCS365 requires an ABC transporter and a large secreted protein. Mol Microbiol 49:905–918. doi:10.1046/j.1365-2958.2003.03615.x

Hirschhausen N, Block D, Bianconi I, Bragonzi A, Birtel J, Lee JC, Dübbers A, Küster P, Kahl J, Peters G, Kahl BC (2013) Extended *Staphylococcus aureus* persistence in cystic fibrosis is associated with bacterial adaptation. Int J Med Microbiol 303:685–692. doi:10.1016/j.ijmm.2013.09.012

Høiby N, Ciofu O, Bjarnsholt T (2010) *Pseudomonas aeruginosa* biofilms in cystic fibrosis. Future Microbiol 5:1663–1674. doi:10.2217/fmb.10.125

Holá V, Ruzicka F, Horka M (2010) Microbial diversity in biofilm infections of the urinary tract with the use of sonication techniques. FEMS Immunol Med Microbiol 59:525–528. doi:10.1111/j.1574-695X.2010.00703.x

http://www.ukmarinesac.org.uk/activities/recreation/r03_03.htm

Huma N, Shankar P, Kushwah J, Bhushan A, Joshi J, Mukherjee T, Raju SC, Purohit HJ, Kalia VC (2011) Diversity and polymorphism in AHL-lactonase gene (*aiiA*) of *Bacillus*. J Microbiol Biotechnol 21:1001–1011. doi:10.4014/jmb.1105.05056

Hussain M, Steinbacher T, Peters G, Heilmann C, Becker K (2015) The adhesive properties of the *Staphylococcus lugdunensis* multifunctional autolysin AtlL and its role in biofilm formation and internalization. Int J Med Microbiol 305:129–139. doi:10.1016/j.ijmm.2014.11.010

Incani V, Omar A, Prosperi-Porta G, Nadworny P (2014) Ag5IO6: novel antibiofilm activity of a silver compound with application to medical devices. Int J Antimicrob Agents S0924–8579(14):00295–00297. doi:10.1016/j.ijantimicag.2014.09.008

Jimenez JC, Federle MJ (2014) Quorum sensing in group a *Streptococcus*. Front Cell Infect Microbiol 4:127. doi:10.3389/fcimb.2014.00127

Kalia VC (2013) Quorum sensing inhibitors: an overview. Biotechnol Adv 31:224–245. doi:10.1016/j.biotechadv.2012.10.004

Kalia VC (2014a) Microbes, antimicrobials and resistance: the battle goes on. Indian J Microbiol 54:1–2. doi:10.1007/s12088-013-0443-7

Kalia VC (2014b) In search of versatile organisms for quorum-sensing inhibitors: acyl homoserine lactones (AHL)-acylase and AHL-lactonase. FEMS Microbiol Lett 359:143. doi:10.1111/1574-6968.12585

Kalia VC, Kumar P (2015a) Potential applications of quorum sensing inhibitors in diverse fields. In: Kalia VC (ed) Quorum sensing vs quorum quenching: a battle with no end in sight. Springer, New Delhi, pp 359–370. doi:10.1007/978-81-322-1982-8_29

Kalia VC, Kumar P (2015b) The Battle: quorum-sensing inhibitors versus evolution of bacterial resistance. In: Kalia VC (ed) Quorum sensing vs quorum quenching: a battle with no end in sight. Springer, New Delhi, pp 385–391. doi:10.1007/978-81-322-1982-8_31

Kalia VC, Purohit HJ (2011) Quenching the quorum sensing system: potential antibacterial drug targets. Crit Rev Microbiol 37:121–140. doi:10.3109/1040841X.2010.532479

Kalia VC, Raju SC, Purohit HJ (2011) Genomic analysis reveals versatile organisms for quorum quenching enzymes: acyl-homoserine lactone-acylase and -lactonase. Open Microbiol J 5:1–13. doi:10.2174/1874285801105010000

Kalia VC, Wood TK, Kumar P (2014) Evolution of resistance to quorum-sensing inhibitors. Microb Ecol 68:13–23. doi:10.1007/s00248-013-0316-y

Kalia VC, Kumar P, Pandian SK, Sharma P (2015) Biofouling control by quorum quenching. In: Kim SK (ed) Hb25_handbook of marine biotechnology, vol 15. Springer, New Delhi, India, pp 431–440. doi: 10.1007/978-3-642-53971-8_15

Karatan E, Watnick P (2009) Signals, regulatory networks, and materials that build and break bacterial biofilms. Microbiol Mol Biol Rev 73:310–347. doi:10.1128/MMBR.00041-08

Karube I, Tadashi M, Shinya T, Shuichi S (1977) Biochemical cells utilizing immobilized cells of *Clostridium butyricum*. Biotechnol Bioeng 19:1727–1733. doi:10.1002/bit.260191112

Kim BH, Kim HJ, Hyun MS, Park DH (1999) Direct electrode reaction of Fe (III) reducing bacterium, *Shewanella putrefacience*. J Microbiol Biotechnol 9:127–131

Kim SR, Oh HS, Jo SJ, Yeon KM, Lee CH, Lim DJ, Lee CH, Lee JK (2013) Biofouling control with bead-entrapped quorum quenching bacteria in membrane bioreactors: physical and biological effects. Environ Sci Technol 47:836–842. doi:10.1021/es303995s

Kjelleberg S, McDougald D, Rasmussen TB, Givskov M (2008) Quorum-sensing inhibition. In: Winans SC, Bassler BL (eds) Chemical communication among bacteria. ASM Press, Washington, D.C., pp 393–416

Kumar P, Patel SKS, Lee JK, Kalia VC (2013) Extending the limits of *Bacillus* for novel biotechnological applications. Biotechnol Adv 31:1543–1561. doi:10.1016/j.biotechadv.2013.08.007

Kumar P, Koul S, Patel SKS, Lee JK, Kalia VC (2015) Heterologous expression of quorum sensing inhibitory genes in diverse organisms. In: Kalia VC (ed) Quorum sensing vs quorum quenching: a battle with no end in sight. Springer, New Delhi, pp 343–356. doi:10.1007/978-81-322-1982-8_28

Laverty G, Gorman SP, Gilmore BF (2014) Biomolecular mechanisms of *Pseudomonas aeruginosa* and *Escherichia coli* biofilm formation. Pathogens 3:596–632. doi:10.3390/pathogens3030596

Logan BE, Hamelers B, Rozendal R, Schröder U, Keller J, Freguia S, Aelterman P, Verstraete W, Rabaey K (2006) Microbial fuel cells: methodology and technology. Environ Sci Technol 40:5181–5192. doi:10.1021/es0605016

Love RM (2010) Biofilm – substrate interaction: from initial adhesion to complex interactions and biofilm maturity. Topics 22:50–57. doi:10.1111/j.1601-1546.2012.00280.x

Lovley DR (2006) Bug juice: harvesting electricity with microorganisms. Nat Rev Microbiol 4:497–508. doi:10.1038/nrmicro1442

Lye DCB, Hughes A, Brien D, Athan E (2005) *Candida glabrata* prosthetic valve endocarditis treated successfully with fluconazole plus caspofungin without surgery: a case report and literature review. Eur J Clin Microbiol Infect Dis 24:753–755. doi:10.1007/s10096-005-0038-2

Lynch MJ, Swift S, Kirke DF, Keevil GW, Dodd CER, Williams P (2002) The regulation of biofilm development by quorum sensing in *Aeromonas hydrophila*. Environ Microbiol 4:18–28. doi:10.1128/JB.186.3.692-698.2004

Lyon GJ, Wright JS, Muir TW, Novick RP (2002) Key determinants of receptor activation in the agr autoinducing peptides of *Staphylococcus aureus*. Biochemistry 41:10095–10104. doi:10.1021/bi026049u

Mancl K (2002) Model for success in on-site wastewater management. J Environ Health 64:29–31

Mitchell A (2001) Quorum quenching. Nat Rev Mol Cell Biol 2:488. doi:10.1038/35080025

Molin S, Tolker-Nielsen T (2003) Gene transfer occurs with enhanced efficiency in biofilms and induces enhanced stabilisation of the biofilm structure. Curr Opin Biotechnol 14:255–261. doi:10.1016/S0958-1669(03)00036-3

Monnet V, Juillard V, Gardan R (2014) Peptide conversations in Gram-positive bacteria. Crit Rev Microbiol 2:100–105. doi:10.3109/1040841X.2014.948804

Mor R, Sivan A (2008) Biofilm formation and partial biodegradation of polystyrene by the actinomycete *Rhodococcus ruber*: biodegradation of polystyrene. Biodegradation 19:851–858. doi:10.1007/s10532-008-9188-0

Mounier J, Camus A, Mitteau I, Vaysse PJ, Goulas P, Grimaud R, Sivadon P (2014) The marine bacterium *Marinobacter hydrocarbonoclasticus* SP17 degrades a wide range of lipids and hydrocarbons through the formation of oleolytic biofilms with distinct gene expression profiles. FEMS Microbiol Ecol 90:816–831. doi:10.1111/1574-6941.12439

Nguyen KT, Piastro K, Gray TA, Derbyshire KM (2010) Mycobacterial biofilms facilitate horizontal DNA transfer between strains of *Mycobacterium smegmatis*. J Bacteriol 192:5134–5142. doi:10.1128/JB.00650-10

Nguyen PQ, Botyanszki Z, Tay PK, Joshi NS (2014) Programmable biofilm-based materials from engineered curli nanofibres. Nat Commun 5:4945. doi:10.1038/ncomms5945

Nwodo UU, Green E, Okoh AI (2012) Bacterial exopolysaccharides: functionality and prospects. Int J Mol Sci 13:14002–14015. doi:10.3390/ijms131114002

O'Loughlin CT, Miller LC, Siryaporn A, Drescher K, Semmelhack MF, Bassler BL (2013) A quorum-sensing inhibitor blocks *Pseudomonas aeruginosa* virulence and biofilm formation. Proc Natl Acad Sci 110:17981–17986. doi:10.1073/pnas.1316981110

Orr IG, Hadar Y, Sivan A (2004) Colonization, biofilm formation and biodegradation of polyethylene by a strain of *Rhodococcus ruber*. Appl Microbiol Biotechnol 65:97–104. doi:10.1007/s00253-004-1584-8

Otsuka R, Imai S, Murata T, Nomura Y, Okamoto M, Tsumori H, Kakuta E, Hanada N, Momoi Y (2015) Application of chimeric glucanase comprising mutanase and dextranase for prevention of dental biofilm formation. Microbiol Immunol 59:28–36. doi:10.1111/1348-0421.12214

Pant D, Van Bogaert G, Diels L, Vanbroekhoven K (2010) A review of the substrates used in microbial fuel cells (MFCs) for sustainable energy production. Bioresour Technol 101:1533–1543. doi:10.1016/j.biortech.2009.10.017

Park HS, Kim HS (2000) Identification and characterization of the nitrobenzene catabolic plasmids pNB1 and pNB2 in *Pseudomonas putida* HS12. J Bacteriol 82:573–580. doi:10.1128/JB.182.3.573-580.2000

Park HS, Lim SJ, Chang YK, Livingston AG, Kim HS (1999) Degradation of chloronitrobenzenes by a coculture of *Pseudomonas putida* and a *Rhodococcus* spp. Appl Environ Microbiol 65:1083–1091

Pei R, Lamas-Samanamud GR (2014) Inhibition of biofilm formation by T7 bacteriophages producing quorum-quenching enzymes. Appl Environ Microbiol 80:5340–5348. doi:10.1128/AEM.01434-14

Peyyala R, Ebersole JL (2013) Multispecies biofilms and host responses: "Discriminating the trees from the forest.". Cytokine 61:15–25. doi:10.1016/j.cyto.2012.10.006

Pham CA, Jung SJ, Phung NT, Lee J, Chang IS, Kim BH, Yi H, Chun J (2003) A novel electrochemically active and Fe(III)-reducing bacterium phylogenetically related to *Aeromonas hydrophila*, isolated from a microbial fuel cell. FEMS Microbiol Lett 223:129–134. doi:10.1016/S0378-1097(03)00354-9

Piola RF, Hopkins GA (2012) Thermal treatment as a method to control transfers of invasive biofouling species via vessel sea chests. Mar Pollut Bull 64:1620–1630. doi:10.1016/j.marpolbul.2012.05.028

Potter MC (1911) Electrical effects accompanying the decomposition of organic compounds. Proc R Soc B Biol Sci 84:260–276

Pous N, Puig S, Coma M, Balaguer MD, Colprim J (2013) Bioremediation of nitrate-polluted groundwater in a microbial fuel cell. J Chem Technol Biotechnol 88:1690–1696. doi:10.1002/jctb.4020

Rabei MG, Gad-Elrab SMF, Abskharon RNN, Hassan SHA, Shoreit AAM (2009) Biosorption of hexavalent chromium using biofilm of *E. coli* supported on granulated activated carbon. World J Microbiol Biotechnol 25:1695–1703. doi:10.1007/s11274-009-0063-x

Raghukumar C, Vipparty V, David JJ, Chandramohan D (2001) Degradation of crude oil by marine cyanobacteria. Appl Microbiol Biotechnol 57:433–436. doi:10.1007/s002530100784

Reeser RJ, Medler RT, Billington SJ, Jost BH, Joens LA (2007) Characterization of *Campylobacter jejuni* biofilms under defined growth conditions. Appl Environ Microbiol 73:1908–1913. doi:10.1128/AEM.00740-06

Riedel DJ, Weekes E, Forrest GN (2008) Addition of rifampin to standard therapy for treatment of native valve infective endocarditis caused by *Staphylococcus aureus*. Antimicrob Agents Chemother 52:2463–2467. doi:10.1128/AAC.00300-08

Robles-Price A, Wong TY, Sletta H, Valla S, Schiller NL (2004) AlgX is a periplasmic protein required for alginate biosynthesis in *Pseudomonas aeruginosa*. J Bacteriol 186:7369–7377. doi:10.1128/JB.186.21.7369-7377.2004

Rodrigues DF, Elimelech M (2009) Role of type 1 fimbriae and mannose in the development of *Escherichia coli* K12 biofilm: from initial cell adhesion to biofilm formation. Biofouling 25:401–411. doi:10.1080/08927010902833443

Rosen R, Bachrach G, Bronshteyn M, Gedalia I, Steinberg D (2005) The role of fructans on dental biofilm formation by *Streptococcus sobrinus*, *Streptococcus mutans*, *Streptococcus gordonii* and *Actinomyces viscosus*. FEMS Microbiol Lett 195:205–210. doi:10.1016/S0378-1097(01)00009-x

Rosenhahn A, Schilp S, Kreuzer HJ, Grunze M (2010) The role of "inert" surface chemistry in marine biofouling prevention. Phys Chem Chem Phys 12:4275–4286. doi:10.1039/c001968m

Saeidi N, Wong CK, Lo TM, Nguyen HX, Ling H, Leong SSJ, Poh CL, Chang MW (2011) Engineering microbes to sense and eradicate *Pseudomonas aeruginosa*, a human pathogen. Mol Syst Biol 7:521. doi:10.1038/msb.2011.55

Scott JA, Karanjkar AM, Rowe DL (1995) Biofilm covered granular activated carbon for decontamination of streams containing heavy metals and organic chemicals. Miner Eng 8:221–230. doi:10.1016/0892-6875(94)00115-S

Sharma VK, Bearson SM, Bearson BL (2010) Evaluation of the effects of *sdiA*, a luxR homologue, on adherence and motility of *Escherichia coli* O157:H7. Microbiology 156:1303–1312. doi:10.1099/mic.0.034330-0

Shantaram A, Beyenal H, Raajan R, Veluchamy A, Lewandowskiz Z (2005) Wireless sensors powered by microbial fuel cells. Environ Sci Technol 39(13): 5037–5042. doi:10-1021/es0480668

Shimada K, Itoh Y, Washio K, Morikawa M (2012) Efficacy of forming biofilms by naphthalene degrading *Pseudomonas stutzeri* T102 toward bioremediation technology and its molecular mechanisms. Chemosphere 87:226–233. doi:10.1016/j.chemosphere.2011.12.078

Sio CF, Otten LG, Cool RH, Diggle SP, Braun PG, Bos R, Daykin M, Cámara M, Williams P, Quax WJ (2006) Quorum quenching by an N-acyl-homoserine lactone acylase from *Pseudomonas aeruginosa* PAO1. Infect Immun 74:1673–1682. doi:10.1128/IAI.74.3.1673-1682.2006

Soini SM, Koskinen KT, Vilenius MJ, Puhakka JA (2002) Effects of fluid-flow velocity and water quality on planktoniç and sessile microbial growth in water hydraulic system. Water Res 3:3812–3820. doi:10.1016/S0043-1354(02)00099-4

Stapper AP, Narasimhan G, Ohman DE, Barakat J, Hentzer M, Molin S, Kharazmi A, Høiby N, Mathee K (2004) Alginate production affects *Pseudomonas aeruginosa* biofilm development and architecture, but is not essential for biofilm formation. J Med Microbiol 53:679–690. doi:10.1099/jmm.0.45539-45540

Starner TD, Zhang N, Kim G, Apicella MA, McCray PB Jr (2006) *Haemophilus influenzae* forms biofilms on airway epithelia: implications in cystic fibrosis. Am J Respir Crit Care Med 174:213–220. doi:10.1164/rccm.200509-1459OC

Steidler L, Hans W, Schotte L, Neirynck S, Obermeier F, Falk W, Fiers W, Remaut E (2000) Treatment of murine colitis by *Lactococcus lactis* secreting interleukin-10. Science 289:1352–1355. doi:10.1126/science.289.5483.1352

Sundar K, Sadiq IM, Mukherjee A, Chandrasekaran N (2011) Bioremoval of trivalent chromium using *Bacillus* biofilms through continuous flow reactor. J Hazard Mater 30:44–51. doi:10.1016/j.jhazmat.2011.08.066

Suseela MR, Toppo K (2007) Algal biofilms on polythene and its possible degradation. Curr Sci 92:285–287

Sutherland I (2001) Biofilm exopolysaccharides: a strong and sticky framework. Microbiology 147:3–9

Tao J, Mancl KM, Tuovinen OH (2010) Attenuation of pollutants in sanitary sewer overflow: comparative evaluation of treatment with fixed media bioreactors. Bioresour Technol 101:1781–1786. doi:10.1016/j.biortech.2009.10.038

Tchouaffi-Nana F, Ballard TE, Cary CH, Macdonald TL, Sifri CD, Hoffman PS (2010) Nitazoxanide inhibits biofilm formation by *Staphylococcus epidermidis* by blocking accumulation on surfaces. Antimicrob Agents Chemother 54:2767–2774. doi:10.1128/AAC.00901-09

Teitzel GM, Parsek MR (2003) Heavy metal resistance of biofilm and planktonic *Pseudomonas aeruginosa*. Appl Environ Microbiol 69:2313–2320. doi:10.1128/AEM.69.4.2313-2320.2003

Toner B, Manceau A, Marcus MA, Millet DB, Sposito G (2005) Zinc sorption by a bacterial biofilm. Environ Sci Technol 39:8288–8294. doi:10.1021/es050528+

Torlak E, Sert D (2013) Combined effect of benzalkonium chloride and ultrasound against *Listeria monocytogenes* biofilm on plastic surface. Lett Appl Microbiol 57:220–226. doi:10.1111/lam.12100

Trautner BW, Darouiche RO (2004) Role of biofilm in catheter-associated urinary tract infection. Am J Infect Control 32:177–183. doi:10.1016/j.ajic.2003.08.005

Tunkel AR, Mandell GL (1992) Infecting microorganisms. In: Kaye D (ed) Infective endocarditis, 2nd edn., pp 85–97

Uhlich GA, Gunther NW 4th, Bayles DO, Mosier DA (2009) The CsgA and Lpp proteins of an *Escherichia coli* O157:H7 strain affect HEp-2 Ccll invasion, motility, and biofilm formation. Infect Immun 77:1543–1552. doi:10.1128/IAI.00949-08

Vazquez V, Liang X, Horndahl JK, Ganesh VK, Smeds E, Foster TJ, Hook M (2011) Fibrinogen is a ligand for the *Staphylococcus aureus* microbial surface components recognizing adhesive matrix molecules (MSCRAMM) bone sialoprotein-binding protein (Bbp). J Biol Chem 286:29797–29805. doi:10.1074/jbc.M110.214981

Verma P, Brown JM, Nunez VH, Morey RE, Steigerwalt AG, Pellegrini GJ, Kessler HA (2006) Native valve endocarditis due to *Gordonia polyisoprenivorans*: case report and review of literature of bloodstream infections caused by *Gordonia* species. J Clin Microbiol 44:1905–1908. doi:10.1128/JCM.44.5.1905-1908.2006

Vivant AL, Garmyn D, Gal L, Piveteau P (2014) The Agr communication system provides a benefit to the populations of *Listeria monocytogenes* in soil. Front Cell Infect Microbiol 4:160. doi:10.3389/fcimb.2014.00160

Wagner M, Loy A (2002) Bacterial community composition and function in sewage treatment systems. Curr Opin Biotechnol 13:218–227. doi:10.1016/S0958-1669(02)00315-4

Walters M, Sperandio V (2006a) Autoinducer-3 and epinephrine signaling in the kinetics of locus of enterocyte effacement gene expression in enterohemorrhagic *Escherichia coli*. Infect Immun 74:5445–5455. doi:10.1128/IAI.00099-06

Walters M, Sperandio V (2006b) Quorum sensing in *Escherichia coli* and *Salmonella typhimurium*. Int J Med Microbiol 296:125–131. doi:10.1016/j.ijmm.2006.01.041

Wang X, Lünsdorf H, Ehrén I, Brauner A, Römling U (2010) Characteristics of biofilms from urinary tract catheters and presence of biofilm-related components in *Escherichia coli*. Curr Microbiol 60:446–453. doi:10.1007/s00284-009-9563-z

Wang S, Liu X, Liu H, Zhang L, Guo Y, Yu S, Wozniak DJ, Ma LZ (2015) The exopolysaccharide Psl-eDNA interaction enables the formation of a biofilm skeleton in *Pseudomonas aeruginosa*. Environ Microbiol Rep 7:330–340. doi:10.1111/1758-2229.12252

Waters CM, Bassler BL (2005) Quorum sensing: cell-to-cell communication in bacteria. Annu Rev Cell Biol 21:319–346. doi:10.1146/annurev.cellbio.21.012704.131001

Whitehead NA, Barnard AM, Slater H, Simpson NJ, Salmond GP (2001) Quorum-sensing in gram-negative bacteria. FEMS Microbiol Rev 25:365–404. doi:10.1016/S0168-6445(01)00059-6

Wiens JR, Vasil AI, Schurr MJ, Vasil ML (2014) Iron-regulated expression of alginate production, mucoid phenotype, and biofilm formation by *Pseudomonas aeruginosa*. mBio 5:e01010–e01013. doi:10.1128/mBio.01010-13

Yakimov MM, Golyshin PN, Lang S (1998) *Alcanivorax borkumensis* gen. nov., sp. nov., a new, hydrocarbon-degrading and surfactant-producing marine bacterium. Int J Syst Bacteriol 48:339–348. doi:10.1099/00207713-48-2-339

Yang S, Yang F, Fu Z, Wang T, Lei R (2010) Simultaneous nitrogen and phosphorus removal by a novel sequencing batch moving bed membrane bioreactor for wastewater treatment. J Hazard Mater 175:551–557. doi:10.1016/j.jhazmat.2009.10.040

Yebra DM, Kiil S, Johansen KD (2004) Antifouling technology-past, present and future steps toward efficient and environmentally friendly antifouling coatings. Prog Org Coat 50:75–104. doi:10.1016/j.porgcoat.2003.06.001

Yeon KM, Lee C-H, Kim J (2009) Magnetic enzyme carrier for effective biofouling control in the membrane bioreactor based on enzymatic quorum quenching. Environ Sci Technol 43:7403–7409. doi:10.1021/es9001323k

Zhong C, Gurry T, Cheng AA, Downey J, Deng SCM, Lu TK (2014) Strong underwater adhesives made by self-assembling multi-protein nanofibres. Nat Nanotechnol 9:858–866. doi:10.1038/nnano.2014.199

Zobell CE (1943) The effect of solid surfaces on bacterial activity. J Bacteriol 46:39–56

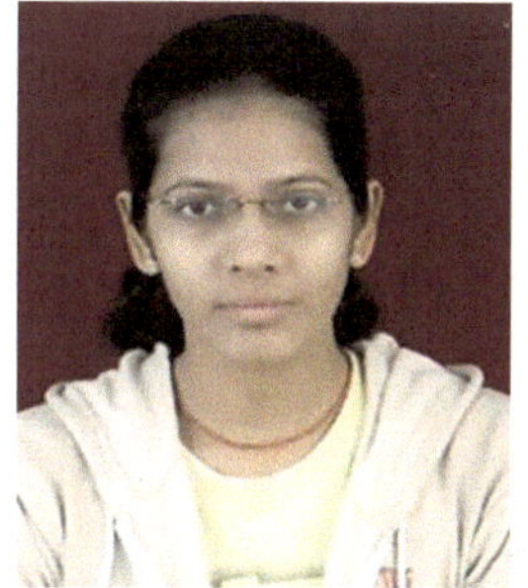

Neha Dubey did her postgraduation in Biotechnology from Jiwaji University, Gwalior, in 2011. Currently, she is enrolled as a Ph.D. scholar at Department of Zoology, Delhi University, under the supervision of Dr. ShamilaBasu-Modak and co-supervision of Dr. Yogendra Singh. Her area of research focuses on the host-pathogen interactions of the tuberculosis pathogenesis caused by Mycobacterium tuberculosis.

Raja Singh received his M.Sc. degree in Biotechnology from Jiwaji University, Gwalior, in 2011. Currently, he is working as a Senior Research Fellow at Special Center for Molecular Medicine (SCMM), Jawaharlal Nehru University, under the supervision of Dr. Vibha Tandon. His current research work focuses on validation of different synthetic inhibitors against topoisomerase family of protein.

Aditya K. Sharma has done his graduation in biomedical sciences from University of Delhi and Master's in Biotechnology from Jamia Millia Islamia. Currently, he is pursuing his doctoral degree from Institute of Genomics and Integrative Biology under supervision of Dr. Yogendra Singh. For his Ph.D., he is working on regulation of kinase and its substrates via phosphorylation in Mycobacterium tuberculosis.

Dr. Sharmila Basu-Modak currently holds the position of Associate Professor at the Department of Zoology, University of Delhi, since 2006. She is a life member of Indian Society of Cell Biology. Her lab is interested in studying the mechanisms of molecular genetics, molecular phylogeny, and gene expression during mouse embryogenesis and development.

Yogendra Singh received his Ph.D. degree from Vallabhbhai Patel Chest Institute, Delhi University, and postdoctoral training at National Institute of Health (NIH), Bethesda. Currently, he is Chief Scientist at CSIR-IGIB and Professor of AcSIR. He is a Fellow of National Academy of Sciences, India, Fellow of Indian National Science Academy, and Fellow of Indian Academy of Sciences, India. He has been honored with Moselio Schaechter Distinguished Service Award in 2014 by American Society for Microbiology. He serves as an Editorial Board Member of *Journal of Biological Chemistry*. His current research interest is to understand the host response in pulmonary/extra-pulmonary tuberculosis and to decipher the role of eukaryotic like Ser/Thr kinases in *Mycobacterium tuberculosis* and *Bacillus anthracis*.

Synthetic Biology in Aid of Bioactive Molecules

12

Shilpi Jain and Swati Shalini

Abstract
Synthetic biology emerged to understand the basic biological processes by designing the novel metabolic systems. The microorganisms are being engineered for the production of fuel, complex chemical compounds, and potential pharmaceutical drugs by using cheaper substrates. Synthetic biology provides essential components needed for engineering cellular metabolism. The well-characterized gene expression, chemical synthesis of metabolites, computer-aided modeling of desired metabolic pathway, and novel mechanism for biological product formation are the main tools of synthetic biology, which ultimately are used for the production of biomolecules. In this chapter, we will discuss different synthetic biology approaches for the production of various valuable biomolecules.

12.1 Introduction

"Synthetic biology," the term was first used by Barbara Hobom to describe genetically engineered bacteria (Benner and Sismour 2005). The consensus definition of it is "synthesis of complex biologically based systems, which displays functions that do not exist in nature." Synthetic biology is basically the union of engineering and biological systems (Serrano 2007). Synthetic biologists assemble the biological components in the organism that are not native to it, but yet it is able to perform its natural function like reproduction and evolution (Benner and Sismour 2005). They aim to achieve the following two goals using synthetic biology:

1. To understand natural processes by mimicking them in the artificial environment
2. To assemble the biological components in an artificial manner to synthesize useful bioactive molecules efficiently and easily (Purnick and Weiss 2009)

The genetic engineering for the production of a particular biomolecule can be done at the various levels vis-à-vis transcriptional, translational,

S. Jain (✉)
RNA II Laboratory, Department of Biochemical Engineering and Biotechnology, Indian Institute of Technology, 110016 Delhi, India
e-mail: shilpi.jain@dbeb.iitd.ac.in

S. Shalini
Plant Cell Culture Laboratory, Department of Biochemical Engineering and Biotechnology, Indian Institute of Technology, 110016 Delhi, India
e-mail: swati.schone@gmail.com

V.C. Kalia (ed.), *Microbial Factories*, DOI 10.1007/978-81-322-2595-9_12

and posttranslational levels depending on the biomolecule or metabolite involved. Since transcription is the first level of gene expression, it is the most common adaptive method for the genetic engineering. It includes modification in the promoter region, RNA polymerase, and the other transcriptional machinery of the gene involved (Khalil and Collins 2010). The hierarchy of synthetic biology is being inspired by the computer engineering (Andrianantoandro et al. 2006) and named accordingly cellular engineering, pathway engineering, metabolic engineering, and protein engineering.

12.1.1 Cellular Engineering

Cellular engineering uses the engineering approach in the cell to solve the basic biological problems and apply them to produce molecules. It includes both the tissue-level modification and the biological designing of the cells. The tissue engineering uses whole cells in the development of other alternatives, whereas bioprocess engineering involves the application of intact cells for the production of biomolecules (Nerem 1991). Cellular engineering includes pathway engineering, metabolic engineering, and protein engineering.

12.1.2 Pathway Engineering

For successful commercial production of bioactive compounds, combined approach of metabolic engineering along with optimization of fermentation procedure is essential. Pathway engineering enables us to increase product yield, enhance carbon use efficiency, and improve the product purity (Chotani et al. 2000). Metabolic pathway engineering includes following points:

1. Modification of native metabolic pathway by altering the regulatory mechanism of transcription.
2. Other pathways using similar substrate are genetically blocked.
3. Enhancing the carbon supply towards biosynthetic pathway of desired product.
4. Increasing the availability of enzymatic cofactors.

12.1.3 Metabolic Engineering

Metabolic engineering involves the manipulation of biochemical reactions in vivo for the production of a certain metabolite (Keasling 2008). If the reaction involved is the part of a certain pathway, then it can also be called as pathway engineering; examples of the metabolic and pathway engineering will be described in detail later in the text.

12.1.4 Protein Engineering

Recombinant versions of natural proteins are used as therapeutics. Protein engineering involves the rational designing of protein structure and its functions. It can be done by modifying natural proteins (glycosylation patterns), making Fc fusion protein and conjugation of protein to the polymer like polyethylene glycol (PEG). These methods are employed to increase the clinical potential by increasing their serum half-life and preventing the rapid clearance from the body.

12.1.5 Bioactive Molecules

Bioactive compounds are natural products having potential use as pharmaceuticals (Nielsen et al. 2013). These bioactive molecules vary from simple, chemical metabolites to complex natural molecules and their derivatives. Usually the useful biomolecules involved in the production of therapeutic proteins, drugs, or biofuels are produced in very minute amounts from their natural sources. Because of this limitation, the production of these molecules form their natural environment is not feasible. Using synthetic biology approach, they can be produced using microorganisms (*E. coli* and yeast) from simple and cost-effective raw materials like mevalonate (for biosynthesis of antimalarial drug).

For the industrial production of bioactive molecules using the tools of synthetic biology, some standard characteristics of the microbial factories are needed, which are mentioned below:

1. The host strain should be well characterized and genetically stable even with all the genetic engineering done in it.
2. The organism should be able to thrive with minimal nutrients to cut down the cost of producing biologically active molecules at the industrial level.
3. Pathways to be modified should be tightly regulated so that it can be switched on or off as required.
4. The enzymes required for the pathway should be coordinated simultaneously to improve the efficiency (Keasling 2008).

Table 12.1 List of terpenoids produced in different organisms using synthetic biology approach

Terpenoids	Organism engineered	Application
Amorphadiene	*E. coli*	Antimalarial
Artemisinic acid	*S. cerevisiae*	Antimalarial
Taxol	*S. cerevisiae*	Anticancer
Lycopene	*S. cerevisiae*	Anticancer
β-Carotene	*E. coli* and *S. cerevisiae*	Anticancer
Sterols	*S. cerevisiae*	For production of vitamin D
Flavonoids	*E. coli* and *S. cerevisiae*	Anticancer

12.2 Synthetic Biology for the Production of Bioactive Molecules

12.2.1 Terpenoids

Terpenoids, derivatives of terpenes or isoprenes, are secondary metabolites produced by plants, animals, and microbial species. They mainly play role in defense and repair mechanisms by acting as toxins or repellants. There are more than 60,000 terpenes which were isolated from their natural sources (Keasling 2008). Chemically, terpenoids are made up of five carbon isoprene units (C_5H_8) and can be classified based on the number of these isoprene units as monoterpenes, diterpenes, triterpenes, and so on.

Traditional Indian and Chinese medicines use extracts of plants for treatment of various diseases like cancer (Lu et al. 2012), diabetes (Indumathi et al. 2014), malaria (Klayman 1985), and bacterial (Jasmine et al. 2011) and viral infections. The main components of these plant extracts were found to be terpenoids. The anticancer drug Taxol and antimalarial drug artemisinin are the two most popular terpenoid-based drugs.

As these terpenoids are produced as secondary metabolites, they are often produced in very small amounts because of which the isolation of terpenoids in the pure form is cumbersome and costly. The terpenoids can be synthesized in large amounts using the synthetic biology approach in the easily culturable microbes (Table 12.1). Since the artemisinin production by synthetic biology has been well studied, we are going to discuss its production in detail.

12.2.1.1 Artemisinin

Artemisinin and its derivatives such as artesunate, artemether, and dihydro-artemisinin are the potent antimalarial drugs. These are used for the artemisinin-based combination therapies (ACTs) against the drug-resistant parasite *Plasmodium falciparum*. Artemisinin is naturally produced in the plant *Artemisia annua*, which was traditionally used for treating malaria in China (Klayman 1985). In 1980s, artemisinin and its derivatives were isolated in China from this plant. The recovery yield of artemisinin from the plant is 5 Kg/1000 Kg of dry leaves (Hale et al. 2007). Using synthetic biology, the whole artemisinin synthesis pathway was engineered in *Saccharomyces cerevisiae* which increased the titer of amorphadiene (precursor of artemisinin) by nearly 100-fold and decreased its cost of production.

Amorphadiene Production in *E. coli* by Mevalonate Pathway

Isoprenoids are synthesized by two precursors, isopentenyl phosphate (IPP) and its isomer dimethylallyl pyrophosphate (DMAPP). There are two pathways to synthesize these two precursors, namely, mevalonate-dependent pathway (MEV) and non-mevalonate pathway (DXP)

(Misawa 2011). In most prokaryotes including *E. coli*, non-mevalonate pathway is involved in the biosynthesis, whereas in eukaryotes, mevalonate pathway is common. Plants use both the pathways. In *E. coli*, the two precursors are condensed to form geranyl diphosphate (GPP) which is then converted to farnesyl pyrophosphate (FPP), precursor of all terpenoids. The chemical diversity of terpenoids is dependent on the functional diversity of terpene synthases. In 2003, the *Saccharomyces cerevisiae* mevalonate-dependent pathway was genetically engineered in *E. coli* while bypassing its native DXP pathway (Martin et al. 2003). The mevalonate pathway of yeast was divided into two synthetic operons and is expressed in the two plasmids. It was observed that in the presence of mevalonate, growth was inhibited because of overproduction of terpenoids and sterols. To produce amorphadiene, amorphadiene synthase (ADS) from the plant *Artemisia annua* was codon optimized and co-expressed in *E. coli*. It was observed that in the presence of ADS, growth inhibition in the presence of mevalonate was relieved and amorphadiene production was proportional to the amount of mevalonate added to the medium.

Artemisinic Acid Production in *Saccharomyces cerevisiae*

The already existing mevalonate pathway in *Saccharomyces cerevisiae* was genetically engineered to produce artemisinic acid in the following ways (Ro et al. 2006) (Fig. 12.1):

1. Increased the production of FPP by overexpressing truncated soluble enzyme, 3-OH-3-methyl-glutaryl-coA reductase
2. Downregulated *ERG9* gene, which encodes squalene synthase, responsible for usage of FPP in sterol biosynthetic pathway
3. Introducing the amorphadiene synthase gene from the plant *Artemisia annua* to convert FPP to amorphadiene
4. Incorporating a cytochrome P450 from *Artemisia annua* which converts amorphadiene to artemisinic acid

All of the above mentioned steps were done by chromosomal integration. Using synthetic biology, artemisinic acid was being successfully produced in yeast. The advantage of using yeast is that the same amount of artemisinic acid can be produced in 4–5 days as compared to months from plants.

12.2.2 Shikimic Acid

Shikimic acid (tri hydroxy-1-cyclohexene-1-carboxylic acid) is used as the starting material for commercial production of oseltamivir which is an orally active inhibitor of the essential neuraminidase of influenza virus. Oseltamivir is potentially used in influenza pandemics such as "bird flu" and "swine flu" and is available in the market in the form of commercial drug TAMIFLU (Bochkov et al. 2011). Currently a major source of shikimic acid is through solvent extraction from the flower of a slow-growing plant "star anise" which is available in China only. However, the huge requirement of shikimic acid for the drug production cannot be met just by the solvent extraction (Ghosh et al. 2012). Alternative technologies are needed so that production of

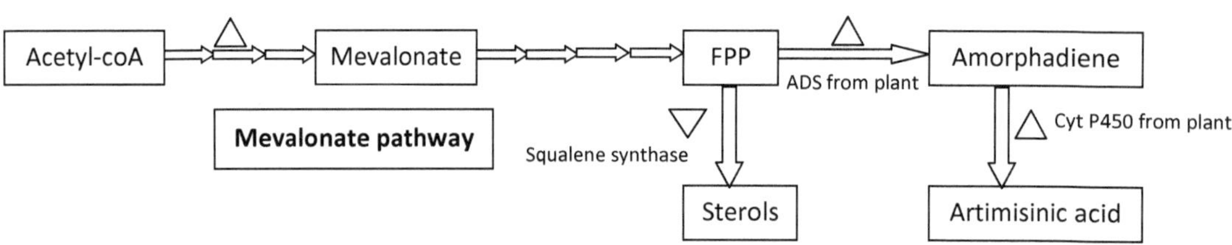

Fig. 12.1 Modified mevalonate pathway in *Saccharomyces cerevisiae* for the production of artemisinic acid. Reverse triangles indicate downregulation of gene and triangles indicate upregulation of gene

shikimic acid can be enhanced to cater the growing needs of the world market.

The pathway for biosynthesis of shikimic acid (cyclohexenecarboxylic acid) is present in all autotrophic microorganisms and plants, since it is the substrate for production of tyrosine, tryptophan, phenylalanine, and other essential aromatic metabolite (Johansson and Lidén 2006). The two initial components of shikimate pathway are phosphoenolpyruvate (PEP), product of glycolytic pathway, and erythrose-4-phosphate (E4P), component of pentose phosphate pathway of glucose oxidation. A combination of these compounds produces DAHP which after cyclization produces dehydroquinic acid (DHQ). DHQ loses water and gets converted into 3-dehydroshikimic acid (DHS). The shikimate dehydrogenase enzyme further converts it into shikimic acid (SA). Shikimic acid gets further phosphorylated to produce shikimic-3-phosphate (S3P) (Fig. 12.2).

Chandran et al. (2003) studied the effect of increasing phosphoenolpyruvate concentration on shikimic acid production in the following ways:

1. PTS facilitates the entry of glucose inside the cytoplasm and add phosphoryl group to glucose for formation of glucose -6-phosphate (initial step of glycolysis). PTS system consumes PEP and limits its supply towards DAHP synthase for shikimic acid biosynthesis. In order to increase the availability of phosphoenolpyruvate, *pps*A, the gene which codes for phosphoenolpyruvate synthase, was overexpressed (which convert the pyruvic acid obtained from PTS-mediated glucose transport into phosphoenolpyruvate).
2. To prevent the PEP consumption by glucose transport, *glf* (glucose facilitator) obtained from *Zymomonas mobilis* was expressed in *E. coli* K-12.
3. Plasmid-containing localized inserts aro F^{FBR} (feedback inhibition insensitive 3-deoxy-Darabino-heptulosonic acid-7-phoshphate), *tkt*A (transketolase) were also inserted in *E. coli* in order to increase the avilability of E4P and PEP towards shikimic acid biosynthesis. The activity of mutant DAHP synthase was decreased by limited presence of E4P and PEP inside the host cell of *Escherichia coli*. Transketolase, *tkt*A, expression enhances the substrate access to the enzyme.

It was found that the highest yield of shikimic acid was obtained when *glf* (glucose facilitator) and *glk* (glucose kinase) were expressed in *E. coli* from *Zymomonas mobilis* in strain lacking PTS system of glucose transport. At 10 l fermentation, this system synthesized 87 g/L shikimic acid during log phase.

Another mechanism which employed to increase the yield of shikimic acid is by reducing the contaminants or byproducts of shikimic acid pathway. Quinic acid is a byproduct of shikimic acid pathway which remains present as contaminant during crystallization of shikimic acid. In order to improve the yield and reduce the quinic acid formation during fermentation, quinate dehydrogenase encoded by *qad* locus obtained from *Klebsiella pneumoniae* was expressed in *E. coli* (Draths et al. 1999).

12.2.3 Therapeutic Proteins

Production of around 140 approved therapeutic proteins at reasonable cost is possible only due to the progress in synthetic biology (Gerngross 2004). Protein-based therapeutics include anticoagulants, antibody-based drugs, blood factors, hormones, interferons, interleukins, and thrombolytics (Andersen and Krummen 2002). The therapeutic proteins can be classified into two parts: proteins that undergo posttranslational modifications like glycosylation and others that do not undergo glycosylation. Aglycosylated proteins can easily be expressed in prokaryotes but for the glycosylated proteins, eukaryotic expression system is needed, as prokaryotes lack the machinery for the glycosylation. *E. coli* and yeasts like *Saccharomyces cerevisiae* and *Pichia pastoris* are the organisms of choice for the production of aglycosylated and glycosylated proteins, respectively. The other problem for the production of therapeutic proteins in microbial

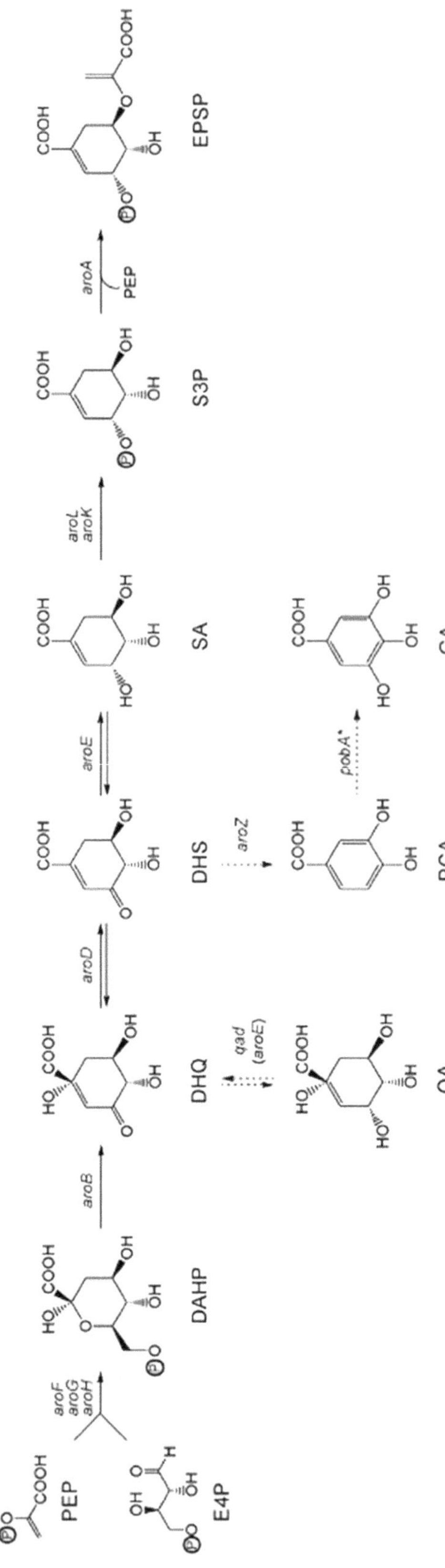

Fig. 12.2 Biosynthetic pathway of shikimic acid

Table 12.2 List of therapeutic proteins produced in different organisms using synthetic biology approach

Therapeutic proteins	Organisms involved
Insulin	*E. coli*
Human growth hormone	*E. coli*
Il-2	*E. coli*
Hirudin	*S. cerevisiae*
Insulin	*S. cerevisiae*
Angiostatin	*P. pastoris*
Elastase inhibitor	*P. pastoris*

factories is that mammalian proteins usually form disulfide linkages. Since cytoplasm of prokaryotes has a reducing environment, disulfide linkages cannot be formed there (Kamionka 2011). This problem has been circumvented by expressing the protein in periplasmic space where oxidation can take place. Using this approach, human growth hormone (hGH) was expressed in *E. coli.* Many therapeutic proteins have been expressed in different organisms by employing different genetic engineering methods (Table 12.2).

12.2.4 Aromatic Amino Acid Production

Aromatic amino acids are one of the first biomolecules synthesized using microorganisms. Intermediates of aromatic amino acid biosynthetic pathways are also valuable products, which can be used as precursors for biosynthesis of other pharmaceuticals. For the production of aromatic amino acids, pathway engineering is usually the employed method. For this, pathways responsible for usage of intermediates are blocked and the different promoters are used for expression of the regulatory genes involved (Chotani et al. 2000). Many derivatives of amino acids are involved in the production of pharmaceuticals like melanin and indigo dye. We are going to discuss production of indigo dyes briefly (Ito et al. 1990).

12.2.4.1 Indigo Dyes

Many different processes of pathway engineering are required for microbial production of textile-grade indigo dye. From *Pseudomonas*, a dioxygenase gene was cloned into *E. coli* for indigo production which was not produced by the host strain naturally. Indigo dye is formed using indole, the intermediate in the tryptophan biosynthetic pathway. Indole is usually bound to an enzyme called tryptophan synthase. This enzyme was engineered to release indole freely in the cytoplasm of *E. coli*. This modification along with dioxygenase is used for the synthesis of indigo.

12.2.5 Succinic Acid

Succinic acid can be used to produce many industrially important biomolecules. Succinic acid is produced as an intermediate in many metabolic pathways of microorganisms. In *E. coli*, the glycolytic product phosphoenolpyruvate (PEP) is being used as a precursor for other pathways. The PEP carboxylase gene was overexpressed to increase succinic acid formation.

12.2.6 Lactic Acid

Lactic acid is used in food industries and pharmaceutical industries. Homofermentative bacteria produce lactic acid under non-limiting nutrient conditions. To produce lactic acid at larger scale, bovine LDH-A gene was cloned in crab-tree-negative strain of *Kluyveromyces lactis* so that pyruvate can be used to synthesize lactic acid only.

12.2.7 Ascorbic Acid

L-Ascorbic acid is used as an antioxidant in pharmaceutical formulations, food, and cosmetics. Till now ascorbic acid was produced by pure chemical method using biocatalyst. By metabolic selection method, *Candida blankii* and *Cryptococcus dimmnae* can convert 2-keto-L-GULONIC acid to ascorbic acid. Microalgae strains were also developed that can produce ascorbic acid.

12.3 Protein Engineering Employed for Pharmaceutical Drugs Production

Large-scale production of drugs from their natural sources is hindered because of their lower concentration and restricted availability of source. Total chemical synthesis of drugs is costly and complicated since they have complex molecular structure. Therefore, there is an urgent need to transfer the drug biosynthetic capabilities of natural hosts to other easily proliferating microorganisms. Next-generation genome sequencing projects had provided us the information about entire metabolic pathways. New technologies of gene synthesis allow the production of whole gene and their regulatory elements in an inexpensive manner.

Proteins having different activities evolve to acquire specific biological functions. This is naturally done by the substitution of some amino acids by the divergent evolution method which involves a number of processes like gene amplification and mutations. This natural process of adaptation is used in many protein engineering methods to design the proteins with specific biological activity. The amino acid residues which can be modified, are identified by structural and biochemical analysis. Keasling and coworkers (Paddon and Keasling 2014) probed these amino acid residues in the active sites of sesquiterpene synthase γ humulene synthase. They developed different mutants having different enzymatic activities and narrowed down to four amino acids which changes catalytic activity significantly. These residues were then *in silico* analyzed for creation of novel terpene synthase catalyzing the synthesis of sesquiterpenes. They developed seven different enzymes capable of producing different products using different reactions (Yoshikuni et al. 2006).

Computer-assisted modeling is used to perform directed evolution of enzymes which could catalyze unnatural chemical reactions that are not possible in intracellular environment. For this, many algorithms were developed. One of the examples of the algorithm used is PRODA-MATCH, an algorithm developed for *in silico* designing of enzymes (Lei et al. 2011). Baker et al. were also able to engineer eight enzymes, which are responsible for the catalysis of Kemp elimination reaction by using computer-aided simulations. This results in rate enhancement up to 10^5 and multiple turnovers. Catalysis depends upon the designed catalytic site confirmed by mutational analysis, and the atomic accuracy of design is measured by studying high-resolution crystal structure (Röthlisberger et al. 2008).

The diversity of natural proteins can be increased in vivo by incorporating synthetic amino acids. These amino acids can also be incorporated genetically into protein with high efficiency using modified aminoacylated tRNA synthetases-tRNA pair. Till now, around 40 synthetic amino acids have been added in vivo in different organisms. The proteins having synthetic amino acids acquire novel properties. These designed amino acids could be used for protein modification, structure and function detection of proteins, and identification of regulating activity of protein (Neumann et al. 2010).

12.4 Conclusion

For engineering of microorganisms, it is necessary to use the well-characterized biological components, for example, identifying the minimal number of genes required by microorganism for sustaining life. With these biological components and tools of recombinant technology, production of therapeutic proteins and useful biomolecules can be done efficiently.

12.5 Opinion

Synthetic biology has come up with many different approaches for the existing crisis of the pharmaceutically important biomolecules and therapeutic proteins. Synthetic biology simplifies and reduces the steps involved in the biosynthesis of complex biomolecules by using engineered microbes. This enables to produce the complex biological intermediates in a cost-effective manner at the industrial level. But to further reduce the cost and time of production of the biomolecules and increase the efficacy of drugs, it is necessary to develop these methods continuously in a rational manner. As mentioned in the text above, many pharmaceutical drugs have been synthesized by the recombinant engineering using the traditional knowledge of medicines. Had it been that the Chinese scientists did not use their traditional knowledge to isolate artemisinin from the plant extracts, it would be impossible to cure many patients affected by resistant malarial parasite *Plasmodium falciparum*. This fact emphasizes that if we combine the traditional knowledge and the modern science for the human benefits, we can come up with the more nature-friendly approach for synthesis of useful biomolecules.

Acknowledgment The authors wish to thank CSIR-HRDG for the research fellowship granted, Dr. V. C. Kalia for giving us the opportunity to write this chapter and Ms. Neha Dhasmana for inspiration.

References

Andersen DC, Krummen L (2002) Recombinant protein expression for therapeutic applications. Curr Opin Biotechnol 13:117–123. doi:10.1016/S0958-1669(02)00300-2

Andrianantoandro E, Basu S, Karig DK, Weiss R (2006) Synthetic biology: new engineering rules for an emerging discipline. Mol Syst Biol 2:0028. doi:10.1038/msb4100073

Benner SA, Sismour AM (2005) Synthetic biology. Nat Rev Genet 6:533–543. doi:10.1038/nrg1637

Bochkov DV, Sysolyatin SV, Kalashnikov AI, Surmacheva IA (2011) Shikimic acid: review of its analytical, isolation, and purification techniques from plant and microbial sources. J Chem Biol 5:5–17. doi:10.1007/s12154-011-0064-8

Chandran SS, Yi J, Draths KM, von Daeniken R, Weber W, Frost JW (2003) Phosphoenolpyruvate availability and the biosynthesis of shikimic acid. Biotechnol Prog 19:808–814. doi:10.1021/bp025769p

Chotani G, Dodge T, Hsu A, Kumar M, LaDuca R, Trimbur D, Sanford K (2000) The commercial production of chemicals using pathway engineering. Biochim Biophys Acta Protein Struct Mol Enzymol 1543:434–455. doi:10.1016/S0167-4838(00)00234-X

Draths KM, Knop DR, Frost JW (1999) Shikimic acid and quinic acid: replacing isolation from plant sources with recombinant microbial biocatalysis. J Am Chem Soc 121:1603–1604

Gerngross TU (2004) Advances in the production of human therapeutic proteins in yeasts and filamentous fungi. Nat Biotechnol 22(11):1409–1414. doi:10.1038/nbt1204-1589e

Ghosh S, Chisti Y, Banerjee UC (2012) Production of shikimic acid. Biotechnol Adv 30(6):1425–1431. doi:10.1016/j.biotechadv.2012.03.001

Hale V, Keasling JD, Renninger N, Diagana TT (2007) Microbially derived artemisinin: a biotechnology solution to the global problem of access to affordable antimalarial drugs. Am J Trop Med Hyg 77:198–202

Indumathi C, Durgadevi G, Nithyavani S, Gayathri PK (2014) Estimation of terpenoid content and its antimicrobial property in Enicostemma littorale. Int J ChemTech Res 6(9):4264–4267

Ito H, Sato K, Enei H, Hirose Y (1990) Improvement in microbial production of L-tyrosine by gene dosage effect of aroL gene encoding shikimate kinase. Agric Biol Chem 54(3):823–824. doi:10.1271/bbb1961.54.823

Jasmine R, Selvakumar B, Daisy P (2011) Investigating the mechanism of action of terpenoids and the effect of interfering substances on an Indian medicinal plant extract demonstrating antibacterial activity. Int J Pharm Stud Res 2:19–24

Johansson L, Lidén G (2006) Transcriptome analysis of a shikimic acid producing strain of Escherichia coli W3110 grown under carbon and phosphate-limited conditions. J Biotechnol 126(4):528–545. doi:10.1016/j.jbiotec.2006.05.007

Kamionka M (2011) Engineering of therapeutic proteins production in *Escherichia coli*. Curr Pharm Biotechnol 12:268–274. doi:10.2174/138920111794295693

Keasling JD (2008) Synthetic biology for synthetic chemistry. ACS Chem Biol 3(1):64–76. doi:10.1021/cb7002434

Khalil AS, Collins JJ (2010) Synthetic biology: applications come of age. Nat Rev Genet 11(5):367–379. doi:10.1038/nrg2775

Klayman DL (1985) Qinghaosu (artemisinin): an antimalarial drug from China. Science 228(4703):1049–1055. doi:10.1126/science.3887571

Lei Y, Luo W, Zhu Y (2011) A matching algorithm for catalytic residue site selection in computational enzyme design. Protein Sci 20:1566–1575. doi:10.1002/pro.685

Lu JJ, Dang YY, Huang M, Xu WS, Chen XP, Wang YT (2012) Anti-cancer properties of terpenoids isolated from Rhizoma Curcumae – a review. J Ethnopharmacol 143(2):406–411. doi:10.1016/j.jep.2012.07.009

Martin VJJ, Pitera DJ, Withers ST, Newman JD, Keasling JD (2003) Engineering a mevalonate pathway in *Escherichia coli* for production of terpenoids. Nat Biotechnol 21(7):796–802. doi:10.1038/nbt833

Misawa N (2011) Pathway engineering for functional isoprenoids. Curr Opin Biotechnol 22(5):627–633. doi:10.1016/j.copbio.2011.01.002

Nerem RM (1991) Cellular engineering. Ann Biomed Eng 19:529–545. doi:10.1007/BF02367396

Neumann H, Wang K, Davis L, Garcia-Alai M, Chin JW (2010) Encoding multiple unnatural amino acids via evolution of a quadruplet-decoding ribosome. Nature 464:441–444. doi:10.1038/nature08817

Nielsen AZ, Ziersen B, Jensen K, Mu L, Olsen CE, Møller BL, Jensen PE (2013) Redirecting photosynthetic reducing power toward bioactive natural product synthesis. J Am Chem Soc Synth Biol. doi:10.1021/sb400136f

Paddon CJ, Keasling JD (2014) Semi-synthetic artemisinin: a model for the use of synthetic biology in pharmaceutical development. Nat Rev Microbiol 12:355–367. doi:10.1038/nrmicro3240

Purnick PEM, Weiss R (2009) The second wave of synthetic biology: from modules to systems. Nat Rev Mol Cell Biol 10:410–422. doi:10.1038/nrm2698

Ro DK, Paradise EM, Ouellet M, Fisher KJ, Newman KL, Ndungu JM, Keasling JD (2006) Production of the antimalarial drug precursor artemisinic acid in engineered yeast. Nature 440:940–943. doi:10.1038/nature04640

Röthlisberger D, Khersonsky O, Wollacott AM, Jiang L, DeChancie J, Betker J, Baker D (2008) Kemp elimination catalysts by computational enzyme design. Nature 453:190–195. doi:10.1038/nature06879

Serrano L (2007) Synthetic biology: promises and challenges. Mol Syst Biol. doi:10.1038/msb4100202

Yoshikuni Y, Ferrin TE, Keasling JD (2006) Designed divergent evolution of enzyme function. Nature 440:1078–1082. doi:10.1038/nature04607

Shilpi Jain graduated in Biochemistry in 2008 from University of Delhi. She received her M.Sc. degree in Biochemistry in 2010 from Hamdard University. She is currently enrolled in Ph.D. with IIT Delhi. She is studying the phenomenon of bistability in *Escherichia coli* lactose and melibiose operons.

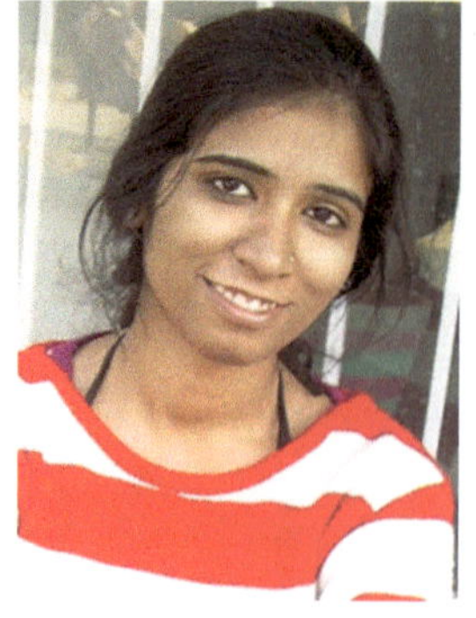

Swati Shalini graduated in Biotechnology in 2008 from Banasthali Vidyapith. She received her M.Sc. degree in Biotechnology in 2010 from CCS Haryana Agriculture University. She is currently enrolled in Ph.D. with IIT Delhi. She is studying the shikimic acid production using plant culture.

Biotechnology Implications of Extremophiles as Life Pioneers and Wellspring of Valuable Biomolecules

13

Ilaria Finore, Licia Lama, Annarita Poli, Paola Di Donato, and Barbara Nicolaus

Abstract

Studies on extremophiles, microorganisms able to survive in extreme environments, are very helpful for the comprehension of life evolution; in fact they are the unique organisms of the Earth at the origin of life. They lie into the three domains of life (*Archaea*, *Bacteria*, and *Eukarya*) and can be found in environmental niches on Earth such as in hydrothermal vents and springs, in salty lakes, in halite crystals, in polar ice and lakes, in volcanic areas, in deserts, or under anaerobic conditions. The existence of life forms beyond the Earth requires an extension of the classical limits of life: the resistance of extremophilic organisms to harsh conditions in terms of temperature, salinity, pH, pressure, dryness, and desiccation makes these living organisms good putative candidates to assess the habitability of other planets. The ability to survive and proliferate in extreme conditions (pH, temperature, pressure, salt, and nutrients) produces a variety of biotechnologically useful molecules such as lipids, enzymes, polysaccharides, and compatible solutes that are employed in several industrial processes. There are many extremophilic enzymes and also endogenous compounds that are used with success for food industry, for preparation of the detergents, for pharmacological applications, and also for genetic studies. In particular enzymes that derive from thermophiles, and for this reason called thermozymes, represent an excellent sources of new catalysts of interest in industrial sectors.

I. Finore • L. Lama • A. Poli • B. Nicolaus (✉)
Council of National Research (C.N.R.), Institute of Biomolecular Chemistry (I.C.B.), Via Campi Flegrei 34, 80078 Pozzuoli (Na), Italy
e-mail: ilaria.finore@icb.cnr.it; llama@icb.cnr.it; apoli@icb.cnr.it; bnicolaus@icb.cnr.it

P. Di Donato
Council of National Research (C.N.R.), Institute of Biomolecular Chemistry (I.C.B.), Via Campi Flegrei 34, 80078 Pozzuoli (Na), Italy

University of Naples Parthenope, Centro Direzionale Isola C4, 80100 Naples, Italy
e-mail: paola.didonato@uniparthenope.it

V.C. Kalia (ed.), *Microbial Factories*, DOI 10.1007/978-81-322-2595-9_13

13.1 Introduction

13.1.1 Origin of Life and Extreme Environments

From several decades, extremophilic microorganisms represent a subject matter deeply investigated. The increasing interest with regard to these life forms firstly lies in their capability to thrive in niches certainly inhospitable with respect to the conventional schemes. Indeed, extremophiles have been isolated from samples coming from the most disparate and unthinkable sites, discovering complete ecosystems, whose existence was unimaginable. The extremophilic microorganisms are classified in diverse groups that take the name from the main stress that affects the living environment. In addition, the microbes are divided in *stress* obligate or *stress* tolerate according to their necessity or tolerance to the stress factor to survive.

The most studied extreme environments are certainly those affected by high temperature, where the microorganisms are named *thermophiles* (>45 °C) or *hyperthermophiles* (>70 °C up to 113 °C). Hot places with so high temperature are hot springs, deep oceanic vents, and deep subsurface.

Hot springs are distributed all over the terrestrial crust, where the warm geothermal water emerges. In zones closest to volcanic sites, the water temperature can reach the boiling point and so it come out in the form of steam (*fumaroles*). Moreover, the groundwater transport causes the enrichment in mineral contents of water and because of the high temperature gives therapeutic properties to the springs. Some famous hot springs in which extremophiles have been isolated are the "Frying Pan Lake" in the Echo Crater of the Waimangu Volcanic Rift Valley, New Zealand; the "Grand Prismatic Spring" in the Yellowstone National Park, USA; and the "Boiling Lake" in the Morne Trois Pitons National Park, in Dominica. Deep oceanic vents are hydrothermal water emissions on the deep ocean floor. The emerging water shows a temperature ranging between 60 and 400 °C; the minerals transported by the groundwater in contact with the sea cold water form accumulating precipitates responsible for building a chimney structure that can be white or black smoker vents, according to the nature of minerals: barium, calcium, silicon, or sulfides. These niches are characterized by close cold and hot water, acid water (*acidophiles*), no solar energy, high pressure (*barophiles*), and no oxygen (*anaerobe*). In 1977, during a research expedition in the Galapagos Rift, the presence of the first hydrothermal vent was revealed. Deep subsurface represents an ubiquitous environment highly reducing, wherein the colonizing microbial communities are able to thrive under elevated temperature and pressure, acidity and alkalinity, low energy and nutrients availability, radioactivity, and high metals concentration (Reith 2011) (Fig. 13.1).

Another category of habitat is that characterized by high salt concentrations and the living microorganisms are named *halophiles*. Salty environments are considered those containing a salt concentration five times higher than that observed in the sea; they can be both of natural origin and artificial origin. Some examples are the salt lakes, internal seas, evaporating pool of seawater, curing brines, salted food products, and saline soils. All those ecosystems contain high amount of salt tending to the saturation; usually the pH ranges between the neutrality and the alkalinity (*halo-alkaliphiles*). Well-known salty lakes are the Owens Lake in California, the Great Salt Lake in Utah, and the Dead Sea; sources of halophilic microorganisms are fermented salty food, soy sauce, Chinese fermented beans, salted cod, anchovies, sauerkraut, etc. (Fig. 13.1). In opposition to hot places, there are colder and gelid sites, distributed all over the word, such as polar ice and glaciers, the alpine soil, deep ocean waters, and high-latitude places. The living microorganisms are named *psychrophiles* or *cryophiles*.

Since these environments are the most similar to the primordial Earth, it has suggested that probably the extremophiles could represent the unique possible primordial life forms capable to survive under those unfavorable conditions. Therefore, the study of the extremophilic microorganisms, looking for their optimal growth

Fig. 13.1 Examples of extreme environments: (**a**) XXI Italian Expedition on Mount Melbourne in Antarctica, (**b**) Mount Melbourne sampling site, ($\mathbf{b_1}$) *Geobacillus thermantarcticus* microscopy, (**c**) samples collected by Dr. Annarita Poli (ICB-CNR) on Mount Rittmann (Antarctica), ($\mathbf{c_1}$) *Anoxybacillus amylolyticus* microscopy, (**d**) Solfatara volcano sampling site (Italy), ($\mathbf{d_1}$) *Sulfolobus solfataricus* microscopy

conditions up to investigate them in terms of cell-wall shape and structures, metabolic pathways, enzymatic reactions, etc., is also important to reproduce at lab scale the chemico-physical parameters that affected the globe when the life appeared for the first time. Recently, it is defining the profile of *poly-extremophile*, that is, a microorganism needing at least three different stress parameters for growth, such as high salt concentration, high temperature, and alkaline pH value (Bowers et al. 2009). In addition, the genetic studies have reinforced the microbiology evidences; the comparison of a highly conservative 16S rRNA gene permitted to build a phylogenetic tree, that is a diagram wherein the length of each branch indicates the level of evolution of each microorganism with respect to the closest node and therefore with respect to a last universal common ancestor (LUCA). It results that the most extreme microorganisms are mainly distributed in the branch of *Archaea* domain, which is the less distant from LUCA. Hence, genetic evidence supports the theory on why the study of extremophilic microorganisms allows the knowledge of the primordial vital forms that have colonized the planet 3.8 billion years ago.

13.1.2 The Tree of Life Forms

The term extremophile (lover of extremes) was introduced by MacElroy (1974). Gradually, it has been used to describe microorganisms that live in environments not suitable for mammalian survival. Conditions that may be "extreme" to a kind of organism result to be necessary for the other's survival; therefore, it is possible to say the extremophily concept only in relative terms. Extremophiles occur in every domain of life: *Archaea*, *Bacteria*, and *Eukarya* (Fig. 13.2).

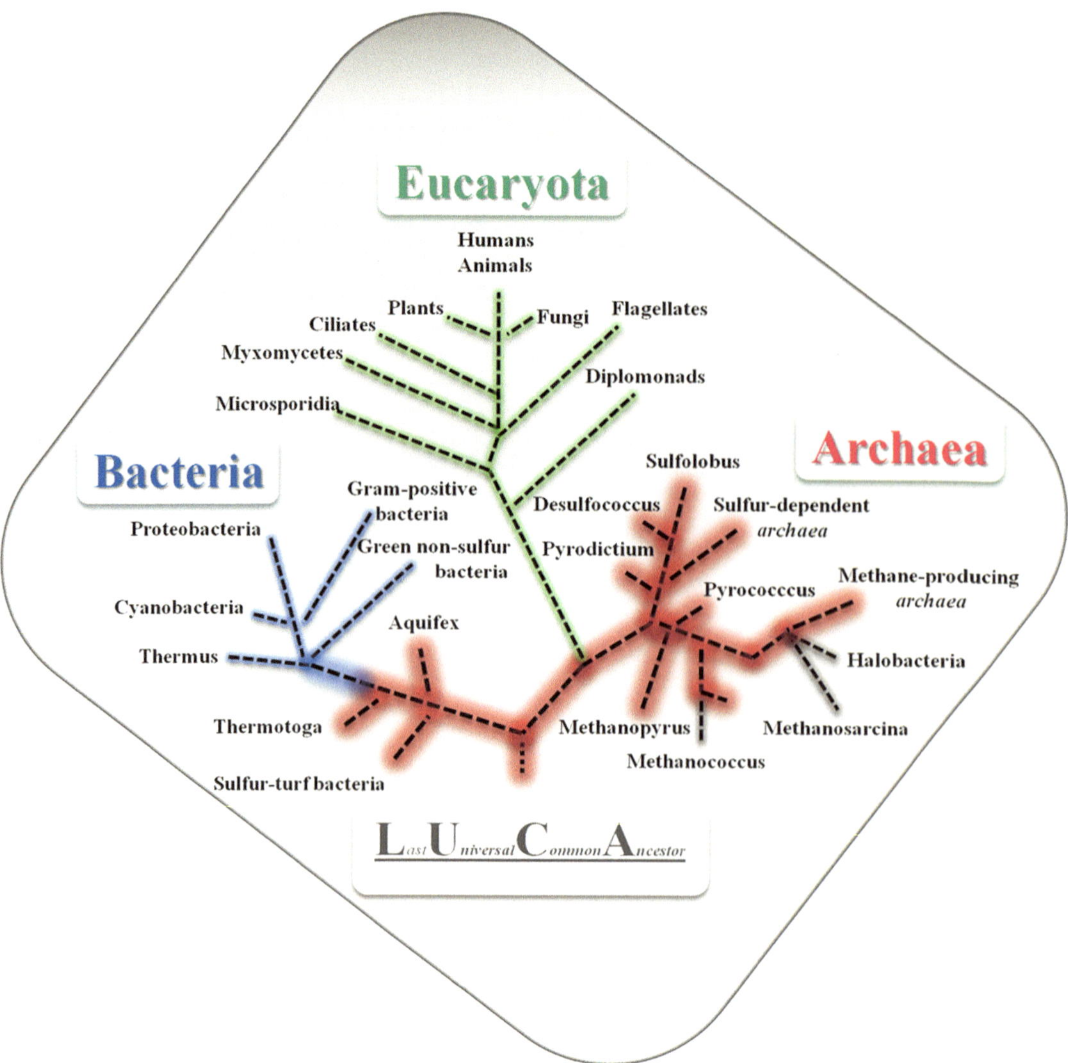

Fig. 13.2 Phylogenetic tree of the three domains

Extremophiles represent an important guide for both biochemical and structural biodiversity; in addition a vast potential lies in this kind of organisms as sources of diverse biological materials with applications in several fields, such as biotechnology and medicine. Therefore, they are the object of investigation and exploitation by research and biotech companies, respectively (Irwin 2010).

The comprehension of extremophiles has really increased our perception of mechanism protein folding and relation structure function. The search of more "extreme" extremophiles has permitted the detection of numerous novel microorganisms so enhancing information about the phylogeny. Nowadays, according to a commercial approach, extremophile products can have many potential applications (Poli et al. 2011). Extremophilic microorganisms can be direct source of enzymes; or as alternative, their structures, on the basis of a comparison to the homologous one from a mesophilic microorganism, can be utilized to build new engineered enzymes appropriately thermostable or with specific catalytic behavior by request to be applied in industrial field (Antranikian et al. 2005; Mesbah and

Wiegel 2012). Regarding to the extremophiles, the major complexity lies in growing them by means of reproducing optimally their environmental conditions both at lab scale and industrial scale; therefore, many efforts are dedicated to the development of more suitable methods for thermophilic and hyperthermophilic cultivation such as continuous fermentation bioprocess. Gene sequences belonging to thermophilic, hyperthermophilic and psychrophilic microorganisms have been successfully engineered by expression in a mesophilic host – *Escherichia coli* – and concerning thermophiles, the low thermostability of proteins belonging to the host has been exploited to isolate the more thermostable protein (Rothschild and Mancinelli 2001; Irwin and Baird 2004; Finore et al. 2014a).

The high stability of biomolecules recovered by extreme microorganisms increased the scientific and engineering interests for the build of biocatalysts with industrial applications. Identification of molecules responsible for the stability under extreme environmental conditions is also having an important impact on our comprehension of survival cellular mechanisms.

13.1.3 The Diversity of Extremophiles

Conventionally, extremophiles are classified according to temperature (both very high and low), pH, salinity, pressure, presence of radiations, oxygen, resistance to heavy metals, high toxic gas concentrations, and tolerance to desiccation (Pikuta et al. 2007).

13.1.3.1 Temperature

A class of extremophiles are those adapted to thrive under extreme temperatures. They are the thermophiles and the hyperthermophiles which include archaea and bacteria, which grow at temperatures over 80 °C, and the psychrophiles, microorganisms resistant to cold temperature. The utmost known extreme of temperature for growth take place into *Archaea* domain. Actually, *Pyrolobus fumarii* is reported to be the most extreme thermophile microorganism; it colonizes the chimney surfaces of the hydrothermal black smoker vents, by surviving in a temperature range between 90 and 113 °C. At the opposite temperature values, microbial communities have been found active up to −20 °C in sea ice (Nicolaus et al. 1993; Manca et al. 1996; Deming 2002), even though bacteria are stored in liquid nitrogen at −196 °C.

One of the most attractive diversity among the extremophiles is represented by the anaerobic thermophilic microorganisms; it is still difficult to comprehend how they are able not only to live at high temperatures but also to thrive in hot water. During the last decades, anaerobic thermophiles have captured the interest of researcher communities to discover new hyperthermophilic *Archaea*, and the understanding that anaerobic thermophiles can be utilized as valuable sources of thermostable biocatalysts was the motivating force for increasing research on thermophiles (Canganella and Wiegel 2014).

13.1.3.2 pH

Acidophilic and alkaliphilic microbes take the name from the extreme environmental parameters of both low and elevated pH wherein they thrive, respectively. An example of an "acid-tolerant" microorganism is *Helicobacter pylori*, which causes peptic ulcer illness both in humans and animals; it secretes urease, responsible for NH_4^+ production that provides barrier for the hydrochloric acid present in the stomach. An example of acidophilic species is *Picrophilus oshimae* that prefers for growth pH values as low as 0.7. For alkaline pH value, several microorganisms able to thrive at pH 10–11 are commonly isolated from soda lakes. These extremophiles resist extreme pH values by maintaining their internal pH nearly to neutrality by pumping H^+ from side to side of their cell membranes (Poli et al. 2011). Some organisms tolerate more stresses at the same time, and therefore they are defined multi-extremophiles, like *Sulfolobus acidocaldarius* which is an archaean able to grow at pH 3 and 80 °C (Nicolaus et al. 1992).

13.1.3.3 Salinity

The inhabitants of salt flats and marine hypersaline basins are really able to tolerate a very high concentrations of salt. The Dead Sea with its content of 30 % salt represents the habitat of some species belonging to the *Halobacteriaceae*, halophilic archaea, and *Dunaliella salina*, a halophilic alga. When the concentration of salts present in the environments is lower, it is possible that some species of bacteria, unicellular eukaryotes, cyanobacteria, and some green algae are able to live.

An ample assortment of saline niches such as coastal areas, salt wetland, mud and salt flats, and inland deserts represents the ideal place where some salt-tolerant plants are found. Since these extreme environments undergo with frequent changes in salinity level, halophiles have developed several adaptive strategies such as the synthesis of active biomolecules (Poli et al. 2011). These halophilic species are greatly of interest for their bioactive metabolites, such as essential oils (terpenes), carotenoids, glycosides, phenolic compounds, vitamins, sterols, polyunsaturated fatty acids, and polysaccharides. These compounds show many biological activities as antitumoral, antimicrobial, and anti-inflammatory suggesting their use for the prevention of some disorder, for example, cancer, atherosclerosis, and cardiovascular disease (Oren 2006; Poli et al. 2009; Ksouri et al. 2012).

13.1.3.4 Radiation

Deinococcus radiodurans is a radiation-resistant extremophile microorganism that survives at 1.5 Mrad, 3000 times more intense with respect to that borne by human cells. Moreover, this microorganism is tolerant to desiccation process and heavy metal presence (Gabani and Singh 2013; Özdemir et al. 2013). In general, the resistance to desiccation is possible using several mechanisms involving the spore formation, exopolysaccharide production, and the synthesis of intracellular solutes. Examples of microorganisms resistant to UV A, B, and C radiations (254–385 nm) are *Natranaerobius thermophilus* and *Natronovirga wadinatrunensis* that show a survival between 40 and 80 % after 28 h of radiation exposure. On the contrary, in the same test conditions, any survival for *E. coli* after 2 h of exposure is not recorded. This resistance could be related to putative DNA repair systems for *N. thermophilus*: in fact, homologs for the genes *recFOR*, *radA*, and *radC* and those coding for DNA mismatch repair proteins (MutS and MutN) have been found in the genome sequence of *Natranaerobius thermophilus* (Mesbah and Wiegel 2012).

13.2 Lipids

Bacteria and *Archaea*, living in peculiar niches of sea environments, have developed adaptation strategies to resist really severe stress parameters. For example, some of them are able to modify the fatty acid composition of cell membranes to guarantee their fluidity under high pressure, low/high temperatures, salinities, and pH. In addition, the bacterial microorganisms can include polyunsaturated fatty acids in the membrane cells to regulate its fluidity influenced by temperature and pressure for the protection of oxidative stress (van de Vossenberg et al. 1998).

Typical of extreme environments are *Archaea*; their cytoplasmic membrane is characterized by unique ether lipids, not easily degradable. They are resistant to high temperature and high salt concentration and are also mechanically resistant. Furthermore, in thermo-acidophilic *Archaea*, membranes showed tetraether lipids organized as a film monolayer membrane that became almost impermeable to ions. All these characteristics enable archaeal to survive under extreme environmental conditions with respect to the ester-type lipids found in *Bacteria* or *Eukarya* (van de Vossenberg et al. 1998). *Archaea* possess a single-layer lipid with ether linkages, whereas the hydrocarbon chains are composed of repeating isoprene units. Their membranes include tetraethers and diethers that became the main factor responsible for the capability of hyperthermophilic microorganisms to resist to a conventional temperature in contrast to membranes made by bilayer structure composed of glycerol and fatty acids (Nicolaus et al. 1990). Furthermore, another factor which helps *Archaea* to preserve the cell by temperature, pH, or salinity stresses is the surface (S) layer of glycoproteins, which envelope the cells and protect them.

Lipids of archaeal cells are radically dissimilar from those of other membranes; the core is made up of archaeal phospholipids – diether of diphytanylglycerol (Kates 1978; Sprott 1992). In particular they are made of isoprenoid (phytanyl) – 15–40 C chains. Chains are connected via ether linkages; the glycerol of archaeal cells, 2,3-di-*O*-*sn*-glycerol, shows the reverse stereochemistry with respect to bacteria (de Carvalho and Caramujo 2012). The cell membranes of haloarchaea consist mainly of bilayer diether lipids, and in archaeal thermophilic organisms, the membranes are mostly constituted of tetraether lipids forming monolayer membranes (Albers et al. 2006; Corcelli and Lobasso 2006; Boucher 2007). As revealed for other organisms, the phospholipids represent the most abundant component in membranes of *Archaea*. In the halophilic archaea, about 10 % of total lipids are composed of neutral lipids – bacterioruberin (Corcelli and Lobasso 2006). The core structure in haloarchaeal lipids is the archaeol, a 2,3-di-*O-phytanyl-sn-glycerol* having C20 isoprenoid chains (Kates 1978). The main type lipid of extreme thermophilic archaea *Sulfolobus* sp. is macrocyclic tetraether – caldarchaeol (Gliozzi et al. 1983; Boucher 2007). So the lipid membrane structures are modulated according to the stress conditions of life, and also *Archaea* microorganisms follow this rule (Corcelli 2009).

In Bacteria and Eukarya domains, the common membrane lipids are composed of an sn-glycerol-3-phosphate (snG-3-P) backbone esterified to linear fatty acids. While the archaeal membrane lipids are characterized by an sn-glycerol-1-phosphate (snG-1-P) backbone etherified to linear isoprenoids, the lipids in *Archaea* are made of saturated chains with methyl branches, which are linked to glycerol through ether linkages. Here the stereochemistry in 2-position of the glycerol is contrary to the mesophilic lipids. Particularly, it underlines the atypical bipolar tetraether lipids occurring in thermoacidophilic and methanogenic species as a mix of regioisomers (Table 13.1).

The unusual bipolar lipids represent an important matter in the *Archaea* survival to extreme environments by improving membrane properties as a direct response to the environmental conditions. This divide membrane lipid is considered evolutionary very significant and implicated in the differentiation of *Archaea* and *Bacteria*. It is not known why and how this differentiation occurred and whether the two phospholipid biosynthesis pathways are originated independently or from an ancestral cell with heterochiral membrane lipid (Lombard et al. 2012a; Jain et al. 2014). The unique structure of their membrane lipids is believed to be vital for the adaptation of these organisms to the extreme environmental conditions (Koga and Morii 2007), but this basic lipid architecture is found in all *Archaea* including the mesophilic Thaumarchaeota. Among *Archaea*, a great diversity of lipids exists, derived from the basic diether structure of archaeol. As in *Sulfolobus* sp. also in the mesophilic *Thaumarchaeota caldarchaeol* is present (Matsumi et al. 2011; Lombard et al. 2012b). However, certain key enzymes (Villanueva et al. 2014) have not yet been identified, precluding a complete reconstitution of the pathway. Very recently, in vitro studies clarified the identification and characterization of the key enzyme CDP-archaeol synthase of the ether lipid biosynthesis pathway that is ubiquitously present in *Archaea*. In conjunction with the other enzymes of this pathway, the synthesis of CDP-archaeol could be reconstituted in vitro using purified enzymes and simple building blocks (Jain et al. 2014).

Since lipids are heat sensitive, the hyperthermophile microorganisms possess special lipids. For example, *Thermotoga maritima*, which lives at 90 °C, has a glycerol ether lipid named 15,16-dimethyl-30-glyceryloxy-triacontanedioic acid. It is responsible of its resistance to high temperatures (De Rosa et al. 1989). Archaeal lipids, characterized by the tetraether lipid, allow the life under harsh conditions typical of *Archaea*. The studies of archaeal tetraether lipids have determined some interesting applications such as new lubricants, system for gene carrier, matrix made of monolayer lipid for sensor devices, and

Table 13.1 Examples of lipids from extremophilic microorganisms

Source	Lipid	Application	References
Archaea			
Thermoplasma acidophilum	Bipolar tetraether lipid with a phosphoglycerol and a β-L-glucopyranose as head groups and a cyclopentane ring per aliphatic chain	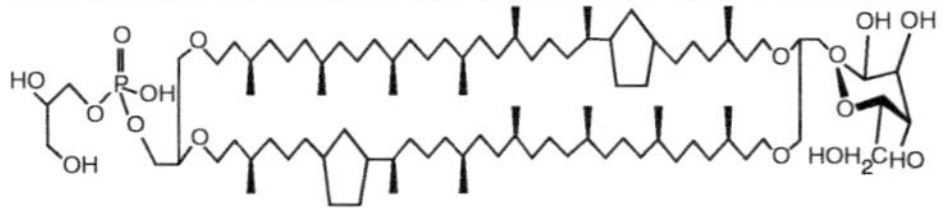 Anticancer therapeutic	Nicolas 2005
Halobacterium salinarum	2,3-Diphytanyl-*sn-glycerol*-1-phospho-3′-*sn-glycerol-1′*-methylphosphate	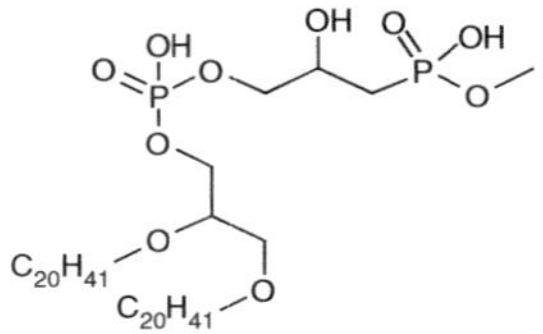 Drug delivery	Kates et al. 1993
Sulfolobus solfataricus	GDGT-4 glycerol dialkyl glycerol tetraethers, with four cyclopentane moieties	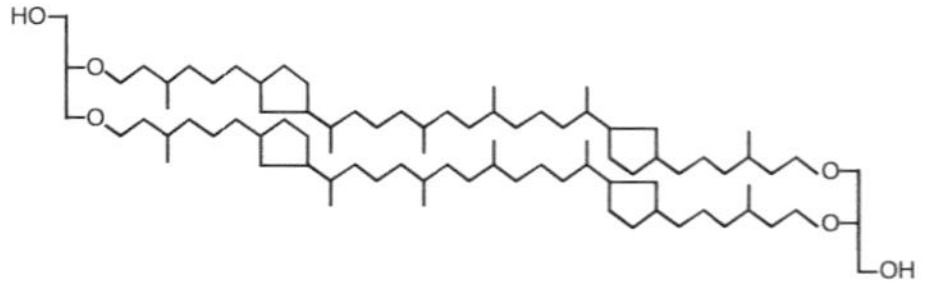 Archaeal ether lipid liposomes (*archaeosomes*)	De Rosa et al. 1989; Knappy et al. 2011.
Methanococcus jannaschii	GMD glycerol monoalkyl diether, macrocyclic archaeol	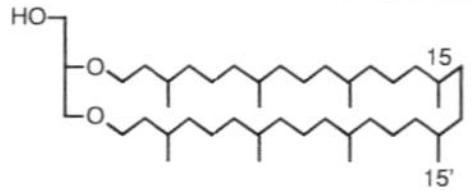	Knappy et al. 2011
Ignisphaera aggregans	GMGT glycerol monoalkyl glycerol tetraether with C80 isoprenoid hydrocarbons	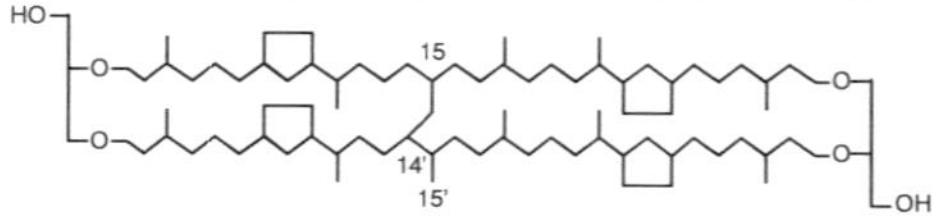	Knappy et al. 2011
Bacteria			
Thermotoga maritima	15,16-Dimethyl-30-glyceryloxy-triacontanedioic acid	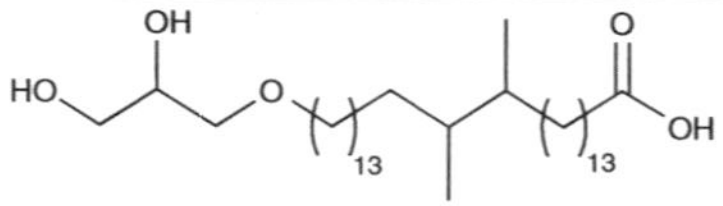 Not exploited yet	De Rosa et al. 1989
Psychrophiles	Polyunsaturated fatty acids	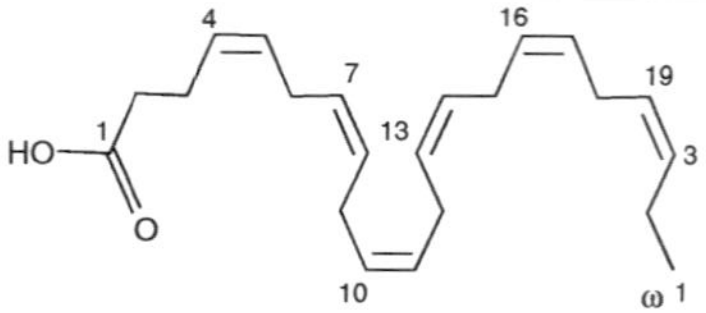 Food supplement	Hamamoto et al. 1999

stabilizer for protein molecules. Hanford and Peeples (2002) reported some applications of archaeal membrane lipids, while Jacquemet et al. (2009) reported the structure of natural and synthetic tetraether lipid and their applications in several fields for drug and gene delivery, vaccines, and liposomes or for formation of films. Another example of biological applications of archaeal lipid is the coated nanoporous aluminum oxide membranes that use stratified natural tetraethers as ultrathin films in order to improve the filtration characteristics of these membranes that resulted with lower permeability and suitable for sterilization process (Muller et al. 2006). The archaeal lipids find another interesting application for the formulation of ultrathin layers for biosensor design (Meister and Blume 2007). Membranes of *Archaea* are impermeable to water and ions and are able to hold liquid in the cell in spite of the environmental temperatures that could be ranging between 0 and 100 °C. Thanks also to the lipid properties, *Archaea* possess high potential bionanotechnology applications. Chugunov et al. (2014) used the molecular dynamics simulations to evaluate the chemical structure and dynamics of archaeal membranes containing tetraether motif and "branched" hydrophobic tails. The hypothesis was that the branched assembly is responsible for a dense packing that ensures low water permeability of these membranes and a liquid-crystalline state that is essential for the life of cells. These kind of studies clarified the structure-function correlation between the chemical type and the physicochemical behavior. Nevertheless, to expand archaeal lipid biotechnological applications, further studies are still necessary to understand their functions (Jacquemet et al. 2009).

13.3 Extremozymes

Extremophiles are considered a scientific curiosity. In the course of about 30 years, both the understanding of the evolution and the origins of life were studied, and only more recently they have been studied for their potential applications in biotechnology, because it is in recent years that the biotechnology has received a concrete response to the need for the development of a more environmentally friendly industry. The interest for the thermostable enzymes has increased drastically, and their thermostability has become an essential property for their use in many industrial processes. To survive and thrive in such adverse conditions, these microorganisms have developed special properties, such as particular structures and chemical components (such as the structure of the membrane), specific metabolic pathways, and extremely stable biomolecules (proteins, nucleic acids, and lipids).

The unique characteristics of extremozymes have allowed their use in all processes where the conditions were widely denaturing, replacing their mesophilic counterparts, thus allowing to broaden the operating processes. The major part of the enzymes have been obtained from mesophiles. These enzymes cease to function when exposed to high temperature or other extreme conditions, and therefore, in the industrial processes that are based on their action, they should be used with special precautions. In recent years, it has therefore intensified the search and characterization of novel enzymatic activities that to be industrially interesting must function in drastic conditions with the prospect of eliminating the need for precautionary measures, increasing efficiency, and reducing the costs.

For this purpose, extremozymes are those that arouse greater interest, due to their ability to extend the range of applications in industrial processes (Elleuche et al. 2014).

The biotechnology is omnipresent and has a greater impact than previously expected on various industrial sectors, such as feed and food production, biofuel and energy generation, as well as sustainable production of high-value chemical compounds. Conditions in an industrial process are often far from conventional biocatalyst's properties. Hence, there is considerable demand for a new generation of stable enzymes that are able to reach this goal to be replaced or integrated to the traditional chemical processes (Woodley 2013).

New applications will be possible with obtaining new enzymes through the isolation of new species of extremophiles and characterization of

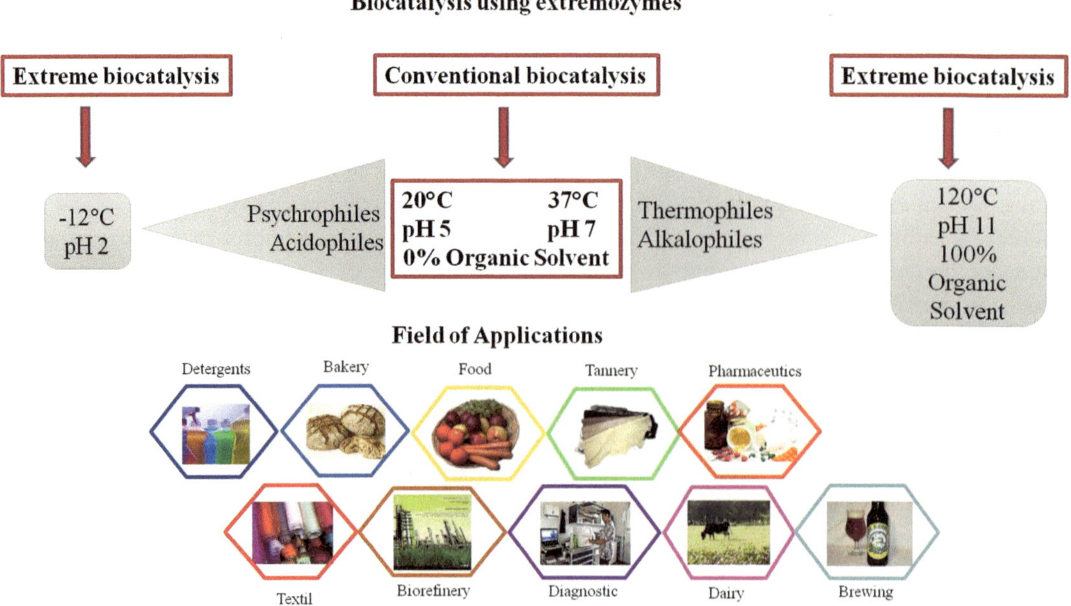

Fig. 13.3 Conventional and extreme biocatalysis

genomic sequences (Hough and Danson 1999; Demirjian et al. 2001; Eichler 2001; Vieille and Zeikus 2001; van den Burg 2003). Until now, on Earth, only a small part of the microorganisms were used. New strategies to produce extremophiles, along with the developments of the cloning and expression of genes in appropriate hosts, are likely to increase the enzymatic transformations. Such trials will allow to prepare products for the industry in the most disparate application sectors (sanitary, agricultural, feeding, energy, etc.), in an alternative way to the proper methodologies of the conventional chemistry (Fig. 13.3). Equally important, the direct evolutions offer strategies to enhance enzyme stability and change the specificity that is not found in the natural world.

Each group of the extremophiles has unique characteristics, which can be exploited to obtain enzymes with a large range of potential applications (Cavicchioli et al. 2002; Haki and Rakshit 2003; Cherry and Fidantsef 2003; Georlette et al. 2004; Wiegel and Kevbrin 2004). Enzyme technology was favored by the spectacular advances in molecular and computational biology, combinatorial methodologies, and biochemical engineering of single and multicomponent enzyme systems. The changes in molecular biology are conducting to the development of thousands of new uses of enzyme technology, feeding strong growth in this multibillion-dollar sector.

Many of these microorganisms belong to the evolutionary line of *Archaea*. The ancient evolutionary segregation of which were subject to the *Archaea* justifies the large biochemical diversity of these organisms compared to all other forms of life. Of particular interest are enzymes isolated from these organisms: for their basic research, experimental models are useful for understanding the stabilization of biomolecules in conditions normally prohibitive, while applied research studies them for their possible use in biotechnology. They are therefore of great interest not only for the exceptional stability of the enzymes but also because they are classes of enzymes with novel features. Therefore, with the advancement of modern biotechnology and protein engineering, it is possible to introduce or modify the capability of the genes that are important for the production of these novel enzymes.

For example, thermostability of engineered pectinases, used in cotton fabric processing, was constructed (Solbak et al. 2005). A xylanase engineered from the fungus *Trichoderma reesei* has been designed for increasing the alkaline stability (Fenel et al. 2006). Furthermore, structural data and computational modeling have provided the mechanisms of protein thermostability (Berezovsky and Shakhnovich 2005) and promise to conduct thermostability to predictive protein (Korkegian et al. 2005). About 80 % of current industrial enzymes are hydrolases, and the most important biotechnological applications are in the cleaning sector, in the textile and tanning, and in the food and pharmaceutical. In particular, proteases represent one of the most used enzymes for their extensive employment in the detergent and dairy industries. The second largest group, used in industries of the starch, textile, detergent, and baking industries, is represented primarily from amylases and cellulases (Gupta et al. 2014).

The global market for industrial enzymes is evaluated at 3.3 billion dollars in 2010, and it is expected to reach more than 4 billion dollars by 2015. In Table 13.2, the classes of enzyme from extremophiles and their related applications used in industrial biotransformations are reported. Thermozymes are often used when the process conditions require high temperatures associated with an higher resistance to chemical denaturant, such as thermostable polymer-degrading enzymes (amylase, pullulanases, xylanase, protease, and cellulase), which have a main role in the food, chemical, pharmaceutical, paper, and waste treatment industries (Table 13.2). The exploitation of thermostable and active enzymes in the industrial processes is noticeable. Their thermal stability makes it possible to operate at high temperatures, allowing an increase of the reaction rate, the solubility of the products and reagents, a decrease in the viscosity of the medium, and, last but not least, reducing microbial contamination. The presence of extra salt bridges, and hydrophobic interactions, or hydrogen bounds could give a further degree of stabilization with respect to mesophilic enzymes (van den Burg 2003).

Psychrophilic enzymes have become interesting for industrial applications, in part for the continuous efforts to decrease energy consumption. A characteristic of psychrophilic enzymes is their high catalytic activity at moderate temperatures and low thermal stability. It has been hypothesized that greater flexibility is related to a reduced stability. In order to reduce the energy consumption and the wear of the textile fibers, cold washing is very advantageous, so the use of hydrolytic enzymes from psychrophiles, in the production of detergents such as protease, lipase, amylase, and cellulase, would be extremely convenient. Furthermore, the industrial processing of leather at low temperatures by using psychrophilic protease or keratinase saves energy and reduces the impact of toxic chemicals used for the hair dying. The cold enzymes have an interesting potential applications: (i) the hydrolysis of lactose in milk using galactosidase, (ii) the extraction and clarification of fruit juice with pectinase, and (iii) the biopolishing and stone washing of textiles with cellulases (Cavicchioli et al. 2002).

Proteins from halophilic to address very high salt concentrations have adapted by having a large number of negatively charged amino acid residues. This prevent the precipitation (e.g., KCl concentrations of 4 M and NaCl concentrations of >5 M) (Demirjian et al. 2001). Instead, the solubility of halophilic enzymes, in an environment with lower salt concentrations, is often very poor and consequently may reduce their applications (Madern et al. 2000). This property has been exploited in nonaqueous media (Klibanov 2001). A remarkable property of acidophilic and alkaliphilic microorganisms is that although their internal pH values are close to neutrality, their extracellular enzymes are able to work at pH extreme. In the production of detergents, where there are reaction conditions strongly acidic or highly alkaline, the use of enzymes, such as amylase, lipase, and protease, which can operate and resist to high pH and high concentrations of chelating agents, may be particularly useful. Proteins that are resistant to the pressure could be used in particular for the production of food, in case of application of high pressure for the treatment and the safety of food (Hayashi 1996).

In general, thermophilic enzymes work in the presence of detergents, chaotropic agents, and organic solvents. Psychrophilic enzymes can be

Table 13.2 Thermozymes and their possible applications

Source	Habitat	Enzyme	Application	References
Thermophiles	Hyperthermophiles (>85 °C)	DNA polymerases	Genetic engineering	Kaledin et al. 1980
	Thermophiles (65–85 °C)	Proteases	Baking, brewing, detergents	Dipasquale et al. 2008
		Amylases	Production of high-glucose, starch conversion	Haki and Rakshit 2003
		Glucoamylases, pullulunase	Hydrolysis of lactose	Haki and Rakshit 2003
		Glycosidases	Synthesis of alkyl glycoside detergents	Haki and Rakshit 2003
		Cellulases	Pulp and paper processing, laundry detergents	Haki and Rakshit 2003
		Xylanases	Paper bleaching	Lama et al. 2004
		Lipase	Detergents, stereo-specific reactions	Klibanov 2001
		Esterase	(e.g., transesterification, organic biosynthesis)	Sellek and Chaudhuri 1999
		Alcohol dehydrogenase	Chemical synthesis	Klibanov 2001
	Moderate thermophiles (45–65 °C)	Xylanases	Paper bleaching	Archana and Satyanarayana 2003
Psychrophile	Low temperature (<15 °C)	Dehydrogenases	Biosensors	Cavicchioli et al. 2002
		Amylases	Detergent and bakery	Cavicchioli et al. 2002
		Cellulases	Detergents, feeds, and textiles	Cavicchioli et al. 2002
		Glycosidases	Hydrolysis of lactose	Cavicchioli et al. 2002
		Lipases	Detergents, food, and cosmetics	Cavicchioli et al. 2002
		Proteases	Detergents and food applications	Cavicchioli et al. 2002
Acidophile	Low pH	Proteases, cellulases	Feed component	Gupta et al. 2014
		Amylases, glucoamylases	Starch processing	Gupta et al. 2014
		Sulfur oxidation	Desulfurization of coal	Norris et al. 2000
		Oxidases	Valuable metal recovery	Norris et al. 2000
Alkalophile	High pH	Cellulases	Polymer degradation in detergents	Wiegel and Kevbrin 2004
		Proteases		Lama et al. 2005
		Xylanases		
Halophile	High salt concentration	Proteases	Peptide synthesis	Madern et al. 2000
		Dehydrogenases	Biocatalysis in organic media	Klibanov 2001
Piezophile	High pressure	Whole microorganism	Formation of gels and starch granules	Hayashi 1996

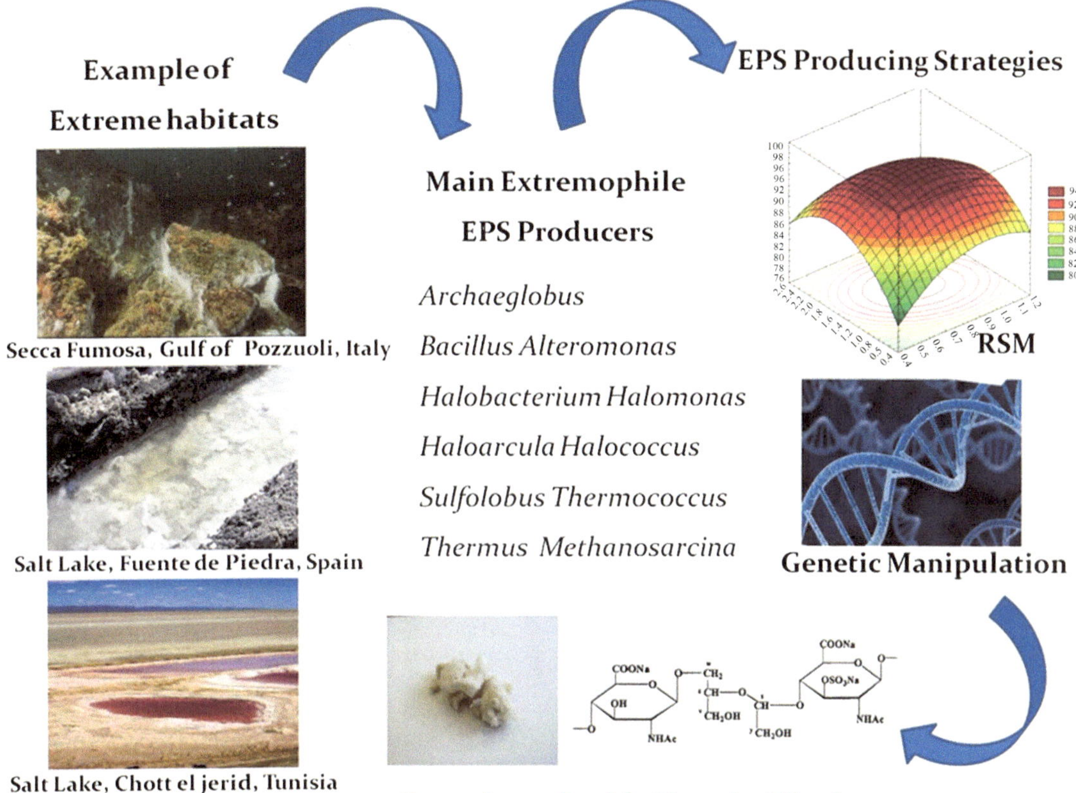

Fig. 13.4 Main extremophile EPS sources and exopolysaccharide study procedures

used to obtain good yields of heat-sensitive products and halophilic enzymes. These are stable at high salt concentrations and can be employed as a model for enzymatic reactions in nonaqueous media (Sellek and Chaudhuri 1999). Heat-stable solvent-tolerant biocatalysts are valuable tools for processes in which, for example, hardly decomposable polymers need to be liquefied and degraded, while cold-active enzymes are of relevance for food and detergent industries. Extremophilic microorganisms are a rich source of naturally tailored enzymes, which are more superior over their mesophilic counterparts for applications at extreme conditions. Extremozymes with unique properties are widely distributed in thermophilic prokaryotes and are of high potential for versatile industrial processes, especially in the processing of lignocellulolytic, amylolytic, and other biomass (Lama et al. 2004, 2005, 2013, 2014; Dipasquale et al. 2008, 2014; Finore et al. 2011, 2014b).

13.4 Extremophilic Biopolymers

Extremophilic microorganisms possess unique ability to resist to several harsh conditions (i.e., extreme pH, temperature, pressure, radiation values): such resistance is often due to the production of several biopolymers, either as extracellular or intracellular macromolecules, that provide defense against environmental stresses (Fig. 13.4). The peculiar features of these biopolymers, which provide a fundamental line of defense for extremophiles, provide them interesting biotechnological properties that in turn make them commercially interesting materials. Among the most remarkable biopolymers are exopolysaccharides (EPS), intracellular polyhydroxyalkanoates (PHA), and polyglutamates (PGA) as summarized in Table 13.3.

Many halophilic microorganisms from the genus *Halomonas* are reported to be producers of EPS that could find application in several indus-

Table 13.3 Selected examples of extremophilic microorganisms producing biopolymers, i.e., exopolysaccharides, polyhydroxyalkanoate, polyglutamic acid, and other molecules with their applications

Microorganism	Biopolymer type/name	Biological/biotechnological properties	References
Halomonas maura	Exopolysaccharide/Mauran	Removal of toxic metals from polluted environments and wastewater; immunomodulator; tissue engineering and drug delivery	Arias et al. 2003; Oren 2010; Raveendran et al. 2013
Halomonas almeriensis	Exopolysaccharide/sulfated mannan	Hydrophobic molecule emulsifier; heavy metal binding	Llamas et al. 2012
Halomonas eurihalina	Exopolysaccharide, glucomannan	Stimulation of human lymphocyte proliferation	Llamas et al. 2012
Halomonas stenophila	Exopolysaccharide/sulfated heteroglucan	Inhibition of human T-lymphocyte tumors	Llamas et al. 2012
Halomonas smyrnensis strain AAD6	Exopolysaccharide/levan	Biocompatibility and affinity with cancerous and noncancerous cell lines	Küçükaşik et al. 2010
Bacillus thermodenitrificans strain B3-72	Exopolysaccharide/mannan	Immunomodulatory and antiviral activities	Poli et al. 2010
Bacillus licheniformis strain B3-15	Exopolysaccharide/mannan	Antiviral activity	Poli et al. 2010
Bacillus licheniformis strain T14	Exopolysaccharide/fructan	Anti-cytotoxic properties	Spanò et al. 2013
Geobacillus thermantarcticus	Exopolysaccharide/mannan	Sulfated polysaccharides	Nicolaus et al. 2004
Pseudoalteromonas strain SM9913	Exopolysaccharide/glucan	Flocculation, biosorption	Poli et al. 2010
Sulfolobus solfataricus strains MT4 and MT3	Exopolysaccharide/sulfated glucomannan	Biofilm	Poli et al. 2011
Halomonas alkaliantarctica	Exopolysaccharide/mannan and a xylomannan	Not exploited yet	Poli et al. 2004
Haloferax mediterranei	Exopolysaccharide/polyhydroxyalkanoate/PHBV	Bioplastics	Bhattacharyya et al. 2014
Halomonas boliviensis	Polyhydroxyalkanoate/PHB	Bioplastics	Oren 2010
Halogeometricum boriquense	Polyhydroxyalkanoate/PHB	Bioplastics	Salgaonkar et al. 2013
Haloarcula marismortui	Polyhydroxyalkanoate/PHB	Bioplastics	Han et al. 2007
Halobiforma haloterrestris strain 135	Polyhydroxyalkanoate/PHB	Bioplastics	Poli et al. 2011

Halopiger aswanensis strain 56	Polyhydroxyalkanoate/PHB	Bioplastics	Poli et al. 2011
Haloterrigena hispanica	Polyhydroxyalkanoate/PHB	Bioplastics	Poli et al. 2011
Pseudomonas strain CT13	Polyhydroxyalkanoate/PHB	Osmolyte/chemical chaperon	Soto et al. 2012
Pseudomonas strain 14-3	Polyhydroxyalkanoate/PHB	Bioplastics	Ayub et al. 2009
Bacillus licheniformis strain WX-02	Polyglutamic acid	Sequestration of toxic metal ions	Wei et al. 2010
Natrialba strain 40	Polyglutamic acid	Protection against dehydration	Hezayen et al. 2000
Natrialba aegyptiaca	Polyglutamic acid	Protectant/stabilizer agent for labile enzymes	Yamasaki et al. 2010
Alkaliphiles	Antibiotics	Treatment of infections	Horikoshi 1999
Halophiles	Compatible solutes	Pharmaceuticals	Lai et al. 1999
Halophiles	Carotene	Food additive	Borowitzka 1999
Psychrophiles	Cells	Bioremediation of oil spills	Lo Giudice et al. 2010
Radiation resistant	Cells	Bioremediation of radioactive waste	Lloyd and Renshaw 2005

trial fields. A remarkable example is mauran produced from *Halomonas maura* (Arias et al. 2003; Oren 2010; Raveendran et al. 2013) that is made up of galactose, glucose, and glucuronic acid, respectively, in the relative proportions of 14/29.3/21.9 %w/w and that also contains 1.3 % w/w phosphate and 6.5 % w/w sulfate. This EPS has been investigated for its environmental applications and for its pharmacological properties. Another interesting example is represented by the EPS from *Halomonas almeriensis* (Llamas et al. 2012), composed by mannose, glucose, and rhamnose in the relative proportions of 72/27.5/0.5 %w/w and containing 1.4 % w/w sulfate: this EPS showed to be able to act as biodetoxifier (thanks to its heavy metal binding capacity) and as emulsifying agent.

Other sulfate- and phosphate-containing EPSs are produced by *Halomonas eurihalina* (Llamas et al. 2012) and by *Halomonas stenophila* (Llamas et al. 2012): it is noteworthy to underline that the presence of such chemical groups is usually associated with the biomedical properties. Indeed EPS from *H. eurihalina* can stimulate the proliferation of human lymphocytes, while EPSs from *Halomonas stenophila* are able to inhibit the growth of human T-lymphocyte tumors. Another interesting EPS from halophiles is represented by levan produced by *Halomonas smyrnensis* sp. AAD6 (Küçükaşik et al. 2010): this exopolysaccharide was produced by using as fermentation medium some cheap substrates such as sugar beet molasses (a waste of sugar beet processing for the sucrose production) or starch molasses (residual from dextrose manufacture from corn) and showed to be biocompatible as confirmed by HeLa and L929 cell line-based tests. *Halomonas alkaliantarctica* strain CRSS produced an heteropolysaccharide when grown either with complex or defined media (Poli et al. 2004). EPSs are produced also by microorganisms belonging to other genera such as *Bacillus thermodenitrificans* strain B3-72 and *Bacillus licheniformis* strain B3-15 (Poli et al. 2010), both isolated form hot marine environments.

The EPS from *B. thermodenitrificans* strain B3-72, mainly composed by mannose and glucose in molar ratio 1:0.2, has been shown to act as an immunomodulator besides being able to restore the immunological disorders determined by viral infections. On the other hand, *B. licheniformis* strain B3-15 exerts an immunomodulatory action by enhancing the production of some key cytokines modulating the immune response to viral infection. Also other *B. licheniformis* strain has been identified as EPS producers: a very recent example is represented by *B. licheniformis* strain T14 (Spanò et al. 2013) that produces an exopolysaccharide composed by fructose, fucose, and glucose in the molar ratio of 1.0:0.75:0.28 with traces of galactosamine and mannose. Such an EPS exhibited anti-cytotoxic properties as confirmed by brine shrimp assay. Other bacillus producers can be found also in terrestrial environments such as *Geobacillus thermantarcticus* (Nicolaus et al. 2004), a thermophilic bacterium isolated from Mount Melbourne in Antarctica. This microorganisms have been identified as a producer of sulfated EPSs, i.e., two main EPSs both made up mainly of glucose and mannose as monomer sugars.

Other examples of EPS-producing strains from marine environment belong to the genus *Pseudoalteromonas* such as *Pseudoalteromonas* strain SM9913 (Poli et al. 2010). This psychrotolerant microorganism produces an EPS mainly composed by glucose with smaller amounts of arabinose, galactose, and xylose: such EPS has a potential ecological role, thanks to its flocculation behavior and to its biosorption capacity.

Finally, interesting examples of extremophilic species producing EPSs can be found in *Archaea* domain like *Sulfolobus solfataricus* strains MT4 and MT3 (Poli et al. 2011), two extreme thermoacidophiles that produce sulfated exopolysaccharides mainly made of glucose, mannose, glucosamine, and galactose as monomer sugars. Such biopolymers are able to form complex biofilms that possess carpet-like structures, as shown by analysis by confocal laser scanning microscopy.

Another interesting class of biopolymers from extremophiles is represented by the polyhydroxyalkanoates (PHA) that are produced by the cells during unbalanced growth conditions (absence of essential nutrients such as nitrogen, phosphorus, and magnesium) and that serve as carbon and energy sources and storage. Thanks to their mechanical properties and biodegradability, such microbial biopolymers are the object of great attention as potential substitutes of petroleum-derived plastics.

Several halophilic species (e.g., *Haloferax*, *Halomonas*, *Halogeometricum*, *Haloarcula*, *Halobiforma*, *Halopiger*, *Haloterrigena*) have been described as PHA producers. For example, this is the case of *Haloferax mediterranei* (Bhattacharyya et al. 2014) that, during the growth under phosphate limitation, can produce PHA up to 65 % of its dry weight. In a very recent report, it has been shown that, by using the residues from of industrial rice conversion to ethanol as fermentation medium, *H. mediterranei* produced poly-3-(hydroxybutyrate-co-hydroxyvalerate) (PHBV) up to a final amount of 16.42 ± 0.02 g/l (Han et al. 2015). Other halophilic producers of PHA are *Halomonas boliviensis* (Oren 2010) that produces high amounts of PHA (88 % w/w) by using as carbon sources acetate, butyrate, or sucrose and *Halogeometricum borinquense* (Salgaonkar et al. 2013), an extremely halophilic archaeal strain that produces, under a variety of harsh conditions, significant amounts of polyhydroxybutyrate (PHB) with interesting thermal resistance. The extreme halophile *Haloarcula marismortui* (Han et al. 2007) produced large amounts of PHB (21 % w/w after 180 h fermentation, as insoluble granules) when grown on minimal medium containing glucose. *Halobiforma haloterrestris* (strain 135^T) (Poli et al. 2011) and *Halopiger aswanensis* (strain 56) (Poli et al. 2011) produce PHB when cultured on yeast extract and casamino acids, reaching higher yields up to 40 % w/w by using yeast extract and butyric acid in the fermentation medium; on the other hand, PHB production reaches a value of 34 % w/w for *H. aswanensis* when grown on acetate and butyric acid. Finally, *Haloterrigena hispanica* strain FP1 (Poli et al. 2011) accumulates intracellular PHB when grown on complex standard medium, but also when fermentation is carried out by using a cheap substrate like wastes from industrial processing of vegetables (i.e., carrot).

Also some species belonging to the genus *Pseudomonas* have been studied for the ability to produce PHAs. The halotolerant bacterium *Pseudomonas* sp. CT13 (Soto et al. 2012), for example, produces PHB that has been investigated for its action as molecular chaperon. Indeed, such biopolymer is able to stabilize citrate synthase by acting on the protein's hydration shell thus preventing its thermal aggregation. The Antarctic bacterium *Pseudomonas* sp. 14-3 (Ayub et al. 2009) has also been reported to be a PHB producer, with a level of polymer of 18 % w/w at 28 °C that increased up to 24 % w/w at 10 °C. This biopolymer showed to be essential to the survival of the bacterium in the oxidative stress conditions developed in extremely cold conditions.

Another kind of biopolymer produced by extremophiles is represented by polyglutamic acid (PGA) that can find applications in several fields, thanks to its biodegradability, for example, in the food drug carrier or in the pharmaceutical industry. PGA is produced predominantly by bacteria belonging to genus *Bacillus* and by halophilic archaebacteria such as *Natrialba* species. One example of PGA-producing species is represented by *Bacillus licheniformis* WX-02, an halotolerant bacterium (Wei et al. 2010). The amount of NaCl in the culture medium affected the molecular weight and productivity of γ-PGA; indeed the molecular weight decreased at increasing salt concentrations, while the production was enhanced at higher NaCl concentrations. *Bacillus licheniformis* has been mainly used for fermentative production of PGA whose biological role is exploited in the defense against stress environment by sequestering toxic metals. *Natrialba* species are also other extremophilic species able to produce PGA. The *Natrialba* strain 40, discovered in Aswan (Egypt) in hypersaline soil, is able to produce significant quantities of PGA when grown on solid medium; in these conditions, it is possible to obtain an almost pure polymer whose main feature is a strong water-binding capacity that affords a significant protective effect against dehydration (Hezayen et al. 2000). Extracellular poly-γ-glutamate synthesized by an extremely halophilic archaeon *Natrialba aegyptiaca* has been investigated for its extremolyte-like applicability (Yamasaki et al. 2010). Interestingly, this biopolymer afforded protection against freeze thawing and proteolysis for a labile DNA ligase that also gained thermostability and alkalotolerance in the presence of such biopolymer.

It is noteworthy to underline that besides the abovementioned biopolymers, extremophiles produce a very large spectrum of valuable compounds that possess diverse chemical structures, biological properties, and biotechnological applications. Some remarkable examples are represented by phenazine antibiotics produced under alkaline conditions by some alkaliphiles *Nocardiopsis* species (Horikoshi 1999), compatible solutes like betaine produced by halophilic microorganisms (Lai et al. 1999), and carotenoids like carotene that is produced by some halophiles and that is exploited in food industry as natural additive (Borowitzka 1999). Finally, also the extremophilic cells can be used as such for other applications like bioremediation of polluted environments: this is the case of some psychrophiles that are able to degrade polluting hydrocarbons (Lo Giudice et al. 2010) or of some *Nocardia* or *Rhodococcus* species that are able to colonize radioactive environments and to transform radioactive wastes (Lloyd and Renshaw 2005).

13.5 Genome

Enzymes and in general bioactive compounds constitute the biological resources of natural products potentially employed in the bioprospecting concept based on the research finding, study, and subsequent distribution on market of active products (Akondi and Lakshmi 2013). An environmental sample potentially contains a very huge amount of different microorganisms, but nothing more than 1–2 % of microbial cells are able to be cultured in the laboratory. This means that the biodiversity that is present in an environmental sample remains unexplored and unfortunately lost (Vester et al. 2015). The exploitation of biodiversity could be realized through different methods. The first one is represented by the improvement of cultivation techniques comprising the traditional culturing steps that bring to the isolation and eventually the screening of some new isolates with interesting enzyme activities. Recently, the metagenomic methods have been developed: this technique is based on the recovery of genetic materials directly from an environmental sample and their subsequent sequencing and bioinformatic analysis. Moreover, a functional expression of metagenomic libraries could be also possible with an ultimate aim to identify interesting genes or gene clusters for potential biotechnology applications (Vester et al. 2013). Metagenomics offers a powerful lens for viewing the whole microbial world starting from an environmental sample (Kim et al. 2013). If this latter came from extreme habitats, this scenario appears more complicated for the poor presence of forms of life and their slowing cellular growth, for the problematic typology of matrices for DNA availability, and for the complexity to reach some restricted area. Moreover, also the choice of a suitable vector system and subsequently an appropriate expression host may be complex: nowadays the number of expression hosts for genes that derived from extremophilic microorganisms is very limited. Sequence-based metagenomics are able to identify many genes encoding biotechnologically useful enzymes even if there is the subsequent problem that these genes should be then expressed as active enzymes in the appropriate hosts. Recently, the development of more multiform vectors and also the new engineered strains has enhanced the functional metagenomics (Ekkers et al. 2012; Cheng et al. 2014; Liebl et al. 2014), and several enzymes such as lipase, cellulase, xylanase, and amylase are present in literature as examples of metagenomic tools to obtain new enzymes from uncultured microorganisms that live in extreme environments (Couto et al. 2010; Jeon et al. 2011; Bhat et al. 2013; Biver et al. 2013; Fu et al. 2013; Vester et al. 2014). The production of exopolysaccharide (EPS) as described for the enzyme production is a process controlled by genes. The main step is the discovery of the genes responsible of the EPS biosynthesis, subsequently the comprehension of the mechanisms in which they are involved.

Starting from a knowledge of a EPS-producer microorganism genome, it is feasible to elucidate the EPS biosynthesis mechanisms, to enhance the microbial productivity via strain improvement strategies, or to modify physicochemical and/or rheological properties of the biopolymer by changing its composition, length, or degree of branching. For example, *Zunongwangia profunda* strain SM-A87, as resulted also in other microorganisms isolated from deep seas, was able to synthesize exopolysaccharide (Liu et al. 2011), and it was the first genome sequenced in the *Bacteroidetes* (Qin et al. 2010). Two polysaccharide biosynthesis gene clusters were found in its genome. The knowledge of strain SM-A87 genome revealed other interesting information related, for example, to its versatility to live in deep-sea habitats and a possible role in the degradation of deposit sediment of organic nitrogen origins. When the whole genome sequence of the EPS producer microorganism is not available, sequence data of a taxonomically close species could also be used for systems-based studies. This approach has been used for the improvement of levan production by the halophilic strain *Halomonas smyrnensis* strain AAD6^T, for which, for the first time, the available whole genome sequence of a taxonomically close microorganism, *Chromohalobacter salexigens* DSM 3043, was used to construct a comprehensive genome-scale metabolic model (Ateş et al. 2011), and then this model was recruited and adopted to the producer strain considering the phenotypic, physiological, and biochemical properties of *H. smyrnensis* AAD6^T. With metabolic system analysis of this generic metabolic model, significant improvement in levan yields was obtained (Ateş et al. 2013).

The knowledge of genome sequence should be a starting point also for other kind of compounds of biotechnological interest. Recently, Maida et al. (2014) described the phenotypic and genomic characterization of the antarctic bacterium *Gillisia* sp. CAL575, a producer of antimicrobial compounds. Sequencing and analysis of the whole genome of *Gillisia* sp. CAL575 revealed that it includes genes that are involved in secondary metabolite productions, adaptation to cold conditions, and different metabolic pathways for the production of energy. The willingness of the genome sequence of *Gillisia* sp. CAL575 provided the possibility to have also an insight on the strategy adopted by this strain for cold adaptation making this strain a possible tool for biotechnology.

Acknowledgments This work has been implemented in the frame of the project PON01_01966 "Integrated agro-industrial chains with high energy efficiency for the development of eco-compatible processes of energy and biochemicals production from renewable sources and for the land valorization" funded by MIUR.

References

Akondi KB, Lakshmi VV (2013) Emerging trends in genomic approaches for microbial bioprospecting. OMICS 17:61–70. doi:10.1089/omi.2012.0082

Albers SV, Konings WN, Driessen AJM (2006) Membranes of thermophiles and other extremophiles. In: Rainey FA, Oren A (eds) Methods in microbiology: extremophiles, vol 35. Elsevier, New York, pp 161–171. doi:10.1016/S0580-9517(08)70010-2

Antranikian G, Vorgias CE, Bertoldo C (2005) Extreme environments as a resource for microorganisms and novel biocatalysts. Adv Biochem Eng Biotechnol 96:219–262. doi:10.1007/b135786

Archana A, Satyanarayana T (2003) Purification and characterization of a cellulase-free xylanase of a moderate thermophile *Bacillus licheniformis* A99. World J Microbiol Biotechnol 19:53–57. doi:10.1023/A:1022527702400

Arias S, Del Moral A, Ferrer M, Tallon R, Quesada E, Bejar V (2003) Mauran, an exopolysaccharides produced by the halophilic bacterium *Halomonas maura*, with a novel composition and interesting properties. Extremophiles 7:319–326. doi:10.1007/s00792-003-0325-8

Ateş Ö, Toksoy Öner E, Arga KY (2011) Genome-scale reconstruction of metabolic network for a halophilic extremophile, *Chromohalobacter salexigens* DSM 3043. BMC Syst Biol 5:12. doi:10.1186/1752-0509-5-12

Ateş Ö, Arga KY, Toksoy Öner E (2013) The stimulatory effect of mannitol on levan biosynthesis: lessons from metabolic systems analysis of *Halomonas smyrnensis* AAD6(T). Biotechnol Prog 29:1386–1397. doi:10.1002/btpr.1823

Ayub N, Tribelli P, López N (2009) Polyhydroxyalkanoates are essential for maintenance of redox state in the Antarctic bacterium *Pseudomonas* sp. 14-3 during low temperature adaptation. Extremophiles 13:59–66. doi:10.1007/s00792-008-0197-z

Berezovsky IN, Shakhnovich EI (2005) Physics and evolution of thermophilic adaptation. Proc Natl Acad Sci U S A 102:12742–12747. doi:10.1073/pnas.0503890102

Bhat A, Riyaz-Ul-Hassan S, Ahmad N, Srivastava N, Johri S (2013) Isolation of cold-active, acidic endocellulase from Ladakh soil by functional metagenomics. Extremophiles 17:229–239. doi:10.1007/s00792-012-0510-8

Bhattacharyya A, Saha J, Haldar S, Bhowmic A, Mukhopadhyay U, Mukherjee J (2014) Production of poly-3-(hydroxybutyrate-co-hydroxyvalerate) by *Haloferax mediterranei* using rice-based ethanol stillage with simultaneous recovery and re-use of medium salts. Extremophiles 18:463–470. doi:10.1007/s00792-013-0622-9

Biver S, Steels S, Portetelle D, Vandenbol M (2013) *Bacillus subtilis* as a tool for screening soil metagenomic libraries for antimicrobial activities. J Microbiol Biotechnol 23:850–855. doi:10.4014/jmb.1212.12008

Borowitzka MA (1999) Commercial production of microalgae: ponds, tanks, tubes and fermenters. J Biotechnol 70:313–321. doi:10.1016/S0079-6352(99)80123-4

Boucher Y (2007) Lipids: biosynthesis, function, and evolution. In: Cavicchioli R (ed) Archaea: molecular and cellular biology. ASM Press, Washington, DC, pp 341–353. ISBN 978-1-55581-391-8

Bowers KJ, Mesbah NM, Wiegel J (2009) Biodiversity of poly-extremophilic bacteria: does combining the extremes of high salt, alkaline pH and elevated temperature approach a physico-chemical boundary for life? Saline Syst 5:9. doi:10.1186/1746-1448-5-9p

Canganella F, Wiegel J (2014) Anaerobic thermophiles; thermal ecosystems; extremophiles; deep-sea; taxonomy; biotechnology. Life 4:77–104. doi:10.3390/life4010077

Cavicchioli R, Siddiqui KS, Andrews D, Sowers KR (2002) Low-temperature extremophiles and their applications. Curr Opin Biotechnol 13:253–261. doi:10.1016/S0958-1669(02)00317-8

Cheng J, Pinnell L, Engel K, Neufeld JD, Charles TC (2014) Versatile broad-host-range cosmids for construction of high quality metagenomic libraries. J Microbiol Methods 99:27–34. doi:10.1016/j.mimet.2014.01.015

Cherry JR, Fidantsef AL (2003) Directed evolution of industrial enzymes: an update. Curr Opin Biotechnol 14:438–443. doi:10.1016/S0958-1669(03)00099-5

Chugunov AO, Volynsky PE, Krylov NA, Boldyrev IA, Efremov RG (2014) Liquid but durable: molecular dynamics simulations explain the unique properties of archaeal-like membranes. Sci Rep 12(4):7462. doi:10.1038/srep07462

Corcelli A (2009) The cardiolipin analogues of *Archaea*. Biochim Biophys Acta 1788(10):2101–2106. doi:10.1016/j.bbamem.2009.05.010

Corcelli A, Lobasso S (2006) Characterization of lipids of halophilic *Archaea*. In: Rainey FA, Oren A (eds) Methods in microbiology: extremophiles. Elsevier, New York, pp 585–613. ISBN 978-0-12-521537-4

Couto GH, Glogauer A, Faoro H, Chubatsu LS, Souza EM, Pedrosa FO (2010) Isolation of a novel lipase from a metagenomic library derived from mangrove sediment from the south Brazilian coast. Genet Mol Res 9:514–523. doi:10.4238/vol9-1gmr738

de Carvalho CC, Caramujo MJ (2012) Lipids of prokaryotic origin at the base of marine food webs. Mar Drugs 10:2698–2714. doi:10.3390/md10122698

De Rosa M, Gambacorta A, Huber R, Lanzotti V, Nicolaus B, Stetter KO, Trincone A (1989) Microbiology of extreme environments and its potential for biotechnology. In: da Costa MS, Duarte JC, Williams RAD (eds) Springer, New York, pp 167–173. ISBN 978-1-85166-361-3

Deming JW (2002) Psychrophiles and polar regions. Curr Opin Microbiol 5:301–309. doi:10.1016/S1369-5274(02)00329-6

Demirjian DC, Moris-Varas F, Cassidy CS (2001) Enzymes from extremophiles. Curr Opin Chem Biol 5:144–151. doi:10.1016/S1367-5931(00)00183-6

Dipasquale L, Calandrelli V, Romano I, Nicolaus B, Gambacota A, Lama L (2008) Purification and characterisation of a highly thermostable extracellular protease from *Bacillus thermantarcticus*, strain M1. Ann Microbiol 58(2):253–259. doi:10.1007/BF03175325

Dipasquale L, Romano I, Picariello G, Calandrelli V, Lama L (2014) Characterization of a native cellulase activity from an anaerobic thermophilic hydrogen-producing bacterium *Thermosipho* sp., strain 3. Ann Microbiol. doi:10.1007/s13213-013-0792-9

Eichler J (2001) Biotechnological uses of archaeal extremozymes. Biotechnol Adv 19:261–278. doi:10.1016/S0734-9750(01)00061-1

Ekkers DM, Cretoiu MS, Kielak AM, Elsas JD (2012) The great screen anomaly—a new frontier in product discovery through functional metagenomics. Appl Microbiol Biotechnol 93:1005–1020. doi:10.1007/s00253-011-3804-3

Elleuche S, Schröder C, Sahm K, Antranikian G (2014) Extremozymes – biocatalysts with unique properties from extremophilic microorganisms. Curr Opin Biotechnol 29:116–123. doi:10.1016/j.copbio.2014.04.003

Fenel F, Zitting AJ, Kantelinen A (2006) Increased alkali stability in *Trichoderma reesei* endo-1, 4-b-xylanase II by site directed mutagenesis. J Biotechnol 121:102–107. doi:10.1016/j.jbiotec.2005.07.010

Finore I, Kasavi C, Poli A, Romano I, Toksoy Oner E, Kirdar B, Dipasquale L, Nicolaus B, Lama L (2011) Purification, biochemical characterization and gene sequencing of a thermostable raw starch digesting α-amylase from *Geobacillus thermoleovorans* subsp. *stromboliensis* subsp. *Nov*. World J Microbiol Biotechnol 27:2425–2433. doi:10.1007/s11274-011-0715-5

Finore I, Di Donato P, Mastascusa V, Nicolaus B, Poli A (2014a) Fermentation technologies for the optimization of marine microbial exopolysaccharide production. Mar Drugs 12:3005–3024. doi:10.3390/md12053005

Finore I, Di Donato P, Poli A, Kirdar B, Kasavi C, Oner Toksoy E, Nicolaus B, Lama L (2014b) Use of agro waste biomass for alpha-amylase production by *Anoxybacillus amylolyticus*. Purification and properties. J Microb Biochem Technol 6:320–326. doi:10.4172/1948-5948.1000162

Fu J, Leiros HK, de Pascale D, Johnson KA, Blencke HM, Landfald B (2013) Functional and structural studies of a novel cold-adapted esterase from an Arctic intertidal metagenomic library. Appl Microbiol Biotechnol 97:3965–3978. doi:10.1007/s00253-012-4276-9

Gabani P, Singh OV (2013) Radiation-resistant extremophiles and their potential in biotechnology and therapeutics. Appl Microbiol Biotechnol 97:993–1004. doi:10.1007/s00253-012-4642-7

Georlette D, Blaise V, Collins T, D'Amico S, Gratia E, Hoyoux A, Marx JC, Sonan G, Feller G, Gerday C (2004) Some like it cold: biocatalysis at low temperatures. FEMS Microbiol Rev 28:25–42. doi:10.1016/j.femsre.2003.07.003

Gliozzi A, Rolandi R, de Rosa M, Gambacorta A (1983) Monolayer black membranes from bipolar lipids of archaebacteria and their temperature-induced structural changes. J Membr Biol 75(1):45–56. doi:10.1007/BF01870798

Gupta GN, Srivastava S, Khare SK, Prakash V (2014) Extremophiles: an overview of microorganism from extreme environment. Int J Agric Environ Biotechnol 7(2):371–380. doi:10.5958/2230-732X.2014.00258.7

Haki GD, Rakshit SK (2003) Developments in industrially important thermostable enzymes: a review. Bioresour Technol 89:17–34. doi:10.1016/S0960-8524(03)00033-6

Hamamoto T, Takata N, Kudo T, Horikoshi K (1999) Characteristic presence of polyunsaturated fatty acids in marine psychrophilic vibrios. FEMS Microbiol Lett 129(1):51–56. doi:10.1016/0378-1097(95)00134-Q

Han J, Lu Q, Zhou L, Zhou J, Xiang H (2007) Molecular characterization of the phaECHm genes, required for biosynthesis of poly(3-hydroxybutyrate) in the extremely halophilic archaeon *Haloarcula marismortui*. Appl Environ Microbiol 73:6058–6065. doi:10.1128/AEM.00953-07

Han J, Wu LP, Hou J, Zhao D, Xiang H (2015) Biosynthesis, characterization and hemostasis potential of tailor-made poly(3-hydroxybutyrate-co-3-hydroxyvalerate) produced by *Haloferax mediterranei*. Biomacromolecules. doi:10.1021/bm5016267

Hanford MJ, Peeples TL (2002) Archaeal tetraether lipids: unique structures and applications. Appl Biochem Biotechnol 97(1):45–62. doi:10.1385/ABAB:97:1:45

Hayashi R (1996) Use of high pressure in bioscience and biotechnology. In: Hayashi R, Balny C (eds) High pressure bioscience and biotechnology. Elsevier, New York, pp 1–6. ISBN 0080544614/9780080544618

Hezayen F, Rehm B, Eberhardt R, Steinbuchel A (2000) Polymer production by two newly isolated extremely halophilic archaea: application of a novel corrosion-resistant bioreactor. Appl Microbiol Biotechnol 54:319–325. doi:10.1007/s002530000394

Horikoshi K (1999) Alkaliphiles: some applications of their products for biotechnology. Microbiol Mol Biol Rev 6(4):735–750, ISSN: 1098-5557

Hough DW, Danson MJ (1999) Extremozymes. Curr Opin Chem Biol 3:39–46. doi:10.1016/S1367-5931(99)80008-8

Irwin JA (2010) Extremophiles and their application to veterinary medicine. Environ Technol 31(8-9):857–869. doi:10.1080/09593330.2010.484073

Irwin JA, Baird AW (2004) Extremophiles and their application to veterinary. Ir Vet J 57(6):348–354. doi:10.1186/2046-0481-57-6-348

Jacquemet A, Barbeau J, Lemiègre L, Benvegnu T (2009) Archaeal tetraether bipolar lipids: structures, functions and applications. Biochimie 91(6):711–717. doi:10.1016/j.biochi.2009.01.006

Jain S, Caforio A, Fodran P, Lolkema JS, Minnaard AJ, Driessen AJM (2014) Identification of CDP-archaeol synthase, a missing link of ether lipid biosynthesis in archaea. Chem Biol 21:1392–1401. doi:10.1016/j.chembiol.2014.07.022

Jeon JH, Kim JT, Lee HS, Kim SJ, Kang SG, Choi SH, Lee JH (2011) Novel lipolytic enzymes identified from metagenomic library of deep-sea sediment. Evid Based Complement Alternat Med 2011:271419. doi:10.1155/2011/271419

Kaledin A, Sliusarenko A, Gorodetski S (1980) Isolation and properties of DNA polymerase from extreme thermophilic bacteria *Thermus aquticus* YT-1. Biokhimiia 45:644–651, ISSN: 0006-2979

Kates M (1978) The phytanyl ether-linked polar lipids and isoprenoid neutral lipids of extremely halophilic bacteria. Prog Chem Fats Other Lipids 15(4):301–342. doi:10.1016/0079-6832(77)90011-8

Kates M, Moldoveanu N, Stewart LC (1993) On the revised structure of the major phospholipid of *Halobacterium salinarium*. Biochim Biophys Acta 1169(1):46–53. doi:10.1016/0005-2760(93)90080-S

Kim M, Lee KH, Yoon SW, Kim BS, Chun J, Yi H (2013) Analytical tools and databases for metagenomics in the next-generation sequencing era. Genomics Inform 11:102–113. doi:10.5808/GI.2013.11.3.102

Klibanov AM (2001) Improving enzymes by using them in organic solvents. Nature 409:241–246. doi:10.1038/35051719

Knappy CS, Nunn CEM, Morgan HW, Keely BJ (2011) The major lipid cores of the archaeon *Ignisphaera aggregans*: implications for the phylogeny and biosynthesis of glycerol monoalkyl glycerol tetraether isoprenoid lipids. Extremophiles 15:517–528. doi:10.1007/s00792-011-0382-3

Koga Y, Morii H (2007) Biosynthesis of ether-type polar lipids in *archaea* and evolutionary considerations. Microbiol Mol Biol Rev 71:97–120. doi:10.1128/MMBR.00033-0

Korkegian A, Black ME, Baker D, Stoddard BL (2005) Computational thermostabilization of an enzyme. Science 308:857–860. doi:10.1126/science.1107387

Ksouri R, Ksouri WM, Jallali I, Debez A, Magné C, Hiroko I, Abdelly C (2012) Medicinal halophytes:

potent source of health promoting biomolecules with medical, nutraceutical and food applications. Crit Rev Biotechnol 32(4):289–326. doi:10.3109/07388551.2011.630647

Küçükaşik F, Kazak H, Güney D, Finore I, Poli A, Yenigün O, Nicolaus B, Oner E (2010) Molasses as fermentation substrate for levan production by *Halomonas* sp. Appl Microbiol Biotechnol 89(6):1729–1740. doi:10.1007/s00253-010-3055-8

Lai MC, Yang DR, Chuang MJ (1999) Regulatory factors associated with synthesis of the osmolyte glycine betaine in the halophilic methanoarchaeon *Methanohalophilus portucalensis*. Appl Environ Microbiol 65(2):828–833, ISSN: 1098-5336

Lama L, Calandrelli V, Gambacorta A, Nicolaus B (2004) Purification and characterization of thermostable xylanase and β-xylosidase by the thermophilic bacterium *Bacillus thermantarcticus*. Res Microbiol 155:283–289. doi:10.1016/j.resmic.2004.02.001

Lama L, Romano I, Calandrelli V, Nicolaus B, Gambacorta A (2005) Purification and characterization of a protease produced by an aerobic haloalkaliphilic species belonging to the *Salinivibrio* genus. Res Microbiol 156:478–484. doi:10.1016/j.resmic.2004.12.004

Lama L, Poli A, Nicolaus B (2013) *Geobacillus thermantarcticus* as source of biotechnological thermozymes and exopolysaccharides: a review. Curr Trends Microbiol 8:1–12, ISSN: 0972-7736

Lama L, Tramice A, Finore I, Anzelmo G, Calandrelli V, Pagnotta E, Tommonaro G, Poli A, Di Donato P, Nicolaus B, Fagnano M, Mori M, Impagliazzo A, Trincone A (2014) Degradative actions of microbial xylanolytic activities on hemicelluloses from rhizome of *Arundo donax*. AMB Express 4:55. doi:10.1186/s13568-014-0055-6

Liebl W, Angelov A, Juergensen J, Chow J, Loeschcke A, Drepper T, Classen T, Pietruzska J, Ehrenreich A, Streit WR, Jaeger KE (2014) Alternative hosts for functional (meta)genome analysis. Appl Microbiol Biotechnol 98:8099–8109. doi:10.1007/s00253-014-5961-7

Liu SB, Qiao LP, He HL, Zhang Q, Chen XL, Zhou WZ, Zhou BC, Zhang YZ (2011) Optimization of fermentation conditions and rheological properties of exopolysaccharide produced by deep-sea bacterium *Zunongwangia profunda* SM-A87. PLoS One 6:e26825. doi:10.1371/journal.pone.0026825

Llamas L, Amjres H, Mata J, Quesada E, Béjar V (2012) The potential biotechnological applications of the exopolysaccharide produced by the halophilic bacterium *Halomonas almeriensis*. Molecules 17:7103–7120. doi:10.3390/molecules17067103

Lloyd JR, Renshaw JC (2005) Bioremediation of radioactive waste: radionuclide–microbe interactions in laboratory and field-scale studies. Curr Opin Biotechnol 16(3):254–260. doi:10.1016/j.copbio.2005.04.012

Lo Giudice A, Bruni V, De Domenico M, Michaud L (2010) Psychrophiles – cold adapted hydrocarbon – degrading microorganisms in: handbook of hydrocarbon and lipid microbiology, pp 1897–1921. doi:10.1007/978-3-540-77587-4_139

Lombard J, Ló pez-García P, Moreira D (2012a) The early evolution of lipid membranes and the three domains of life. Nat Rev Microbiol 10:507–515. doi:10.1038/nrmicro2815

Lombard J, Ló pez-García P, Moreira D (2012b) Phylogenomic investigation of phospholipid synthesis in archaea. Archaea 630910. doi:10.1155/2012/630910

MacElroy RD (1974) Some comments on the evolution of extremophiles. Biosystems 6:74–75. doi:10.1016/0303-2647(74)90026-4

Madern D, Ebel C, Zaccai G (2000) Halophilic adaptation of enzymes. Extremophiles 4:91–98. doi:10.1007/s007920050142

Maida I, Fondi M, Papaleo MC, Perrin E, Orlandini V, Emiliani G, de Pascale D, Parrilli E, Tutino ML, Michaud L, Lo Giudice A, Romoli R, Bartolucci G, Fani R (2014) Phenotypic and genomic characterization of the Antarctic bacterium *Gillisia* sp. CAL575, a producer of antimicrobial compounds. Extremophiles 18:35–49. doi:10.1007/s00792-013-0590-0

Manca MC, Lama L, Esposito E, Improta R, Gambacorta A, Nicolaus B (1996) Chemical composition of two exopolysaccharides from *Bacillus thermoantarcticus*. Appl Environ Microbiol 62(9):3265–3269, ISSN: 1098-5336

Matsumi R, Atomi H, Driessen AJM, van der Oost J (2011) Isoprenoid biosynthesis in *Archaea*—biochemical and evolutionary implications. Res Microbiol 162:39–52. doi:10.1016/j.resmic.2010.10.003

Meister A, Blume A (2007) Self-assembly of bipolar amphiphiles. Curr Opin Colloid Interface Sci 12:138–147. doi:10.1016/j.cocis.2007.05.003

Mesbah NM, Wiegel J (2012) Life under multiple extreme conditions: diversity and physiology of the halophilic alkalithermophiles. Appl Environ Microbiol 78(12):4074. doi:10.1128/AEM.00050-12

Muller S, Pfannmoller M, Teuscher N, Heilmann A, Rothe U (2006) New method for surface modification of nanoporous aluminum oxide membranes using tetraether lipid. J Biomed Nanotechnol 2:16–22. doi:10.1166/jbn.2006.003

Nicolas JP (2005) A molecular dynamics study of an archaeal tetraether lipid membrane: comparison with a dipalmitoylphosphatidylcholine lipid bilayer. Lipids 40:1023–1030. doi:10.1007/s11745-005-1465-2

Nicolaus B, Trincone A, Esposito E, Vaccaro MR, Gambacorta A, De Rosa M (1990) Calditol tetraether lipids of the archaebacterium *Sulfolobus solfataricus*. Biosynth Stud Biochem J 266:785–791, ISSN Electronic: 1470-8728

Nicolaus B, Trincone A, Lama L, Palmieri G, Gambacorta A (1992) Quinone composition in *Sulfolobus solfataricus* grown under different conditions. Syst Appl Microbiol 15:18–20. doi:10.1016/S0723-2020(11)80131-1

Nicolaus B, Manca MC, Romano I, Lama L (1993) Production of an exopolysaccharide from two thermophilic archaea belonging to the genus *Sulfolobus*. FEMS Microbiol Lett 109:203–206. doi:10.1111/j.1574-6968.1993.tb06168.x

Nicolaus B, Schiano Moriello V, Lama L, Poli A, Gambacorta A (2004) Polysaccharides from extremophilic microorganisms. Orig Life Evol Biosph 34:159–169. doi:10.1023/B:ORIG.0000009837.37412.d3

Norris PR, Burton NP, Faulis NAM (2000) Acidophiles in bioreactor mineral processing. Extremophiles 4(2):71–76. doi:10.1007/s007920050139

Oren A (2006) Life at high salt conditions. In: Dworkin M, Falkow S, Rosenberg E, Schleifer K-H, Stackebrandt E (eds) The Prokaryotes, a handbook on the biology of bacteria: ecophysiology and biochemistry. Springer, New York, pp 263–282. ISBN 978-0-387-30742-8

Oren A (2010) Industrial and environmental applications of halophilic microorganisms. Environ Technol 31:8–9. doi:10.1080/09593330903370026

Özdemir S, Kilink E, Nicolaus B, Poli A (2013) Resistance and bioaccumulation of Cd, Cu, Co and Mn by thermophilic bacteria, *Bacillus thermantarcticus* and *Anoxybacillus amylolyticus*. World J Microbiol Biotechnol 28(1):155–6363. doi:10.1007/s13213-013-0598-9

Pikuta EV, Hoover RB, Tang J (2007) Microbial extremophiles at the limits of life. Crit Rev Microbiol 33:183–209. doi:10.1080/10408410701451948

Poli A, Schiano Moriello V, Esposito E, Lama L, Gambacorta A, Nicolaus B (2004) Exopolysaccharide production by a new *Halomonas* strain CRSS isolated from saline lake Cape Russell in Antarctica growing on complex and defined media. Biotechnol Lett 26:1635–1638. doi:10.1007/s10529-004-3187-y

Poli A, Kazak H, Gurleyendag B, Tommonaro G, Pieretti G, Toksoy Oner E, Nicolaus B (2009) High level synthesis of levan by a novel *Halomonas* species growing on defined media. Carbohydr Polym 78:651–657. doi:10.1016/j.carbpol.2009.05.031

Poli A, Anzelmo G, Nicolaus B (2010) Bacterial exopolysaccharides from extreme marine habitats: production, characterization and biological activities. Mar Drugs 8:1779–1802. doi:10.3390/md8061779

Poli A, Di Donato P, Abbamondi GR, Nicolaus B (2011) Synthesis, production and biotechnological applications of exopolysaccharides and polyhydroxyalkanoates by *Archaea*. Review article in Archaea, Article ID 693253. doi:10.1155/2011/693253

Qin QL, Zhang XY, Wang XM, Liu GM, Chen XL, Xi BB, Dang HY, Zhou BC, Yu J, Zhang YZ (2010) The complete genome of *Zunongwangia profunda* SM-A87 reveals its adaptation to the deep-sea environment and ecological role in sedimentary organic nitrogen degradation. Genomics 11:247. doi:10.1186/1471-2164-11-247

Raveendran S, Dhandayuthapani B, Nagaoka Y, Yoshida Y, Maekawa T, Kumar D (2013) Biocompatible nanofibers based on extremophilic bacterial polysaccharide, Mauran from *Halomonas Maura*. Carbohydr Polym 92:1225–1233. doi:10.1016/j.carbpol.2012.10.033

Reith F (2011) Life in the deep subsurface. Geology 39:287–288. doi:10.1130/focus032011.1

Rothschild LJ, Mancinelli RL (2001) Life in extreme environments. Nature 409:1092–1101. doi:10.1038/35059215

Salgaonkar B, Mani K, Braganca J (2013) Accumulation of polyhydroxyalkanoates by halophilic archaea isolated from traditional solar salterns of India. Extremophiles 17:787–795. doi:10.1007/s00792-013-0561-5

Sellek GA, Chaudhuri JB (1999) Biocatalysis in organic media using enzymes from extremophiles. Enzyme Microb Technol 25:471–48228. doi:10.1016/S0141-0229(99)00075-7

Solbak AI, Richardson TH, McCann RT, Kline KA, Bartnek F, Tomlinson G, Tan X, Parra-Gessert L, Frey GJ, Podar M, Luginbühl P, Gray KA, Mathur EJ, Robertson DE, Burk MJ, Hazlewood GP, Short JM, Kerovuo J (2005) Discovery of pectin-degrading enzymes and directed evolution of a novel pectate lyase for processing cotton fabric. J Biol Chem 11:9431–9438. doi:10.1074/jbc.M411838200

Soto G, Setten L, Lisi C, Maurelis C, Mozzicafreddo M, Cuccioloni M, Angeletti M, Ayub N (2012) Hydroxybutyrate prevents protein aggregation in the halotolerant bacterium *Pseudomonas* sp. CT13 under abiotic stress. Extremophiles 16:455–462. doi:10.1007/s00792-012-0445-0

Spanò A, Gugliandolo C, Lentini V, Maugeri T, Anzelmo G, Poli A, Nicolaus B (2013) A novel EPS-producing strain of *Bacillus licheniformis* isolated from a shallow vent off Panarea Island (Italy). Curr Microbiol 67:21–29. doi:10.1007/s00284-013-0327-4

Sprott GD (1992) Structures of archaebacterial membrane lipids. J Bioenerg Biomembr 24(6):555–566. doi:10.1007/BF00762348

van de Vossenberg JL, Driessen AJ, Konings WN (1998) The essence of being extremophilic: the role of the unique archaeal membrane lipids. Extremophiles 2(3):163–170. doi:10.1007/s007920050056

van den Burg B (2003) Extremophiles as a source for novel enzymes. Curr Opin Microbiol 6:213–218. doi:10.1016/S1369-5274(03)00060-2

Vester JK, Glaring MA, Stougaard P (2013) Improving diversity in cultures of bacteria from an extreme environment. Can J Microbiol 59:581–586. doi:10.1139/cjm-2013-0087

Vester JK, Glaring MA, Stougaard P (2014) Discovery of novel enzymes with industrial potential from a cold and alkaline environment by a combination of functional metagenomics and culturing. Microb Cell Fact 13:72. doi:10.1186/1475-2859-13-72

Vester JK, Glaring MA, Stougaard P (2015) Improved cultivation and metagenomics as new tools for bioprospecting in cold environments. Extremophiles 19.17–29. doi:10.1007/s00792-014-0704-3

Vieille C, Zeikus GJ (2001) Hyperthermophilic enzymes: sources, uses, and molecular mechanisms for thermostability. Microbiol Mol Biol Rev 65:1–43. doi:10.1128/MMBR.65.1.1-43

Villanueva L, Damsté JSS, Schouten S (2014) A re-evaluation of the archaeal membrane lipid biosynthetic pathway. Nat Rev Microbiol 12:438–448. doi:10.1038/nrmicro3260

Wei X, Ji Z, Chen S (2010) Isolation of halotolerant *Bacillus licheniformis* WX-02 and regulatory effects of sodium chloride on yield and molecular sizes of poly-γ-glutamic acid. Appl Biochem Biotechnol 160:1332–1340. doi:10.1007/s12010-009-8681

Wiegel J, Kevbrin VV (2004) Alkalithermophiles. Biochem Soc Trans 32:193–198, ISSN Electronic: 1470-8752

Woodley JM (2013) Protein engineering of enzymes for process applications. Curr Opin Chem Biol 17:310–316. doi:10.1016/j.cbpa.2013.03.017

Yamasaki D, Minouchi Y, Ashiuchi M (2010) Extremolyte-like applicability of an archaeal exopolymer, poly-γ-L-glutamate. Environ Technol 31(10):1129–1134. doi:10.1080/09593331003592279

Ilaria FINORE received her M.Sc. and Ph.D. degree in Environmental Sciences in 2006 from "Parthenope" University, Naples, Italy, and Ph.D. degree in "Environment, Resources and Sustainable Development" with Label of "Doctor Europaeus." She is currently working at CNR-Institute of Biomolecular Chemistry, and her interests are extremophilic microorganisms, bioenergy, and biopolymers.

Licia Lama She is currently a Researcher at CNR, Institute of Biomolecular Chemistry, in Pozzuoli (NA). She is expert on identification, purification, and physicochemical characterization of the extremophilic enzymes for developing of new and innovative products and processes more economical and environmental friendly. Her current interests are extremophiles, extremozyme, and biorefinery.

Annarita Poli, M.Sc. and Ph.D. She is responsible of the work package "Studies of biochemical properties of natural biomolecules from macro- and microorganisms" of C.N.R. She is a Member of the Scientific Board of the International Ph.D. "Environment, Resources, and Sustainable Development" of Parthenope University, Italy. Her current interests are extremophiles and their molecules, biopolymers, and reuse of vegetable wastes.

Paola Di Donato, M.Sc. and Ph.D. She is Associate Professor of Biochemistry at Parthenope University of Naples, Italy. She is a Member of the Scientific Board of the International Ph.D. "Environment, Resources, and Sustainable Development" of Parthenope University. Her current interests include valorization of vegetable waste biomass as sources of value-added products including polyphenols, carotenoids, polysaccharides, and the study of biological and biotechnological properties of such biomolecules.

Barbara Nicolaus received her M.Sc. degree in Science Biology from Federico II University of Naples Italy. She is presently Dirigent of Research at CNR, Institute of Biomolecular Chemistry, ICB, and Active Director of ICB-CNR (June 2013–May 2014). She is a Member of the Scientific Board of the International Ph.D. "Environment, Resources, and Sustainable Development" of Parthenope University, Italy. She is Scientific Responsible of International Project. Her current interests are extremophiles, biopolymers, and biorefinery.

Microbial CRISPR–Cas System: From Bacterial Immunity to Next-Generation Antimicrobials

14

Alka Mehra

Abstract

Microbes live in multi-microbial communities called microbiome. Discoveries that can help in the regulation of the composition of the microbiome are likely to impact diverse functions of microbes from health, environment, to biotechnology. Antimicrobials offer such regulatory potential and are slowly but surely evolving for the benefit of human health and biotechnology. Antibiotics are the first discovered antimicrobials which are low molecular weight natural microbial products that inhibit the growth of other microbes. However, emergence of microbial resistance to conventional antibiotics has presented an urgent need for novel antimicrobials. Here, we describe another native microbial machinery, CRISPR ("clustered regularly interspaced short palindromic repeats")–Cas ("CRISPR associated") system, that confers adaptive immunity to microbes by employing CRISPR RNAs to recognize and destroy complementary nucleic acids of invasive foreign genetic elements. Further, sequence-based targeting by CRISPR–Cas system has been leveraged for the development of sequence-specific novel antimicrobials, genome editing, and genome regulation tools.

14.1 Introduction

14.1.1 A Microbial Immune System

Microbes need to survive under a constantly changing environment. They can adapt to the environment by acquiring new traits which confer selective advantage leading to genome evolution. Therefore, there is exchange of genetic material between microbes by horizontal gene transfer (HGT) (Nakamura et al. 2004). HGT can occur by transformation of DNA from the

A. Mehra (✉)
Allergy and Infectious Diseases, CSIR-Institute of Genomics and Integrative Biology (CSIR-IGIB), 110007 Delhi, India
e-mail: alkam08@gmail.com

V.C. Kalia (ed.), *Microbial Factories*, DOI 10.1007/978-81-322-2595-9_14

environment or uptake of bacteriophages by transduction and foreign mobile genetic elements like transposons and plasmids by conjugation. Majority of HGT events are harmful for the recipient, and therefore, microbes have systems for protection from the invading DNA (Thomas and Nielsen 2005). Restriction–modification enzyme systems from bacteria have been known for long to help eliminate the invaders by restriction based on differential methylation of the host and invading DNA (Kusano et al. 1995). Many microbes, archaea, and bacteria also harbor CRISPR (Clustered Regularly Interspaced Short Palindromic Repeats)–Cas ("CRISPR associated") systems that use a novel mechanism of action along the lines of eukaryotic RNA interference (RNAi) (Makarova et al. 2006). However, while prokaryotic CRISPR–Cas systems target DNA (Barrangou et al. 2007; Marraffini and Sontheimer 2008), eukaryotic RNAi targets RNA. CRISPER–Cas system is a microbial adaptive immune system since it is directed by sequence specific for the invading DNA conferring acquired resistance to major routes of HGT – transformation, transduction, and conjugation.

14.1.2 Genomic Architecture and Discovery of CRISPR–Cas System

CRISPRs are arrays of DNA direct repeats interspersed with unique spacers. Main feature of CRISPER locus is a A/T-rich leader sequence (L), followed by an array of variable number of conserved short DNA repeat (R) sequences of 21–48 base pairs interspaced by stretches of variable sequences called spacers (S) of 26–72 bp (see Fig. 14.1). Cas proteins are encoded by putative operons adjacent to CRISPR locus (Jansen et al. 2002a; Makarova et al. 2006) (see Fig. 14.1). CRISPRs were first identified in 1987 in upstream of *iap* gene, responsible for alkaline phosphatase isozyme conversion in *Escherichia coli* (Ishino et al. 1987). Over the next 10 years, as more microbial genomes were sequenced, these genetic elements were identified in several prokaryotes like in *Haloferax mediterranei*, *Streptococcus pyogenes*, *Anabaena*, and *Mycobacterium tuberculosis* (Groenen et al. 1993; Mojica et al. 1995; Masepohl et al. 1996; Hoe et al. 1999). They were defined as a novel class of repeats regularly spaced by intervening sequences of equal length, and they were referred to as short regularly spaced repeats (SRSRs) (Mojica et al. 2000). Later, Jansen et al. 2002b carried out more analysis of these repeats across multiple genomes and found that intervening spacers have no similarity with each other unlike similarity between repeats organized in same orientation and called them as Spacers Interspersed Direct Repeats (SPIDRs) (Jansen et al. 2002b). Thereafter, Mojica et al. and Jansen et al. 2002a revised its nomenclature to the acronym "CRISPR" which they believed to truly reflect organization of its structural elements. At the same time, a group of genes were discovered adjacent to CRISPR loci and called as CRISPR-associated (*cas*) genes which indicated a functional relationship between the two (Jansen et al. 2002a). Since its discovery, CRISPR now is a widely distributed family of repeats in prokaryotes. However, the function of the CRISPR–Cas system was unknown for long. An interesting insight came into their function by systematic analysis of unique spacer sequences suggesting an extrachromosomal origin of this DNA element mainly from phages or plasmids (Bolotin et al. 2005; Mojica et al. 2005). Along this was the observation of a reciprocal relationship between the presence of a spacer in the cell and ability of the foreign genetic material like plasmids or viruses to be transferred to the cell (Mojica et al. 2005). Additionally, CRISPRs were found to be transcribed as long precursors and processed into small monomers (Tang et al. 2002; Lillestol et al. 2006). Subsequently, the transcribed and processed CRISPR RNA was shown to complex with Cas proteins to target foreign DNA conferring immunity by interference. Hence CRISPR-Cas system was discovered to be a novel adaptive immune mechanism in prokaryotes (described below).

14.1.3 Mechanism of Immunity

Microbes have evolved nucleic acid-based "immunity" system whereby resistance is con-

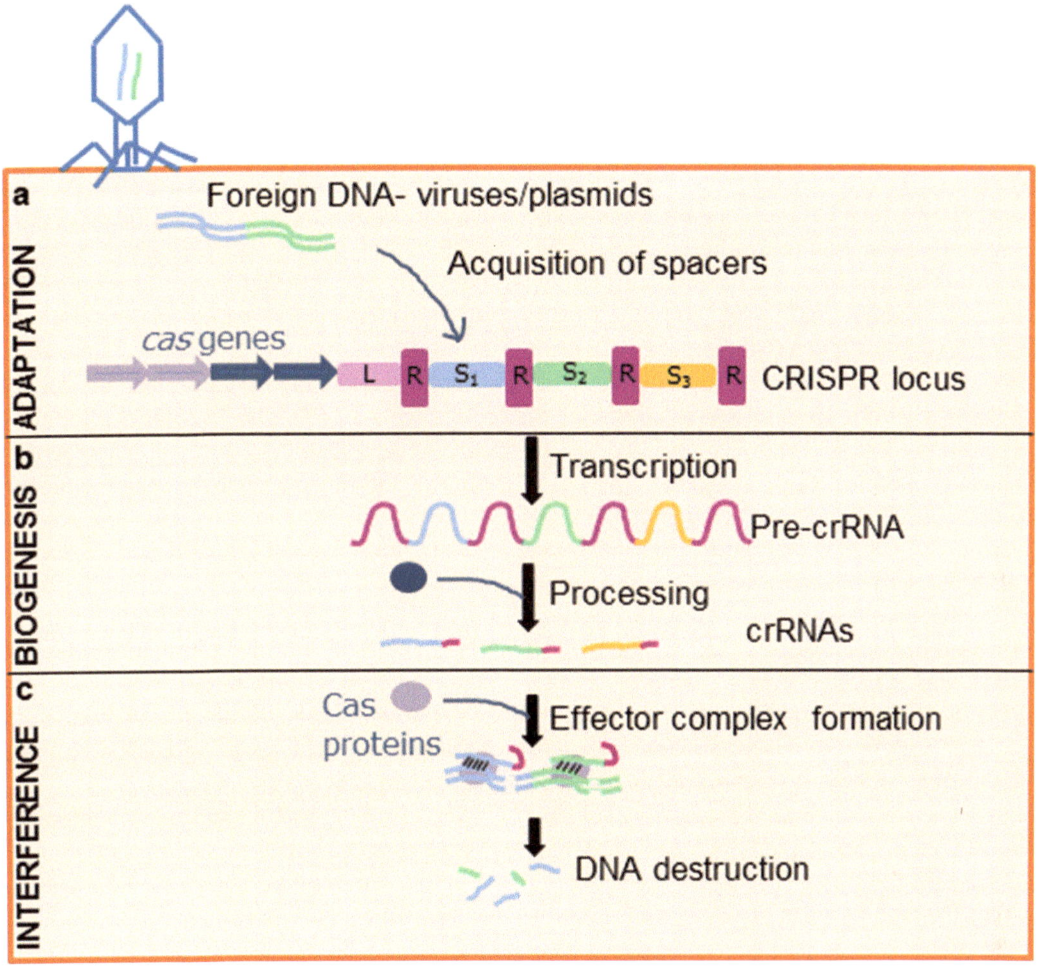

Fig. 14.1 Mechanism of adaptive immunity by microbial CRISPR–Cas system. (**a**) Adaptation: spacer sequences (e.g. S_1, S_2, S_3) are acquired from the invading foreign genetic element and incorporated into the CRISPR locus. (**b**) Biogenesis: transcription of the locus generates longer precursor pre-crRNA which is processed into set of shorter crRNAs containing a short repeat sequence and the variable spacer sequence (guide) with sequence identity to invading nucleic acids. (**c**) Interference: crRNA forms an effector complex with the Cas proteins and guides it to region of sequence complementary on the invading nucleic acid for cleavage and/or degradation

ferred by Cas enzymatic machinery in spacer content-defined manner. Additionally, some *cas* gene-encoded proteins do not directly provide resistance but are involved in the insertion of CRISPR spacers and repeats at the CRISPR loci. There are three distinct steps in this immunity: (1) CRISPR adaptation, (2) CRISPR biogenesis, and (3) interference or silencing (see Fig. 14.1). During adaptation, Cas proteins recognize and integrate short sequences of the invasive DNA into the CRISPR as spacers (Swarts et al. 2012; Yosef et al. 2012). Barrangou and colleagues for the first time experimentally demonstrated that CRISPR–Cas system-based immunity provided resistance to *Streptococcus thermophilus* against phage attack. They showed that new spacers were acquired during phage attack bearing sequence identity to phage sequence, leading to resistance to the phage (Barrangou et al. 2007). New spacers acquired from the invading DNA are added at the leader (L) side of CRISPR (Barrangou et al. 2007; Swarts et al. 2012; Yosef et al. 2012). Highly conserved Cas1 and Cas2 proteins are thought to play a role in spacer acquisition, but the exact mechanism remains to be elucidated. The polarized acquisition and integration of the spacer is suggestive of CRISPR with historical records of previous encounters with foreign DNA. Immunologically, adaptation step is like immunization or vaccination of bacteria with spacer DNA which leads to memorization of the invader. This acquired trait reflects Lamarckian-based mechanism of acquired immunity in the microbes (Koonin and Wolf 2009). CRISPR array is expressed into a long precursor CRISPR RNA (pre-crRNA) which is further processed into small CRISPR RNA (crRNA), consisting of one conserved repeat and one variable spacer sequence called the guide sequence. The guide sequence is complementary to a region of the invasive DNA called the protospacer (Brouns et al. 2008). CRISPR transcripts were first identified in the *Archaeoglobus fulgidus* (Tang et al. 2002) and *Sulpholobus solfataricus* (Tang et al. 2005). The first biochemical evidence of the CRISPR processing came from *E. coli* strain K12 transcript analysis (Brouns et al. 2008) and was subsequently shown for *Pyrococcus furiosus* (Hale et al. 2008), *Sulfolobus acidocaldarius* (Lillestol et al. 2009), and *Xanthomonas oryzae* (Semenova et al. 2009). The leader (L) sequence harbors the promoter that drives the transcription of the CRISPR locus. The Cas proteins bound to crRNA form an effector complex that recognizes target sequences in invasive nucleic acids based on complementarity of guide sequence to double-stranded DNA (Garneau et al. 2010) or single-stranded mRNA (Hale et al. 2009) leading to sequence-specific cleavage and/or degradation of foreign genetic elements. Immunologically, the interference phase correlates with an immune response of host vaccinated to invasive foreign DNA by incorporation of the invader-derived spacer (Datsenko et al. 2012).

14.2 CRISPR–Cas System Classification

The CRISPR–Cas system occurs in 40 % of bacteria and 90 % of archaea genomes (Kunin et al. 2007). The system exhibits diversity across microbes, in that, the number of CRISPER loci per genome and number of repeat-spacer units per CRISPER array are variable (Bult et al. 1996). Based on sequence similarities and stem–loop-forming ability, CRISPRs are classified into 12 categories (Grissa et al. 2007). Additional diversity comes from *cas* genes located as a cluster within 1 kb of CRISPR array. *Cas* genes encode 65 sets of orthologous Cas proteins which are classified into 25 gene families (Haft et al. 2005; Makarova et al. 2006). Currently, the diverse CRISPER–Cas systems fall into three major types (I–III) which are further subdivided into subtypes based on content and order of the *cas* gene families, repeat length, and sequences at CRISPR loci. The core *cas1* and *cas2* gene families are present in all CRISPR loci. Three major types of CRISPR–Cas systems have different organization features and mechanisms to target DNA. See Table 14.1 for summarization of comparative features of the three types.

Different modes of DNA interference of three different major classes of CRISPR–Cas systems are explained below (see Fig. 14.2).

In the *type I system*, the transcribed precursor crRNA (pre-crRNA) forms hairpin structures due to partial palindromic nature of the repeat units leading to recognition and processing by Cas6 or Cas5d endoribonucleases (Carte et al. 2008; Nam et al. 2012). Mature crRNAs include a spacer flanked on 5′- and 3′ ends by 8-nucleotides (nts) and 21-nts, respectively, due to cleavage of pre-crRNA within repeat stems (Gesner et al. 2011; Sashital et al. 2011). CrRNA is then incorporated into a large multi-Cas protein complex called CRISPR-associated complex for antiviral defense (CASCADE). crRNA–CASCADE complex is guided by sequence complementarity to the target DNA in the foreign genetic element called protospacer. Additionally, CASCADE complex scans for the correct protospacer adjacent motif (PAM) in the noncomplementary/noncoding strand of the duplex DNA upstream of the protospacer and unwinds the DNA creating R loop (Jore et al. 2011). Now Cas3 protein binds the single-stranded DNA which leads to activation of its ATPase/helicase activity leading to cleavage of the DNA in the protospacer region (Sinkunas

Table 14.1 Types of microbial CRISPER–Cas systems

Features	Type I	Type II	Type III
Prevalence	Bacteria and archaea	Bacteria	Archaea
CRISPER–Cas machinery components	*Cas genes*: multiple Cas proteins; Cas3 and CASCADE (group of Cas proteins) *CRISPR array*: repeats are partially palindromic	*Cas genes*: minimal set of cas genes *tracrRNA*: 25 nt stretch of tracrRNA is complementary to CRISPR repeats *CRISPR array*: repeats are partially palindromic	*Cas genes*: multiple Cas proteins; Cas6, Cas10, and Cmr complex *CRISPR array*: repeats are non-palindromic except *Staphylococcus epidermis* of III A subtype
Target	DNA	DNA	DNA and RNA
crRNA processing	Cas6 or Cas5d endoribonuclease	tracrRNA and host RNase III	Cas6 endoribonuclease, Csm4, and/or Cas10
Signature Cas protein	Cas3	Cas9	Cas10
PAM	2–3 nt sequence upstream of protospacer differs between subtypes and organisms	2–5 nt PAM (NGGNG) immediately downstream of protospacer, differs between different organisms	Not required
crRNA seed sequence	6–12 nts at 5′ end	12–15 nts at 3′ end	5′ 1–8 nts
Subtypes	I-A to I-F	II A, II B, and II C	III A and III B

et al. 2011). Thereafter, Cas3 protein translocates in 3′ to 5′ direction on the noncoding strand and degrades it (Westra et al. 2012).

The Type II system has the simplest organization with minimal set of *cas* genes, CRISPR array, and trans-encoded small RNA (called a trans-activating CRISPR RNA or tracrRNA). The transcribed tracrRNA binds to the pre-crRNA through 24-nt region of complementarity. This complex is stabilized by Cas9 and then cleaved by host RNase III in the repeat units of pre-crRNA (Deltcheva et al. 2011) generating mature crRNAs. A mature crRNA is shorter than type I crRNA since it only has a 3′ extended 22-nt handle without the 5′ handle which is cleaved by unknown nuclease. The Cas9–crRNA–tracrRNA complex locates and binds to a protospacer sequence in the target DNA in a PAM-dependent process (Chylinski et al. 2013; Karvelis et al. 2013). Binding of crRNA to complementary strand and displacement of the noncomplementary strand of the target upstream of the PAM results in an R loop structure. Further, Cas9 cleaves with RuvC domain on the noncoding strand and with HNH domain on the coding strand upstream of PAM to generate blunt ends (Gasiunas et al. 2012).

In the type III system, expressed pre-crRNAs have repeats which are non-palindromic (Kunin et al. 2007) with exception of *Staphylococcus epidermis* where they are partially palindromic (Hatoum-Aslan et al. 2011). Nonetheless, the pre-crRNA is processed first by Cas6 endoribonuclease which cuts in the repeat unit to yield an intermediate which is subsequently cleaved at 3′-end (Carte et al. 2008; Wang et al. 2011). This yields crRNAs with an 8-nt 5′-handle from the repeat region, but the spacer region is trimmed at 3′-end (Carte et al. 2010). A group of proteins encoded within *cas* operon called Cmr proteins bind to the crRNA forming Cmr complex that cleaves the target sequence (DNA or RNA) (Zhang et al. 2012; Deng et al. 2013) as defined by crRNA (Hale et al. 2009).

14.3 Microbial Genome Targeting by CRISPR–Cas System Is Lethal

While majority of the spacers have targets in phages and plasmids, Stern and colleagues found that around 0.4 % of the spacers have targets on chromosomal DNA of microbes and therefore are

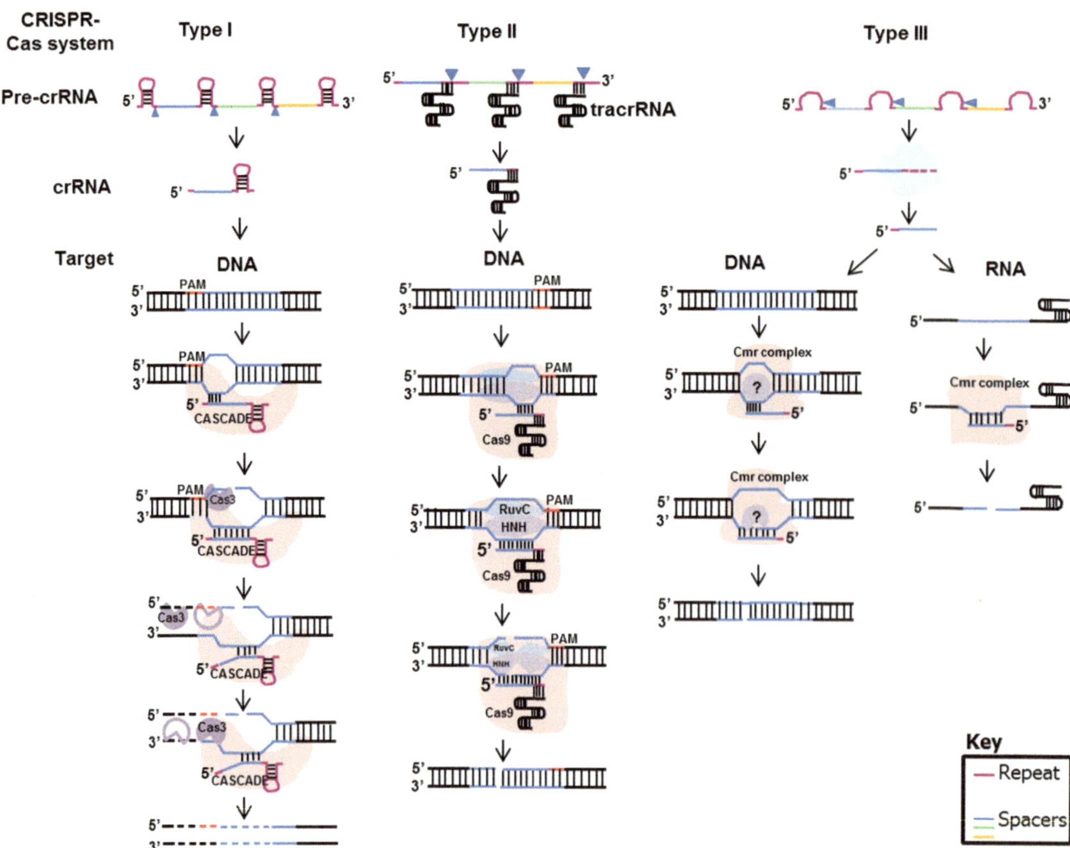

Fig. 14.2 Types of CRISPR–Cas systems. Schematic representation of the different structures of pre-crRNAs, mature crRNAs, and different effector complexes mediating targeted cleavage and/or degradation of nucleic acids (*thin black lines* represent regions of base pairing between two strands, but the numbers do not exactly correspond to the number of nt base pairing as mentioned in the text in Sect. 14.2. CRISPER–Cas system classification). In the type I system, crRNA with 5′ and 3′ handle in complex with CASCADE binds to the protospacer adjacent to 5′ PAM and unwinds it. Cas3 joins this complex, cleaves first the noncomplementary strand, and then translocates in 3′ to 5′ direction to cleave the targeted complementary strand. In the type II system, the crRNA with 3′ handle binds to tracrRNA by base pairing with nts in 3′ handle. This is bound by a large protein Cas9 with RuvC and HNH active sites, and the effector complex is targeted to protospacer with flanking 3′ PAM. The RuvC site cuts the noncomplementary strand, whereas the HNH active site cuts the complementary strand of DNA to generate the blunt ends. In the type III system, the crRNA has 5′ handle with trimmed 3′ end which binds with Cmr/Csm complex of proteins to form an effector complex. This complex can cleave both RNA and DNA without the requirement of PAM in region of the strand complementary to crRNA

self-targeting (Stern et al. 2010). It was thought to be possibly useful for regulation of these targeted genes. However, these self-targeting spacers with incomplete to full match with endogenous DNA lacked conservation, i.e., each spacer-target pair existed only in one organism. Hence, this suggests that these spacers do not have any functional role that could confer any selective advantage to the host for their perpetuation. Self-targeting spacers are suggested to originate by leaky/accidental processing of endogenous DNA by CRISPR–Cas system. Most of the spacers matched to regions of genome encoding functionally important molecules as DNA polymerase, 16S rRNA, and tRNA synthetases, and therefore, if functional, they can have lethal consequences for the organism. Bioinformatic analysis further showed that they are majorly located close to the leader at 1–2 position of the CRISPR array suggestive of a recent acquisition which probably renders the array nonfunctional. Often, these spacers are associated with degraded CRISPR system, i.e., either spacers or repeats are inactivated by mutation, have *cas* pseudogenes, or the target/ PAM sequences are mutated in the genome conferring protection to the organism. In the absence of these protective changes, CRISPRs with self-targeting spacers can compromise essential functions of the microbe leading to loss of these microbial populations by purifying selection. Paez-Espino D and colleagues cocultured *S. thermophilus* with the pathogenic phage and monitored the expansion of active CRISPR loci over 15 days by deep sequencing of the PCR amplicons. The bacteria mainly acquired spacers corresponding to phage DNA with an adjacent conserved PAM sequences. Acquisition of self-targeting spacers was rare but would lead to cell death (Paez-Espino et al. 2013). Heat-stable nucleoid-structuring (H-NS) encoded by *hns* gene is a repressor of CRISPR–Cas system in *E. coli* (Pul et al. 2010). *E. coli* Δ*hns* lysogenized with lambda (λ) phage showed considerably reduced transformation efficiencies for plasmids having spacers targeting λ prophage integrated in the chromosome (Yosef et al. 2011). All these features suggested that chromosomal targeting by CRISPR–Cas system is lethal for the microbe, and this correlates with an autoimmune response.

In the eukaryotic systems, type II CRISPER–Cas9 system has been extensively used for genome editing by trans-expression of Cas9 along with tracrRNA and CRISPR array with spacers designed to target the desired region of DNA (Cong et al. 2013; Jiang et al. 2013; Mali et al. 2013). The double-stranded break so introduced into the genomic DNA is repaired by homologous recombination or nonhomologous end joining (NHEJ). The prokaryotes rely mainly on homologous recombination for DNA repair, and recently NHEJ machinery has also been discovered but is limited or poor. There are instances where the microbes adapt to this phenomenon under strong selective pressure for survival by rapid mutation. For instance, in *Pectobacterium atrosepticum*, a plant pathogen, transformation of plasmids with engineered CRISPR harboring single spacer targeting chromosomally encoded *expI* gene (coding for N-acyl homoserine lactone synthase) whose protein product is required for generating quorum-sensing signals resulted in no viable colonies relative to control plasmid. However, this phenotype was rescued in Δ*cas* mutant. Similarly when *lacZ* was targeted with a single spacer, it resulted in growth inhibition which was again rescued in Δ*cas* mutant demonstrating *cas*-dependent cytotoxicity when chromosomal regions are targeted (Vercoe et al. 2013). Natural strains of *P. atrosepticum* bear spacer 6 in their CRISPR 2 with exact match to *eca0560* gene within ~ 100 kb horizontally acquired island 2 (HAI2). However, the PAM corresponding to the protospacer does not match the consensus PAM for type I-F, and this mismatch subverts self-targeting and toxicity through crRNA corresponding to the self-targeting spacer. Therefore, *P. atrosepticum* transformation with plasmid bearing an engineered CRISPR with spacer that could target *eca0560* gene near the type I-F PAM led to growth inhibition. Despite toxicity, few cells continued to grow and showed filamentous morphology indicative of cellular stress in response to DNA damage and SOS response (Huisman and D'Ari 1981). In the few spontaneous survivors that arose, most of the survivors lost the whole pathogenicity island indicative of rapid genome evolution for survival (Vercoe et al. 2013). In the hyperthermophilic archaeon, *Sulfolobus solfataricus*, targeting of host-encoded nonessential gene *β-galactosidase* with a synthetic mini-CRISPR array having regions of homology to native CRISPR locus decreased the growth of bacteria under selective conditions. Clones of these bacteria were subjected to two rounds of selection during which they resumed growth in selective media relative to controls with nontargeting spacers. Subsequent analysis of the extrachromosomal DNA (on which mini-CRISPR array is harbored) from these clones showed increased size of the CRISPR locus matching that of native CRISPR locus. This suggested loss of the deleterious CRISPR locus and replacement with the chromosomal CRISPR locus by recombination for survival (Manica et al. 2011). In each of these experiments, survivors acquire protection either by marked genomic deletions comprising the targeted region or alteration of the CRISPER locus with the self-targeting spacer.

In a systematic analysis of chromosomal targeting-associated toxicity, Gomaa AA and colleagues (Gomaa et al. 2014) designed plasmid-borne CRISPR spacers *targeting different regions* of *E. coli* K12 genome (harbors type I CRISPR–Cas system) – positive or negative strands, template and untemplate strands of genes, and non-transcribed regions. They were transformed into *E. coli* K12 harboring two plasmids for inducible expression of *cas3* and *casA-E* operon in Δ*hns* mutant (since H-NS represses *cas* operon). In all cases, chromosomal targeting was lethal as long as the targeted protospacer region bore a 5′ adjacent PAM. This suggested that chromosomal injury leads to cytotoxicity which is not location or gene specific. They further tested how they could use this to program removal of bacteria by targeting specific chromosomal sequences as desired. Selective elimination of a specific strain among closely related bacteria, e.g., two strains of *E. coli*, *E. coli* K12, and *E. coli* B sharing 99 % sequence homology, can be challenging with currently available antimicrobials. Therefore, the possibility of using CRISPR–Cas system was tested to distinguish between these two bacteria using genomic sequence information. The CRISPR array was designed with spacers bearing identity to PAM-flanking target regions unique to *E. coli* K12 (within *fucP* gene) or *E. coli* B (within *ogr* gene). Targeting protospacer in *fucP* or ogr genes in pure populations of bacteria bearing plasmids expressing *casA-E* and *cas3* selectively eliminated *E. coli* K12 or *E. coli* B, respectively. Likewise, pathogenic bacteria, *Salmonella enterica*, and a commensal bacteria inhabiting the gut, *E. coli* K12, share~ 70 % genomic sequence homology. CRISPR spacers were designed to target PAM-flanking regions specific for *E. coli* (within the *arp* gene), *S. enterica* (within the *mviM* gene), and both (within *groL* gene). These constructs were transformed into pure cultures of either *E. coli* or *S. enterica* harboring plasmids expressing *casA-E* or *cas3*. Targeting *arp* killed *E. coli*, *mviM* killed *S. enterica*, and targeting *groL* killed both bacteria.

14.4 Sequence-Specific Killing of Bacteria

14.4.1 Microbial Pathogenicity and Antibiotics

The term “superbugs” includes microbes showing increased resistance to specific antibiotics prescribed for treating associated infections. Use of conventional antibiotics promotes emergence of pathogenic bacteria that are resistant to antibiotics making it difficult to treat the infection. This has led to increased morbidity and mortality associated with diseases caused by superbugs since known antibiotics are no longer effective leaving lesser alternative therapeutic options. The virulence and antibiotic resistance determinants are encoded on mobile genetic elements, native conjugative plasmids, plasmids, or phages. Superbugs show increased transmittance of these pathogenicity determinants through HGT (Davies and Davies 2010). Additionally, unwanted targeting like with broad spectrum antibiotics can lead to indiscriminate killing of even beneficial or commensal microbes along with pathogenic microbial population in the niche (Dethlefsen et al. 2008). Eradication of commensal bacteria in the niche by antibiotic treatment gives an uncompleted opportunity for the proliferation of antibiotic-resistant pathogenic bacteria and spreading of resistance to nonpathogenic species which further complicates the treatment. Thus, antibiotic treatment leads to drastic alteration of the microbiome which is important for health (Mazmanian et al. 2008). Recently, CRISPR–Cas system based antimicrobials have been developed to specifically target and kill antibiotic-resistant *Staphylococcus aureus* and Enterobacteriaceae (Bikard et al. 2014; Citorik et al. 2014). Pathogenic Gram-positive bacteria, *Staphylococcus aureus*, and group of enteric Gram-negative bacteria called Enterobacteriaceae are examples of antibiotic-resistant superbugs which are very difficult to treat with currently available antibiotics. The most notorious superbug, *S. aureus*, is a versatile human pathogen. It causes superficial skin and soft tissue infection, organ abscesses, and toxic shock syndrome.

Penicillin was initially used to control *S. aureus* infections, but soon the microbe produced penicillinase leading to resistance. The first synthetic antibiotic methicillin targeting penicillinase offered respite before the resistance to methicillin emerged. Now, methicillin-resistant *S. aureus* (MRSA) is one of the most important causes of the hospital-acquired *S. aureus* infection which is resistant to multiple antibiotics. MRSA strain USA300 is also the leading cause of community-acquired *S. aureus* infections in the USA causing severe septicemia, necrotizing pneumonia, and necrotizing fasciitis (DeLeo and Chambers 2009). Genomic sequence analysis of USA300 revealed that most of the virulence and antibiotic resistance-encoding genes are located on mobile genetic elements and three plasmids pUSA01-03 which are highly transmissible (Diep et al. 2006). These possibly account for diverse pathogenicity and colonization associated with *S. aureus*. Chromosomally integrated mobile genetic element, staphylococcal chromosomal cassette mec (SSCmec), that encodes for *mecA* gene confers resistance to β-lactam antibiotics, while pUSA02 carries *tetK* gene which confers resistance to tetracycline (Malachowa and DeLeo 2010). Likewise, Enterobacteriaceae, intestinal gram-negative bacteria, which give rise to opportunistic infections have developed resistance to carbapenems, class of beta-lactam antibiotics that target the cell wall of the bacteria. β-lactamase-encoding genes, $bla_{\text{SHV-18}}$ and $bla_{\text{NDM-1}}$, confer extended and pan-resistance to β-lactam antibiotics, respectively. The $bla_{\text{NDM-1}}$ is harbored along with other resistant determinants on mobile plasmids which greatly facilitates transfer of resistance to other Enterobacteriaceae (Nordmann et al. 2012).

14.4.2 Generation and Delivery of Targeted CRISPER–Cas Antimicrobial to Bacteria

The different types of CRISPR–Cas systems offer different design rules for DNA targeting. For generation of CRISPR–Cas antimicrobials, two groups (Bikard et al. 2014; Citorik et al. 2014) have used the simplest and the most well-studied type II CRISPR–Cas9 system (see Fig. 14.2 and Table 14.1) of *Streptococcus pyogenes* for sequence-specific killing of microbes.

Bikard Dand colleagues designed ΦNM1-based phagemid system for sequence-specific killing of *S. aureus* strains:

1. Generation of the cloning plasmid: *S. pyogenes cas9* and tracrRNA and a minimal CRISPR array (optimized for one-step cloning) were put in staphylococcal vector, pC194 (Horinouchi and Weisblum 1982), to generate pDB114. 20 bp spacers targeting specific genes were cloned between the repeat units of pDB114 to generate pDB114::gene.
2. Fragment of~ 2 kb size containing *rinA*, *terS*, and *terL* genes and a packaging site from ΦNM1 phage were cloned into staphylococcal vector, pC194, with chloramphenicol resistance marker generating phagemid, pDB91.
3. CRISPR sequences from pDB114 and pDB114::gene are cloned into pDB91 to yield phagemids, pDB121 and pDB121::gene, respectively.
4. Phagemids, pDB121 and pDB121::gene, were transformed into wild-type *S. aureus* strain, RN4220, and then infected with phage, ΦNM1, leading to lysis of the cells. This lysate has both wild-type ΦNM1 phage and phage with phagemid packaged in the capsid generating CRISPR–Cas9-based antimicrobial phage. Therefore, this lysate is used to infect cells like RN^{Φ} and RNK^{Φ} which are RN4220 and RNK (*S. aureus* with chromosomally encoded kanamycin resistance gene, *aph*) lysogenized with ΦNM1, respectively, so that these cells are resistant to superinfection with wild-type phage and only phagemid transduction can be detected. These transducing phagemids were called *the gene-targeted CRISPR–Cas9 antimicrobial.*

Citorik RJ and colleagues designed M13-based phages for sequence-specific killing of Enterobacteriaceae strains:

1. Cloning plasmid: *S. pyogenes cas9* and tracrRNA were ligated into plasmid backbone

containing chloramphenicol resistance marker making plasmid, p-RGN (RNA-guided nuclease). CRISPER loci were assembled by oligo annealing and moved from intermediate cloning vector into p-RGN into which the desired spacer sequence targeting the desired gene can be cloned to give plasmid called p-RGN*gene*.

2. The tracrRNA, *cas9*, and CRISPR with the spacer sequence were amplified from pZB-RGN gene and ligated to phagemid, pZEf-*gfp* (Lutz and Bujard 1997), to give pZE-*gene-gfp* harboring kanamycin resistance and F1 ori for packaging into M13 phage particles.
3. Phagemids encoding RGN, pZf-*gene-gfp*, and those without a targeting spacer, pZE-*gfp*, were transformed into *E. coli* DH5αPro along with M13 helper phage to produce M13 phagemid particles. The supernatants of the culture were filtered and concentrated by PEG-6000 precipitation, resuspended, and solubilized for treatment of microbes and were called *ΦRGN gene antimicrobial*.

14.4.3 Targeting In Situ

Targeting the Chromosomal Genes The spacers in CRISPR–Cas9 antimicrobials, pDB121::*aph* and pDB121::*mecA*, were designed to target either the chromosomally encoded kanamycin resistance gene (*aph-3*) or chromosomally integrated methicillin and β-lactam antibiotic resistance gene (*mecA*). These phagemids were transduced into ΦNM1 lysogenized *S. aureus* wild-type strain, RN4220$^{\Phi}$, *S. aureus* strain bearing *aph-3* in the chromosome, RNK$^{\Phi}$, and the clinical isolate USA300$^{\Phi}$ bearing the staphylococcal cassette chromosomal mec (SCCmec) encoding *mecA* integrated in its chromosome (Malachowa and DeLeo 2010). Treatment of mixed populations of RN4220$^{\Phi}$and RNK$^{\Phi}$ with pDB121::*aph-3* selectively eliminated RNK$^{\Phi}$ relative to pDB121 treated cells. Similarly, treatment of mixed populations of RN4220$^{\Phi}$and USA300$^{\Phi}$ with pDB121::*mecA* led to selective killing of USA300$^{\Phi}$ relative to pDB121 treated cells. Treatment of USA300$^{\Phi}$ with a CRISPR–Cas9 pDB121::*mecA sek* phagemid harboring multiplex spacers targeting more than one region of chromosome, namely, *mecA* gene or *sek* gene, encoding superantigen enterotoxin was effective in killing UAS300$^{\Phi}$ (Bikard et al. 2014). Likewise, Citorik RJ and colleagues targeted mix populations of wild-type *E. coli* and *E. coli* harboring single-nucleotide mutation in DNA gyrase (*gyrA*$_{D87G}$) making it resistant to quinolone antibiotics with RNA-guided nucleases (RGNs) antimicrobial, ΦRGN*gyrA*$_{D87G}$. The antimicrobial killed *E. coli* harboring *gyrA*$_{D87G}$ but not the wild-type *E. coli*. Apart from this, RGN targeting *eae* gene encoding for a virulence factor, intimin, in enterohemorrhagic *E. coli* O157:H7 (EHEC) also resulted in cytotoxicity (Citorik et al. 2014). Therefore, antimicrobials based on the principle of CRISPR–Cas system mediated chromosomal cytotoxicity exhibit single-nucleotide precision and can be programmed to specifically eliminate desired microbe from the microbiome.

Targeting the Plasmid Treatment of *S. aureus* USA300$^{\Phi}$ with CRISPR–Cas9 antimicrobial pDB121::*tetK* targeting tetracycline resistance gene, *tetK*, on pUSA02 made the strain sensitive to tetracycline by *plasmid curing* without killing the bacteria. On the contrary, wild-type RN4220 cells treated with CRISPR–Cas9 antimicrobial pDB121::*tetK* prior to transduction with ΦNM1 phage harboring pUSA02 did not become tetracycline resistant (Bikard et al. 2014). Likewise, Citorik RJ and colleagues transferred the native plasmid bearing *bla*$_{NDM-1}$ gene, p-NDM1, present in clinical isolate to wild-type *E. coli*, EMG2, through conjugation yielding strain, EMG2 p-NDM1. Interestingly, transduction of EMG2 p-NDM1 with ΦRGN targeting *bla*$_{NDM-1}$ was lethal but not for *E. coli* bearing the same gene on cloning vector. The lethal phenotype was due to derepression of the toxin activity by RGN targeted loss of the native plasmid, p-NDM1, that co-harbors toxin–antitoxin system. The labile antitoxin quenches stable toxin, and so loss of p-NDM1 by ΦRGN targeting *bla*$_{NDM-1}$ led to accumulation of the toxin which kills the bacteria. Thus, targeting native plasmids may compromise associated functions which can prove lethal.

Since most of the plasmids bearing antibiotic resistance genes can spread through HGT, plasmid curing by CRISPR–Cas9 offers a strategy to remove unwanted genetic elements from a population and can resensitize microbes to antibiotics. Prior treatment of nonresistant bacteria with the antimicrobial acts like a vaccine to immunize the bacteria against plasmid transfer and development of antibiotic resistance.

14.4.4 Targeting In Vivo

***S. aureus* Mouse Skin Colonization Model**

S. aureus colonizes skin, and therefore, mouse skin can be injected with *S. aureus* and subsequently used for testing of antimicrobials for decolonization of the bacteria. Back of the mice was injected with equimolar mixture of RN4220$^{\Phi}$ and green fluorescent protein (GFP) expressing RNK$^{\Phi}$ (strains described above) to colonize the skin. Thereafter, groups of mice were treated topically with either CRISPR–Cas9 antimicrobial pDB121::*aph*, ΦNM1 phage alone, or mupirocin (used for decolonization of staphylococci). Ratio of GFP colony-forming units (CFUs) to total CFU at 24 h posttreatment suggested that treatment of mice with antimicrobial pDB121::*aph* significantly reduced RNK$^{\Phi}$ burden relative to ΦNM1, PBS, or mupirocin-treated mice (Bikard et al. 2014).

***Galleria mellonella* Larvae Infection Model**

G. mellonella or wax moth is recently gaining attention as an alternative insect infection model for several human pathogenic Gram-positive and Gram-negative bacteria and fungus to aid the search of novel antimicrobials (Ramarao et al. 2012). Its utility comes in light of findings that invertebrate (wax moth) immune system is functionally similar to human innate immune system and often identical virulence factors of pathogens are required by the microbe for infection in human and in insect larvae. Usage is not subject to ethical issues and offers easier handling and maintenance option to bridge the gap between in vitro and in vivo expensive animal study for determining efficacy and toxicity of new antimicrobials. Using *G. mellonella* larvae, Citorik RJ and colleagues injected two groups with either PBS or enterohemorrhagic *E. coli* O157:H7 (EHEC) treated either with antibiotics or ΦRGN*eae* and monitored their survival. Treatment with ΦRGN*eae* significantly increased the survival of the larvae relative to antibiotic treatment to which EHEC bacteria are resistant (Citorik et al. 2014).

Thus, in vivo targeting shows promising utility of these antimicrobials for treatment of diseases caused by microbes, but this will require further testing.

14.5 Beyond Killing and Escape: Precision Genome Editing and Genome Regulation

14.5.1 CRISPRing Recombination

In order to understand functions of genomic sequences, it is important to precisely alter them and monitor the associated gain/loss of function phenotypes. Such genome alterations require editing tools to insert, delete, and mutagenize nucleotides at specific loci. As described above, the cytotoxicity associated with CRISPR–Cas targeting of chromosomal DNA is circumvented by occasional escape mutants with mutations in the target genes. This observation led to the exciting idea of using CRISPR–Cas system along with recombineering to introduce mutations in host chromosome. The existing popular mode of altering or mutagenizing prokaryotic chromosome is based on homologous recombination (Sung et al. 2001; Bae and Schneewind 2006) or phage-based recombineering methods (Sharan et al. 2009). However, in the absence of selection of the recombinants, it becomes a tedious task to screen for recombinants that arise at a low frequency.

Among Streptococcus genus, *S. pneumoniae* does not have any CRISPR system. Jiang W and colleagues (Jiang et al. 2013) integrated a simplified Cas9-based CRISPR locus from *S. pyogenes* into non-capsulated *S. pneumoniae* to make a *S. pneumoniae* crR6M strain. This strain carried a spacer in the CRISPR array with a region of perfect match to prophage (Φ) 8232.5 but with mismatches to Φ370.1. The *S. pyogenes* prophages, Φ8232.5 and Φ370.1, were inserted into *S. pneumoniae* at locus of nonessential gene, *srtA*, to generate strains, $R6^{\Phi 8232.5}$*and* $R6^{\Phi 370.1}$, respectively. Transformation of *S. pneumoniae* crR6M genomic DNA into $R6^{\Phi 8232.5}$ gave tenfold lesser transformation efficiency than into $R6^{\Phi 370.1}$ due to active CRISPR interference. However, transformation of *crR6M genomic DNA along with wild-type srtA gene into* $R6^{\Phi 8232.5}$ led to increased transformation and editing efficiency relative to just crR6M transformation alone. Transformants showed replacement of endogenous *srtA* into which Φ8232.5 is integrated with wild-type *srtA* DNA. This suggested that concomitant presence of *srtA* DNA as an editing template during transformation promoted recombination with *srtA* sequences flanking Φ8232.5 DNA rather than Cas9-mediated interference by targeted cleavage of the Φ8232.5 DNA.

Further, introduction of point mutations by CRISPR-mediated genome editing requires editing templates that not only derives homologous recombination but also abolishes Cas9 nuclease activity. This possibility would be achieved by using editing templates with mutations in the PAM or the protospacer sequences. This was tested using *S. pneumoniae* β-galactosidase (*bga*) gene whose protein product can easily be monitored. A spacer corresponding to the region close to PAM (TGG) and encompassing the active site of the β-galactosidase enzyme was cloned into the CRISPR targeting construct with kanamycin resistance. Editing templates were created for *bga* gene-(i) template bearing R>A mutation in the protospacer seed region and (ii) template bearing a mutation in PAM and NE>AA double-point mutation at 218 nucleotides downstream of the protospacer region. Co-transformation of the CRISPR targeting construct and editing templates generated ten times more kanamycin-resistant clones than co-transforming with wild-type *bga* template and led to incorporation of the desired mutations in *bga* gene. This was further confirmed by measuring the loss of β-galactosidase activity. Likewise, in order to introduce multiple mutations at different loci like deletion of *bga* and *srtA* gene, one can engineer the spacers targeting each gene into the targeting construct. This can be co-transformed with two editing templates directed for deletion of *bga* and *srtA* genes.

Thus, co-transformation of the editing template for the chromosomal region against which the CRISPR–Cas system bears a spacer leads to homologous recombination-based gene replacement, deletion, or editing by CRISPR–Cas system.

14.5.2 Kill and Escape Goes On

In order to accurately assess genome editing efficiency, *S. pneumonia* strain was constructed in which erythromycin resistance gene (*ermAM*) was introduced into *srtA* locus with a premature stop codon, *ermAM (stop)*, rendering the bacteria erythromycin sensitive. Repairing of *ermAM* gene would rescue erythromycin sensitivity, and this was used to assess the contribution of CRISPR system to editing of the gene. These cells were transformed with editing template of wild-type *ermAM* allele along with either a kanamycin-resistant CRISPR targeting construct (spacer targeting either the *ermAM (stop)* or non-targeting spacer). Co-selection for kanamycin and erythromycin resistance gave ten times more fraction of edited transformants than selecting for erythromycin resistance alone. This suggested that a fraction of cells is more active for transformation and/or recombination independent of CRISPR targeting under co-selection pressure. This fraction was further enhanced by CRISPR targeting relative to nontargeting CRISPR since the subpopulations that do not edit the *ermAM (stop)* gene are killed by CRISPR targeting of *ermAM* gene, and thus purifying selection increases the fraction of edited cells to 99 %. The co-transformants with CRISPR construct targeting *ermAM (stop)* had twofold more colonies than CRISPR construct with nontargeting spacer suggesting that Cas9-mediated double-strand cleavage which stimulates DNA repair may be stimulating recombination (Jiang et al. 2013). Thus, a combination of co-selection, CRISPR-mediated targeting of non-edited cells and Cas9 cleavage stimulated recombination contribute to CRISPR-based genome editing. A fraction of non-edited cells that may not have received editing template during co-transformation escape killing by CRISPR system and are kanamycin resistant. These colonies harbored mutations like deletions of the spacers or Cas9 inactivating mutations. Such background colonies are limitations of this method, and so a recombination frequency above this background is required to recover clones with desired mutations in their genomes. While CRISPR-based editing is benefited by CRISPR targeting and killing which selects for and recovers desired mutation with high efficiency in the cells, alternatives to limit the generation of escapers can further improve it.

14.5.3 Synthetic Gene Regulation

Targeted genome regulation offers an opportunity to program genetic repression and/or expression. In this way, it can give insight into gene function without editing DNA. This is particularly useful for organisms in which DNA manipulation methods are not well developed. It also offers to functionally map genetic regulatory modules like promoters. Cas9 is a double-stranded endonuclease and binds and cleaves dsDNA in gRNA-dependent manner with its two different domains – RuvC and HNH (see Fig. 14.2). Mutation of key catalytic residues, D10A and H840A, in RuvC and HNH domains, respectively, renders it catalytically inactive, but it can still bind DNA and is called a dead Cas9 or "dCas9" (Jinek et al. 2012). Targeting the chromosomal regions with dCas9 alleviates the lethality associated with Cas9-mediated DNA cleavage (Qi et al. 2013), and scientists have used this dCas9 to program transcription regulation.

Bikard and colleagues (Bikard et al. 2013) designed a GFP reporter assay to monitor dCas9-mediated gene regulation in *E. coli*. Gene *gfp-mut2* was cloned downstream of a promoter with many NGG PAM sequences in the plasmid on both strands of DNA. Plasmid bearing a CRISPR array with many spacers targeting different regions of promoter was introduced along with plasmid bearing dCas9 and tracrRNA into *E. coli*. Region targeting −35 and −10 promoter elements and Shine–Dalgarno sequence led to greater than 100-fold reduction in GFP fluorescence relative to nontargeting spacer control. This suggested targeting of dCas9 to promoter prevents transcription initiation perhaps by sterically hindering RNA polymerase (RNAP) binding. Targeting the spacers to both the noncoding and coding strands of the GFP open reading frame (ORF) resulted in reduction in GFP fluorescence to varying degrees. Thus, dCas9 can be targeted to gene promoter or ORF to inhibit transcription initiation or elongation, respectively, for repression of gene expression. Similarly, when the *bga* gene of *S. pneumoniae* was targeted in the promoter and the ORF region, there was reduction in β-galactosidase activity. However, targeting of *bga* promoter, which is regulated in response to lactose metabolism, showed different pattern of regulation by dCas9 than *gfp-mut2* promoter. This suggested targeting of ORFs by dCas9 consistently leads to decrease in gene expression with targeting of noncoding strand showing more efficiency of gene repression than coding strand targeting.

CRISPR–Cas9 system from *S. pyogenes* is a simple minimal naturally machinery requiring tracrRNA, Cas9, and CRISPR array (see Fig. 14.2 and Table 14.1). Maturation of pre-crRNA requires base pairing of tracrRNA with repeats of the CRISPR transcript which is processed by host RNase III. Qi and colleagues (Qi et al. 2013) engineered single guide RNA (sgRNA) that mimics crRNA/tracrRNA duplex in that the crRNA is joined to the tracrRNA by a hairpin and a 20 bp region of complementarity of crRNA to the target directs Cas9 cleavage. This simplifies the CRISPR–Cas9 system functional requirement to co-expression of Cas9 and sgRNA only. The CRISPR–Cas9 system was further repurposed for knocking down gene expression using dCas9 and sgRNA in what is called as CRISPR interference (CRISPRi). *E. coli* MG1655 strains co-expressing dCas9 and sgRNAs targeting different promoter and ORF regions of the template and non-template strands of the gene were made. Targeting of the promoter region on either strand was effective in gene repression, but unlike Bikard and colleagues, only targeting of non-template regions of ORF was effective in gene repression. This was thought to be due to collision of RNAP with dCas9/sgRNA complex on the target leading to block in transcription. Further, dCas9 and targeting sgRNA were put under control of an inducible promoter. In the presence of the inducer, the target protein levels decreased, and thereafter, on removal of the inducer, the levels of the target protein returned back to original levels. This suggested CRISPRi could be designed to knockdown gene expression in an inducible and reversible manner. Whole-transcriptome shotgun sequencing (RNA-seq) analysis of the cells showed considerable specificity of sgRNA guided knockdown with no significant off-target effects. Therefore, CRISPRi offers a valuable programmable facile tool for knockdown of gene expression in prokaryotic systems. Further, targeting of dCas9, fused to omega (ω) subunit of RNAP, to the promoters resulted in activation of gene. Gene activation showed a positive correlation with increasing distance of binding from −35 region of the promoter (up to 59 nt) of dCas9-ω fusion protein and decreasing strength of the promoter (Bikard et al. 2013). Gene regulation studies require expensive and laborious genome and protein engineering efforts to modify cis-elements that can bind to desired regulators like small molecules or modular proteins generated from proteins like Zinc finger DNA-binding proteins and transcription activator-like effectors (Zhang et al. 2011; Cong et al. 2012). However, CRISPR–dCas9 modular system requires only designing of sgRNA and can be targeted to any cis-acting motifs in gene regulatory elements, and dCas9 can block the binding of endogenous trans-acting factors to facilitate such studies.

14.6 Conclusion

CRISPR–Cas is a microbial machinery that can be programmed to target DNA in sequence-specific manner. In a natural environment, microbes use this machinery to target invading DNA conferring protection against harmful genetic material received by HGT. Thus, it is called the adaptive immune system of the microbe. Though rarely but when the machinery targets self-DNA in the microbe, it mainly leads to cell death with a minor population surviving by mutation of either the target DNA or the CRISPR–Cas system itself. Both these features of self-targeting have led to development of diverse tools that broadly come under the categories described above – antimicrobials, genome editing, and regulation tools. All these tools have started to contribute in novel ways to specifically control microbial populations, understanding of gene functions and treat diseases.

14.7 Opinion

Undoubtedly, the CRISPR–Cas system is programmable to any target sequence with immensely wide applications. However, certain issues need to be addressed for general applicability and robustness of the method. The mode of delivery of CRISPR–Cas antimicrobials described here is based on HGT by bacteriophages. It is a natural way to administer the antimicrobial to maximum number of microbes though it is restricted by the host range specificity of the bacteriophage. It is useful for many biotechnological applications and ecological purposes. However, other modes of delivery like nanoparticles need to be developed to make it safer for treatment of human diseases caused by microbes. Like antibiotics, the CRISPR–Cas system originates from microbes and likewise microbes can also develop resistance to CRISPR–Cas antimicrobials which should be actively monitored in the future. On the upside, the genome editing features of CRISPR–Cas systems have revolutionized the way genome engineering is being done across three domains of life. Most of the microbial genomes have been sequenced, and CRISPR–Cas system can utilize this sequence information to generate tools that examine novel genes with unknown functions. Additionally, they can be used to activate cryptic gene clusters in microbial genomes that could function as secondary metabolic pathways producing novel small molecule-based traditional antibiotics.

Acknowledgements Alka Mehra thanks Dr. V. C. Kalia and Dr. Yogendra Singh of CSIR-Institute of Genomics and Integrative Biology (IGIB) for providing this opportunity.

References

Bae T, Schneewind O (2006) Allelic replacement in *Staphylococcus aureus* with inducible counter-selection. Plasmid 55:58–63. doi:10.1016/j.plasmid.2005.05.005

Barrangou R, Fremaux C, Deveau H, Richards M, Boyaval P, Moineau S, Romero DA, Horvath P (2007) CRISPR provides acquired resistance against viruses in prokaryotes. Science 315:1709–1712. doi:10.1126/science.1138140

Bikard D, Jiang W, Samai P, Hochschild A, Zhang F, Marraffini LA (2013) Programmable repression and activation of bacterial gene expression using an engineered CRISPR-Cas system. Nucleic Acids Res 41:7429–7437. doi:10.1093/nar/gkt520

Bikard D, Euler CW, Jiang W, Nussenzweig PM, Goldberg GW, Duportet X, Fischetti VA, Marraffini LA (2014) Exploiting CRISPR-Cas nucleases to produce sequence-specific antimicrobials. Nat Biotechnol 32:1146–1150. doi:10.1038/nbt.3043

Bolotin A, Quinquis B, Sorokin A, Ehrlich SD (2005) Clustered regularly interspaced short palindrome repeats (CRISPRs) have spacers of extrachromosomal origin. Microbiology 151:2551–2561. doi:10.1099/mic.0.28048-0

Brouns SJ, Jore MM, Lundgren M, Westra ER, Slijkhuis RJ, Snijders AP, Dickman MJ, Makarova KS, Koonin EV, van der Oost J (2008) Small CRISPR RNAs guide antiviral defense in prokaryotes. Science 321:960–964. doi:10.1126/science.1159689

Bult CJ, White O, Olsen GJ, Zhou L, Fleischmann RD, Sutton GG, Blake JA, FitzGerald LM, Clayton RA, Gocayne JD, Kerlavage AR, Dougherty BA, Tomb JF, Adams MD, Reich CI, Overbeek R, Kirkness EF, Weinstock KG, Merrick JM, Glodek A, Scott JL, Geoghagen NS, Venter JC (1996) Complete genome sequence of the methanogenic archaeon, *Methanococcus jannaschii*. Science 273:1058–1073

Carte J, Wang R, Li H, Terns RM, Terns MP (2008) Cas6 is an endoribonuclease that generates guide RNAs for invader defense in prokaryotes. Genes Dev 22:3489–3496. doi:10.1101/gad.1742908

Carte J, Pfister NT, Compton MM, Terns RM, Terns MP (2010) Binding and cleavage of CRISPR RNA by Cas6. RNA 16:2181–2188. doi:10.1261/rna.2230110

Chylinski K, Le Rhun A, Charpentier E (2013) The tracrRNA and Cas9 families of type II CRISPR-Cas immunity systems. RNA Biol 10:726–737. doi:10.4161/rna.24321

Citorik RJ, Mimee M, Lu TK (2014) Sequence-specific antimicrobials using efficiently delivered RNA-guided nucleases. Nat Biotechnol 32:1141–1145. doi:10.1038/nbt.3011

Cong L, Zhou R, Kuo YC, Cunniff M, Zhang F (2012) Comprehensive interrogation of natural TALE DNA-binding modules and transcriptional repressor domains. Nat Commun 3:968. doi:10.1038/ncomms1962

Cong L, Ran FA, Cox D, Lin S, Barretto R, Habib N, Hsu PD, Wu X, Jiang W, Marraffini LA, Zhang F (2013) Multiplex genome engineering using CRISPR/Cas systems. Science 339:819–823. doi:10.1126/science.1231143

Datsenko KA, Pougach K, Tikhonov A, Wanner BL, Severinov K, Semenova E (2012) Molecular memory of prior infections activates the CRISPR/Cas adaptive bacterial immunity system. Nat Commun 3:945. doi:10.1038/ncomms1937

Davies J, Davies D (2010) Origins and evolution of antibiotic resistance. Microbiol Mol Biol Rev 74:417–433. doi:10.1128/MMBR.00016-10

DeLeo FR, Chambers HF (2009) Reemergence of antibiotic-resistant *Staphylococcus aureus* in the genomics era. J Clin Invest 119:2464–2474. doi:10.1172/JCI38226

Deltcheva E, Chylinski K, Sharma CM, Gonzales K, Chao Y, Pirzada ZA, Eckert MR, Vogel J, Charpentier E (2011) CRISPR RNA maturation by trans-encoded small RNA and host factor RNase III. Nature 471:602–607. doi:10.1172/JCI38226

Deng L, Garrett RA, Shah SA, Peng X, She Q (2013) A novel interference mechanism by a type IIIB CRISPR-Cmr module in *Sulfolobus*. Mol Microbiol 87:1088–1099. doi:10.1111/mmi.12152

Dethlefsen L, Huse S, Sogin ML, Relman DA (2008) The pervasive effects of an antibiotic on the human gut microbiota, as revealed by deep 16S rRNA sequencing. PLoS Biol 6, e280. doi:10.1111/mmi.12152

Diep BA, Gill SR, Chang RF, Phan TH, Chen JH, Davidson MG, Lin F, Lin J, Carleton HA, Mongodin EF, Sensabaugh GF, Perdreau-Remington F (2006) Complete genome sequence of USA300, an epidemic clone of community-acquired meticillin-resistant *Staphylococcus aureus*. Lancet 367:731–739. doi:10.1016/S0140-6736(06)68231-7

Garneau JE, Dupuis ME, Villion M, Romero DA, Barrangou R, Boyaval P, Fremaux C, Horvath P, Magadan AH, Moineau S (2010) The CRISPR/Cas bacterial immune system cleaves bacteriophage and plasmid DNA. Nature 468:67–71. doi:10.1038/nature09523

Gasiunas G, Barrangou R, Horvath P, Siksnys V (2012) Cas9-crRNA ribonucleoprotein complex mediates specific DNA cleavage for adaptive immunity in bacteria. Proc Natl Acad Sci U S A 109:E2579–E2586. doi:10.1073/pnas.1208507109

Gesner EM, Schellenberg MJ, Garside EL, George MM, Macmillan AM (2011) Recognition and maturation of effector RNAs in a CRISPR interference pathway. Nat Struct Mol Biol 18:688–692. doi:10.1038/nsmb.2042

Gomaa AA, Klumpe HE, Luo ML, Selle K, Barrangou R, Beisel CL (2014) Programmable removal of bacterial strains by use of genome-targeting CRISPR-Cas systems. mBio 5:e00928-13. doi:10.1128/mBio.00928-13

Grissa I, Vergnaud G, Pourcel C (2007) The CRISPRdb database and tools to display CRISPRs and to generate dictionaries of spacers and repeats. BMC Bioinf 8:172. doi:10.1186/1471-2105-8-172

Groenen PM, Bunschoten AE, van Soolingen D, van Embden JD (1993) Nature of DNA polymorphism in the direct repeat cluster of *Mycobacterium tuberculosis*; application or strain differentiation by a novel typing method. Mol Microbiol 10:1057–1065

Haft DH, Selengut J, Mongodin EF, Nelson KE (2005) A guild of 45 CRISPR-associated (Cas) protein families and multiple CRISPR/Cas subtypes exist in prokaryotic genomes. PLoS Comput Biol 1, e60. doi:10.1371/journal.pcbi.0010060

Hale C, Kleppe K, Terns RM, Terns MP (2008) Prokaryotic silencing (psi)RNAs in *Pyrococcus furiosus*. RNA 14:2572–2579. doi:10.1261/rna.1246808

Hale CR, Zhao P, Olson S, Duff MO, Graveley BR, Wells L, Terns RM, Terns MP (2009) RNA-guided RNA cleavage by a CRISPR RNA-Cas protein complex. Cell 13:945–956. doi:10.1016/j.cell.2009.07.040

Hatoum-Aslan A, Maniv I, Marraffini LA (2011) Mature clustered, regularly interspaced, short palindromic repeats RNA (crRNA) length is measured by a ruler mechanism anchored at the precursor processing site. Proc Natl Acad Sci U S A 10:21218–21222. doi:10.1073/pnas.1112832108

Hoe N, Nakashima K, Grigsby D, Pan X, Dou SJ, Naidich S, Garcia M, Kahn E, Bergmire-Sweat D, Musser JM (1999) Rapid molecular genetic subtyping of serotype M1 group A *Streptococcus* strains. Emerg Infect Dis 5:254–263. doi:10.3201/eid0502.990210

Horinouchi S, Weisblum B (1982) Nucleotide sequence and functional map of pC194, a plasmid that specifies inducible chloramphenicol resistance. J Bacteriol 150:815–825

Huisman O, D'Ari R (1981) An inducible DNA replication-cell division coupling mechanism in *E. coli*. Nature 290:797–799

Ishino Y, Shinagawa H, Makino K, Amemura M, Nakata A (1987) Nucleotide sequence of the *iap* gene, responsible for alkaline phosphatase isozyme conversion in *Escherichia coli*, *iap* gene, responsible for alkaline phosphatase isozyme conversion in *Escherichia coli*, and identification of the gene product. J Bacteriol 169:5429–5433

Jansen R, Embden JD, Gaastra W, Schouls LM (2002a) Identification of genes that are associated with DNA repeats in prokaryotes. Mol Microbiol 43:1565–1575

Jansen R, van Embden JD, Gaastra W, Schouls LM (2002b) Identification of a novel family of sequence repeats among prokaryotes. Omics J Int Biol 6:23–33

Jiang W, Bikard D, Cox D, Zhang F, Marraffini LA (2013) RNA-guided editing of bacterial genomes using CRISPR-Cas systems. Nat Biotechnol 31:233–239. doi:10.1038/nbt.2508

Jinek M, Chylinski K, Fonfara I, Hauer M, Doudna JA, Charpentier E (2012) A programmable dual-RNA-guided DNA endonuclease in adaptive bacterial immunity. Science 337:816–821. doi:10.1126/science.1225829

Jore MM, Lundgren M, van Duijn E, Bultema JB, Westra ER, Waghmare SP, Wiedenheft B, Pul U, Wurm R, Wagner R, Beijer MR, Barendregt A, Zhou K, Snijders AP, Dickman MJ, Doudna JA, Boekema EJ, Heck AJ, van der Oost J, Brouns SJ (2011) Structural basis for CRISPR RNA-guided DNA recognition by cascade. Nat Struct Mol Biol 18:529–536. doi:10.1038/nsmb.2019

Karvelis T, Gasiunas G, Miksys A, Barrangou R, Horvath P, Siksnys V (2013) crRNA and tracrRNA guide Cas9-mediated DNA interference in *Streptococcus thermophilus*. RNA Biol 10:841–851. doi:10.4161/rna.24203

Koonin EV, Wolf YI (2009) Is evolution Darwinian or/and Lamarckian? Biol Direct 4:42. doi:10.1186/1745-6150-4-42

Kunin V, Sorek R, Hugenholtz P (2007) Evolutionary conservation of sequence and secondary structures in CRISPR repeats. Genome Biol 8:R61

Kusano K, Naito T, Handa N, Kobayashi I (1995) Restriction-modification systems as genomic parasites in competition for specific sequences. Proc Natl Acad Sci U S A 92:11095–11099

Lillestol RK, Redder P, Garrett RA, Brugger K (2006) A putative viral defence mechanism in archaeal cells. Archaea 2:59–72. doi:10.1155/2006/542818

Lillestol RK, Shah SA, Brugger K, Redder P, Phan H, Christiansen J, Garrett RA (2009) CRISPR families of the crenarchaeal genus *Sulfolobus*: bidirectional transcription and dynamic properties. Mol Microbiol 72:259–272. doi:10.1111/j.1365-2958.2009.06641

Lutz R, Bujard H (1997) Independent and tight regulation of transcriptional units in *Escherichia coli* via the LacR/O, the TetR/O and AraC/I1-I2 regulatory elements. Nucleic Acids Res 25:1203–1210

Makarova KS, Grishin NV, Shabalina SA, Wolf YI, Koonin EV (2006) A putative RNA- interference-based immune system in prokaryotes: computational analysis of the predicted enzymatic machinery, functional analogies with eukaryotic RNAi, and hypothetical mechanisms of action. Biol Direct 1:7. doi:10.1186/1745-6150-1-7

Malachowa N, DeLeo FR (2010) Mobile genetic elements of *Staphylococcus aureus*. Cell Mol Life Sci 67:3057–3071. doi:10.1007/s00018-010-0389-4

Mali P, Yang L, Esvelt KM, Aach J, Guell M, DiCarlo JE, Norville JE, Church GM (2013) RNA-guided human genome engineering via Cas9. Science 339:823–826. doi:10.1126/science.1232033

Manica A, Zebec Z, Teichmann D, Schleper C (2011) In vivo activity of CRISPR-mediated virus defence in a hyperthermophilic archaeon. Mol Microbiol 80:481–491. doi:10.1111/j.1365-2958.2011.07586.x

Marraffini LA, Sontheimer EJ (2008) CRISPR interference limits horizontal gene transfer in staphylococci by targeting DNA. Science 322:1843–1845. doi:10.1126/science.1165771

Masepohl B, Gorlitz K, Bohme H (1996) Long tandemly repeated repetitive (LTRR) sequences in the filamentous cyanobacterium *Anabaena* sp. PCC 7120. Biochim Biophys Acta 1307:26–30

Mazmanian SK, Round JL, Kasper DL (2008) A microbial symbiosis factor prevents intestinal inflammatory disease. Nature 453:620–625. doi:10.1038/nature07008

Mojica FJ, Ferrer C, Juez G, Rodriguez-Valera F (1995) Long stretches of short tandem repeats are present in the largest replicons of the Archaea *Haloferax mediterranei* and *Haloferax volcanii* and could be involved in replicon partitioning. Mol Microbiol 17:85–93

Mojica FJ, Diez-Villasenor C, Soria E, Juez G (2000) Biological significance of a family of regularly spaced repeats in the genomes of Archaea, Bacteria and mitochondria. Mol Microbiol 36:244–246

Mojica FJ, Diez-Villasenor C, Garcia-Martinez J, Soria E (2005) Intervening sequences of regularly spaced prokaryotic repeats derive from foreign genetic elements. J Mol Evol 60:174–182. doi:10.1007/s00239-004-0046-3

Nakamura Y, Itoh T, Matsuda H, Gojobori T (2004) Biased biological functions of horizontally transferred genes in prokaryotic genomes. Nat Genet 36:760–766. doi:10.1038/ng1381

Nam KH, Haitjema C, Liu X, Ding F, Wang H, DeLisa MP, Ke A (2012) Cas5d protein processes pre-crRNA and assembles into a cascade-like interference complex in subtype I-C/Dvulg CRISPR-Cas system. Structure 20:1574–1584. doi:10.1016/j.str.2012.06.016

Nordmann P, Dortet L, Poirel L (2012) Carbapenem resistance in Enterobacteriaceae: here is the storm! Trends Mol Med 18:263–272. doi:10.1016/j.molmed.2012.03.003

Paez-Espino D, Morovic W, Sun CL, Thomas BC, Ueda K, Stahl B, Barrangou R, Banfield JF (2013) Strong bias in the bacterial CRISPR elements that confer immunity to phage. Nat Commun 4:1430. doi:10.1038/ncomms2440

Pul U, Wurm R, Arslan Z, Geissen R, Hofmann N, Wagner R (2010) Identification and characterization of *E. coli* CRISPR-cas promoters and their silencing by H-NS. Mol Microbiol 75:1495–1512. doi:10.1111/j.1365-2958.2010.07073.x

Qi LS, Larson MH, Gilbert LA, Doudna JA, Weissman JS, Arkin AP, Lim WA (2013) Repurposing CRISPR as an RNA-guided platform for sequence-specific control of gene expression. Cell 152:1173–1183. doi:10.1016/j.cell.2013.02.022

Ramarao N, Nielsen-Leroux C, Lereclus D (2012) The insect *Galleria mellonella* as a powerful infection model to investigate bacterial pathogenesis. J Visual Exp JoVE: e4392. doi:10.3791/4392

Sashital DG, Jinek M, Doudna JA (2011) An RNA-induced conformational change required for CRISPR RNA cleavage by the endoribonuclease Cse3. Nat Struct Mol Biol 18:680–687. doi:10.1038/nsmb.2043

Semenova E, Nagornykh M, Pyatnitskiy M, Artamonova II, Severinov K (2009) Analysis of CRISPR system function in plant pathogen *Xanthomonas oryzae*. FEMS Microbiol Lett 296:110–116. doi:10.1111/j.1574-6968.2009.01626.x

Sharan SK, Thomason LC, Kuznetsov SG, Court DL (2009) Recombineering: a homologous recombination-based method of genetic engineering. Nat Protoc 4:206–223. doi:10.1038/nprot.2008.227

Sinkunas T, Gasiunas G, Fremaux C, Barrangou R, Horvath P, Siksnys V (2011) Cas3 is a single-stranded DNA nuclease and ATP-dependent helicase in the CRISPR/Cas immune system. EMBO J 30:1335–1342. doi:10.1038/emboj.2011.41

Stern A, Keren L, Wurtzel O, Amitai G, Sorek R (2010) Self-targeting by CRISPR: gene regulation or autoimmunity? Trends Genet 26:335–340. doi:10.1016/j.tig.2010.05.008

Sung CK, Li H, Claverys JP, Morrison DA (2001) An rpsL cassette, janus, for gene replacement through negative selection in *Streptococcus pneumoniae*. Appl Environ Microbiol 67:5190–5196. doi:10.1128/AEM.67.11.5190-5196.2001

Swarts DC, Mosterd C, van Passel MW, Brouns SJ (2012) CRISPR interference directs strand specific spacer acquisition. PLoS One 7, e35888

Tang TH, Bachellerie JP, Rozhdestvensky T, Bortolin ML, Huber H, Drungowski M, Elge T, Brosius J, Huttenhofer A (2002) Identification of 86 candidates for small non-messenger RNAs from the archaeon *Archaeoglobus fulgidus*. Proc Natl Acad Sci U S A 99:7536–7541. doi:10.1073/pnas.112047299

Tang TH, Polacek N, Zywicki M, Huber H, Brugger K, Garrett R, Bachellerie JP, Huttenhofer A (2005) Identification of novel non-coding RNAs as potential antisense regulators in the archaeon *Sulfolobus solfataricus*. Mol Microbiol 55:469–481. doi:10.1111/j.1365-2958.2004.04428.x

Thomas CM, Nielsen KM (2005) Mechanisms of, and barriers to, horizontal gene transfer between bacteria. Nat Rev Microbiol 3:711–721. doi:10.1038/nrmicro1234

Vercoe RB, Chang JT, Dy RL, Taylor C, Gristwood T, Clulow JS, Richter C, Przybilski R, Pitman AR,

Fineran PC (2013) Cytotoxic chromosomal targeting by CRISPR/Cas systems can reshape bacterial genomes and expel or remodel pathogenicity islands. PLoS Genet 9, e1003454. doi:10.1371/journal.pgen.1003454

Wang R, Preamplume G, Terns MP, Terns RM, Li H (2011) Interaction of the Cas6 riboendonuclease with CRISPR RNAs: recognition and cleavage. Structure 19:257–264. doi:10.1016/j.str.2010.11.014

Westra ER, van Erp PB, Kunne T, Wong SP, Staals RH, Seegers CL, Bollen S, Jore MM, Semenova E, Severinov K, de Vos WM, Dame RT, de Vries R, Brouns SJ, van der Oost J (2012) CRISPR immunity relies on the consecutive binding and degradation of negatively supercoiled invader DNA by Cascade and Cas3. Mol Cell 46:595–605. doi:10.1016/j.molcel.2012.03.018

Yosef I, Goren MG, Kiro R, Edgar R, Qimron U (2011) High-temperature protein G is essential for activity of the *Escherichia coli* clustered regularly interspaced short palindromic repeats (CRISPR)/Cas system. Proc Natl Acad Sci U S A 108:20136–20141. doi:10.1073/pnas.1113519108

Yosef I, Goren MG, Qimron U (2012) Proteins and DNA elements essential for the CRISPR adaptation process in *Escherichia coli*. Nucleic Acids Res 40:5569–5576. doi:10.1093/nar/gks216

Zhang F, Cong L, Lodato S, Kosuri S, Church GM, Arlotta P (2011) Efficient construction of sequence-specific TAL effectors for modulating mammalian transcription. Nat Biotechnol 29:149–153. doi:10.1038/nbt.1775

Zhang J, Rouillon C, Kerou M, Reeks J, Brugger K, Graham S, Reimann J, Cannone G, Liu H, Albers SV, Naismith JH, Spagnolo L, White MF (2012) Structure and mechanism of the CMR complex for CRISPR-mediated antiviral immunity. Mol Cell 45:303–313. doi:10.1016/j.molcel.2011.12.013

Alka Mehra graduated in Biochemistry from Delhi University. She did her Master's and Ph.D. at the School of Life Sciences, Jawaharlal Nehru University. Her Ph.D. is in the field of Molecular Parasitology. Further, she has done postdoctoral studies from USA on type VII secreted proteins of *Mycobacterium tuberculosis* (Mtb) and antimicrobial host defense pathways. She is a Guest Researcher at CSIR-IGIB, Delhi, where she is interested in host-pathogen interactions in innate immunity to tuberculosis.

Photorhabdus: A Microbial Factory of Insect-Killing Toxins

15

Jyoti Kushwah and Vishal Singh Somvanshi

Abstract

The overuse of chemical pesticides to meet the production and productivity goals in modern agriculture is causing a number of unintended side effects and destruction of the environment. Eco-friendly pest management techniques and strategies are urgently needed. *Photorhabdus* spp. are Gram-negative gamma-proteobacteria of the family *Enterobacteriaceae*, found exclusively in symbiotic association with nematodes of the genus *Heterorhabditis*. *Heterorhabditis* nematodes are widely used as a biological control agent for insect-pests of crops. These nematodes carry the symbiont bacteria in their gut and release them in insect hemocoel upon infection of new insect host. Inside the insect hemocoel, *Photorhabdus* multiplies and releases a multitude of insecticidal toxins and secondary metabolites resulting in death of the insect by septicemia and toxemia. Some of these toxins are highly specific to their target species, while others are generalists. Stand-alone formulation of *Photorhabdus* bacteria is reported to be selling well in markets for insect management. *Photorhabdus* toxins are considered next to Bt toxins in their potential for use in insect-pest management in agriculture.

15.1 Introduction

Photorhabdus is a bioluminescent bacterium belonging to the family Enterobacteriaceae that lives in a symbiotic relationship with insect parasitic nematodes of family *Heterorhabditis*. This genus comprises of three species – *Photorhabdus luminescens*, *P. temperata*, and *P. asymbiotica*. All the three species are found as symbionts of nematodes, whereas *P. asymbiotica* has also been isolated from human wounds (Waterfield et al. 2009).

J. Kushwah (✉)
EPN Genomics Lab, Division of Nematology, Indian Agricultural Research Institute (IARI), 110012 New Delhi, India
e-mail: jyotiskushwah@gmail.com

V.S. Somvanshi
#314, Division of Nematology, ICAR-Indian Agricultural Research Institute, Pusa Campus, 110012 New Delhi, India
e-mail: vssomvanshi@iari.res.in

V.C. Kalia (ed.), *Microbial Factories*, DOI 10.1007/978-81-322-2595-9_15

Infective juvenile stage of *Heterorhabditis* contains *Photorhabdus* spp. as intestinal symbionts. The bacterium is mutualistic with the nematodes but is a deadly pathogen of insects (Clarke 2008). *Heterorhabditis* attacks the insects and releases its bacteria in the insect hemocoel and kills the insects within 48 h by causing toxemia and septicemia (Forst et al. 1997). Several enzymes and secondary metabolites are produced by bacteria, which are used for bioconversion of insect cadaver into a nutrient soup (Munch et al. 2008). The nematodes develop inside the insect cadaver and complete few generations, and after depleting the food in the cadaver, the IJs emerge and disperse in the search of new host. Formulations of *Heterorhabditis* and *Steinernema* nematodes are being used worldwide for the management of many dangerous soil insect-pests, e.g., *Agrotis ipsilon* (black cutworm) of the order Lepidoptera, *Diabrotica* spp. and *Diaprep* ssp. and *Otiorhynchus* sp. root weevils of the Coleoptera, and the cabbage root fly *Delia radicum* and fungus gnat (*Sciaridae*) of the Diptera. Experiments suggest that foliar application of *P. luminescens* on cabbage butterfly *Pieris brassiae* showed direct toxicity against the insects (Mohan et al. 2003). This suggests that*Photorhabdus* could be used as biopesticide in agriculture on its own, a fact supported by availability of standalone formulations of *Photorhabdus* that are used as spray-able insecticides (Lacey and Shapiro-Ilan 2008).

A special feature of *Photorhabdus* biology is the phenomenon of phase variation. The bacterium is found in two phenotypic phases. Primary phase is associated with IJ stage of nematodes and is characterized by production of dyes, array of antibiotics, proteases, lipases, and bioluminescence. Secondary phase arises after prolonged incubation of primary variant, probably as a result of environmental stress, and this stage lacks or has reduced level of all the characteristics of primary phase as described above (Volgyi et al. 1998). Recently, the cell variant of *Photorhabdus* forming mutualistic relationship with the nematodes has been described as "M-form," which is a small-sized cell variant and does not produce toxins or secondary metabolites. M-form is not the same as secondary phase and can revert back to primary-phase cells (Somvanshi et al. 2012).

15.2 Toxins

Analysis of the *P. luminescens* genome has resulted in the identification of more putative toxin genes than had been found in any other bacteria (Duchaud et al. 2003). This bacterium contains many toxins which were characterized and grouped as: (i) the toxin complexes (TCs), (ii) "Makes caterpillars floppy" (McF) toxins, and (iii) the *Photorhabdus* Virulence Cassettes (PVC), insect-related toxins (Pir), etc. (Bowen et al. 1998; ffrench-Constant et al. 2007; Rodou et al. 2010).

15.2.1 Toxin Complexes

The first class of *Photorhabdus* toxins, the toxin complexes (tc toxins), are ABC-type toxins which are high-molecular weight protein complexes consisting of many subunits: Tca, Tcb, Tcc, and Tcd. Injectable toxicity is found in all the four groups whereas Tca and Tcd exhibit oral toxicity as well. The TcC proteins were found as the main functional components that have the ADP ribosyltransferases activity and cause the cell death by causing defects in phagocytosis mechanism of cell and actin clustering. The genes present in these four loci show significant similarity and have been grouped into tcdA-like [A], the tcaC-like [B], and the tccC-like [C] subunits. All the three – [A], [B], and [C] – subunits are required for complete toxicity. Induction of actin clustering, defects in phagocytosis and cell death-like characters were caused by the [C] subunit, which was identified as the main functional component, whereas the [A] and [B] proteins make a transmembrane pore/channel that helps translocate the [C] proteins through an injection-like mechanisms (Gatsogiannis et al. 2013; Meusch et al. 2014; Aktories et al. 2015).

Genome sequencing of some isolates of *Bacillus thuringiensis* revealed the presence of an operon that encodes for Tca similar to *Photorhabdus* (Blackburn et al. 2011). *Serratia entomophila*, which causes "amber" disease in New Zealand grass-grub larvae, also contains locus for three tc-like genes that encode for ABC components (Hurst et al. 2000). *Yersinia pestis*, the etiological agent of plague, and *Paenibacillus* bacteria, the causative agent of American foulbrood of honey bee disease are two other insect-associated bacteria that show presence of tc-like genes. Apart from this, some bacteria with no insect association have also been found having amino acid similarity with tc-like toxins; these are *Pseudomonas syringae* pv. *tomato, Pseudomonas fluorescens*, and *Fibrobacter succinogenes*. Presence of tc-like genes in insect-associated and noninsect-associated bacteria shows that it has wider range of hosts as well as applications.

15.2.1.1 Caterpillars Floppy (Mcf) Toxins

The second class of *Photorhabdus* toxins, the Mcf toxins, are members of a family of potent high-molecular mass toxins that encode differing effector domains at their N-termini and are active upon injection. These toxins were first identified by screening genomic DNA library of the *Photorhabdus* strain W14 infecting the insect *Manduca sexta*. The injections made the caterpillars loose the body turgor; hence these toxins were named Mcf for "*M*akes *C*aterpillars *F*loppy" (Daborn et al. 2002). Two variants of these toxins are encoded in *P. luminescens* genome: Mcf1 and Mcf2. These two genes display limited homology to the cytotoxin B of *Clostridium difficile*. Mcf1 is a high-molecular weight toxin associated with persistence in the host and insect death. The predicted structure of Mcf1 and Mcf2 includes an N-terminal proapoptotic BH3 domain. Upon exposure to Mcf1, changes in cell morphology of hemocytes and midgut epithelial cells occur as a result of toxin-triggered apoptosis (programmed cell death). The second Mcf toxin (Mcf2) shows homology to HopA1 gene of *Pseudomonas syringae* – a type-III effector protein (Waterfield et al. 2003; Dowling et al. 2004, 2007).

15.2.2 Other *Photorhabdus* Toxins

The *Photorhabdus* insect-related toxins (PirAB) are binary toxins exhibiting injectable and oral toxicity against mosquitos and lepidopterans, displaying similarity to δ-endotoxins from *Bacillus thuringiensis*. PirA does not show much similarity to known proteins, although PirB shows high homology with Cry2A insecticidal toxin (ffrench-Constant et al. 2007). A similar novel binary toxin of the class mono-ADP-ribosyltransferases (mARTs), named Photox, was isolated from *Photorhabdus*. This toxin ADP-ribosylates the α-skeletal actin and non-muscle β- and the γ-actin at Arg^{177}. This results in the inhibition of the polymerization of actin filaments (Visschedyk et al. 2010). Another prominent toxin group is the *Photorhabdus* Virulence Cassettes (PVCs), which are homologues of the prophage-like loci in entomopathogen *Serratia entomophila*. This locus consists of approximately 15 phage-like genes immediately upstream of a varying number of effector-coding sequences, like known toxins such as LopT, Cif, and PhxA and domains of CNF and MCF. They display no antibacterial activity and instead are active against insect hemocytes, causing actin cytoskeleton condensation (Yang et al. 2006). Additionally, eight genes (four coding sequences and four pseudogenes) have been identified in the *Photorhabdus* showing the homology with RtxA toxins of *Vibrio cholerae*. RTX (repeats-in-toxins) proteins are known to exhibit cytolytic metalloprotease and lipase activity (Bowen et al. 2003). One uncharacterized toxin is *Photorhabdus* insecticidal toxin (Pit) which showed 30 % amino acid sequence similarity to part of the Cry protein of *B. thuringiensis*. It shows injectable toxicity, but no oral toxicity (Rodou et al. 2010). It also produces many other secondary metabolites like derivatives of furan stilbene, genistein derivatives, glidobactin/luminmycin, and phenol. This is indicative of the fact that the biosynthesis of these compounds could be a factor of insecticidal activity along with the toxins (Ullah et al. 2014, 2015) (Table 15.1).

15.3 Perspective

The nematode hosts of *Photorhabdus* are used worldwide for the biological management of insect-pests of agricultural crops. *Photorhabdus* are known to produce a range of bioactive compounds having insecticidal, nematicidal, antifungal, and antibacterial properties. Presently, these

Table 15.1 Major insecticidal toxin genes and their products encoded in the *Photorhabdus luminescens* ssp. *laumondii* TTO1 genome (Duchaud et al. 2003; Wilkinson et al. 2009)

Locus	Gene product
plu0404-plu0419	Fimbrial proteins, adhesins
plu0525-plu0541, plu0545-plu0549, plu0634-plu0643, plu1408-plu1413	Hemolysins
plu1030-plu1059	Pili operon
plu0802-plu0808, plu0961-plu0968	Tc insecticidal toxins
plu04165-plu04175	Tc insecticidal toxins TccB1,TccA1
plu4182-plu4186	Tc insecticidal toxins TccC6
plu4488	Tc insecticidal toxins TccC7
plu1536-plu1537	Similar to Bt insecticidal toxin
plu2213-plu2223	Nematicidal protein
plu3124-plu3129	Protein involved in toxin secretion
plu3667-plu3669, plu1341, plu1344, plu3217, plu3324, plu1336/plu1337, plu1339/plu1340, plu1342/plu1343, plu3209/ plu3207	RTX toxins/rtxA genes
plu4187- plu4197	Anthraquinone biosynthesis
plu2631,plu2632-plu2637, plu2850-plu2853, plu4621-fepG, plu2315-plu2324, hemR, plu0739, plu2715, plu3513, plu3519, plu3838	Iron/hemin/siderophore uptake system
plu0822	Photox
plu4141-plu4246/plu3111-plu3140	Mcf/Mcf-like toxins

toxins are the most sought after insect killer toxins next to delta-endotoxins from *Bacillus thuringiensis*, and efforts are on to make insect-resistant transgenic plants using *Photorhabdus* toxins. Combination of *B. thuringiensis* and *P. luminescens* is another strategy to effectively use these bacteria without their nematode host (Benfarhat-Touzri et al. 2014). Using this combination may be cost-effective and a solution to manage development of insect resistance against *B. thuringiensis* (Benfarhat-Touzri et al. 2014). Studies on these bacterial toxins and mechanism involved in their virulence against insects may suggest novel ways to use these toxins as biopesticides for managing a wide range of insect-pests and diseases of crop plants (ffrench-Constant and Bowen 1999).

References

Aktories K, Schmidt G, Lang AE (2015) *Photorhabdus luminescens* toxins TccC3 and TccC5: insecticidal ADP-ribosyltransferases that modify threonine and glutamine. Curr Top Microbiol Immunol 384:53–67. doi:10.1007/82_2014_382

Benfarhat-Touzri D, Amira AB, Benkhedher S, Givaudan A, Jaoua S, Tounsi S (2014) Combinatorial effect of *Bacillus thuringiensis* kurstaki and *Photorhabdus luminescens* against *Spodoptera littoralis* (Lepidoptera: Noctuidae). J Basic Microbiol 54:1160–1165. doi:10.1002/jobm.201300142

Blackburn MB, Martin PA, Kuhar D, Farrar RR Jr, Gundersen-Rindal DE (2011) The occurrence of photorhabdus-like toxin complexes in *Bacillus thuringiensis*. PLoS One 6, e18122. doi:10.1371/journal.pone.0018122

Bowen D, Rocheleau TA, Blackburn M, Andreev O, Golubeva E, Bhartia R, ffrench-Constant RH (1998) Insecticidal toxins from the bacterium *Photorhabdus luminescens*. Science 280:2129–2132. doi:10.1126/science.280.5372.2129

Bowen DJ, Rocheleau TA, Grutzmacher CK, Meslet L, Valens M, Marble D, Dowling A, ffrench-Constant R, Blight MA (2003) Genetic and biochemical characterization of PrtA, an RTX-like metalloprotease from *Photorhabdus*. Microbiol 149:1581–1591. doi:10.1099/mic.0.26171-0

Clarke DJ (2008) *Photorhabdus*: a model for the analysis of pathogenicity and mutualism. Cell Microbiol 10:2159–2167.doi:10.1111/j.1462-5822.2008.01209.x

Daborn PJ, Waterfield N, Silva CP, Au CP, Sharma S, Ffrench-Constant RH (2002) A single *Photorhabdus* gene, makes caterpillars floppy (mcf), allows *Escherichia coli* to persist within and kill insects. Proc Natl Acad Sci U S A 99:10742–10747. doi:10.1073/pnas.102068099

Dowling AJ, Daborn PJ, Waterfield NR, Wang P, Streuli CH, ffrench-Constant RH (2004) The insecticidal toxin makes caterpillars floppy (Mcf) promotes apoptosis in mammalian cells. Cell Microbiol 6:345–353. doi:10.1016/S0378-1097(03)00846-2

Dowling AJ, Waterfield NR, Hares MC, Le Goff G, Streuli CH, ffrench-Constant RH (2007) The Mcf1 toxin induces apoptosis via the mitochondrial pathway and apoptosis is attenuated by mutation of the BH3-like domain. Cell Microbiol 9:2470–2484. doi:10.1111/j.1462-5822.2007.00974.x

Duchaud E, Rusniok C, Frangeul L, Buchrieser C, Givaudan A, Taourit S, Bocs S, Boursaux-Eude C, Chandler M, Charles JF, Dassa E, Derose R, Derzelle S, Freyssinet G, Gaudriault S, Medigue C, Lanois A, Powell K, Siguier P, Vincent R, Wingate V, Zouine M, Glaser P, Boemare N, Danchin A, Kunst F (2003) The genome sequence of the entomopathogenic bacterium *Photorhabdus luminescens*. Nat Biotechnol 21:1307–1313. doi:10.1038/nbt886

ffrench-Constant R, Bowen D (1999) Photorhabdus toxins: novel biological insecticides. Curr Opin Microbiol 2(3):284–288. doi:10.1016/S1369-5274(99)80049-6

ffrench-Constant RH, Dowling A, Waterfield NR (2007) Insecticidal toxins from *Photorhabdus* bacteria and their potential use in agriculture. Toxicon 49:436–451. doi:10.1016/j.toxicon.2006.11.019

Forst S, Dowds B, Boemare N, Stackebrandt E (1997) *Xenorhabdus* and *Photorhabdus* spp.: bugs that kill bugs. Annu Rev Microbiol 51:47–72. doi:10.1146/annurev.micro.51.1.47

Gatsogiannis C, Lang AE, Meusch D, Pfaumann V, Hofnagel O, Benz R, Aktories K, Raunser S (2013) A syringe-like injection mechanism in *Photorhabdus luminescens* toxins. Nature 495:520–523. doi:10.1038/nature11987

Hurst MR, Glare TR, Jackson TA, Ronson CW (2000) Plasmid-located pathogenicity determinants of *Serratia entomophila*, the causal agent of amber disease of grass grub, show similarity to the insecticidal toxins of *Photorhabdus luminescens*. J Bacteriol 182:5127–5138. doi:10.1128/JB.182.18.5127-5138.2000

Lacey LA, Shapiro-Ilan DI (2008) Microbial control of insect pests in temperate orchard systems: potential for incorporation into IPM. Annu Rev Entomol 53:121–144. doi:10.1146/annurev.ento.53.103106.093419

Meusch D, Gatsogiannis C, Efremov RG, Lang AE, Hofnagel O, Vetter IR, Aktories K, Raunser S (2014) Mechanism of Tc toxin action revealed in molecular detail. Nature 508:61–65. doi:10.1038/nature13015

Mohan S, Raman R, Gaur HS (2003) Foliar application of *Photorhabdus luminescens*, symbiotic bacteria from entomopathogenic nematode *Heterorhabditis indica*, to kill cabbage butterfly *Pieris brassicae*. Curr Sci 84:1397

Munch A, Stingl L, Jung K, Heermann R (2008) *Photorhabdus luminescens* genes induced upon insect infection. BMC Genomics 9:229. doi:10.1186/1471-2164-9-229

Rodou A, Ankrah DO, Stathopoulos C (2010) Toxins and secretion systems of *Photorhabdus luminescens*. Toxins 2:1250–1264. doi:10.3390/toxins2061250

Somvanshi VS, Sloup RE, Crawford JM, Martin AR, Heidt AJ, Kim KS, Clardy J, Ciche TA (2012) A single promoter inversion switches *Photorhabdus* between pathogenic and mutualistic states. Science 337:88–93. doi:10.1126/science.1216641

Ullah I, Khan AL, Ali L, Khan AR, Waqas M, Lee IJ, Shin JH (2014) An insecticidal compound produced by an insect-pathogenic bacterium suppresses host defenses through phenoloxidase inhibition. Molecules 19:20913–20928. doi:10.3390/molecules191220913

Ullah I, Khan AL, Ali L, Khan AR, Waqas M, Hussain J, Lee IJ, Shin JH (2015) Benzaldehyde as an insecticidal, antimicrobial, and antioxidant compound produced by *Photorhabdus temperata* M1021. J Microbiol 53:127–133. doi:10.1007/s12275-015-4632-4

Visschedyk DD, Perieteanu AA, Turgeon ZJ, Fieldhouse RJ, Dawson JF, Merrill AR (2010) Photox, a novel actin-targeting mono-ADP-ribosyltransferase from *Photorhabdus luminescens*. J Biol Chem 285:13525–13534. doi:10.1074/jbc.M109.077339

Volgyi A, Fodor A, Szentirmai A, Forst S (1998) Phase variation in *Xenorhabdus nematophilus*. Appl Environ Microbiol 64:1188–1193

Waterfield NR, Daborn PJ, Dowling AJ, Yang G, Hares M, ffrench-Constant RH (2003) The insecticidal toxin makes caterpillars floppy 2 (Mcf2) shows similarity to HrmA, an avirulence protein from a plant pathogen. FEMS Microbiol Lett 229:265–270. doi:10.1016/S0378-1097(03)00846-2

Waterfield NR, Ciche T, Clarke D (2009) *Photorhabdus* and a host of hosts. Annu Rev Microbiol 63:557–574. doi:10.1146/annurev.micro.091208.073507

Wilkinson P, Waterfield NR, Crossman L, Carton C, Sanchez-Contreras M, Vlisidou I, Barron A, Bignell A, Clark L, Ormond D, Mayho M, Banson N, Smith M, Churcher C, Harris D, Thompson NR, Quail M, Parkhill J, ffrench-Constant RH (2009) Comparative genomics of the emerging human pathogen *Photorhabdus asymbiotica* with the insect pathogen *Photorhabdus luminescens*. BMC Genomics 10:302

Yang G, Dowling AJ, Gerike U, ffrench-Constant RH, Waterfield NR (2006) *Photorhabdus* virulence cassettes confer injectable insecticidal activity against the wax moth. J Bacteriol 188:2254–2261. doi:10.1128/JB.188.6.2254-2261.2006

Ms. Jyoti Kushwah received her M.Sc. degree in Microbiology from Jiwaji University, Gwalior, Madhya Pradesh, India. She is currently working as Senior Research Fellow at EPN Genomics Lab, Division of Nematology, Indian Agricultural Research Institute (IARI). Her current research focuses on characterization of symbiotic bacteria *Photorhabdus* sp. associated with indigenous entomopathogenic nematodes *Heterorhabditis* sp.

Vishal Singh Somvanshi received his M.Sc. Ag. degree (Plant Pathology) from G. B. Pant University of Agriculture and Technology, Pantnagar (Uttarakhand), and Ph.D. degree (Nematology) from IARI, New Delhi. He did his postdoctoral training in Israel and USA from 2005 to 2012 and is presently working as a Senior Scientist at ICAR-IARI. His research interests are molecular genetics of nematode-bacterium symbiosis and nematode biosystematics.

Microbial Vesicles: From Ecosystem to Diseases

16

Shashank S. Kamble, Nancy Garg, Brijendra Kumar Tiwari, Lalit K. Singh, Neha Dhasmana, and Yogendra Singh

Abstract

The production of outer membrane vesicles (OMVs) is conserved in eukaryotes and prokaryotes. OMVs are double-layered structures with contents from outer membrane, periplasmic space, and even cytosol. Some of them have been shown to contain nucleic acids as well, which explains the specialized system for the packaging of these vesicles. OMVs mediate essential processes such as transport of nutrients, antigens, and virulence factors, etc., which help the microorganisms in communication as well as in killing of other microbial cells. The biogenesis of OMVs differs in bacteria and archaea. The archaeal OMV biogenesis is similar to eukaryotes involving ESCRT machinery, while in gram-negative bacteria, it occurs either due to broken OM-PG interaction or due to increased periplasmic pressure. OMVs containing antigens have been recently explored for use as vaccine which provides another dimension for its applications.

16.1 Introduction

Since the revelation of microbial existence back in the sixteenth century, several fascinating queries have intrigued researchers worldwide about the way microorganisms communicate. Production of outer membrane vesicles (OMVs) by microbes has answered many of these questions in great detail. OMVs have been shown to be crucial for intercellular communication in the natural environment as well as inside the host. This way of communication has been shown to facilitate complex processes such as horizontal gene transfer, transport of nutrients, and secretion of virulence factors.

S.S. Kamble (✉) • N. Garg • B.K. Tiwari • L.K. Singh • N. Dhasmana
Allergy and Infectious Diseases, CSIR-Institute of Genomics and Integrative Biology, 110007 Delhi, India
e-mail: shashank.kamble@igib.in; brijendratiwari@gmail.com; b.tiwari@igib.in; lalit.singh@igib.in; neha.dhasmana@igib.in

Y. Singh
Allergy and Infectious Diseases, CSIR-Institute of Genomics and Integrative Biology, Room No. 208, Mall Road, 110007 Delhi, India
e-mail: ysingh@igib.res.in

V.C. Kalia (ed.), *Microbial Factories*, DOI 10.1007/978-81-322-2595-9_16

Release of extracellular material was first observed in lysine auxotrophs of *Escherichia coli* which were found to secrete a complex consisting of lipopolysaccharides (LPSs) and lipoproteins (Lpp) in the media continuously under lysine-limiting conditions. Electron microscopy of lysine-limited cells harvested at regular intervals revealed the accumulation of globular and hollow spheres which started to accumulate after 12 h of growth (Knox et al. 1966). Another report indicated the secretion of lipoglycoconjugate-like material by a strain of *E. coli* (Municio et al. 1963; Bishop and Work 1965). The C^{14} labeled OMVs were later shown to be released upon the interaction of *E. coli* cells with T4 phage as early as 2 min postinfection (Loeb 1974). Production of OMVs also showed to occur during normal growth conditions of *E. coli*. The report also suggested that vesicles arise at a site where the linkage between the outer membrane and underlying peptidoglycan layer is severed either due to breakage of existing linkage or due to insertion of LPS or other molecules instead of lipoproteins, which may occur during cell division (Hoekstra et al. 1976). These early reports opened the exciting area of membrane vesicles for research. In this chapter, we will discuss the recent research on OMVs and its physiological relevance.

16.2 OMV Biogenesis

Early reports suggested that the OMVs arise from the outer membrane of bacteria at a region wherein the outer membrane-peptidoglycan (OM-PG) linkage is severed. This was proposed owing to the fact that the outer membrane of gram-negative bacteria is covalently attached to the underlying murein layer by dense interactions of Braun's lipoprotein. Hence, the release of OMVs is possible only when these interactions are either poorly formed or get broken. In the former case, OMVs will be mainly composed of newly inserted proteins, which failed to form interactions with murein layer, while in the later case, OMVs will be consisting mainly of older outer membrane which got replaced with newly inserted proteins. Using dual labeling, *E. coli* JC411 was shown to produce OMVs with 7.5-fold higher content of newly synthesized proteins. This report indicated that indeed the OMVs originate from the outer membrane at a site where newly synthesized proteins get inserted (Mug-Opstelten and Witholt 1978). The attempt to insert newer proteins, however, could create a pressure on older proteins which are covalently attached to the murein layer. Older membrane proteins resist this newer force, thereby pushing the newer proteins to bulge outward. If the bulge is smaller, newer proteins still get a chance to establish the covalent or non-covalent interactions with murein layer resulting in the disappearance of the bulge. However, if the bulge grows beyond a size, it becomes difficult for the newer proteins to establish these interactions with murein which ultimately leads to its release in the form of OMV (Mug-Opstelten and Witholt 1978) as has been depicted in Fig. 16.1.

In the case of archaea, the process of vesicle biogenesis seems to be of primitive nature and is related to eukaryotic exosome release. In eukaryotes, endosomal sorting complex required for transport (ESCRT) is involved in the vesicle release from the cell surface. There are five types of ESCRT complexes: ESCRT-0, ESCRT-I, ESCRT-II, ESCRT-III, and Vps34. The ESCRT system is highly conserved and is essential for processes such as multivesicular body (MVB) pathway, cytokinesis, and the budding of human immunodeficiency virus (HIV). Together, these complexes are required for membrane budding, bending, and scission that ultimately leads to vesicle release (Schmidt and Teis 2012). ESCRT-III- and Vps34-like proteins have also been found in OMVs released from some of the archaeal species (Ellen et al. 2009). In these species, release of OMVs has been shown to involve outward bulging and pinching off of the vesicle from the cell surface. One interesting aspect of this machinery is that it cleaves off the membranous neck from the

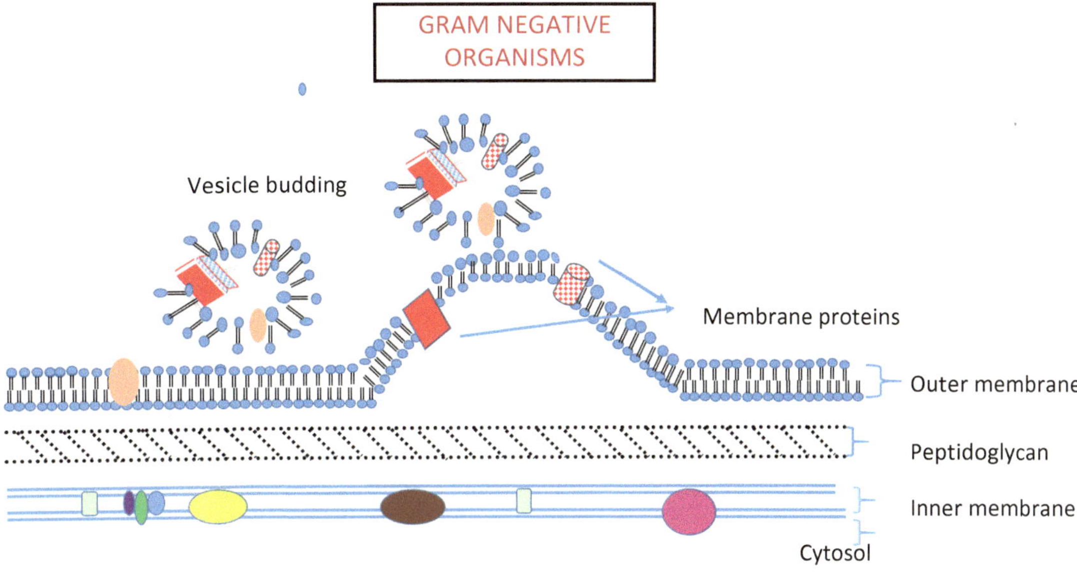

Fig. 16.1 Biogenesis of outer membrane vesicles (OMVs) in gram-negative organisms which involves outward bulging of outer membrane (OM) due to broken OM-murein linkage

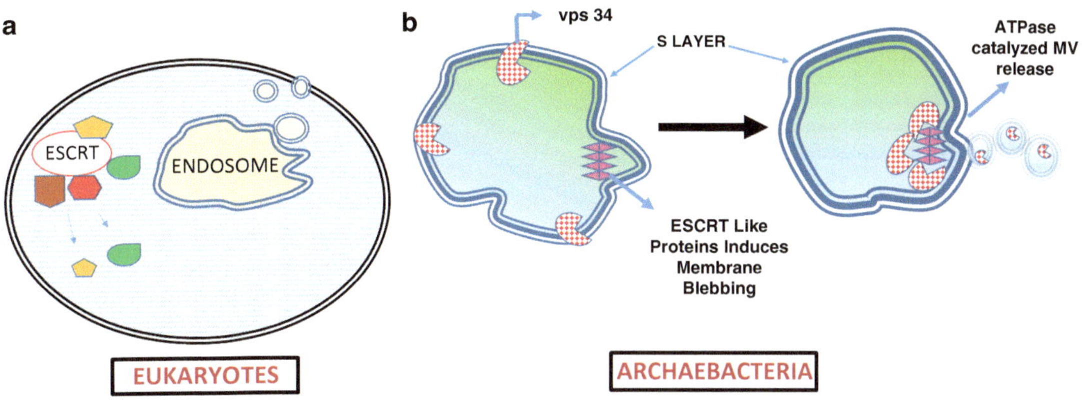

Fig. 16.2 Biogenesis of OMVs in eukaryotes (**a**) involves endosomal sorting complex required for transport (ESCRT) essential for budding of vesicle, as well as its release from organelles such as endosomes and in archaea (**b**) involves ESCRT-like machinery involving ESCRT-III- and Vps34-like proteins

inside of the vesicle rather than constricting it from the outside. ESCRT proteins are present in a monomeric state in cytosol and get activated upon conformational change. Activated proteins in turn get assembled at the site of vesicle biogenesis in the form of multi-subunit complex. Eventually, this complex gets released from the site as a result of dissociation by Vps34 ATPases (Deatherage and Cookson 2012). OMV biogenesis in eukaryotes and archaea has been depicted in Fig. 16.2.

16.3 Distribution of OMVs in Bacteria, Archaea, Fungi, and Parasites

Microbial vesicle (MV) formation is a conserved process across diverse microbial life spanning both prokaryotes and eukaryotes. Membrane vesicles can be categorized into different types based on their size: In bacteria and archaebacteria, these structures are often referred to as membrane vesicles, whereas fungal and parasitic vesicles are known as exosomes or microvesicles (shedding microvesicles) (Mathivanan et al. 2010). The diameter of microbial vesicles derived from gram-positive bacteria (*Bacillus* spp.) are found to be of size similar (20–200 nm) to that of gram-negative bacteria. OMVs released from archaebacteria such as *Sulfolobus* spp. have a diameter between 90 and 230 nm. Microbial vesicles released from parasites and fungi include two distinct populations, i.e., exosomes (40–100 nm) and shedding microvesicles (100–1000 nm). Eukaryotes also release membrane vesicles under normal conditions as well upon infection with pathogenic microbial strains. The production of exosomes has been depicted in (Fig. 16.3) human monocyte-differentiated macrophages infected with a virulent strain of *M. tuberculosis*. The process of OMV production seems to be similar across the microbial life, but the cellular machinery participating is completely different (Oliveira et al. 2010).

MVs derived from gram-positive as well as gram-negative bacteria have related constituents. It has been reported that *Staphylococcus aureus-derived* MVs contain membrane lipids, virulence factors, coagulase, adhesins, and various other toxins. The superantigens present inside these vesicles are known to induce sepsis (Lee et al. 2009). Protein profiling of OMVs derived from *E. coli* showed the presence of various virulence factors and other toxic compounds (Lee et al. 2007). Also OMVs derived from different genera like *Neisseria, E. coli,* and *Haemophilus* are known to contain RNA and DNA. Vesicles from parasites such as *Leishmania, Cryptococcus*, and *Trypanosoma* are known to contain factors that promote the dissemination of pathogen (Silverman et al. 2010; Huang et al. 2012). It has been demonstrated that vesicles from fungal microorganisms like *Cryptococcus neoformans* are known to act as virulent bags, i.e., they may contain concentrated toxins which enhance the invasion into the host cell (Rodrigues et al. 2008). MVs are thought to be a stress response mechanism for organism living in extreme conditions such as *Sulfolobus* and *Thermococcus* spp. of archaebacteria (Ellen et al. 2009). These OMVs are known to be composed of many protein sorting complexes.

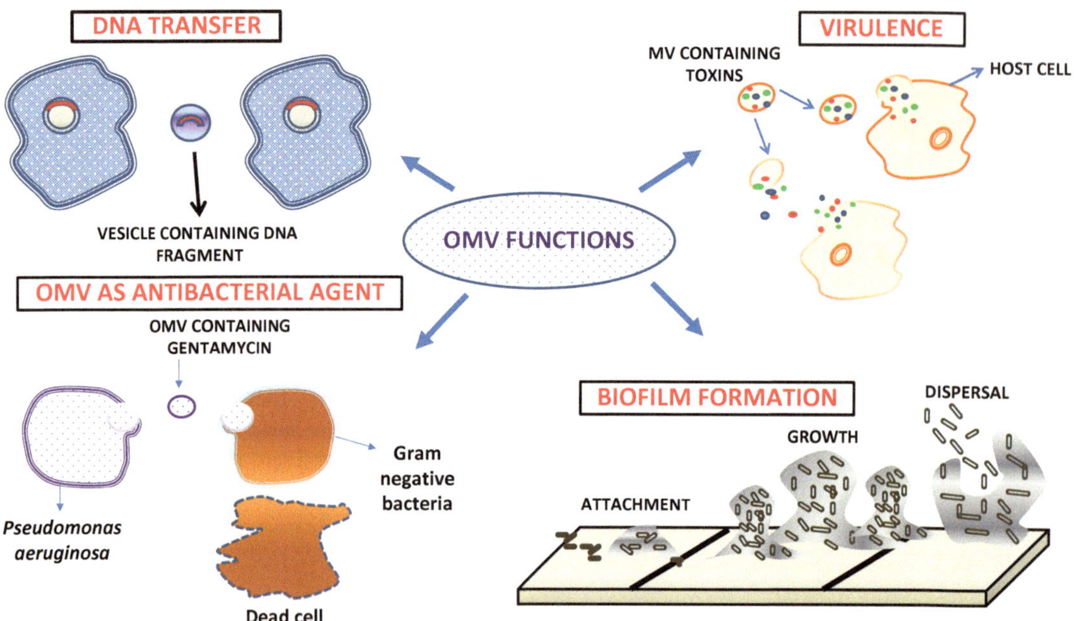

Fig. 16.3 Various functions of outer membrane vesicles (OMVs) such as (1) transfer of DNA, (2) transfer of virulence factors, (3) transport of antimicrobial agents to neighboring cells, and (4) formation of biofilms

Microbes from all the branches of life share the feature of OMV release, which contributes to its pathogenesis and cellular functioning. OMVs may modify the cell surface so as to avoid immune response, facilitate cell to cell communication, interaction between microbes and host cells and are mainly responsible for the secretion of virulence factors. For example, *Salmonella-derived* MVs are released during intracellular growth in epithelial cells (Garcia-del Portillo et al. 1997), and virulence factors are released from OMVs derived from *H. pylori* (Fiocca et al. 1999). OMVs from *Neisseria meningitidis* contribute to sepsis through toll-like receptors (TLRs) and nod-like receptors (NLRs) (Hellerud et al. 2008). Similarly, on interaction with host cells, some eukaryotic microbes such as *Leishmania donovani* also release OMVs during macrophage infection (Silverman et al. 2008). Proteins present in the OMVs facilitate the transfer of other proteins like β-lactamase and thereby help to retain the functionality of these proteins in receiving cells (Gurung et al. 2011). Many properties and functions are common between vesicles derived from different microbial life forms. Therefore, it can be concluded that vesicle secretion is an evolutionary conserved process.

16.4 Modulation of OMV Production

OMV production by microorganisms is a conserved process and plays a crucial role in their inhabitation. Hence, it is essential to understand the processes that may modulate the production of OMV. Since many gram-negative bacteria have been shown to produce OMVs, they represent a better model to study alterations in OMV production (Kulkarni and Jagannadham 2014). In the gram-negative bacteria, peptidoglycan layer of the periplasmic space is tethered to the outer membrane by lipoproteins (Kuehn and Kesty 2005). In order for a vesicle to bud out of OM, it is important that the interaction between PG and OM is transiently disrupted. Since PG layer is subject to modulation during cell growth and growth phase transitions, bacteria may use this to alter OMV production in accordance with the environment they encounter. Vesicle biogenesis has been shown to depend on the structure and content of the cellular envelope. It varies with either the accumulation of proteins in the periplasm or the concentration of lipoprotein in the OM. The concentration of bound Lpp correlates inversely with the OMV production; although in strains wherein OMV production occurs due to the increase in "periplasmic pressure" with the accumulation of proteins, this does not hold true (Schwechheimer et al. 2014).

Pseudomonas aeruginosa is the causative agent of nosocomial infections and an opportunistic pathogen. Previously reports have shown the production of OMVs in naturally growing and gentamicin exposed strain of *P. aeruginosa*. The OMV production increased threefold after exposure to gentamicin, and these vesicles contained more DNAs and LPS as compared to naturally growing bacteria (Kadurugamuwa and Beveridge 1995). *P. aeruginosa* is being studied extensively for its ability to form biofilm on multiple surfaces as well as the secretion of signaling molecules in extracellular millieu to coordinate activities between the cells. The later is termed as quorum-sensing (QS) system (Kalia 2013, 2014a, 2015; Kalia and Purohit 2011). *P. aeruginosa* secrete numerous small molecules that diffuse into neighboring cells in order to sense the density and to coordinate behavior of the population (Harmsen et al. 2010). These include *Pseudomonas* quinolone signal (PQS) and hydroxyquinone (HHQ). PQS was shown to be packaged preferentially in OMVs over other quinolones such as butyryl homoserine lactone, while mutants with deletion in *pqsH*, a gene that catalyzes the final step in PQS biosynthesis, had severely impaired OMV formation. OMVs containing PQS showed antimicrobial activity against actively dividing cells of *Staphylococcus epidermidis* (Mashburn and Whiteley 2005). Later studies showed that the formation of OMVs required the presence of PQS. PQS stimulates OMV formation by interacting with the LPS of *P. aeruginosa*. LPS consists of lipid A, oligosaccharides, and polysaccharide O antigen. PQS has been shown to interact with the lipid A portion of

LPS thereby lowering the membrane fluidity and thus to provide the curvature necessary for outward bulging of OMVs. Thus, PQS acts as a molecule with dual nature being involved in quorum sensing as well in regulation of OMV formation (Warren et al. 2008).

The OMVs of *Mycobacterium tuberculosis* carry LpqH, a lipoprotein that is also TLR2 agonist, and thus elicit immune response in hosts. Recently, a protein encoded by the gene rv0431 was shown to regulate the release of OMV and thereby regulating the immunostimulatory potential of the pathogen in host. The gene was termed "vesiculogenesis and immune response regulator" or *virR* (Rath et al. 2013). *M. tuberculosis* contains a family of 11 eukaryotic-like serine/threonine kinases, and many of which have been shown to play an important role during pathogenesis (Koul et al. 2004; Sajid et al. 2011; Kusebauch et al. 2014). It will be interesting to see if these kinases do play any role during OMV production in *M. tuberculosis*.

In the case of *E. coli*, mutation in a gene *degP* which encodes a serine protease has been shown to hamper OMV production. *degP* encodes a periplasmic protein with a dual nature, thus acting as a protease as well as molecular chaperones. The dual activity of this protein helps maintain the protein homeostasis in the periplasmic space by preventing the accumulation of misfolded proteins. Mutations in *degP* lead to accumulation of misfolded proteins thus causing increased "periplasmic pressure." This in turn causes outward bulging of the outer membrane leading to the release of outer membrane vesicle. A similar locus in *P. aeruginosa* termed as mucD, however, has a less severe effect on OMV production (Schwechheimer and Kuehn 2013).

16.5 Biological Significance

OMVs mediate processes such as horizontal gene transfer, intercellular communication, and transport of virulence factors as well as seem to play essential roles in other ecosystems as depicted in Fig. 16.4.

16.5.1 Horizontal Gene Transfer

The most well-known function of OMV is to mediate horizontal gene transfer, thus allowing cell to cell communication in complex microbial community. OMVs are used extensively by various microorganisms to transfer genetic information such as DNA and RNA in order to adapt to environmental cues. *P. aeruginosa* is known to produce OMVs under natural conditions. *P. aeruginosa* strain PAO1 containing a plasmid pAK1900 has exhibited packaged plasmid DNA in released OMVs. However, these OMVs showed a very little ability to efficiently transform parent PAO1 strain as well as *E. coli* DH5-α, although various conditions can be used to recheck the transformation ability of these vesicles. Two reasons can be accounted for the packaging of DNA into vesicles: (1) trafficking of DNA to periplasm which then would be released in OMVs along with other periplasmic components, although this seems to be possible for plasmid DNA but difficult for chromosomal DNA and (2) uptake of exogenous DNA by OMV before their release from the cell surface. It is believed that a combination of above routes might be responsible for DNA packaging of OMVs in most of the cases (Renelli et al. 2004).

The association of DNA with OMVs was earlier shown in *N. gonorrhoeae*. OMVs produced by this pathogen also termed as "gonococcal blebs" were shown to contain both chromosomal and plasmid DNA. The DNA contained in these blebs was resistant to denaturation by extensive DNase treatment. Interestingly, some of these blebs were shown to carry plasmids encoding penicillinase-resistant gene. *N. gonorrhoeae* strains susceptible to penicillin treatment are shown to acquire penicillin resistance upon incubation with the blebs from wild-type strain, and this transfer occurred even in the presence of DNase which showed the DNA protective role of these blebs. The blebs were shown to contain DNA up to 36Kbp (conjugative plasmid) and small amounts of RNA as well. This showed that OMVs serve as a very effective way of DNA transfer between bacterial cells (Dorward et al. 1989). Later, the presence of DNA-binding pro-

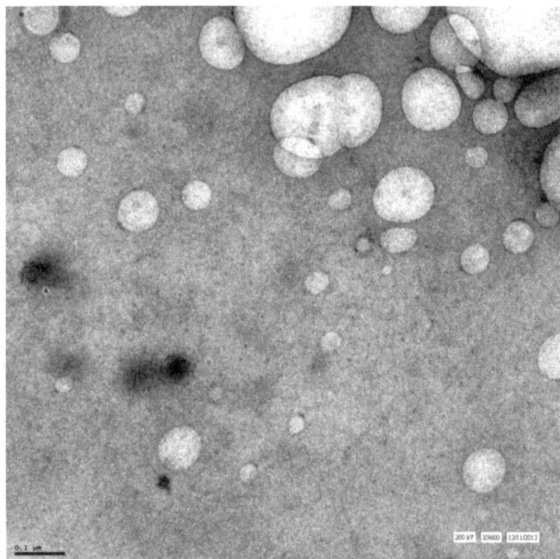

Fig. 16.4 Eukaryotic membrane derived vesicles isolated from human monocyte differentiated macrophages infected with *M. tuberculosis*

teins was revealed in OMVs from *N. gonorrhoeae* wherein these protein-DNA complexes were identified using a nonradioactive assay by probing blotted proteins for their DNA-binding activity. Proteins such as SSB (single-stranded DNA-binding protein), RecA (recombinase A), and Cyt C (cytochrome C) from whole cell and bleb fractions were shown to have significant DNA-binding activity (Dorward et al. 1989).

Similarly, *Haemophilus influenzae* has been studied extensively for natural transformation. Competent cells of these bacteria were observed to produce numerous membranous extensions. These extensions help in binding and uptake of DNA by bacterial cells, also providing physical protection against cellular nucleases prior to integration (Kahn et al. 1982). Specialized structures for DNA uptake and transfer have been termed as "transformasomes" in *H. influenzae*. Linear DNA was shown to be in these structures for short time (less than 5 min) after which it appears in the chromosomes, while plasmid DNA required more time to exit. The composition of transformasomes revealed the presence of many proteins associated with competence as well as other outer membrane proteins of *H. influenzae* (Kahn et al. 1983).

16.5.2 Intercellular Communication: Bacterial Biofilms

During adverse conditions that the microbes encounter, a system is required to aid in bacterial survival and persistence by reducing the magnitude of the stress (Kalia et al. 2014, 2015). MVs secreted during in vitro and in vivo growth provide such a system that is widely used by microorganisms. Biofilms are microbial community embedded in self-synthesized extracellular polysaccharides. The ability to form biofilm is essential for the survival of microorganisms in the ecosystem as well as in the host. The biofilm moves through five distinct phases, each phase being marked by distinct changes in the genotypic and phenotypic characteristics of the cells involved (Ma et al. 2009). The five phases involved are:

Phase 1: Attachment of cells to the surface, which is reversible

Phase 2: Irreversible attachment by the extracellular polysaccharides, wherein cells lose their flagella-driven motility to minimize ATP loss

Phase 3: Early maturation phase with developed biofilm architecture

Phase 4: Final maturation phase with fully mature biofilms characterized by the complex biofilm structure

Phase 5: Active dispersion/ Passive erosion of cells from biofilm matrix to colonize the new site

A mature biofilm resembles a multicellular organism with specialized ways of communication and transport of secretory molecules between biofilm cells. Many reports show that MVs play a role in the initial stages and during maturation of biofilm.

Myxococcus xanthus is a soil-dwelling scavenger with a characteristic gliding motility. Upon nutrient starvation, it forms a multicellular fruiting body containing spores, which requires coordinated gliding motility of all the cells. How this process is regulated inside complex biofilm communities of *M. xanthus* is incompletely understood. Recent study showed the presence of OMV-like structures within the biofilm with the help of 3D electron tomography. OMVs were shown to act as a tether which connects cell-cell and cell-substratum. This technique also revealed the presence of other appendage-like structures such as flagella and pili within the mature biofilm (Palsdottir et al. 2009). Previously it was shown that the cells in the biofilm of *M. xanthus* exchange outer membrane proteins such as CglB and Tgl that are required for the gliding motility of the organism (Nudleman et al. 2005). The authors also report the rescue of gliding motility defects in cells of *M. xanthus* upon contact with cells having intact gliding motility. A recent study shows direct visualization of the outer membrane exchange mediated protein transfer using a type II signal sequence for outer membrane localization fused to mCherry fluorescent reporter. When these cells were mixed with GFP-labeled recipient bacteria, fluorescence of mCherry reporter could be visualized in the recipient cells. The outer membrane localization of signal sequence was sufficient to initiate protein exchange between cells. This study also showed that proper cell-cell alignment is required for the lipoprotein exchange (Wei et al. 2011). This exchange of phenotype without having the need to transfer nucleic acid provides a means of intercellular communication within complex microbial communities such as biofilms.

H. pylori is a gastrointestinal pathogen that colonizes gastric mucosa. It has been shown previously to form biofilm in various conditions which indicate the fact that the pathogen may reside in gastric mucosa in the form of a biofilm. The differential ability of clinical isolates and reference strains of *H. pylori* to form biofilm has been linked to the difference in the ability to produce OMVs. Scanning electron microscopy demonstrated the presence of many OMV-like structures tethered to bacterial surfaces in *H. pylori* TK1402 biofilm. The amount of OMV released correlated with fetal calf serum concentration in the medium. The OMVs were mainly seen at the interface of bacterium and the substratum and, interestingly, as a means of cell to cell attachment within the biofilm. The addition of OMV fraction from these cells exogenously enhanced biofilm-forming ability in dose-dependent manner. Thus, OMVs secreted by this strain of *H. pylori* may contain factors that enhance cell aggregation and thus biofilm formation (Yonezawa et al. 2009). Further study is needed to support this notion.

Cell membrane of a gram-negative bacterial cell has a difficult task of combating the stress directly and also of relaying this stress onset to other cellular organelles so that appropriate response can be mounted. Therefore, in order to maintain the integrity of membrane, its composition changes with respect to the environment. Membrane properties of *Pseudomonas putida* change in response to multiple stressors such as long-chain fatty-acid alcohol such as 1-octanol, osmotic stress such as NaCl, and EDTA as well as heat shock. It was observed that in comparison to control cells where little or no vesicles were seen, *P. putida* strains treated with above stresses induced the formation of OMVs with different structures within as early as 10 min. Though the vesicles were of the same size irrespective of the stress applied, the zeta potential varied. The degree of saturation of these OMVs increased with increased osmotic stress due to enrichment of stearic acid. The release of MVs in response to these stresses increased the cell surface hydrophobicity, which in turn enhanced their biofilm-forming ability (Baumgarten et al. 2012).

16.5.3 OMVs as Antimicrobial Agents

Whether OMVs produced by a microorganism can differentiate between self and nonself is controversial. It has been shown in previous reports that the MVs may contain periplasmic and cytosolic components, including enzymes such as phospholipase C, proteases, phosphatases, etc. OMVs from the cultures of gram-negative bacteria can kill other bacteria having different PG chemotypes. The MVs isolated from 15 strains of gram-negative bacteria containing the periplasmic enzyme peptidoglycan hydrolase showed the ability to lyse the PG layer of other gram-negative and gram-positive bacteria. The use of bacteria with different PG chemotypes showed the bacteriolysis range of MVs produced (Li et al. 1998).

Kadurugamuwa and Beveridge in their reports (Kadurugamuwa and Beveridge 1995, 1996, 1999) showed that the OMVs from *P. aeruginosa* may mediate killing of other gram-negative and gram-positive bacteria. The study showed that *P. aeruginosa* produce OMVs during natural growth cycle. However, their production increased upon treatment with gentamicin. OMVs produced during normal growth cycle contained in addition to other factors autolysin, a major virulence factor. OMVs produced upon gentamicin treatment also contained traces of gentamicin along with autolysin. These gentamicin-containing OMVs could lyse isolates of *P. aeruginosa* with resistance to gentamicin due to reduced permeability of the antibiotic. Use of immunogold electron microscopy revealed that OMVs containing PG hydrolase fused and released the content into the periplasmic space of gram-negative bacteria, while these OMVs adhered to the surface of gram-positive bacteria including those containing the S layer proteins. Although the recipient cells lysed upon the release of PG hydrolase and autolysin, the capture of actual PG breakdown could provide a new dimension to this study.

16.5.4 OMV Release during Microbial Infections: Transport of Virulence Factors

In an ever changing environment that a microbe encounters, it's important for the microbe to sense environmental cues and respond accordingly. The most efficient way of responding is through the release of secreted vesicles. Production of membrane vesicles provides an efficient means of targeted delivery of the content to its destination without having the need for close proximity between donor and acceptor membranes (Ducret et al. 2013). Although soluble material can be secreted directly by the pathogen in the environment, MVs transport it specifically to the target site without diluting the signal moiety as may happen in the case of direct secretion. MVs produced by pathogenic bacteria may carry toxins, virulence factors, and membrane proteins as well as enzymes. The transfer of various molecules through MVs such as proteins, lipids, or nucleic acids helps the microbes to alter variety of host processes such as cytoskeletal rearrangement and cellular signaling, including cell proliferation and cell death pathways and thereby making the host environment favorable for its own persistence. These MVs also act as a mean of cell to cell communication inside the host, thereby contributing to the pathogenesis (Kulkarni and Jagannadham 2014). Varieties of mechanisms have been proposed for the targeted delivery of vesicular content: (1) spontaneous release of contents upon vesicular lysis inside the host cell and (2) interaction of vesicles with host cell surface, internalization, and fusion.

In peridontitis, a common gum infection of *Aggregatibacter actinomycetemcomitans* causes alveolar bone resorption and systemic infection. Recent reports indicate the crucial role of OMVs in the pathogenesis of this oral pathogen. This mechanism of toxin delivery represents a way of exposing pathogen-associated molecular patterns (PAMPs) to host pattern recognition receptors (PRRs) thereby inducing immune response. The membrane vesicles produced by the pathogen carry the cytolethal-distending toxin (CDT) and deliver it to the healthy cells of peridontium. The resulting induction of NOD-1- and NOD-2-(cytosolic peptidoglycan sensors)mediated innate immune response targeting oral bacteria in turn causes alveolar bone resorption, which may result in tooth loss (Thay et al. 2014).

OMV production from gram-negative bacteria is a well-known fact; however, little is known about the release of MVs from pathogenic gram-positive bacteria during infection. Recent studies reveal the production of MVs from the outer layer of the bacterial cell (50–150 nm) in the supernatant of *Bacillus anthracis* cultures. The release of MVs also occurs inside the macrophages with the vesicles carrying toxins, including protective antigen, lethal factor, edema factor, and anthrolysin. Mice experiments show the immunogenic potential of these vesicles thus inducing a strong immune response against anthrax toxins. Thus, these vesicles carrying anthrax toxins may prove to be valuable in developing vaccine against *B. anthracis* (Rivera et al. 2010).

One of the major physiological symptoms of the infection due to intestinal pathogens is diarrhea. Intestinal pathogens such as *E. coli* and *Vibrio cholerae* secret toxins which increase the efflux of electrolytes into the small intestine thereby leading to watery diarrhea. The heat labile enterotoxin (LT) secreted by enterotoxigenic *E. coli,* which is around 80 % homologues to cholera toxin, has been shown to be associated with bacterial outer membrane vesicles. OMVs carrying the toxin get internalized into the intestinal epithelial cells and delivers toxin inside the cytosol. Indeed, this process has been found to be directed by the toxin itself via interaction with GM1, and the receptor for LT is a soluble receptor present in lipid rafts (Kesty et al. 2004). Other pathogens such as *P. aeruginosa* (Kadurugamuwa and Beveridge 1995; Renelli et al., 2004; Bauman and Kuehn 2006; MacDonald and Kuehn 2013), *H. pylori* (Ismail et al. 2003; Parker et al. 2010), *Shigella* (Menard et al. 1996) have also been shown to secrete membrane vesicles carrying toxins, immunomodulators, and virulence factors.

16.5.5 Release of OMVs in Ecosystem

Although the role of membrane vesicles during in vitro and in vivo growth have been well documented, the significance of these vesicles in natural ecosystem or rather in nonpathogenic organisms is not known. Recently for the first time, the role of OMVs of metal-reducing bacteria has been shown in enzymatic transformation of heavy metals and radionuclides. The MVs isolated from these bacteria under normal growth conditions showed the ability to reduce Fe [III] (with hydrogen as electron donor) which then accumulated as precipitates on the surface of these vesicles. Inconsistent with this, OMVs from *Shewanella* spp. demonstrated the ability to reduce uranium and technetium under anaerobic conditions with hydrogen as electron donor. The ability to reduce metals owed to the presence of a variety of proteins as well as cytochrome in the lumen of these OMVs. These proteins could couple the reduction of metals to the oxidation of hydrogen (Gorby et al. 2008). This can have biogeochemical applications as these OMVs can reduce many other metals which may act as terminal electron acceptor.

However, some questions remain regarding the transfer of electrons from bacteria to the vesicles. Studies have shown the ability of metal-reducing bacteria to use external redox-reactive compounds which might transfer electron from bacteria to metal oxides. *Shewanella* spp. indeed has been shown to produce redox-reactive metabolites which are thought to serve as platform for transfer of electron from bacteria to metal oxides. Previous study showed that OMVs in *Shewanella* spp. are tethered to outer layer of the cell by tiny appendages called nanowires. Initial studies showed the ability of these nanowires to conduct electricity, which provoked researchers to hypothesize that these nanowires may play an essential role in electron flow from bacteria to OMVs to metal oxides (Gorby et al. 2006). However, more work needs to be done to prove this exciting hypothesis (Pirbadian et al. 2014). Also, these OMVs from *Shewanella* spp. having metal-reducing ability may act as nannofossils with significant paleontological applications.

Production of OMVs by heterotrophs is a well-known fact; however, production of OMVs by a phototrophic organism in natural ecosystem was first reported by a recent study (Biller et al. 2014). Marine cyanobacteria *Prochlorococcus* show continuous production of bilayered OMVs containing various biomolecules such as lipids, proteins, and DNA. Proteome analysis of these vesicles revealed the presence of many proteins such as nutrient transporters, enzymes such as proteases and hydrolases, porins, and many other unknown proteins. Sequencing of the DNA from these vesicles revealed that the DNA mapped to terminal region of the genome. One intriguing aspect of vesicle production in marine ecosystem is that these vesicles may represent a source of dissolved organic carbon (DOC) which in turn can support the growth of other heterotrophs. Looking at the abundance of vesicles produced in the ocean (10^5–10^6 per ml), the global vesicle production by *Prochlorococcus* species could account for nearly 10^4–10^5 tonnes of DOC per day (Biller et al. 2014). This can in turn play a significant role in carbon recycling in the ocean. We can say that *Prochlorococcus* spp. may be able to move fixed carbon down the food web of the ocean. Since *Prochlorococcus* is abundant in the ocean, production of OMVs by these organisms may have a significant impact on the environment.

16.5.6 Membrane Vesicles as Vaccines

Infectious diseases remain a major cause of death among the world population. According to World Health Organization's (WHO) report in 2012, these diseases kill almost nine million people annually and may also cause lifelong disability. Vaccination is considered to be a most effective measure to reduce the mortality rate associated with infectious diseases. A variety of vaccines have been generated so far to combat infectious pathogens. However, these vaccines have several drawbacks associated with them. For example, the use of attenuated live bacteria is not always safe as they may mutate and revert back to virulent form. Protein subunit vaccines can overcome the safety limitations of live attenuated vaccines but are able to induce a strong immune response only in the presence of adjuvant. Additionally, protein subunit vaccines must be formulated in a controlled release system which makes them economically nonviable.

Bacterial OMV's are small, nonreplicating vesicles with immunogenic properties. They can be easily purified and offer many advantages over conventional vaccines in use currently. They are inexpensive and easily obtainable. Because of their vesicular structure, they are potent to work as antigen carrier and vaccine adjuvant. Hence, they can boost the vaccination effect without the aid of adjuvants. Also, OMVs are nonliving despite of carrying all the multiple putative virulence factors and therefore overcome the limitation of vaccine efficacy of single antigen vaccine.

OMVs from a variety of gram-negative bacteria have been extensively studied to work as a potent acellular vaccine. The only licensed vaccine based on OMVs of *N. meningitidis* is against serogroup B utilized in Chile, Cuba, New Zealand, Norway, Australia, and Brazil (Boslego et al. 1995; Fredriksen et al. 1991). However, this strain does not provide protection against heterologous strain. This is due to high variability of PorA protein (major non-capsular protective Ag) across different isolates of *N. meningitidis* (Holst et al. 2009). Hexavalent and non-hexavalent engineered vaccines, which express different PorA subtypes, deal with this variation. OMV stimulates the production of various cytokines and antibodies in the host. OMVs from *N. meningitidis* induce the production of proinflammatory cytokines and chemokines such as TNF-α, IL-1β, IL-6, and IL-8 (Mirlashari et al. 2001). *E. coli-derived* OMVs have been explored to investigate the mechanism of immune response elicited by gram-negative pathogen derived from OMVs. These OMVs strongly depend on stimulation of T-cell-mediated immunity involving Th1 and Th17 cell response rather than B-cell humoral immunity. IL-17 and IFN- ϒ are one of the most essential factors responsible for facilitating bacterial clearance. It is known that immunization with OMVs derived from *E. coli* are able to prevent lethality induced by bacteria and also the systemic inflammatory response syndrome (SIRS) (Kim et al. 2013).

OMV derived from *Brucella abortus* is known to elicit a protective immune response without using adjuvant. Outer membrane protein 16 is the factor responsible for this effect which is found to be independent of protein lipidation. This is the first vaccine against *Brucella* which is able to induce response similar to that of control live vaccine S19. This self-adjuvating OMV from *Brucella* elicits a Th-1-specific response (Pasquevich et al. 2010). OMVs from *V. cholerae* are potential candidates for vaccine against cholera. It has been studied that a long-lasting, high-titer immune response is induced when female mice are immunized with these OMVs. Regardless of the route of immunization, it is observed that offsprings of immunized female mice are protected against colonization with *V. cholerae* challenge. These studies may play an integral role in the development of OMV-based vaccines against *B. abortus* and several other enteric pathogens (Schild et al. 2008).

After extensive studies on OMVs derived from *V. cholerae*, *N. meningitides*, and *H. pylori*, it is evident that OMVs are safe candidates to be used as vaccines against various microbial diseases. Also OMVs derived from *Shigella boydii* type IV strains are found to be safe and less reactogenic. It induces IgA, IgG, and IgM response in an adult mice and 75 % average passive immunity against heterologous *Shigella* strains in neonatal mice. Complete, passive immunity is difficult to achieve due to serotype diversity in *Shigella* strains. Based on these studies, OMVs can prove to be a next-generation efficient vaccine against shigellosis (Mitra et al. 2013).

Gram-positive bacteria due to the absence of outer membrane were overlooked for the secretion of OMVs. However, recent studies reported that certain gram-positive bacteria such as *Staphylococcus aureus* and *Bacillus subtilis* secrete MVs into the extracellular milieu (Lee et al. 2009). Similar to gram-negative bacteria derived from OMVs, these MVs are enriched with several virulence proteins. These findings provide a hint on the role of these vesicles as vaccine candidates (Kim et al. 2013). *S. aureus-derived* vesicles are composed of several virulence factors, including adhesins, toxins, proteolysin, coagulase, and other related enzymes that have pathological effects in the human body (Gurung et al. 2011). Extracellular vesicles from *B. anthracis* have been reported to be immunogenic and show a protective response to vaccine, suggesting a possibility that these vesicular preparations may develop into effective vaccines (Rivera et al. 2010).

OMV vaccines have been widely explored for the development of vaccine against bacterial infections. It would be interesting to explore the idea of the development of an OMV-based TB vaccine from a nonpathogenic bacterial strain carrying potent toxins of tuberculosis. It has been shown previously that CD8+ T-cell response is induced upon infection of macrophages with apoptotic vesicles from dendritic cells (DCs) infected with *M. tuberculosis*. Whether similar response is produced by DCs upon internalization of OMVs from *M. tuberculosis* needs to be checked (Schaible et al. 2003).

16.6 Summary and Conclusion

Membrane vesicles constitute an essential part of microbial existence. The fact that in some ecosystems, the number of released vesicles exceed the total number of microorganisms shows the dependence of microbes on these vesicles for a variety of physiological functions. The metabolic expenditure that an organism has to invest in releasing membrane vesicles is enormous which again states how crucial this process is for microbial life. Hence, it is very essential to understand the pathways involved in vesicle biogenesis, contents of membrane vesicles, and processes that modulate their production and release across different forms of life.

Membrane vesicles from different organisms serve a variety of functions including transport of nutrients, toxins, virulence factors, etc. They are also involved in adhesion of a cell with neighboring cells as well as with the substratum that is involved in biofilm formation. These functions can be explored for a variety of purposes, including selective killing of microbes, or inhibition of biofilm-forming pathogens (Huma et al. 2011; Kalia 2014b; Kalia et al. 2011; Kalia and Kumar 2015a, b). The other functions can be the targeted delivery of the molecules of interest in case of vaccine development as well as to understand the formation of complex microbial communities such as biofilm.

Acknowledgments The authors wish to thank the Director of CSIR-Institute of Genomics and Integrative Biology (IGIB), Government of India for providing the necessary funds, and facilities. Authors are also thankful to Academy of Scientific and Innovative Research (AcSIR), New Delhi. SSK is a Senior Research Fellow supported by UGC, India. NG is a project assistant funded from BSC0123 (DRDO, India). BKT is a research associate supported by CSIR, India. LKS is a Senior Research Fellow supported by University Grant Commission, India. ND is a Shyama Prasad Mukherjee-Senior Research Fellow supported by CSIR, India. We highly acknowledge Neha Dubey and Dr. V. C. Kalia from CSIR-IGIB, Delhi, India, for inspiration and critical comments on the manuscript.

References

Bauman SJ, Kuehn MJ (2006) Purification of outer membrane vesicles from *Pseudomonas aeruginosa* and their activation of an IL-8 response. Microbes Infect 8:2400–2408. doi:10.1016/j.micinf.2006.05.001

Baumgarten T, Sperling S, Seifert J, von Bergen M, Steiniger F, Wick LY, Heipieper HJ (2012) Membrane vesicle formation as a multiple-stress response mechanism enhances *Pseudomonas putida* DOT-T1E cell surface hydrophobicity and biofilm formation. Appl Environ Microbiol 78:6217–6224. doi:10.1128/AEM.01525-12

Biller SJ, Schubotz F, Roggensack SE, Thompson AW, Summons RE, Chisholm SW (2014) Bacterial vesicles in marine ecosystem. Science 343:183–186. doi:10.1126/science.1243457

Bishop DG, Work E (1965) An extracellular glycolipid produced by *Escherichia coli* grown under lysine-limiting conditions. Biochem J 96:567–576, PMC1207076

Boslego J, Garcia J, Cruz C, Zollinger W, Brandt B, Ruiz S, Martinez M, Arthur J, Underwood P, Silva W, Moran E, Hankins W, Gilly J, Mays J (1995) Efficacy, safety, and immunogenicity of a meningococcal group B (15:P1.3) outer membrane protein vaccine in Iquique, Chile. Vaccine 13:821–829. doi:10.1016/0264-410X(94)00037-N

Deatherage BL, Cookson BT (2012) Membrane vesicle release in bacteria, eukaryotes and archaea: a conserved yet underappreciated aspect of microbial life. Infect Immun 80:1948–1957. doi:10.1128/IAI.06014-11

Dorward DW, Garon CF, Judd RC (1989) Export and intracellular transfer of DNA via membrane blebs of *Neisseria gonorrhoeae*. J Bacteriol 171:2499–2505

Ducret A, Fleuchot B, Bergam P, Mignot T (2013) Direct live imaging of cell-cell protein transfer by transient outer membrane fusion in *Myxococcus xanthus*. Elife 2:e00868. doi:10.7554/eLife.00868

Ellen AF, Albers SV, Huibers W, Pitcher A, Hobel CF, Schwarz H, Folea M, Schouten S, Boekema EJ, Poolman B, Driessen AJ (2009) Proteomic analysis of secreted membrane vesicles of archaeal *Sulfolobus* species reveals the presence of endosome sorting complex components. Extremophiles 13:67–79. doi:10.1007/s00792-008-0199-x

Fiocca R, Necchi V, Sommi P, Ricci V, Telford J, Cover TL, Solcia E (1999) Release of *Helicobacter pylori* vacuolating cytotoxin by both a specific secretion pathway and budding of outer membrane vesicles. Uptake of released toxin and vesicles by gastric epithelium. J Pathol 188:220–226. doi:10.1002/(SICI)10969896(199906)188:2<220, AID-PATH307>3.0.CO;2-C

Fredriksen JH, Rosenqvist E, Wedege E, Bryn K, Bjune G, Froholm LO, Lindbak AK, Mogster B, Namork E, Rye U (1991) Production, characterization and control of MenB-vaccine "Folkehelsa": an outer membrane vesicle vaccine against group B meningococcal disease. NIPH Ann 14:67–79

Garcia-del Portillo F, Stein MA, Finlay BB (1997) Release of lipopolysaccharide from intracellular compartments containing *Salmonella typhimurium* to vesicles of the host epithelial cell. Infect Immun 65:24–34, PMID: 8975888

Gorby YA, Yanina S, McLean JS, Rosso KM, Moyles D, Dohnalkova A, Beveridge TJ, Chang IS, Kim BH, Kim KS, Culley DE, Reed SB, Romine MF, Saffarini DA, Hill EA, Shi L, Elias DA, Kennedy DW, Pinchuk G, Watanabe K, Ishii S, Logan B, Nealso KH, Fredrickson JK (2006) Electrically conductive bacterial nanowires produced by *Shewanella oneidensis* strain MR-1 and other microorganisms. Proc Natl Acad Sci U S A 103:11358–11363. doi:10.1073/pnas.0604517103

Gorby Y, McLean J, Korenevsky A, Rosso K, El-Naggar MY, Beveridge TJ (2008) Redox-reactive membrane vesicles produced by *Shewanella*. Geobiology 6:232–241. doi:10.1111/j.1472-4669.2008.00158.x

Gurung M, Moon DC, Choi CW, Lee JH, Bae YC, Kim YC, Lee YC, Seol SY, Cho DT, Kim S, Lee JC (2011) *Staphylococcus aureus* produces membrane-derived vesicles that induce host cell death. PLoS One 6:e27958. doi:10.1371/journal.pone.0027958

Harmsen M, Yang L, Pamp SJ, Tolker-Nielsen T (2010) An update on *Pseudomonas aeruginosa* biofilms formation, tolerance and dispersal. FEMS Immunol Med Microbiol 59:253–268. doi:10.1111/j.1574-695X.2010.00690.x

Hellerud BC, Stenvik J, Espevik T, Lambris JD, Mollnes TE, Brandtzaeg P (2008) Stages of meningococcal sepsis simulated in vitro, with emphasis on complement and toll like receptor activation. Infect Immun 76:4183–4189. doi:10.1128/IAI.00195-08

Hoekstra D, Van der Laan JW, De Leij L, Witholt B (1976) Release of outer membrane fragments from normally growing *Escherichia coli*. Biochim Biophys Acta 455:889–899. doi:10.1016/0005-2736(76)90058-4

Holst J, Martin D, Arnold R, HuegroC C, Oster P, O'Hallahan J, Rosengyiste E (2009) Properties and clinical performance of vaccines containing outer membrane vesicles from *Neisseria meningitides*. Vaccine 27:B3–B12. doi:10.1016/j.vaccine.2009.04.07

Huang SH, Wu CH, Chang YC, Kwin-Chung KJ, Brown RJ, Jong A (2012) *Cryptococcus neoformans* derived microvesicles enhance the pathogenesis of fungal brain infection. PLoS One 7:e48570. doi:10.1371/journal.pone.0048570

Huma N, Shankar P, Kushwah J, Bhushan A, Joshi J, Mukherjee T, Raju SC, Purohit HJ, Kalia VC (2011) Diversity and polymorphism in AHL-lactonase gene (*aiiA*) of *Bacillus*. J Microbiol Biotechnol 21:1001–1011. doi:10.4014/jmb.1105.05056

Ismail S, Hampton MB, Keenan JI (2003) *Helicobacter pylori* outer membrane vesicles modulate proliferation and interleukin-8 production by gastric epithelial cells. Infect Immun 71:5670–5675. doi:10.1128/IAI.71.10.5670-5675.2003

Kadurugamuwa JL, Beveridge TJ (1995) Virulence factors are released from *Pseudomonas aeruginosa* in association with membrane vesicles during normal growth and exposure to gentamicin: a novel mechanism of enzyme secretion. J Bacteriol 177:3998–4008, PMC177130

Kadurugamuwa JL, Beveridge TJ (1996) Bacteriolytic effect of membrane vesicles from *Pseudomonas aeruginosa* on other bacteria including pathogens: conceptually new antibiotics. J Bacteriol 178:2764–2774, PMC178010

Kadurugamuwa JL, Beveridge TJ (1999) Membrane vesicles derived from *Pseudomonas aeruginosa* and *Shigella flexneri* can be integrated into the surfaces of other gram-negative bacteria. Microbiology 145:2051–2060. doi:10.1099/13500872-145-8-2051

Kahn ME, Maul G, Goodgal SH (1982) Possible mechanism for donor DNA binding and transport in *Haemophilus*. Proc Natl Acad Sci U S A 79:6370–6374, PMC347123

Kahn ME, Barany F, Smith HO (1983) Transformasomes: specialized membranous structures that protect DNA during *Haemophilus* transformation. Proc Natl Acad Sci U S A 80:6927–6931, PMID:6316334

Kalia VC (2013) Quorum sensing inhibitors: an overview. Biotechnol Adv 31:224–245. doi:10.1016/j.biotechadv.2012.10.004

Kalia VC (2014a) Microbes, antimicrobials and resistance: the battle goes on. Indian J Microbiol 54:1–2. doi:10.1007/s12088-013-0443-7

Kalia VC (2014b) In search of versatile organisms for quorum-sensing inhibitors: acyl homoserine lactones (AHL)-acylase and AHL-lactonase. FEMS Microbiol Lett 359:143. doi:10.1111/1574-6968.12585

Kalia VC (2015) Microbes: the most friendly beings? In: Kalia VC (ed) Quorum sensing vs quorum quenching: a battle with no end in sight. Springer, pp 1–5. doi:10.1007/978-81-322-1982-8_1

Kalia VC, Kumar P (2015a) Potential applications of quorum sensing inhibitors in diverse fields. In: Kalia VC (ed) Quorum sensing vs quorum quenching: a battle with no end in sight. Springer, pp 359–370. doi:10.1007/978-81-322-1982-8_29

Kalia VC, Kumar P (2015b) The battle: quorum-sensing inhibitors versus evolution of bacterial resistance. In: Kalia VC (ed) Quorum sensing vs quorum quenching: a battle with no end in sight. Springer, New Delhi, pp 385–391. doi:10.1007/978-81-322-1982-8_31

Kalia VC, Purohit HJ (2011) Quenching the quorum sensing system: potential antibacterial drug targets. Crit Rev Microbiol 37:121–140. doi:10.3109/1040841X.2010.532479

Kalia VC, Raju SC, Purohit HJ (2011) Genomic analysis reveals versatile organisms for quorum quenching enzymes: acyl-homoserine lactone-acylase and -lactonase. Open Microbiol J 5:1–13. doi:10.2174/1874285801105010000

Kalia VC, Wood TK, Kumar P (2014) Evolution of resistance to quorum-sensing inhibitors. Microb Ecol 68:13–23. doi:10.1007/s00248-013-0316-y

Kalia VC, Kumar P, Pandian SK, Sharma P (2015) Biofouling control by quorum quenching. In: Kim SK (ed) Handbook of marine biotechnology. Springer, pp 431–440. doi:10.1007/978-3-642-53971-8_15

Kesty NC, Mason KM, Reedy M, Miller SE, Kuehn MJ (2004) Enterotoxigenic *Escherichia coli* vesicles target toxin delivery into mammalian cells. EMBO J 23:4538–4549. doi:10.1038/sj.emboj.7600471

Kim OY, Hong BS, Park KS, Yoon YJ, Choi SJ, Lee WH, Roh TY, Lötvall J, Kim YK, Gho YS (2013) Immunization with *Escherichia coli* outer membrane vesicles protects bacteria-induced lethality via Th1 and Th17 cell responses. J Immunol 190:4092–4102. doi:10.4049/jimmunol.1200742

Knox KW, Vesk M, Work E (1966) Relation between excreted lipopolysaccharide complexes and surface structures of a lysine- limited culture of *Escherichia coli*. J Bacteriol 92:1206–1217, PMC276396

Koul A, Herget T, Klebl B, Ulrich A (2004) Interplay between mycobacteria and host signalling pathways. Nat Rev Microbiol 2:189–202, PMID: 15083155

Kuehn MJ, Kesty NC (2005) Bacterial outer membrane vesicles and host-pathogen interaction. Genes Dev 19:2645–2655. doi:10.1101/gad.1299905

Kulkarni HM, Jagannadham MV (2014) Biogenesis and multifaceted roles of outer membrane vesicles from gram negative bacteria. Microbiol 160:2109–2121. doi:10.1099/mic.0.079400-0

Kusebauch U, Ortega C, Ollodart A, Rogers RS, Sherman DR, Moritz RL, Grundner C (2014) *Mycobacterium tuberculosis* supports protein tyrosine phosphorylation. Proc Natl Acad Sci U S A 111:9265–9270. doi:10.1073/pnas.1323894111

Lee EY, Bang JY, Park GW, Choi DS, Kang JS, Kim HJ, Park KS, Lee JO, Kim YK, Kwon KH, Kim KP, Gho YS (2007) Global proteomic profiling of native outer membrane vesicles derived from *Escherichia coli*. Proteomics 7:3143–3153, PMID 1778703

Lee EY, Choi DY, Kim DK, Kim JW, Park JO, Kim S, Kim SH, Desiderio DM, Kim YK, Kim KP, Gho YS (2009) Gram-positive bacteria produce membrane vesicles: proteomics-based characterization of *Staphylococcus aureus*-derived membrane vesicles. Proteomics 9:5425–5436. doi:10.1002/pmic.200900338

Li Z, Clarke AJ, Beveridge TJ (1998) Gram-negative bacteria produce membrane vesicles which are capable of killing other bacteria. J Bacteriol 180:5478–5483, PMC107602

Loeb MR (1974) Bacteriophage T4-mediated release of envelope components from *Escherichia coli*. J Virol 13:631—641, PMCID: PMC355348

Ma L, Conover M, Lu H, Parsek MR, Bayles K, Wozniak DJ (2009) Assembly and development of the *Pseudomonas aeruginosa* biofilm matrix. PLoS Pathog 5:e1000354. doi:10.1371/journal.ppat.1000354

MacDonald IA, Kuehn MJ (2013) Stress-induced outer membrane vesicle production by *Pseudomonas aeruginosa*. J Bacteriol 195:2971–2981. doi:10.1128/JB.02267-12

Mashburn LM, Whiteley M (2005) Membrane vesicles traffic signals and facilitate group activities in a prokaryote. Nature 437:422–425. doi:10.1038/nature03925

Mathivanan S, Ji H, Simpson RJ (2010) Exosomes: extracellular organelles important in intercellular communication. J Proteomics 73:1907–1920. doi:10.1016/j.jprot.2010.06.006

Menard R, Prevost MC, Gounon P, Sansonetti P, Dehio C (1996) The secreted Ipa complex of *Shigella flexneri* promotes entry into mammalian cells. Proc Natl Acad Sci U S A 93:1254–1258

Mirlashari MR, Hoiby EA, Holst J, Lyberg T (2001) Outer membrane vesicles from *Neisseria meningitidis*: effects on cytokine production in human whole blood. Cytokine 13:91–97, PMID: 11145848

Mitra S, Chakrabarti MK, Koley H (2013) Serotype outer membrane vesicles of *Shigellae* confer passive protection to the neonatal mice against Shigellosis. Vaccine 31:3163–3173. doi:10.1016/j.vaccine.2013.05.001

Mug-Opstelten D, Witholt B (1978) Preferential release of new outer membrane fragments by exponentially growing *Escherichia coli*. Biochim Biophys Acta 508:287–295. doi:10.1016/0005-2736(78)90331-0

Municio AM, Diaz T, Martinez A (1963) The presence of a mucopeptide in the media of an *E. coli* mutant and its relation to the cell wall. Biochem Biophys Res Commun 11:195–200. doi:10.1016/0006-291X(63)90333-4

Nudleman E, Wall D, Kaiser D (2005) Cell-to-cell transfer of bacterial outer membrane lipoproteins. Science 309:125–127. doi:10.1126/science.1112440

Oliveira DL, Nakayasu ES, Joffe LS, Guimarães AJ, Sobreira TJ, Nosanchuk JD, Cordero RJ, Frases S, Casadevall A, Almeida IC, Nimrichter L, Rodrigues ML (2010) Characterization of yeast extracellular vesicles: evidence for the participation of different pathways of cellular traffic in vesicle biogenesis. PLoS One 5:e11113. doi:10.1371/journal.pone.0011113

Palsdottir H, Remis JP, Schaudinn C, O'Toole E, Lux R, Shi W, McDonald KL, Costerton JW, Auer M (2009) Three-dimensional macromolecular organization of cryofixed *Myxococcus xanthus* biofilms as revealed by electron microscopic tomography. J Bacteriol 191:2077–2082. doi:10.1128/JB.01333-08

Parker H, Chitcholtan K, Hampton MB, Keenan JI (2010) Uptake of *Helicobacter pylori* outer membrane vesicles by gastric epithelial cells. Infect Immun 78:5054–5061. doi:10.1128/IAI.00299-10

Pasquevich KA, García Samartino C, Coria LM, Estein SM, Zwerdling A, Ibañez AE, Barrionuevo P, Oliveira FS, Carvalho NB, Borkowski J, Oliveira SC, Warzecha H, Giambartolomei GH, Cassataro J (2010) The protein moiety of *Brucella abortus* outer membrane protein 16 is a new bacterial pathogen-associated molecular pattern that activates dendritic cells *in vivo*, induces a Th1 immune response, and is a promising self-adjuvanting vaccine against systemic and oral acquired brucellosis. J Immunol 184:5200–5212. doi:10.4049/jimmunol.0902209

Pirbadian S, Barchinger SE, Leung KM, Byun HS, Jangir Y, Bouhenni RA, Reed SB, Romine MF, Saffarini DA, Shi L, Gorby YA, Golbeck JH, El-Naggar MY (2014) *Shewanella oneidensis* MR-1 nanowires are outer membrane and periplasmic extensions of the extracellular electron transport components. Proc Natl Acad Sci U S A 111:12883–12888. doi:10.1073/pnas.1410551111

Rath P, Huang C, Wang T, Wang T, Li H, Prados-Rosales R, Elemento O, Casadevall A, Nathan CF (2013) Genetic regulation of vesiculogenesis and immunomodulation in *Mycobacterium tuberculosis*. Proc Natl Acad Sci U S A 110:E4790–E4797. doi: 10.1073/pnas.1320118110

Renelli M, Martias V, Lo RY, Beveridge TJ (2004) DNA-containing membrane vesicles of *Pseudomonas aeruginosa* PAO1 and their genetic transformation potential. Microbiology 150:2161–2169. doi:10.1099/mic.0.26841-0

Rivera J, Cordero RJ, Nakouzi AS, Frases S, Nicola A, Casadevall A (2010) *Bacillus anthracis* produces membrane-derived vesicles containing biologically active toxins. Proc Natl Acad Sci U S A 107:19002–19007. doi:10.1073/pnas.1008843107

Rodrigues ML, Nakayasu ES, Oliveira DL, Nimrichter L, Nosanchuk JD, Almeida IC, Casadevall A (2008) Extracellular vesicles produced by *Cryptococcus neoformans* contain protein components associated with virulence. Eukaryot Cell 7:58–67

Sajid A, Arora G, Gupta M, Upadhyay S, Nandicoori VK, Singh Y (2011) Phosphorylation of *Mycobacterium tuberculosis* Ser/Thr phosphatase by PknA and PknB. PLoS One 6:e17871. doi:10.1371/journal.pone.0017871

Schaible UE, Winau F, Sieling PA, Fischer K, Collins HL, Hagens K, Modlin RL, Brinkmann V, Kaufmann SH (2003) Apoptosis facilitates antigen presentation to T lymphocytes through MHC-1 and CD1 in tuberculosis. Nat Med 9:1039–1046. doi:10.1038/nm906

Schild S, Nelson EJ, Camilli A (2008) Immunization with *Vibrio cholerae* outer membrane vesicles induces protective immunity in mice. Infect Immun 77:472–484. doi:10.1128/IAI.01139-08

Schmidt O, Teis D (2012) The ESCRT machinery. Curr Biol 22:R116–R120. doi:10.1016/j.cub.2012.01.028

Schwechheimer C, Kuehn MJ (2013) Synthetic effect between envelope stress and lack of outer membrane vesicle production in *Escherichia coli*. J Bacteriol 195:4161–4173. doi:10.1128/JB.02192-12

Schwechheimer C, Kulp A, Kuehn MJ (2014) Modulation of bacterial outer membrane vesicle production by envelope structure and content. BMC Microbiol 14:324. doi:10.1186/s12866-014-0324-1

Silverman JM, Chan SK, Robinson DP, Dwyer DM, Nandan D, Foster LJ, Reiner NE (2008) Proteomic analysis of the secretome of *Leishmania donovani*. Genome Biol 9:R35. doi:10.1186/gb-2008-9-2-r35

Silverman JM, Clos J, de'Oliveira CC, Shirvani O, Fang Y, Wang C, Foster LJ, Reiner NE (2010) An exosome-based secretion pathway is responsible for protein export from *Leishmania* and communication with macrophages. J Cell Sci 123:842–852. doi:10.1242/jcs.056465

Thay B, Damm A, Kufer TA, Wai SN, Oscarsson J (2014) *Aggregatibacter actinomycetemcomitans* outer membrane vesicles are internalized in human host cells and trigger NOD1 and NOD2-dependent NF-κB activation. Infect Immun 82:4034–4046. doi:10.1128/IAI.01980-14

Warren LM, Howe J, Garidel P, Richter W, Steiniger F, Roessle M, Brandenburg K, Whiteley M (2008) Interaction of quorum signals with outer membrane lipids: insights into prokaryotic membrane vesicle formation. Mol Microbiol 69:491–502. doi:10.1111/j.1365-2958.2008.06302.x

Wei X, Pathak DT, Wall D (2011) Heterologous protein transfer within structured myxobacteria biofilms. Mol Microbiol 81:315–326. doi:10.1111/j.1365-2958.2011.07710

Yonezawa H, Osaki T, Kurata S, Fukuda M, Kawakami H, Ochiai K, Hanawa T, Kamiya S (2009) Outer membrane vesicles of *Helicobacter pylori* TK1402 are involved in biofilm formation. BMC Microbiol 9:197. doi:10.1186/1471-2180-9-197

Shashank S. Kamble received his M.Sc. degree from University of Mumbai, India. Currently, he is pursuing Ph.D. under the supervision of Dr. Yogendra Singh at CSIR-IGIB, India. He is interested in the host-pathogen cross talk during infection with *M. tuberculosis*. His work focuses on the modulation of host proteases by *M. tuberculosis*.

Nancy Garg obtained her M.Tech. degree from Amity Institute of Biotechnology, Noida, in 2014. Currently, she is working as project assistant at CSIR-IGIB, Delhi, India. Her major interests include microbial communication, biofilm formation, as well as host-pathogen interaction during infection with *M. tuberculosis*.

Brijendra Kumar Tiwari received his M.Tech. degree in Biotechnology from M.P. Technological University, Bhopal, India. He is currently working as a Research Associate, at CSIR-IGIB, Delhi, India. His current interests are microbiology, immunology, and cell signaling. His current research focuses on "Role of microRNA during *Mycobacterium tuberculosis* infection in dendritic cells." He is a member of American Society for Microbiology (ASM) and Indian Immunology Society (IIS).

Lalit K. Singh received the M.Sc. degree in Biotechnology from Rajiv Gandhi Biotechnology Centre, RTM Nagpur University, Nagpur, India. He is currently working as Senior Research Fellow, CSIR-IGIB, Delhi, India. His current research focuses on understanding the role of caseinolytic proteases and Ser/Thr kinases that mediated cellular architecture alterations, cell wall remodeling, and sporulation in *Bacillus*.

Neha Dhasmana graduated in Microbiology in 2008 from University of Delhi. She received her M.Sc. degree in Biochemistry in 2010 from Hamdard University. She is currently enrolled in Ph.D. with Academy of Innovative and Scientific Research (AcSIR) and works at CSIR-Institute of Genomics and Innovative Biology. She is studying the physiological significance of Ser/Thr kinases in *Bacillus anthracis*.

Yogendra Singh received his Ph.D. degree from Vallabhbhai Patel Chest Institute, Delhi University, and postdoctoral training at National Institute of Health (NIH), Bethesda. Currently, he is Chief Scientist at CSIR-IGIB and Professor of AcSIR. He is a Fellow of National Academy of Sciences, India, Fellow of Indian National Science Academy, and Fellow of Indian Academy of Sciences, India. He has been honored with Moselio Schaechter Distinguished Service Award in 2014 by American Society for Microbiology. He serves as an Editorial Board Member of "Journal of Biological Chemistry." His current research interest is to understand the host response in pulmonary/extra-pulmonary tuberculosis and to decipher the role of eukaryotic like Ser/Thr kinases in *Mycobacterium tuberculosis* and *Bacillus anthracis*.

Bacteriophage Diversity in Different Habitats and Their Role in Pathogen Control

17

Nishant A. Dafale, Zubeen J. Hathi, Sarmistha Bit, and Hemant J. Purohit

Abstract

Bacteriophages are selective to their host bacteria and do not have any direct interaction with other members of the indigenous flora of the community. In any niche, the diversity of phage is directly proportional to the associated discriminating bacterial population as a host. Pathogens including MDR bacteria in wastewater are responsible for big outbreaks in many developing nations and the biggest emerging threat to public health. Every effort leading to reduction of pathogens in the environment has to be promoted and implemented for establishing a healthy and sustainable environment. The inefficient functioning of STPs is the significant reservoir of diverse bacterial population, including MDR. A biocontrol capability through phage has been demonstrated in the control of pathogens in food industry, medicine, aquaculture and agriculture and could be implemented to environment. In order to apply phage formulation in the environment, it must be a wild type and should be well characterised; that includes genome annotation, validation of genes of obligate lytic cycle and without the virulence factors. The implementation will demand long-term trials with microbial community. This can be ensured through advances in metagenomics, which will monitor shifts in the community structure after phage interventions. Phages would emerge as eco-friendly biocontrol agents and could be considered as an obvious alternative to chemical disinfectant and preservatives.

17.1 Introduction

Bacteriophages are the bacterial viruses that infest the bacterial cell. Edward Twort (1915) and Felix d'Herelle (1917), nearly a century ago, independently reported isolation of filterable entities competent for killing bacterial population and develop small clear plaques on host lawn. It was Felix d'Herelle, a Canadian, who coined the term *bacteriophage*. Bacteriophages also known

N.A. Dafale (✉) • Z.J. Hathi • S. Bit
Environmental Genomics Division, CSIR-National Environmental Engineering Research Institute (NEERI), Nagpur 440020, Maharashtra, India
e-mail: na_dafale@neeri.res.in; nishant.dafale@gmail.com; hathizubeen@gmail.com

H.J. Purohit
Environmental Genomics Division, CSIR-National Environmental Engineering Research Institute (NEERI), Nehru Marg, Nagpur 440015, India
e-mail: hj_purohit@neeri.res.in; hemantdrd@hotmail.com

V.C. Kalia (ed.), *Microbial Factories*, DOI 10.1007/978-81-322-2595-9_17

as phage are the most abundant and biologically diverse entities with an estimated 10^{31} particles present on Earth (Allen et al. 2013). Bacteriophage is an entity that contains either DNA or RNA surrounded by protein coat (Verheust et al. 2010). Bacteriophages are classified on the basis of morphology and the type of nucleic acid. Till now, nineteen families have been identified which are divided on the basis of their mode of replication in two groups: lytic phage and lysogenic (temperate) phage. Lytic phage progresses instantly after infecting the host and progeny is released by lysis of the host cell, whereas, in case of temperate phages, the genetic material is incorporated into host chromosome depending upon the nature and physicochemical conditions (Verheust et al. 2010).

Phages are host specific as they recognise particular receptor sites present on the host cell and penetrate the genetic material which ultimately leads to the infection (Goyal 1987). Bacteriophage infection consists of the following consecutive stages: adsorption on the host bacteria, penetration of phage nucleic acids, intracellular synthesis of phage components and assembly of virions and release of progeny with lysis of the host cell (Rakhuba et al. 2010). They secrete lytic enzymes which target the integrity of bacterial cell wall and invade one of the four key bonds in peptidoglycan of bacteria. During development process in the host, lytic enzymes accumulate in the cytoplasm and at genetically precise time, holin molecules are infused into cytoplasm membrane which leads to membrane disruption and release of phage progeny (Fischetti 2005).

Phages are natural microflora as they are commonly isolated from different habitats and are remarkably stable in all environments (Kutter and Sulakvelidze 2005; Pirnay et al. 2012; Allen et al. 2013). The lytic bacteriophage disrupts bacterial metabolism after infection and promotes lyses leading to its application as therapy/biocontrol in various fields to eliminate pathogens (Fu et al. 2010). Phages have been detected from a number of food samples and so they are ingested by everyone, every day (Hudson et al. 2005). Hence, phages are considered as a natural alternative to chemical disinfectant and preservatives (McIntyre et al. 2012).

17.2 Classification of Bacteriophage

The very first classification of bacteriophage was done by Australian microbiologist Sir Macfarlane Burnet in 1937 on the basis of phage size and resistance against physiochemical agents. In 1962, Lwoff, Tournier and Home suggested that viruses should be classified on the basis of properties of virion and nucleic acid. In 1965, a provisional committee on nomenclature of viruses was formed which was later constituted as International Committee on Taxonomy of Viruses (ICTV) in 1971. It is the only international body which deals with virus taxonomy. ICTV classification of bacteriophage with example and their author study is mentioned in Table 17.1 (Mc Grath and van Sinderen 2007).

The classification of phages is based on nucleic acid content and morphology which includes 6 orders, 87 families, 19 subfamilies and 348 genera. Among which, two families contain RNA genomes and two have ssDNA genomes. The phages having tail comprise the order Caudovirales with families named as Myoviridae, Podoviridae and Siphoviridae. These families are characterised by contractile or non-contractile and long or short tails represent 96 % of phages. The tailed phages show many facultative features like unusual bases and tail appendages. Ligamenvirales are linear viruses infecting archaea and possess double-stranded DNA genomes having 15.9–56 kb, it consists of two families: Lipothrixviridae and Rudiviridae. Phages belonging to this family are rod-shaped and enveloped, and capsid size is around 24–38 nm in diameter. Rudiviridae members are rigid, non-enveloped, and rod-shaped having three-tail fibres at the end and mostly infect thermophilic archaea *Sulfolobus islandicus*.

Table 17.1 ICTV classification of bacteriophage

Order	Family	Morphology	Nucleic acid	Examples	References
Caudovirales	*Myoviridae*	Non-enveloped, contractile tail	Linear dsDNA	T4 phage	Miller et al. (2003)
	Siphoviridae	Non-enveloped, non-contractile tail (long)	Linear dsDNA	Λ phage, T5 phage	Wang et al. (2005)
	Podoviridae	Non-enveloped, non-contractile tail (short)	Linear dsDNA	T7 phage, T3 phage, P22, P37	Bleackley et al. (2009)
Ligamenvirales	*Lipothrixviridae*	Enveloped, rod-shaped	Linear dsDNA	*Acidianus filamentous* virus 1 (AFV 1)	Bettstetter et al. (2003)
	Rudiviridae	Non-enveloped, rod-shaped	Linear dsDNA	*Sulfolobus islandicus* rod-shaped virus 1 (SIFV 1)	Arnold et al. (2000)
Unassigned	*Ampullaviridae*	Enveloped, bottle-shaped	Linear dsDNA	ABV	Peng et al. (2007)
	Bicaudaviridae	Non-enveloped, lemon-shaped	Circular dsDNA	STSV-1, ATV	Scheele et al. (2011)
	Clavaviridae	Non-enveloped, rod-shaped	Circular dsDNA	APBV-1	Mochizuki et al. (2010)
	Corticoviridae	Non-enveloped, isometric	Circular dsDNA	PM2	Mannisto et al. (1999)
	Cystoviridae	Enveloped, spherical	Segmented dsRNA	*Pseudomonas* phage Φ6 (phi 6)	Ford et al. (2014)
	Fuselloviridae	Non-enveloped, lemon-shaped	Circular dsDNA	SSV-1	Fusco et al. (2014)
	Globuloviridae	Enveloped, isometric	Linear dsDNA	*Pyrobaculum* spherical virus (PSV), TSV1	Schoenfeld et al. (2008)
	Guttaviridae	Non-enveloped, ovoid	Circular dsDNA	SNDV1, SNDV2	Arnold et al. (2000)
	Inoviridae	Non-enveloped, filamentous	Circular ssDNA	Enterobacteria phage f1	Ackermann. (2012)
	Leviviridae	Non-enveloped, isometric	Linear ssRNA	MS2, Qβ	O'Connell et al. (2006)
	Microviridae	Non-enveloped, isometric	Circular ssDNA	ΦX174	Jaschke et al. (2012)
	Plasmaviridae	Enveloped, pleomorphic	Circular dsDNA	Acholeplasma phage L2, (MV-L 1)	Elbreki et al. (2014)
	Tectiviridae	Non-enveloped, isometric	Linear dsDNA	*Bacillus anthracis* phage AP50	Sozhamannan et al. (2008)

17.3 Prevalence of Bacteriophage in Different Ecological Niches

The several diverse types of bacteriophages are present in various ecological niches. However, the abundance and diversity mainly depend on bacterial load. Till recent date, phages have been isolated from various habitats, but most of the studies have been carried out in aquatic environments (Ashelford et al. 1999).

17.3.1 Hot and Cold Deserts

17.3.1.1 Sahara Desert

The Sahara is the largest desert on Earth. Previously, desert has been also studied for the isolation and detection of bacteriophages. In desert, archaea dominate the bacterial population which provides more chance for its phage isolation. The metagenomic approach was also used for the analysis of composition, taxonomy and functional diversity of phage communities from four gueltas *Ilji, Molomhar, Hamdoun and El Berbera* in Sahara (Fancello et al. 2013). On the basis of metagenomic studies, it was found that Caudovirales largely dominate in the gueltas. Metagenome analysis of *Molomhar* desert represented the largest number of reads related to Myoviruses. Phages like Prochlorococcus and Synechococcus infecting photosynthetic bacteria were most abundant in *Ilji*, *El Berbera*, and *Molomhar*.

17.3.1.2 Antarctica

Antarctica has remained geographically cutoff for millions of years nurturing the unexplored ecosystem on Earth. Microorganisms adapted to extreme environmental conditions such as low temperature, months of nearly total darkness during winter and low nutrient levels that are mostly found in this region. Lake Limnopolar (Byers Peninsula) is home to diverse phages distributed across 12 phage families (López-Bueno et al. 2009). During spring, ssDNA phages were found to be present in majority among the phage community. Although some of the phages were connected to Microviruses phylogenetically linked to genus Chlamydiamicrovirus, the phage community contains ssDNA connected to phage families Circovirus, Nanovirus, Geminivirus and Satellites which have not been earlier reported for aquatic environment. Further, the change in phage assemblage before and after ice melt was monitored using five different search approaches which probably indicates shift in host community (López-Bueno et al. 2009). After the ice cap melted in summer, the viral community was dominated by dsDNA viruses belonging mainly to Phycodnaviridae, Caudoviridae and Mimiviridae families. The limited metabolic activity under ice cap promotes the lysogeny in large Antarctic phages (López-Bueno et al. 2009).

The presence of sea ice bacteria and their phages in the Baltic Sea of south-west Finland was also investigated. The phages were screened against 43 purified bacterial cultures isolated from melted sea ice samples. Phages were found against 15 of 43 bacterial cultures at 3 °C of which 8 were selected for morphological and structural protein analysis. Six phages infecting *Shewanella* sp. were Myoviruses and one of the phages was identified as Siphovirus (Luhtanen et al. 2014).

17.3.2 Marine Waters and Hot Spring

Hot Springs Several researchers have explored the hot spring niche for study of phage diversity. The study was conducted at different temperatures ranging from 61 to 94 °C, with pH varying from 2.0 to 6.0 which includes 38 acidic hot springs and mud holes in a solfataric field (Zillig et al. 1993). On the basis of electron microscopic studies, it was indicated that the phage STSV1 is present in such extreme environment (Xiang et al. 2005). Similarly, many authors have used this niche for the isolation and detection of phages.

Marine Water The marine water contains good number of phages and till date, over 5000 bacteriophages have been studied through electron microscopy and can be attributed to 11 viruses' families (Ackermann and Krisch 1997). Among them, about 150 have morphology identical to phage T4 (Tetart et al. 2001). Further, the metagenomic and electron microscopic analysis of marine phage communities of North Sea and North Atlantic indicates the dominance of tailed phages mainly of Myoviridae family in marine environment. The marine bacterial community is accountable for a major portion of primary production and regeneration of nutrients in the marine environment. The genotypic and phenotypic diversity of phage population depend on the host and phage interaction (Wichels et al. 1998). Marine phages play a vital role in maintaining marine bacterial community.

Several bacteriophages have been isolated from marine ecosystem against *Aeromonas* sp. and *Vibrio* sp. previously (Baross et al. 1978). Marine phages isolated are found to be morphologically similar but genetically diverse. Most of the phages possess linear double-stranded DNA (dsDNA) and are divided on the basis of tail morphology into Siphoviridae, Podoviridae and Myoviridae. The Myoviridae has a strong retractable tail and have broad host specificity mostly found in seawater. Podoviridae has non-retractable and short tail with narrow host range. Siphoviridae has usually long tails and relatively weak lytic activity and can be easily isolated from marine environments (Zhang et al. 2011).

17.3.3 Water and Wastewater

Bacteriophage dominates water and wastewater microbial community due to the presence of heavy bacterial load and favourable conditions. Diversity of STPs changes with type of bacterial load and diversity. Earlier, bacteriophages have been isolated from various sources like seawater, fresh water, soil and sewage water system (Das et al. 2009). The detection of bacteriophages was carried out from well water of Nagpur city. In this work, a coal bed concentration method was employed for coliphage detection against *E. coli* strain EC-R8. Several phages were detected in river and well water with this method (Dafale et al. 2007). Previously, many studies have been published for isolation and detection of bacteriophages from water. In few studies, coliphages were used as an indicator for water pollution. The coliphage is easy to detect and resistant to chemical treatment as compared to coliform. The detection of coliphage in water indicates the faecal contamination.

Bacteriophages were isolated against enteric pathogens from sewage water from Jinke Park, Bangalore. Five bacterial cultures *Klebsiella* sp., *Pseudomonas aeruginosa*, *E. coli*, *Shigella* sp. and *Salmonella typhi* were isolated from sampled water (Das et al. 2009). Among the five isolates, *P. aeruginosa*, *E. coli* and *S. typhi* were found to be sensitive to the bacteriophages. According to the study carried out by the researchers, it was found that phages against human pathogens are widespread in the sewage and can be easily isolated from the wastewater. Further, phage PPST1 was detected in hospital wastewater treatment tank located at Thailand against drug-resistant 23 strains of *S. typhi* SSH1. Three strains of *S. typhi* (SSH1, ATCC 19430 and ATCC 19214) were found to be susceptible to the phage. According to the present research studies, it was found that lytic phage PPST1 has broad host specificity which is stable at high temperature and wide range of pH. All these features make phage PPST1 a potential therapeutic agent for treating infections caused by *S. typhi* (Rattanachaikunsopon and Phumkhachorn 2012). Similarly, 30 phages were isolated against *E. coli* from wastewater in Sweden of which 6 phages were selected for further studies (Mirzae et al. 2014). These phages infest gram-negative and gram-positive 'Lactococci' belonged to Podoviridae family.

17.3.4 Human Gut

Gut ecosystem is a very tempting environment for bacteriophages. The high diversity of gut microflora increases the risk of contamination to natural resources. The stool samples are the major source of diverse phage and pathogens in STPs. The metagenomic- and culture-dependent techniques are available for their detection. Culture-dependent techniques permit researchers to isolate intact replicating phage particles in values up to 10^4 PFU/g of dry faeces (Gantzer et al. 2002). These techniques are applicable in isolating phages infecting *Salmonella*, *Bacteroides*, *Escherichia* and *Enterococcus* species from faeces, sewage and wastewater samples (Gantzer et al. 2002). The work was carried out in eastern France in order to determine concentration of *Bacteroides fragilis* phages and somatic coliphages in human faeces. These phages serve as an indicator to estimate the viral contamination of water. Results indicate the presence of somatic coliphages in 68 % of the faecal samples at a mean concentration of 4.3×10^3 PFU/g and *Bacteroides fragilis* phages of 7×10^1 PFU/g in 11 % of the stools (Gantzer et al. 2002). Study reflects that there is no correlation between concentration of phages, age and sex of human subjects.

A virulent phage was isolated from sewage that infects gram-positive *Lactobacillus paracasei* NFBC338 localising in human intestine (Alemayehu et al. 2009). The work indicates that other similar strains of phages might be possibly present in the human distal intestine. Genome sequence of phage Lb338-1 indicated that it has 142 kb of genome and 199 ORF (open reading frames). Among them, 22 % had considerable homology to genes associated with gram-positive phages. Interestingly, Lb338-1 had little or almost no genomic similarity with *Lactobacillus paracasei* NFBC338 host indicating it has not evolved from the host. Similarly, bacterial and phage community in four pairs of healthy female twins and their mothers was analysed at three different time points over a period of one year. The results indicate that there exists a significant degree of similarity among composition of bacterial communities between the twin pair and their mother. However, significant differences were observed in the phage community of an individual irrespective of their relationship. These studies suggest that the phage population is diverse in humans and predator–prey coadaptation rarely occurs in the human gut, as phages are very specific only targeting their host bacteria (Reyes et al. 2010).

17.4 Mechanism of Phage Infection

Phage genome entry into bacteria requires a key step of adsorption on the bacterial cell envelope. The primary step of infectivity is typically coupled with depolarisation of the host cytoplasm membrane and release of intracellular K^+ ions (Poranen et al. 2001; Jakutyte et al. 2012). The period and model of ion leakage differ among phages and yield a fingerprint of infection for each phage species (Jakutyte et al. 2012). Phage infection of host bacteria involves phage adsorption on host cell and penetration of phage nucleic acid into host, synthesis of phage components and assembly of virions, lysis of host bacteria and release of phage progeny (Rakhuba et al. 2010).

17.4.1 Receptors for Phage Adsorption

The bacteriophage adsorption takes place as a result of random phage bacteria collision. The higher the concentration of virions and host cells, the more the number of random collision taking place resulting in a higher adsorption rate. The adsorption rate is influenced by various physical, chemical and physiological factors and culturable conditions of the host cell. The adsorption process takes place in two stages: reversible and irreversible binding. The molecular mechanisms of interaction in case of both stages are very specific and differ from one phage-host system to another.

The outer membrane of gram-negative bacteria is highly permeable as compared to the plasma membrane of gram-positive bacteria due to the presence of greater number of integral protein forming transport channels. Protein present in the outer membrane and various lipopolysaccharide sites serves as bacteriophage receptors. Porins were the first characterised receptor protein of gram-negative bacteria. Protein OmpC acts as a receptor for phage Hy2, ss4, T4 and Tulb. Bacteriophage T4 uses the protein OmpC in association of lipopolysaccharide (LPS) as a receptor. Protein gp37 present in phage T4 is responsible for bacterial receptor OmpC recognition and protein gp38 found in phage T6 is responsible for receptor protein OmpF recognition. Similarly, the LamB act as a receptor for phage λ and proteases OmpT and OmpX serve as a receptor for T-like phages. Receptor protein forms narrow channel through the use of aromatic positively charged amino acid residues which allow the selective transport of maltose and derived polymers (Brussow et al. 2004).

The structure of R-type LPS is limited by lipid A and the core. The core is a negatively charged LPS region, which is required for the gram-negative cell wall rigidity by intermolecular cationic bridging. The core domain has an oligosaccharide moiety attached to lipid A containing heptose and 3-deoxy-D-mannooctulosonic acid as sugars. Disruption in core biosynthesis of LPS occurring at different stages results in incomplete core formation which in turn will severely affect bacteriophage adsorption. Phage F0 lysing wild-type *Salmonella* possesses LPS with complete core while, mutant strain lacking the terminal glucosamine moiety is resistant to phage infection. Thus, phage tends to attack and lyse the wild-type strain and not the mutant. Bacteriophage recognises the R-type LPS of *Shigella* and *Escherichia* which acts as a receptor for T-like phages infection. Phage T3 infecting *S. flexneri* mutant strain has core terminated with glucose bonded to heptose. Mutant strain *E. coli K12* with core terminating in glucose and heptose is infected by phage T3 and T7. Thus, it is observed that the major function in receptor formation is played by spatial arrangement around terminal glycosidic bond in polysaccharide chain of the core (Brussow et al. 2004; Rakhuba et al. 2010).

The key component of bacterial cell wall is peptidoglycan, a heteropolymer consisting of disaccharide monomer made up of N-acetyl glucosamine and N-acetylmuramic acid. Tetrapeptide is attached to hydroxyl group of N-acetylmuramic acid. This mediates covalent bond between peptidoglycan fibres. Teichoic acids are one of the important constituents of gram-positive microorganisms. They contain glycerol or ribitol moieties bonded together by phosphodiester bond. The presence of tetrapeptide is essential for irreversible binding of bacteriophage on host bacteria during phage adsorption in gram-positive bacteria. Irreversible binding of bacteriophages 3C, 52A, 71, 77, 79 and 80 during adsorption on *Staphylococcus aureus* takes place in the presence of tetrapeptide only. Protein GamE acts/serves as a receptor for phage γ on *B. anthracis.* Bacteriophage infecting *Lactobacillus delbrueckii* adsorb on lipoteichoic acid isolated from bacterial cell wall. Rhamnose, glucose and galactose moieties are responsible for initial recognition and attachment of phage virions (Brussow et al. 2004).

17.4.2 Phage Infection and Cell Lysis

Penetration of phage nucleic acid takes place after irreversible adsorption of bacteriophage on host bacteria. Factors which influence the penetration of nucleic acid are electrochemical membrane potential and enzymatic degradation of peptidoglycan layer and ATP molecules. The reversible binding of three or more phage tail fibres on LPS receptor protein of host makes the conformational change of phage tail which is required for DNA penetration. Subsequently, six short phage fibres are generated and result in irreversible adsorption to the host heptose moiety in LPS core. Stellar conformation causes contraction of the tail sheath so that inner hollow tube pierces through the host outer membrane. Baseplate protein gp5 found at the end of hollow tube secretes enzyme lysozyme responsible for peptidoglycan degradation. When the protein comes in contact with phosphatidylglycerol of the host bacteria inner membrane, a signal is produced for the transport of DNA along the tail tube into the host. Bacterial transport protein FhuA of outer membrane is involved in the transport of iron in cell serving as receptor for T5 phage. Phage T5 irreversibly adsorbs to the receptor which triggers the DNA introduction into host cells (Lurz et al. 2001). The introduced viral DNA controls the synthesis of proteins responsible for bacterial DNA degradation. Phage T1 and φ80 use FhuA as receptor but

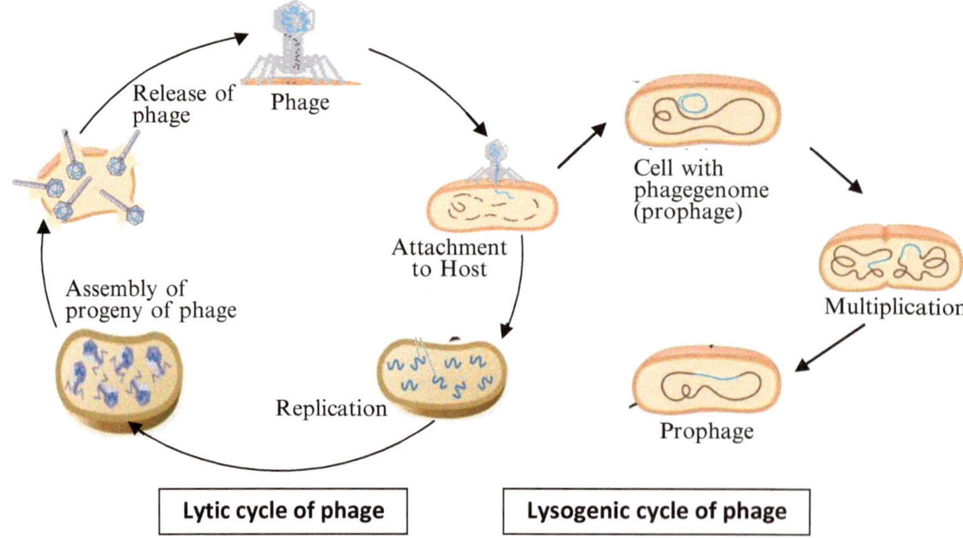

Fig. 17.1 The lytic and lysogenic life cycle of bacteriophage

require energy for adsorption which is provided by electrochemical proton gradient produced on bacterial inner membrane through the electron transport chain (Rakhuba et al. 2010).

17.4.3 Life Cycle of Phage

The life cycle of phages can be categorised in two parts, lytic (virulent) and lysogenic (temperate) cycles. Figure 17.1 depicted the lytic and lysogenic life cycle of bacteriophage. In case of lytic cycle, bacteriophage adsorbs to specific receptor sites on a host bacteria followed by penetration of phage nucleic acid. Host receptor recognition and attachment to the specific receptor determine the host range of bacteriophage. Phage tail penetrates the host cell wall and lytic enzyme secreted by it degrades the cell wall. Inside, the cell-specific enzymes encoded by bacteriophage genome are secreted to divert the host cell machinery towards the production of new phage particles. Phage-encoded structural proteins and enzymes assemble to form new virions and the newly replicated phage genomes are packed into phage heads. At an appropriate time at the end of the cycle, phage-encoded holins form holes in the host cell membrane permitting phage-encoded peptidoglycan hydrolase access to host peptidoglycan.

Temperate phages, in addition to being capable of undergoing lytic life cycle, possess the ability to exist as prophage within the genome of host bacteria (Oyaski and Hatfull 1992). The prophage may enter the lytic cycle in response to environmental stimulus which subsequently leads to lysis of host bacterial cell. Temperate phages are strictly avoided for direct use as therapeutics as they have a tendency to transfer genetic material through transduction from one bacterium to another. They can transfer genes that increase the virulence of the host through a process known as lysogenic conversion and the infected bacteria does not get killed (Endersen et al. 2014).

17.5 Phage–Host Relationship and Ecological Perspective

It is believed that the phages are found where bacteria flourish in the ecosystem. Phage shares a dynamic relationship with activity, abundance and density of the host bacterial populations. Mechanism of phage–bacteria coexistence plays a central role in structuring the host communities. Earlier, two *Pseudomonas species* and their phage parasites were propagated under different conditions and observed for coexistence

(Brockhurst et al. 2006). Phages lead to evolution of host genomes which would play a role in increase of pathogenicity and may develop an antibiotic resistance in the bacterial population. Phages have shown profound effects on various bacterial species in a given community making them ecologically significant. They have opened up the possibilities for horizontal DNA transfer across dynamic microbial population of ecosystem. Thus, it is not surprising that the phages occur even in humans. In a survey, 600 stool samples of healthy adults were analysed for the presence of coliphages. Results indicated that 34 % of the samples were found positive and majority of them had temperate phages (Chibani-Chennoufi et al. 2004). It is postulated that phage number is mostly 10 times higher than the bacterial population in the environment. Further, transmission electron microscopy studies revealed that 10^7/g phages were present in terrestrial ecosystems. In hybridisation experiments with *Pseudomonas* sp. and *Serratia*, it was observed that 5 % of phages actually infected the bacteria.

17.5.1 Phage–Host Interaction

Phage and its bacterial host interact first via random collision. Later, when phage recognises its respective host, it inserts its genetic material into the host. This is usually a three-step process starting from adsorption. It is the key stage in the phage-specific host interaction. The mechanism of adsorption involves various factors such as nature of bacterial receptors, its chemical configuration, spatial arrangement, specific interaction mechanism and structure of phage receptor. The amount and densities of receptors present on bacterial cell surface play an important role. Lipopolysaccharides present on the outer membrane of gram-negative bacteria serve as receptor for some phages. It is made up of monosaccharide and fatty acids and consists mainly of three parts: lipid A, core and O-chain (side chain, O-antigen). LPS is of two types, smooth(S) type which consists of typical LPS structure while rough(R) type lacks O-chain. Bacteriophage recognises the O-antigen through the enzyme present at its tail end, attaching to it leading to subsequent hydrolysis of one of the bond in polysaccharide chain O-antigen. The P22 has tail spikes to adsorb on bacterial surface. This group of phage contains carbohydrates as surface receptors. The adhesin of P22 phage is tail heterotrimeric protein of 215 KDa. P22 phage recognises O-antigen in outer membrane lipopolysaccharide of *Salmonella anatum* and *Salmonella typhimurium*, the pathogenic salmonella species. O-antigen has a trisaccharide α-D-mannose–α-L-rhamnose–α-D-galactose as common recurring units only differing in dideoxyhexose substituent at C3 of mannose. The endorhamnosidase or endoglycosidase activity of P22 hydrolyses O-antigen polysaccharide at rhamnose–galactose α-(1 → 3) glycosidic linkage thereby producing dimers of repeating unit (Baxa et al. 1996). The various types of receptors and proteins at bacterial cell surface are mentioned in Table 17.2.

Lurz 2001 observed that the capsid consists of many copies of several proteins. Phage tail is

Table 17.2 Various types of receptors and protein at bacterial cell surface

Receptor	Type	Phage	References
OmpC	Protein	Hy2,SS4,Tulb,T4	Lu and Breidt (2015)
OmpF	Protein	T2	Black (1988)
LamB	Protein	Phage λ	Rakhuba et al. (2010)
Fhu, TonB	Protein	T7,T1,T5,φ80	Rakhuba et al. (2010)
OmpT, OmpX	Protein	T-like phages	Nakata et al. (1993)
GamR	Protein	Phage γ	Rakhuba et al. (2010)
Vi-antigen	Protein	Phage II	Rakhuba et al. (2010)
Capsid protein A	Protein	P17, M12,fr, Qβ,f2,f4	Rakhuba et al. (2010)
O-antigen	Protein	ε^{15}, P22, Sf6,φ1,	Lander et al. (2006)
YueB	Protein	SPP1	Sao-Jose et al. (2006)

attached to the head via a connector and assembles the dsDNA in capsid (Lurz et al. 2001). It plays the role of gatekeeper and prevents DNA leakage under high pressure. As the signal is transmitted by tail, connector opens and allows the release of DNA into the bacteria when phage gets attached to the bacteria (Plisson et al. 2007). Leiman et al. (2010) observed that the phage tail and its accessory structures play a key role securing the entry of phage DNA into the host at infectivity stage. Tails contain outer appendages at distal end mainly including baseplate and many fibres with tip which possess specificity to membrane receptor of the host. During the short-termed reversible attachments to the host, just after the membrane receptor is discovered, the binding of bacteriophage to host outer membrane makes the attachment irreversible (Sao-Jose et al. 2006).

Research studies indicate that gram-positive phage lytic enzyme has two domain structures. The N-terminal domain with catalytic activity of enzyme will precisely cleave the key bonds of peptidoglycan. The C-terminal domain is mainly attached to carbohydrate found in the host bacterium (Fischetti 2005).As compared to N-terminal, C-terminal is more specific as it binds to substrates which are found in enzyme-sensitive bacteria only. Electron microscopic studies show that lytic enzymes attack the host bacteria by forming holes through peptidoglycan digestion. Interruption in cell wall integrity causes discharge of cytoplasmic membrane which finally leads to hypotonic lysis (Fischetti 2005). Phage lytic enzyme or lysin acts on the cell wall receptor and breaks one of the four main bonds of the peptidoglycan. This can be either endo-β-N-acetylglucosaminidase or N-acetylmuraminidase activity which reacts with the sugar fraction. Endopeptidase and N-acetylmuramoyl-1-alanine amidase reacts with the peptide component and hydrolyses the amide linkage of glycan strand and peptide fraction (Young 1992). Lysin gets collected in the cytoplasm of phage-infected bacterium and at a precise time, holin is inserted that finally leads to membrane disruption. As compared to large, small DNA phage encode proteins to impede the host enzyme for peptidoglycan biosynthesis, causing cell wall lysis (Bernhardt et al. 2001). The lytic enzyme can easily degrade cell wall of gram-positive bacteria, whereas this action is opposed by outer membrane in case of gram-negative bacteria.

17.5.2 Phage Evolution

Phages have evolved over the period of time by responding to changes of bacterial defence mechanisms. Bacteria possess wide array of defence mechanisms like altering: (i) target receptor on outer membrane to which phage binds, (ii) restriction enzyme modification system (CRISPR-CAS) and (iii) apoptotic mechanism which have a role to play upon phage DNA entry in their genome. Bacteriophages have developed counter mechanisms to tolerate host's defensive system and survive by means of generating its progeny inside the host. The toxin antitoxin system is widespread in prokaryotes. Phage ΦTE evolved sequences which mimicked extracellular antidote and kept on actively generating antitoxin without becoming a victim of bacterium's immune system (Blower et al. 2012). Prophages possess virulence factors which confer resistance to host upon its genome integration. The coverage of phage genome being extremely low phages analysed so far shows great variability in their genome.

17.5.3 Role of Phages in Biogeochemical Cycle

It is understood that phages have significant role in ecology as they cycle the organic matter in biosphere in the ecosystem. Biogeochemical cycle is a process by which organic and inorganic constituents circulate through biotic and abiotic forms of ecosystem. Microorganisms play a role in regulating biogeochemical cycle and have an impact on our ecosystem and environment. The effect of microbial community in particular that of phages to regulate biogeochemical cycle and ecosystem has to be thoroughly understood. Bacteriophage can alter many biogeochemical and ecological processes through their manipulation of bacterial community by getting involved in bacterial mortality, evolution, genetic diversity and biodiversity (Fuhrman and Schwalbach 2003). To estimate the impact of bacteriophage on decrease in bacterial number, a quantitative model of a steady state food web has been used. The 27 % rise in bacterial growth along with 37 % reduction in transfer of bacterial carbon to protistan grazers and decrease in 7 % macrozooplankton production were observed. The work indicates significant role of phage activity in shifting dissolved organic carbon towards bacteria and contributing to its increase in respiration (Fuhrman and Noble 1995). Phage–host interaction is a dominant factor in shaping bacterial community and thereby plays an impact on availability of nutrients. The bacteria which are engulfed by phages tend to play a central role in recycling of nutrients. Thus, engulfment of specific bacteria would impact the geochemical cycle of ecosystem (Brussow et al. 2004).

17.5.4 Pathogen Evolution: Role of Prophage

There are mainly two groups of bacterial pathogens with respect to prophage. One consists of pathogens in which no prophage or similar genetic elements are present. Second and the larger group being of pathogens having prophage and phage-like elements present in their genome. Prophages are reported to behave as rampant invaders and thereby contributing to development of new bacterial strain. Prophages sometimes contribute to the development of a new phenotype or change in existing phenotype which is towards pathogenicity, for example, increase in virulence due to production of toxin and improved host colonisation because of resistance to oxidative stress (Wang et al. 2010). In most cases with a few exception, phages encode toxins which confers a new phenotype to bacterial host belonging to this category (Brussow et al. 2004; Tinsley et al. 2006). While a few depend on specific phage-encoded virulence factor to cause a specific disease (*V. cholera, C. botulinum*), others, namely, *S. aureus* and *S. pyogenes*, harbour number of prophages and each prophage-encoded virulence factor contributes to fitness of pathogen.

In addition to self-modification, bacteriophages contribute to significant change in bacterial genome architecture. The prophage modifies the lysogen, a bacterium through the expression of its gene by a process called lysogenic conversion. In most of the cases, the prophage integrated in host chromosome and remains in a dormant state as phage-encoded repressor binds the operators controlling early promoters. Release of mature phage takes place through the removal of repressor and induction of small fraction of lysogen, a process known as spontaneous induction which takes place as a result of DNA damage. The SOS response consigns production of more number of enzymes involved in DNA repair including RecA protein that facilitates auto-cleavage of several phage repressors. Acquisition of prophage-encoded virulence genes includes: *lom* encodes for outer membrane protein concerned with macrophages survival, *sod* encodes superoxide dismutase and tellurite resistance genes contributes for the emergence of food-borne pathogenic *E. coli* O157 (Boyd and Brussow 2002).

17.5.5 Virulence Arsenal of Bacteriophage

Bacteriophages have been observed as a suitable tool to explore the bacterial genetics. Available multiple whole genome sequence of many bacterial species has uncovered the importance of prophage in modification of strains. Genome analysis of different bacterial strains indicates the integration of prophage-containing virulence factor converts natural bacterial population to pathogens. These virulence factors may be situated on morphologically diverse bacteriophage from family Podoviridae (short tail), Myoviridae (contractile tail), Siphoviridae (non-contractile, long flexible tail) and Inoviridae (filamentous) (Busby et al. 2013).

In 1971, it was first revealed that the bacteriophage epsilon from *Salmonella* encodes O-antigen genes. The genes modifying O-antigen have been found in many bacteriophages preying on a number of gram-negative bacteria. The modification of O-antigen generates antigenic variation which alters receptors and helps the phage to escape host immune system. The genes *rfb*, *oac* and *gtr* are responsible for O-antigen modification situated on diverse bacteria in *Shigella* sp. and *E. coli.* The location of genes causing lysogenic conversion was found near to the phage and host chromosomal integration site and indicates fragments of DNA are acquired through imprecise prophage excision from parental host.

17.6 Microscopic and Genomic Insights on Bacteriophage Biocontrol

Bacteriophages are believed as a worthy alternative to antibiotics and disinfectants. Phage therapy is widely used to treat various undesired bacteria. Phage used for therapy should be strictly lytic as virulent phage has a tendency of transferring virulent genes into the host bacteria. Phage cocktails are subjected to a number of tests to check its efficacy as a biocontrol agent.

(a) *Optimal MOI of phage*: The ratio of phage number to host bacteria is called multiplicity of infection (MOI). This is estimated by dividing the phage number to the number of bacteria added. The maximum yield giving MOI was considered as optimal MOI (Augustine et al. 2013). For example, the MOI of phageBp7 was examined using it along with *E. coli* (EC041029A, serotype O78) in initial log phase and at various MOIs (0.001, 0.1, 1 and 10). A bacterial culture without phage was used as control. The optimal MOI was determined by the multiplicity of infection result in maximum phage titre within 4 h and is used for large-scale phage production (Zhang et al. 2013).
(b) *One-step growth curve*: It uses bacterial cells in log phase to examine the growth kinetics of phage. The latent period, amplification and burst size of phage were estimated using growth curve (Augustine et al. 2013).
(c) *Effect of physicochemical parameters on phage infectivity*: The effect of temperature ranging from 50 °C to 100 °C and pH 7–14 on phage propagation were studied (Muller et al. 2005). Further, the studies were also carried out to observe the impact of different NaCl concentration on viability of phage.
(d) *Host range studies*: Phage–host specificity was determined based on their ability to target the different host strains. The characterised phage has to be checked for their host specificity for different targets.
(e) *Screening for virulence genes*: The host and phage genome were screened to detect the presence of virulence genes using PCR-based screening method. The test is carried out against the virulence genes reported in that specific strain in NCBI using multiplex PCR.

17.6.1 Microscopic Views on Bacteriophage

In 1940, the first phage electron micrograph published in Germany proved the particulate nature of bacteriophage. Further, electron microscopy was introduced in the study of structural classification of phages. Previously, the Enterobacteriophage T1, T4, T5 and T7 and *Pseudomonas* phage have been initially identified.

The characteristic feature of bacteriophages consists of capsid which protects the genome and structures which allow interfacing with bacterial membrane for genome release. Researchers observed through cryoelectron microscopy (cryo-TEM) that tailed phages of Caudovirales have dsDNA, capsid and a tail. Phage of this order differ from each other approximately 24–400 nmin length and DNA sequences from 18 to 400 kb. Phage has mainly icosahedral head and envelops the genome in capsid. In 2012, Orlova observed the X-ray and cryo-TEM structure of different phages and described the connector opening and release of genetic material via tail tube into the host cell (Orlova, 2012).

The T4 phage of Myoviridae family is about 200 nm long and 80–100 nm wide and capsid has icosahedral shape with rigid tail of two main layers. The contractile sheath of inner tail tube contracts during phage infection which would support the entry of genome into host (Leiman et al. 2004). Fokine et al., in 2004, observed through TEM that the phage head is connected to tail sheath by neck. The T4 has a massive baseplate with fibres attached which help in locating receptors on the host cell. The capsid is composed of three proteins, gp23, gp24 and gp20, forming the distinct dodecameric portal vertex. T4 tail consists of three types of fibrous proteins: long, short tail fibres and whiskers. The tail fibres are anchored to the baseplate while whiskers extend outwards in a region where the tail is connected to capsid (Kostyuchenko et al. 2003).

In 1974, Marples and Zierdt published the first micrograph of phages infecting *P. acne*. Zierdt observed smaller phages with an isometric head of 42–44 nm whereas Marples identified phages with larger head of 130 nm. Similarly, bacteriophage ϕYS61 was isolated from commercial Chinese cabbage kimchi after 1 week of manufacture (Kleppen et al. 2012). Purified phage samples were negatively stained and examined by TEM. Microscopy (TEM) studies revealed phage ϕYS61 has moderately elongated capsid, short contractile tail, wide baseplate of 28 nm, and 6 appendages with central spike belonging to Podoviridae family of C2 morphotype.

17.6.2 Genomic Diversity of Phage

Bacteriophage genome was first sequenced with 5386 bp single-stranded DNA (ssDNA) phage φX174 in 1977. The first complete sequence of dsDNA lambda phage was 48,506 bp genome size. Despite the elevated phage diversity, population and abundance, its genome is approximately 1 % of bacterial chromosome. The genome size range varies greatly from *Leuconostoc* phage L5 (2435 bp) to *Pseudomonas* phage phi2-1(316,674 bp) and majority of them are larger than 15 kbp. The several genomic tools were used for genome analysis of different phages (Brewer et al. 2014). Similarly, the Phamerator was used to calculate the G/C content and number of ORFs and program PROSITE, Pfam, ProDom, SMART, PRINTS and TIGRFAMS database were used to find protein domain and family (Grose et al. 2014). Program CoreGenes 3.0 predicted the percentage of protein conservation, average nucleotide identity was calculated by k-align and MUSCLE alignment was used for phylogenetic tree construction (Oakley et al. 2011).

Phage microbe dynamic in gut ecosystem has also been studied using metaHIT set which consists of more than 500 billion bases of sequences (Qin et al. 2010). Diversity on ssDNA phages in faecal samples of five healthy individuals using density gradient ultracentrifugation followed by random amplification of DNA through phi29 polymerase along with 454 pyrosequencing gave 400,133 sequences, among which 86.2 % of genomes were not characterised in public database (Kim et al. 2011). The dsDNA podophages (52–74 %), siphophages (11–30 %) and myophages (1–4 %) were reported to be present among previously identified population of phages as they constituted major portion of their genome.

The metagenomic studies that are completed on human gut provide information on diversity and abundance of phage community residing in human intestine (Minot et al. 2011). Genome sequencing revealed the faecal phage community was mostly novel as indicated by the fact that there were no significant hits, while the identified hits belong to dsDNA of Podoviridae, Siphoviridae and Myoviridae family. The infant intestinal phage community indicated less diversity and only 8 viral genotypes were present. The most abundant genotype comprised 43.6 % of the total phage community. The overall phage community composition changed between 2 weeks of age, as indicated by microarray experiments (Breitbart et al. 2008). The phage genomes were observed to share some of the host common nucleotide sequences. Analysis of the complete nucleotide sequence of 27 *Staphylococcus aureus* bacteriophages shows that the class of same genome size often has homology with each other, but not with the class of different genome size (Kwan et al.

2005). Phages with different morphology showed the greater genome diversity and thus, genome architecture imposes restriction on genetic exchange (Hatfull 2008).

17.6.3 Broad Host Range

The host range of a phage is determined by its ability to target the different host strains. The characterised phage has to be checked for their host specificity to different targets. Previously, the host specificity of phageBp7 was checked by spotting 10^5 PFU of phage on freshly prepared lawns of 35 clinical and 4 lab isolates of *E. coli*. This phage Bp7 belonging to family Myoviridae was isolated from chicken faeces. Phage Bp7 was found sensitive to 16 clinical strains and 4 laboratory strains of *E. coli* (Zhang et al. 2013). In another study, two broad host range bacteriophages were analysed with regard to their ability to form plaques on different bacterial species. Nine of ten bacteriophages studied were observed to be broad host range.

17.7 Application of Phage Cocktail as a Biocontrol

Phage therapy can act as the alternative for antibiotic-resistant bacterial infections. Many developed and developing nations have taken the initiative to control waterborne pathogens, including MDR bacteria using phage therapy. Presently, R-plasmid-carrying bacteria and few deadly pathogens are a major cause of serious environmental pollution. Indiscriminate discharge of hospital wastewater in the STPs contributes to the release of deadly pathogens and MDR bugs in the environment. The STPs provide the rich source of diverse phages to target the various undesired bacterial populations in different ecosystems. In order to analyse the application of phage formulation in the environment, it must be well characterised, and genome annotation validation for gene of obligate lytic phage and virulence factors and natural environment must be studied (Strauch et al. 2007; Bourdin et al. 2014). The falling costs of whole genome sequencing (Kisand and Lettieri 2013) should make tracking the evolution and spread of virulence genes easier and more accurate (Didelot et al. 2012). Furthermore, advances in metagenomics may make monitoring the effects of environment on microbial communities feasible and allow researchers to track changes over long time periods.

Researchers have supported the phage therapy for the numbers of undesired bacteria without damaging the natural microflora of system. Phage preparation can be used against various bacterial pathogens. The complete information of the following parameters are essential for using phage lysate:

1. Understanding the biology of phage.
2. Phage suspension should meet all the safety requirements.
3. It should contain infective phage particles.
4. Phage receptors should be identified.
5. Efficacy should be tested in an animal model.

17.7.1 Food Industry

Food is an essential requirement of everyday life, but contamination of food with pathogenic bacteria results in illness and sometimes even death. With the rapid increase in food-borne outbreaks and corresponding cost, maintaining the quality at a prominence has social and financial significance. The rising awareness among consumer has increased the pressure on food industry to apply better manufacturing process, quality assurance and hygienic and safety models (Oliveira et al. 2012). Food-borne pathogens of concern are *Salmonella*, *Campylobacter*, and *Listeria monocytogenes* and pathogenic varieties of *E. coli* can be associated with grave gastrointestinal infection (Endersen et al. 2014). The *Listeria* proves to be a serious health risk to the consumer due to its unique ability to tolerate at high salt concentration and low temperature, making it very challenging. According to the law in most countries, *Listeria monocytogenes* must be absent from 25 g of food sample which is a tough challenge for any detection system. The European Food and Safety Authority and Food

Standards Australia New Zealand have also approved phage-related formulations as additives to organic foods. Several companies are adopting this new technology and start to release the commercial products to the market. Several products such as ListShield (Intralytix, Inc., USA), Listex (MICREOS) for the prevention of *Listeria* contamination on ready-to-eat foods, EcoShieldTM to control *E. coli O157*, and AgriPhage for the treatment of phytopathogens are also available in the market (Sulakvelidze and Barrow 2005; Monk et al. 2010; Klumpp et al. 2013). A study was carried out to estimate an *E. coli* O157:H7 phage cocktail on different foods. A three-phage cocktail known as ECP 100 was used to reduce contamination of three-strain *E. coli* O157:H7 of hard surface establishes in food production facilities. When treated for 5 min with ECP100 at three different concentrations, the treated samples show significant reduction in the number of *E. coli* O157:H7 (Abuladze et al. 2008).

The potential risk associated with chemical preservatives used to enhance the food safety has increased concern in the society and thus, the focus has been moved towards natural antimicrobial alternatives such as endolysins. Endolysins are peptidoglycan hydrolases (PGHs) which are encoded by phage and used to enzymatically disrupt host cell wall in the final phase of reproduction (Oliveira et al. 2012; Schmeler et al. 2012). Complete analysis of endolysins is carried out to check the stability and safety on food items and manufacturing units. Further, researchers have shown that *staphylococcal* phage lysin LysH5 destroys *S. aureus* present in pasteurised milk and it acts synergistically with bacteriocins (Obeso et al. 2008; García et al. 2010). Similarly, the lysins of *Erwinia* phage are used to eradicate *Erwinia amylovora* on fruits (Kim et al. 2007) and Ply118, Ply511 and Ply500 are used as antilisterial agent on iceberg lettuce (Schmeler et al. 2011).

17.7.2 Agriculture and Horticulture

Plant-derived lysins are widely used in agriculture to control the pathogens (During et al. 1993; Lottmann et al. 2000). Similarly, phage has been successfully used to treat bacterial infections in peaches, cabbage and pepper. The phage-based biocontrol against plant pathogens has been effectively tried on *Xanthomonas pruni* and *Xanthomonas campestris*, which causes bacterial spots on peaches and tomatoes, respectively. A bacterial spot of mushrooms occurring due to *Pseudomonas tolaasii* infection has also been treated by phages (Gill et al. 2003). Similarly, the EcoShield™ to control *E. coli O157* and AgriPhage produced by OmniLytics for the treatment of phytopathogens are also available in the market (Sulakvelidze and Barrow 2005; Monk et al. 2010).

17.7.3 Phage Typing

Phage typing is a tool to distinguish between bacterial cultures and is applied as clinical and molecular method to identify and characterise pathogenic strains. With several more sophisticated tools available, including genomic for differentiation, phage typing stands out because of its relative simplicity, speed and cost-effectiveness. Studies on enterohemorrhagic *E. coli* and *Campylobacter* reflect upon the need to involve phage typing, especially because any single method alone cannot succeed to produce all the relevant data pertaining to epidemiological relatedness (Hopkins et al. 2004).

17.7.4 Role in Bacterial Infection Control

Phage therapy has significant impact on immunology, drug discovery, pharmacology and cell biology. The rise of antibiotic-resistant deadly pathogen such as *Pseudomonas aeruginosa*, *Enterococcus faecium*, *Staphylococcus aureus*, *Mycobacterium tuberculosis* and *Acinetobacter baumannii* has raised huge outcry (Hanlon 2007). In 2002, Markoishvili developed a cocktail of phages commercially named as *Pyophage* active against *Staphylococcus*, *P. aeruginosa*, *Proteus* sp., *E. coli* and *Streptococci* that can be used for prophylaxis and treatment of purulent wound infections (Markoishvili et al. 2002). It is successfully used in case of surgery, burns, wounds, osteomyelitis, ear and eye infections and skin infections. This cocktail is used in commercial wound dressing product known as *Phage Bioderm*. It is a novel biodegradable polymer impregnated with both antibiotic ciprofloxacin and phage cocktail. This product is used to treat wounds even those infected with multiple resistant *S. aureus*. Similarly, the specific sequences coding for enzymes of phage have been used previously to target the undesired pathogens. A suitable example of treatment is of *Bacillus anthracis* which causes anthrax. Phage lytic enzymes are highly specific and efficient in destroying pathogenic bacteria. Research studies show that a lysin isolated from γ phage specifically kills *B. anthracis* within few seconds of 100 units of Ply G and decreased the counts of ~10^7 bacilli by thousand folds. Vegetative cells and spore-germinating cells of *B. anthracis* was also found to be sensitive to this lysine. Further, bacteriophage K was used to eliminate *Staphylococcus aureus* biofilm on central venous catheter material. The phage used in this experiment is a polyvalent phage competent of lysing 10 diverse *Staphylococcus epidermidis* strains and nine different *Staphylococcus* sp. (Lungren et al. 2013).

The genetic manipulation of phage capsid protein can increase the efficiency of phage cocktail. The technique uses modification of bacteriophage capsid through foreign proteins and peptides. Gene sequence encoding for particular proteins can be inserted into specific sites of DNA which codes for phage capsid proteins. The altered site produces fusion protein that is incorporated in the new progeny phage (Uchiyama et al. 2005). This method can be used in vivo or in vitro to represent the enormous amount of diverse protein and to create a *phage display library* in a cheap and rapid way where each phage clone displays a different peptide on its surface.

17.7.5 Pathogen Removal from Wastewater

The main objectives of STPs are reduction and removal of COD, BOD and suspended solids along with pathogens which are continued till date. Phages are known to be highly host specific and used to control the pathogenic bacteria in the field of medicine, agriculture, aquaculture, wastewater treatment, food industry and sludge bulking. Phage production at commercial level to target the *E. coli* O157:H7 in manure and to eradicate pathogens from food preparation areas along with carcasses has been already underway (Flynn et al. 2004; Thiel 2004). Nowadays, reduction and removal of waterborne pathogens and MDRs are one of the concerned issues in developing and developed countries (Chitnis et al. 2004; Ekhaise and Omavwoya 2008; Periasamy and Sundaram 2013). In one study, eight different samples of hospital wastewater in Tamil Nadu were analysed for the degree of pollution and isolated the bacteriophage to target the *E. coli*. It was concluded that biocontrol provides a feasible tool to control the pathogens in wastewater (Periasamy and Sundaram 2013). The emerged MDR *Acinetobacter baumannii* is infamous for causing nosocomial wound infections (Davis et al. 2005; Jones et al. 2006; Petersen et al. 2007). It has an ability to form biofilm on environmental surfaces and resistant to a broad range of antibiotics. This bacterium develops resistance very quickly due to its long-term exposure to an antibiotic-producing soil organism (Manchanda et al. 2010). The resistance mechanisms adopted by deadly pathogen include enzymatic inactivation, modification of target sites, and decreased influx of drugs (Peleg et al. 2008).

Recently, the 26 bacteriophages were isolated against different combination of 34 strains of *Acinetobacter baumannii* from hospital wastewater (Merabishvili et al. 2014). Phages Acibel004 and Acibel007 were selected on the basis on genotyping and host range. They were identified as Myovirus and Podovirus. The absence of lysogeny-associated gene in the genome of both the phages presents them as a suitable candidate for use in phage therapy against *A. baumannii* infections. Similarly, the bacteriophages to control the *Salmonella* sp. and other *Enterobacteriaceae* in environmental samples were also studied in vitro at different temperatures. Further, the biocontrol application of predominant filamentous bacterium, *Sphaerotilus natans* and *Haliscomenobacter hydrossis*, notoriously known to cause filamentous bulking in wastewater treatment plants has also been studied (Choi et al. 2011; Petrovski et al. 2011; Turki et al. 2012). Thus, the potential phage cocktail is required to improve effluent quality of wastewater treatment system and sludge disposal into the environment.

17.7.6 Biofilms and Biofouling

Biofilm is a complex, densely packed assembly of microorganisms adhered to living or inert surfaces. Microbes which are adhered to the surface secrete extracellular polysaccharides and attract the other population to form a biofilm. Previously, several studies have been published to control the biofilm formation through phage cocktails. Further, phages have been isolated which are found to have a role in degrading extracellular polysaccharides of various gram-negative biofilm-forming bacteria (Sillankorva et al. 2008). Phage ΦIBB-PF7 was isolated to control the *P. fluorescens* which is key food spoilage organism and typically observed in biofilms. Phage ΦIBBPF7A was observed to greatly reduce the bacterial population present on stainless steel slides after 2 h of infection period. This result was even more pronounced for 4 h; however, complete eradication was not achieved under any given conditions (Sillankorva et al. 2008). Similarly, the foot infections in diabetic patients caused by *S. aureus*, *P. aeruginosa* and *A. baumannii* have been treated using lytic phage cocktail (Mendes et al. 2014). Bacteriophages were identified members of *Podoviridae* or *Myoviridae* and did not harbour any virulence genes. These phages found to be active against all the three bacterial species in both planktonic and biofilm condition at an input multiplicity of 10 (Mendes et al. 2014).

Biofouling is a process of aggregation of microbes on moist surface leading to aggregation of similar communities. Phages intervene in the process of biofouling by lysing bacterial population. Although biofouling is considered as a major concern for environmental microbiologists, the phage treatment has opened the door of its regulation. For example, Lu and Collins (2007) used a bacteriophage to remove the bacterial biofilm of *E. coli*. Further, phages have also been suggested as a way to counter biofouling of condenser tubes of thermal power plant (Goodridge and Abedon 2003). In another study, lytic bacteriophages were used against three bacterial species known to cause biofouling: *P. aeruginosa*, *A. johnsonii* and *B. subtilis*. The phages were found to reduce biofouling activity by about 60 % compared to control. Work suggests the potential use of specific lytic phages in reducing membrane biofouling caused by *P. aeruginosa, A. johnsonii and B. subtilis.* (Goldman et al. 2009).

The current chapter addresses the issue of phage and their products to control the undesired pathogens from various environmental niches. The evaluation and application of phage cocktail formulation as a biocontrol would help in resolving the issue of public health threat. Further, it will help to understand the exact mechanism and the safety of phage cocktail through whole genome analysis which is key part of applications.

References

Abuladze T, Li M, Menetrez Y, Dean T, Senecal A, Sulakvelidze A (2008) Bacteriophages reduce experimental contamination of hard surfaces, tomato, spinach, broccoli, and ground beef by *Escherichia coli* O157:H7. Appl Environ Microbiol 74:6230–6238. doi:10.1128/AEM.01465-08

Ackermann W (2012) Bacteriophage electron microscopy. Adv Virus Res 82:1–32

Ackermann W, Krisch H (1997) A catalogue of T4-type bacteriophages. Arch Virol 142:2329–2345

Alemayehu D, Ross R, O'Sullivan O, Coffey A, Stanton C, Fitzgerald G, McAuliffe O (2009) Genome of a virulent bacteriophage Lb338-1 that lyses the probiotics *Lactobacillus paracasei* cheese strain. Gene 448:29–39. doi:10.1016/j.gene.2009.08.008

Allen K, Levine Y, Looft T, Bandrick M, Casey A (2013) Treatment, promotion, commotion: antibiotic alternatives in food-producing animals. Trends Microbiol 21:114–119. doi:10.1016/j.tim.2012.11.001

Arnold P, Zillig W, Ziese U, Holz I, Crosby M, Utterback T, Fraser M (2000) A novel lipothrixvirus, SIFV, of the extremely thermophilic crenarchaeon *Sulfolobus*. Virology 267:252–266. doi:10.1006/viro.1999.0105

Ashelford K, Fry J, Bailey M, Jeffries A, Day M (1999) Characterization of six bacteriophages of *Serratia liquefaciens* CP6 isolated from the sugar beet phytosphere. Appl Environ Microbiol 65:1959–1965

Augustine J, Louis L, Varghese S, Bhat S, Kishore A (2013) Isolation and partial characterization of ΦSP-1, a *Salmonella* specific lytic phage from intestinal content of broiler chicken. J Basic Microbiol 53:111–120. doi:10.1002/jobm.201100319

Baross A, Liston J, Morita Y (1978) Incidence of *Vibrio parahaemolyticus* bacteriophages and other *Vibrio* bacteriophages in marine samples. Appl Environ Microbiol 36:492–499

Baxa U, Steinbacher S, Miller S, Weintraub A, Huber R, Seckler R (1996) Interactions of phage P22 tails with their cellular receptor, Salmonella O-antigen polysaccharide. Biophys J 71(4):2040. doi:10.1016/S0006-3495(96)79402-X

Bernhardt G, Wang SK, Young R (2001) A protein antibiotic in the phage Qβ virion: diversity in lysis targets. Science 292:2326–2329. doi:10.1126/science.1058289

Bettstetter M, Peng X, Garrett A, Prangishvili D (2003) AFV1, a novel virus infecting hyperthermophilic archaea of the genus *Acidianus*. Virology 315:68–79

Black N (1988) The fadL gene product of *Escherichia coli* is an outer membrane protein required for uptake of long-chain fatty acids and involved in sensitivity to bacteriophage T2. J Bacteriol 170:2850–2854

Bleackley J, Cooper J, Kaminski M, Sandilands S (2009) The reduction of T7 phage adsorption in *Escherichia coli* B23 cells treated with sub-lethal levels of kanamycin. J Exp Microbiol Immunol 13:89–92

Blower R, Evans J, Przybilski R, Fineran C, Salmond P (2012) Viral evasion of a bacterial suicide system by RNA–based molecular mimicry enables infectious altruism. PLoS Genet 8:e10. doi:10.1371/journal.pgen.1003023

Bourdin G, Schmitt B, Guy M, Germond E, Zuber S, Michot L, Reuteler G, Brussow H (2014) Amplification and purification of T4-like *Escherichia coli* phages for phage therapy: from laboratory to pilot scale. Appl Environ Microbiol 80:1469–1476. doi:10.1128/AEM.03357-13

Boyd F, Brussow H (2002) Common themes among bacteriophage-encoded virulence factors and diversity among the bacteriophages involved. Trends Microbiol 10:521–529. doi:10.1016/S0966-842X(02)02459-9

Breitbart M, Haynes M, Kelley S, Angly F, Edwards R, Felts B, Mahaffy J, Mueller J, Nulton J, Rayhawk S, Rodriguez-Brito B, Salamon P, Rohwer F (2008) Viral diversity and dynamics in an infant gut. Res Microbiol 159:367–373. doi:10.1016/j.resmic.2008.04.006

Brewer E, Stroupe E, Jones M (2014) The genome, proteome and phylogenetic analysis of Sinorhizobium meliloti phage ΦM12, the founder of a new group of T4-superfamily phages. Virology 450:84–97. doi:10.1016/j.virol.2013.11.027

Brockhurst A, Buckling A, Rainey B (2006) Spatial heterogeneity and the stability of host-parasite coexistence. J Evol Biol 19(2):374–379

Brussow H, Canchaya C, Hardt D (2004) Phages and the evolution of bacterial pathogens: from genomic rearrangements to lysogenic conversion. Microbiol Mol Biol Rev 68:560–602. doi:10.1128/MMBR.68.3.560-602.200

Busby N, Panjikar S, Landsberg J, Hurst R, Lott S (2013) The BC component of ABC toxins is an RHS-repeat-containing protein encapsulation device. Nature 501:547–550. doi:10.1038/nature12465

Chibani-Chennoufi S, Bruttin A, Dillmann M, Brussow H (2004) Phage-host interaction: an ecological perspective. J Bacteriol 186:3677–3686. doi:10.1128/JB.186.12.3677-3686.2004

Chitnis V, Chitnis S, Vaidya K, Ravikant S, Patil S, Chitnis S (2004) Bacterial population changes in hospital effluent treatment plant in central India. Water Res 38:441–447

Choi J, Kotay M, Goel R (2011) Bacteriophage based biocontrol of biological sludge bulking in wastewater. Bioeng Bugs 2:214–217. doi:10.1016/j.watres.2010.08.038

Dafale N, Lakhe S, Yadav K, Purohit H, Chakrabarti T (2007) Concentration and recovery of coliphages from water with bituminous coal. Water Environ Res 80:282–288. doi:10.2175/106143007X220923

Das S, Biswas A, Dasgupta S, Abraham A (2009) Bacterial foraging optimization algorithm: theoretical foundations, analysis, and applications. Found Comput Intell 3(1):23–55, Springer, Berlin/Heidelberg

Davis A, Moran K, McAllister C, Gray P (2005) Multidrug resistant *Acinetobacter* extremity infections in soldiers. Emerg Infect Dis 11:1218–1224. doi:10.3201/1108.050103

Didelot X, Bowden R, Wilson J, Peto A, Crook W (2012) Transforming clinical microbiology with bacterial genome sequencing. Nat Rev Genet 13:601–612. doi:10.1038/nrg3226

During K, Porsch P, Fladung M, Lorz H (1993) Transgenic potato plants resistant to the phytopathogenic bacterium *Erwinia carotovora*. Plant J 3:587–598. doi:10.1046/j.1365-313X.1993.03040587.x

Ekhaise O, Omavwoya P (2008) Influence of hospital wastewater discharged from University of Benin Teaching Hospital (UBTH), Benin City on its receiving environment. Afr J Appl Zool Environ Biol 10:56–60

Elbreki M, Ross P, Hill C, O'Mahony J, McAuliffe O, Coffey A (2014) Bacteriophages and their derivatives as biotherapeutic agents in disease prevention and treatment. J Viruses 1:1–20. doi:10.1155/2014/382539

Endersen L, Mahony J, Hill C, Ross R, McAuliffe O, Coffey A (2014) Phage therapy in food industry. Ann Rev Food Sci Technol 5:327–349. doi:10.1146/annurev-food-030713-092415

Fancello L, Trape S, Robert C, Boyer M, Popgeorgiev N, Raoult D, Desnues C (2013) Viruses in the desert: a metagenomic survey of viral communities in four perennial ponds of the Mauritanian Sahara. ISME J 7:359–369. doi:10.1038/ismej.2012.101

Fischetti V (2005) Bacteriophage lytic enzymes: novel anti-infectives. Trends Microbiol 13:491–496. doi:10.1016/j.tim.2005.08.007

Flynn G, Ross P, Fitzgerald F, Coffey A (2004) Evaluation of a cocktail of three bacteriophages for biocontrol of *E coli* O157:H7. Appl Environ Microbiol 70:3417–3424. doi:10.1128/AEM.70.6.3417-3424.2004

Fokine A, Chipman R, Leiman G, Mesyanzhinov V, Rao B, Rossmann G (2004) Molecular architecture of the prolate head of bacteriophage T4. Proc Natl Acad Sci 101:6003–6008. doi:10.1073/pnas.0400444101

Ford E, Sun B, Carpino J, Chapler S, Ching J, Choi Y, Dennehy J (2014) Frequency and fitness consequences of bacteriophage Φ6 host range mutations. PLoS One 9:e11. doi:10.1371/journal.pone.0113078

Fu W, Forster T, Mayer O, Curtin J, Lehman M, Donlan M (2010) Bacteriophage cocktail for the prevention of biofilm formation by *Pseudomonas aeruginosa* on catheters in an *in vitro* model system. Antimicrob Agents Chemother 1:397–404. doi:10.1128/AAC.00669-09

Fuhrman A, Noble T (1995) Viruses and protists cause similar bacterial mortality in coastal seawater. Limnol Oceanogr 40:1236–1242. doi:10.4319/lo.1995.40.7.1236

Fuhrman A, Schwalbach M (2003) Viral influence on aquatic bacterial communities. Biol Bull 204:192–195

Fusco S, Aulitto M, Bartolucci S, Contursi P (2014) A standardized protocol for the UV induction of Sulfolobus spindle-shaped virus 1. Extremophiles 1:1–8. doi:10.1007/s00792-014-0717

Gantzer C, Henny J, Schwartzbrod L (2002) *Bacteroides fragilis* and *Escherichia coli* bacteriophages in human faeces. Int J Hyg Environ Heal 205:325–328

García S, De la Torre Á, Morales PB, Tolón J, Domingo S, Picazo J (2010) Clinical outbreak of linezolid-resistant Staphylococcus aureus in an intensive care unit. JAMA 303(22):2260–2264. doi:10.1001/jama.2010.757

Gill J, Svircev M, Smith R, Castle J (2003) Bacteriophages of Erwinia amylovora. Appl Environ Microbiol 69(4):2133–2138. doi:10.1128/AEM.69.4.2133-2138.200

Goldman G, Starosvetsky J, Armon R (2009) Inhibition of biofilm formation on UF membrane by use of specific bacteriophages. J Membr Sci 342:145–152

Goodridge L, Abedon S (2003) Bacteriophage biocontrol and bioprocessing: application of phage therapy to industry. SIM News 53:254–262

Goyal SM (1987) Methods in phage ecology. Wiley Interscience 1:267–287

Grose H, Jensen L, Burnett H, Breakwell P (2014) Genomic comparison of 93 Bacillus phages reveals 12 clusters, 14 singletons and remarkable diversity. BMC Genomics 15:855. doi:10.1186/1471-2164-15-855

Hanlon W (2007) Bacteriophages: an appraisal of their role in the treatment of bacterial infections. Int J Antimicrob Agents 30:118–128. doi:10.1016/j.ijantimicag.2007.04.006

Hatfull F (2008) Bacteriophage genomics. Curr Opin Microbiol 11:447–453. doi:10.1016/j.mib.2008.09.004

Hopkins L, Desai M, Frost A, Stanley J, Logan M (2004) Fluorescent amplified fragment length polymorphism genotyping of Campylobacter jejuni and Campylobacter coli strains and its relationship with host specificity, serotyping, and phage typing. J Clin Microbiol 42(1):229–235. doi:10.1128/JCM.42.1.229-235.2004

Hudson A, Billington C, Carey-Smith G, Greening G (2005) Bacteriophages as biocontrol agents in food. J Food Prot 68:426–437. doi:10.1016/j.copbio.2010.10.008

Jakutyte L, Lurz R, Baptista C, Carballido-Lopez R, Sao-Jose C, Tavares P, Daugelavicius R (2012) First steps of bacteriophage SPP1 entry into *Bacillus subtilis*. Virology 422:425–434. doi:10.1016/j.virol.2011.11.010

Jaschke R, Lieberman K, Rodriguez J, Sierra A, Endy D (2012) A fully decompressed synthetic bacteriophage øX174 genome assembled and archived in yeast. Virology 434:278–284. doi:10.1016/j.virol.2012.09.020

Jones A, Walsh M, Turton J, Livermore D, Pitt T, Green A, Gill M, Mortiboy D (2006) Importation of multidrug-resistant *Acinetobacter* sp. infections with casualties from Iraq. Lancet Infect Dis 6:317–318. doi:10.1016/S1473-3099(06)70471-6

Kisand V, Lettieri T (2013) Genome sequencing of bacteria: sequencing, de novo assembly and rapid analysis using open source tools. BMC Genomics 1:211. doi:10.1186/1471-2164-14-211

Kleppen P, Holo H, Jeon R, Nes F, Yoon S (2012) Novel Podoviridae family bacteriophage infecting *Weissella cibaria* isolated from kimchi. Appl Environ Microbiol 78:7299–7308. doi:10.1128/AEM.00031-12

Klumpp J, Fouts E, Sozhamannan S (2013) Bacteriophage functional genomics and its role in bacterial pathogen detection. Brief Funct Genomics 9(1):8–14. doi:10.1093/bfgp/elt009

Kostyuchenko A, Leiman G, Chipman R, Kanamaru S, van Raaij J, Arisaka F, Rossmann G (2003) Three-dimensional structure of bacteriophage T4 base plate. Nat Struct Mol Biol 10:688–693. doi:10.1038/nsb970

Kutter E, Sulakvelidze A (eds) (2005) Bacteriophages: biology and application. CRC Press/Taylor & Francis Group, Boca Raton, pp 335–380

Kwan T, Liu J, DuBow M, Gros P, Pelletier J (2005) The complete genomes and proteomes of 27 Staphylococcus aureus bacteriophages. Proc Natl Acad Sci U S A 102(14):5174–5179. doi:10.1073/pnas.0501140102

Lander C, Tang L, Casjens R, Gilcrease B, Prevelige P, Poliakov A, Johnson E (2006) The structure of an infectious P22 virion shows the signal for headful DNA packaging. Science 312:1791–1795. doi:10.1126/science.1127981

Leiman G, Arisaka F, van Raaij J, Kostyuchenko A, Aksyuk A, Kanamaru S, Rossman G (2010) Morphogenesis of the T4 tail and tail fibers. Virol J 7(1):355. doi:10.1186/1743-422X-7-355

Leiman G, Chipman R, Kostyuchenko A, Mesyanzhinov V, Rossmann G (2004) Three-dimensional rearrangement of proteins in the tail of bacteriophage T4 on infection of its host. Cell 118:419–429. doi:10.1016/j.cell.2004.07.022

Lima-Oliveira G, Guidi C, Salvagno L, Montagnana M, Rego LG, Picheth G (2012) Is phlebotomy part of the dark side in the clinical laboratory struggle for quality? Lab Med 43(5):172–176. doi:10.1309/LMZ7YARD6ZSDIID

López-Bueno A, Tamames J, Velázquez D, Moy A, Quesada A, Alcamí A (2009) High diversity of the viral community from an Antarctic lake. Science 326:858–861. doi:10.1126/science.1179287

Lottmann J, Heuer H, DeVries J, Mahn A, During K, Wackernagel W, Berg G (2000) Establishment of introduced antagonistic bacteria in the rhizosphere of transgenic potatoes and their effect on the bacterial community. FEMS Microbiol Ecol 33:41–49. doi:10.1111/j.1574-6941.2000.tb00725.x

Lu K, Collins J (2007) Dispersing biofilms with engineered enzymatic bacteriophage. Proc Natl Acad Sci 104(27):11197–11202. doi:10.1073/pnas.0704624104

Lu Z, Breidt F (2015) *Escherichia coli* O157: H7 bacteriophage Φ241 isolated from industrial cucumber fermentation at high acidity and salinity. Front Microbiol 6:67. doi:10.3389/fmicb.2015.00067

Luhtanen M, Eronen-Rasimus E, Kaartokallio H, Rintala M, Autio R, Roine E (2014) Isolation and characterization of phage – host systems from the Baltic Sea ice. Extremophiles 18(1):121–130. doi:10.1007/s00792-013-0604-y

Lungren M, Christensen D, Kankotia R, Falk I, Paxton B, Kim C (2013) Bacteriophage K for reduction of *Staphylococcus aureus* biofilm on central venous catheter material. Bacteriophage 3:4. doi:10.4161/bact.26825

Lurz R, Orlova V, Gunther D, Dube P, Droge A, Weise F, Tavares P (2001) Structural organisation of the head-to-tail interface of a bacterial virus. J Mol Biol 310:1027–1037. doi:10.1006/jmbi.2001.4800

Manchanda V, Sanchaita S, Singh P (2010) Multidrug resistant *Acinetobacter*. J Global Infect Dis 2:291. doi:10.4103/0974-777X.68538

Mannisto H, Kivela M, Paulin L, Bamford H, Bamford K (1999) The complete genome sequence of PM2, the first lipid-containing bacterial virus to be isolated. Virology 262:355–363

Markoishvili K, Tsitlanadze G, Katsarava R, Glenn J, Sulakvelidze A (2002) A novel sustained-release matrix based on biodegradable poly (ester amide) and impregnated with bacteriophages and an antibiotic shows promise in management of infected venous stasis ulcers and other poorly healing wounds. Int J Dermatol 41:453–458. doi:10.1046/j.1365-4362.2002.01451.x

Mc Grath S, van Sinderen D (eds) (2007) Bacteriophage: genetics and molecular biology. Horizon Scientific Press

McIntyre L, Hudson A, Billington C, Withers H (eds) (2012) Biocontrol of food borne bacteria. Novel technologies in food science. Springer, pp 183–204. doi:10.1007/978-1-4419-7880-6_8

Mendes J, Leandro C, Mottola C, Barbosa R, Silva F, Oliveira M, Vilela C, Cristino J, Gorski A, Pimentel A, Sao-Jose C, Cavaco-Silva P, Garcia M (2014) In vitro design of a novel lytic bacteriophage cocktail with therapeutic potential against organisms causing diabetic foot infections. J Med Microbiol 63:1055–1065. doi:10.1099/jmm.0.071753-0

Merabishvili M, Vandenheuvel D, Kropinski M, Mast J, De Vos D, Verbeken G, Pirnay P (2014) Characterization of newly isolated lytic bacteriophages active against Acinetobacter baumannii. doi:10.1371/journal.pone.0104853

Miller S, Heidelberg F, Eisen A, Nelson C, Durkin S, Ciecko A, Fraser M (2003) Complete genome sequence of the broad-host-range vibriophage KVP40: comparative genomics of a T4-related bacteriophage. J Bacteriol 185:5220–5233. doi:10.1128/JB.185.17.5220-5233.2003

Minot S, Sinha R, Chen J, Li H, Keilbaugh S, Wu G, Lewis J, Bushman F (2011) The human gut virome: inter-individual variation and dynamic response to diet. Genome Res 21:1616–1625. doi:10.1101/gr.122705.111

Mirzae K, Eriksson H, Kasuga K, Haggard-Ljungquist E, Nilsson S (2014) Genomic, proteomic, morphological, and phylogenetic analyses of vB_EcoP_SU10, a podoviridae phage with C3 morphology. PLoS One 9:12. doi:10.1371/journal.pone.0116294

Mochizuki T, Yoshida T, Tanaka R, Forterre P, Sako Y, Prangishvili D (2010) Diversity of viruses of the hyperthermophilic archaeal genus *Aeropyrum*, and isolation of the *Aeropyrum pernix* bacilliform virus 1, APBV1, the first representative of the family Clavaviridae. Virology 402:347–354. doi:10.1016/j.virol.2010.03.046

Monk AB, Rees CD, Barrow P, Hagens S, Harper R (2010) Bacteriophage applications: where are we now? Lett Appl Microbiol 51:363–369. doi:10.1111/j.1472-765X.2010.02916.x

Muller-Merbach M, Rauscher T, Hinrichs J (2005) Inactivation of bacteriophages by thermal and high-pressure treatment. Int Dairy J 15:777–784

Nakata N, Tobe T, Fukuda I, Suzuki T, Komatsu K, Yoshikawa M, Sasakawa C (1993) The absence of a surface protease, OmpT, determines the intercellular spreading ability of *Shigella*: the relationship between the *ompT* and *kcpA* loci. Mol Microbiol 9:459–468. doi:10.1111/j.1365-2958.1993.tb01707.x

Oakley B, Talundzic E, Morales C, Hiett K, Siragusa G, Volozhantsev N, Seal B (2011) Comparative genomics of four closely related *Clostridium perfringens* bacteriophages reveals variable evolution among core genes with therapeutic potential. BMC Genomics 12:282. doi:10.1186/1471-2164-12-282

Obeso M, Martínez B, Rodríguez A, García P (2008) Lytic activity of the recombinant staphylococcal bacteriophage ΦH5 endolysin active against Staphylococcus aureus in milk. Int J Food Microbiol 128(2):212–218. doi:10.1016/j.ijfoodmicro.2008.08.010

O'Connell P, Bucher R, Anderson E, Cao J, Khan S, Gostomski V, Valdes J (2006) Real-time fluorogenic Reverse transcription-PCR assays for detection of bacteriophage MS2. Appl Environ Microbiol 72:478–483. doi:10.1128/AEM.72.1.478-483.2006

Oliveira H, Azeredo J, Lavigne R, Kluskens D (2012) Bacteriophage endolysins as a response to emerging foodborne pathogens. Trends Food Sci Technol 28(2):103–115. doi:10.1016/j.tifs.2012.06.016

Orlova V (2012) Bacteriophages and their structural organisation. InTech, 3–30. doi: 10.5772/34642

Oyaski M, Hatfull F (1992) The cohesive ends of mycobacteriophage L5 DNA. Nucleic Acids Res 20:3251. doi:10.1093/nar/20.12.3251

Peleg Y, Seifert H, Paterson L (2008) *Acinetobacter baumannii*: emergence of a successful pathogen. Clin Microbiol 21:538–582. doi:10.1128/CMR.00058-07

Peng X, Basta T, Haring M, Garrett A, Prangishvili D (2007) Genome of the Acidianus bottle-shaped virus and insights into the replication and packaging mechanisms. Virology 364:237–243

Periasamy D, Sundaram A (2013) A novel approach for pathogen reduction in wastewater treatment. J Environ Health Sci Eng 11:12. doi:10.1186/2052-336X-11-12

Petersen K, Riddle M, Danko J, Blazes D, Hayden R, Tasker A, Dunne J (2007) Trauma-related infections in battlefield casualties from Iraq. Ann Surg 245:803–811. doi:10.1097/01.sla.0000251707.32332.c1

Petrovski S, Seviour J, Tillett D (2011) Characterization of the genome of the polyvalent lytic bacteriophage GTE2, which has potential for biocontrol of *Gordonia*, *Rhodococcus*, and *Nocardia* stabilized foams in activated sludge plants. Appl Environ Microbiol 77:3923–3929. doi:10.1128/AEM.00025-11

Pirnay P, Verbeken G, Rose T, Jennes S, Zizi M, Huys I, Lavigne R, Merabishvili M, Vaneechoutte M, Buckling A, De Vos D (2012) Introducing yesterday's phage therapy in today's medicine. Futur Virol 7:379–390. doi:10.2217/fvl.12.24

Plisson C, White E, Auzat I, Zafarani A, Sao-Jose C, Lhuillier S, Orlova V (2007) Structure of bacteriophage SPP1 tail reveals trigger for DNA ejection. EMBO J 26:3720–3728. doi:10.1038/sj.emboj.7601786

Poranen M, Paatero O, Tuma R, Bamford H (2001) Self-assembly of a viral molecular machine from purified protein and RNA constituents. Mol Cell 7:845–854. doi:10.1016/S1097-2765(01)00228-3

Qin J, Li R, Raes J, Arumugam M, Burgdorf K, Manichanh C, Nielsen T, Pons N, Levenez F, Yamada T, Mende DR, Li J, Xu J, Li S, Li D, Cao J, Wang B, Liang H, Zheng H, Xie Y, Tap J, Lepage P, Bertalan M, Batto JM, Hansen T, Le Paslier D, Linneberg A, Nielsen HB, Pelletier E, Renault P, Sicheritz-Ponten T, Turner K, Zhu H, Yu C, Li S, Jian M, Zhou Y, Li Y, Zhang X, Li S, Qin N, Yang H, Wang J, Brunak S, Doré J, Guarner F, Kristiansen K, Pedersen O, Parkhill J, Weissenbach J, Bork P, Ehrlich SD, Wang J, MetaHIT Consortium (2010) A human gut microbial gene catalogue established by metagenomic sequencing. Nature 464:59–65. doi:10.1038/nature08821

Rakhuba V, Kolomiets E, Dey E, Novik G (2010) Receptors, adsorption and penetration of bacteriophage. Pol J Microbiol 59:145–155

Rattanachaikunsopon P, Phumkhachorn P (2012) Bacteriophage PPST1 Isolated from Hospital wastewater, a potential therapeutic agent against drug resistant *Salmonella enterica subsp. Enterica serovar typhi*. In tech open. doi:10.5772/30446

Reyes A, Haynes M, Hanson N, Angly F, Heath A, Rohwer F, Gordon J (2010) Viruses in the fecal microbiota of monozygotic twins and their mothers. Nature 466:334–338

Sao-Jose C, Lhuillier S, Lurz R, Melki R, Lepault J, Santos A, Tavares P (2006) The ectodomain of the viral receptor YueB forms a fiber that triggers ejection of bacteriophage SPP1 DNA. J Biol Chem 281:11464–11470. doi:10.1074/jbc.M513625200

Scheele U, Erdmann S, Ungewickell J, Felisberto-Rodrigues C, Ortiz-Lombardía M, Garrett A (2011) Chaperone role for proteins p618 and p892 in the extracellular tail development of Acidianus two-tailed virus. J Virol 85:4812–4821. doi:10.1128/JVI.00072-11

Schmeler M, Frumovitz M, Ramirez T (2011) Conservative management of early stage cervical cancer: is there a role for less radical surgery? Gynecol Oncol 120(3):321–325. doi:10.1016/j.ygyno.2010.12.352

Schmeler M, Sood K, Bell-McGuinn M, Coleman RL, Frumovitz M, Sonoda Y, Gardner J (2012) Proceedings from the 9th international conference on ovarian cancer. Gynecol Oncol 125(1):5. doi:10.1016/j.ygyno.2011.12.445

Schoenfeld T, Patterson M, Richardson M, Wommack E, Young M, Mead D (2008) Assembly of viral metagenomes from yellowstone hot springs. Appl Environ Microbiol 74:4164–4174. doi:10.1128/AEM.02598-07

Sillankorva S, Neubauer P, Azeredo J (2008) Pseudomonas fluorescens biofilms subjected to phage philBB-PF7A. BMC Biotechnol 8(1):79. doi:10.1186/1472-6750-8-79

Sozhamannan S, McKinstry M, Lentz M, Jalasvuori M, McAfee F, Smith A, Read D (2008) Molecular characterization of a variant of *Bacillus anthracis*-specific phage AP50 with improved bacteriolytic activity. Appl Environ Microbiol 74:6792–6796. doi:10.1128/AEM.01124-08

Strauch E, Hammerl J, Hertwig S (2007) Bacteriophages: new tools for safer food? J Verbr Lebensm 2:138–143. doi:10.1007/s00003-007-0188-5

Sulakvelidze A, Barrow P (2005) Phage therapy in animals and agribusiness. In: Kutter E, Sulakvelidze A (eds) Bacteriophages: biology and applications. CRC Press, Boca Raton, pp 335–380. doi:10.1202/9780203491751.ch13

Tetart F, Desplats C, Kutateladze M, Monod C, Ackermann W, Krisch M (2001) Phylogeny of the major head and tail genes of the wide-ranging T4-type bacteriophages. J Bacteriol 183:358–366. doi:10.1128/JB.183.1.358-366.2001

Thiel K (2004) Old dogma, new tricks–21st century phage therapy. Nat Biotechnol 22:31–36. doi:10.1038/nbt0104-31

Tinsley C, Bille E, Nassif X (2006) Bacteriophages and pathogenicity: more than just providing a toxin?

Microbiol Infect 8:1365–1371. doi:10.1016/j.micinf.2005.12.013

Turki Y, Ouzari H, Mehri I, Ben Ammar A, Hassen A (2012) Evaluation of a cocktail of three bacteriophages for the biocontrol of *Salmonella* of wastewater. Food Res Int 45:1099–1105. doi:10.1128/AEM.70.6.3417-3424.2004

Uchiyama T, Abe T, Ikemura T, Watanabe K (2005) Substrate induced gene-expression screening of environmental metagenome libraries for isolation of catabolic genes. Nat Biotechnol 23:88–93. doi:10.1038/nbt1048

Verheust C, Pauwela K, Mahillon J, Helinski D, Herman P (2010) Contained use of bacteriophages: risk assessment and safety recommendations. App Bio Saf 15:32

Wang J, Jiang Y, Vincent M, Sun Y, Yu H, Wang J, Hu S (2005) Complete genome sequence of bacteriophage T5. Virology 332:45–65

Wang X, Kim Y, Ma Q, Hong S, Pokusaeva K, Sturino J, Wood T (2010) Cryptic prophages help bacteria cope with adverse environments. Nat Commun 1:147. doi:10.1038/ncomms1146

Wichels A, Biel H, Gelderblom H, Brinkhoff T, Muyzer G, Schutt C (1998) Bacteriophage diversity in the North Sea. Appl Environ Microbiol 64:4128–4133

Xiang X, Chen L, Luo Y, She Q, Huang L (2005) *Sulfolobus tengchongensis* spindle shaped virus SSTSV1: virus – host interactions and genomic features. J Virol 79:8677–8686. doi:10.1128/JVI.79.14.8677-8686.2005

Young Y (1992) Bacteriophage lysis: mechanism and regulation. Microbiol Rev 56:430

Zhang C, Li W, Liu W, Zou L, Yan C, Lu K, Ren H (2013) T4-like phage Bp7, a potential antimicrobial agent for controlling drug-resistant *Escherichia coli* in chickens. Appl Environ Microbiol 79:5559–5565. doi:10.1128/AEM.01505-13

Zhang Y, Huang C, Yang J, Jiao N (2011) Interactions between marine microorganisms and their phages. Chin Sci Bull 56(17):1770–1777. doi:10.1007/s11434-011-4503-2

Zillig W, Kletzin A, Schleper C, Holz I, Janekovic D, Hain J, Kristjansson K (1993) Screening for Sulfolobales, their plasmids and their viruses in Icelandic solfataras. Syst Appl Microbiol 16:609–662

Nishant A. Dafale received his M.Sc. and Ph.D. degree in Microbiology from CSIR-NEERI, Nagpur (degree awarded by RTM Nagpur University, Nagpur). He is currently working as a Scientist at EGD, CSIR-NEERI, Nagpur. He has published several papers in SCI journals and significantly contributed in books. He is recognized Ph.D. supervisor for Microbiology and Biotechnology in RTM Nagpur University, Nagpur, and NMU, Jalgaon, MS (India). His current areas of interest are genomics and metagenomics with respect to environment and health issues.

Zubeen J. Hathi graduated in Biochemistry in 2011 from Rashtrasant Tukadoji Maharaj Nagpur University, Nagpur. He received his M.Sc. degree in Biochemistry in 2013 from RTM Nagpur University, Nagpur. He is currently working as Project Assistant-II in project entitled "Microbial Diversity of Human Gut" at CSIR-National Environmental Engineering Research Institute.

Sarmistha Bit received the M.Sc. degree in Biotechnology from RTM Nagpur University, Nagpur, India. She is currently working as Project Assistant-II in project entitled "Microbial diversity of human gut" at CSIR-National Environmental Engineering Research Institute. Her current interests are bacteriophage therapy and clinical pathology.

Hemant J. Purohit is a microbiologist with training in molecular biology working as Head of Environmental Genomics Division, CSIR-NEERI. His current interests include developing deep insight through gene expression and their networks into biofuel production, bioremediation, and stress tolerance through application of omics tools. He is Editor of Indian Journal of Microbiology, Associate Editor of Applied Biochemistry and Biotechnology, and Academic Editor of Plos One.

18 Metagenomics: A Systemic Approach to Explore Microbial World

Manoj Kumar, Jitendra Kumar, and Nar Singh Chauhan

Abstract

Microbes are ubiquitous in nature and have been identified from all possible habitats on earth like water, soil, air, deep-sea sediments, living organisms, etc. Isolation, characterization, and industrial applications of these microbes were acknowledged from previous centuries. However, recent culture-independent microbial diversity studies with small subunit rRNA sequencing have speculated that majority of microbes (~99 %) were still hidden in respective ecosystem. Thus a huge microbial diversity remained untapped. Recently, a molecular biology technique was developed to explore their untapped gene pool, without the need of culturing them, i.e., metagenomics. Metagenomics tried to overcome this impediment by establishing and employing culture-independent approaches. Discoveries from metagenomic techniques have led to the agglomeration of novel gene sequences. This novel genetic information fetched with metagenomic approach has been utilized in biotechnological and pharmaceutical applications, as well as to boast our awareness of the microbial ecology in complex ecosystems. The metagenomic data can help to decode some of the key issues associated with the proper functioning of these complex microbial communities, as well as to analyze the microbial interactions within niche.

18.1 Introduction

Microbiology, a branch of life sciences, to study microorganisms has witnessed a tremendous development in last few decades. In the past, microbiology had started with the discovery of microscopes, followed by microbial classification with Bergey's manuals, role in diseases, and fermentation technology. These laborious techniques with limited scope were major hurdles and hampered the discovery rate in the field of microbiology. However, the development of molecular techniques like automated nucleic acid sequencing, cloning, and polymerase chain reaction technology and their applicability in

M. Kumar • J. Kumar • N.S. Chauhan (✉)
System Biology, Department of Biochemistry, Maharshi Dayanand University, 124001 Rohtak, Haryana, India
e-mail: nschauhan@mdurohtak.ac.in; narsinghc@gmail.com

V.C. Kalia (ed.), *Microbial Factories*, DOI 10.1007/978-81-322-2595-9_18

microbiology to estimate analyze the microbial diversity of various ecosystems like soil, sea water, freshwater, industrial effluents, biological samples and have placed microbiology on the path of tremendous growth. These techniques have performed a huge makeover on the classical face of microbiology and turned it into a very interesting and rapidly growing branch of life sciences.

Even these developments have shifted the microbiologist's view about the mode of exploration of these microbes from various habitats. It leads to the utilization of molecular biology techniques in addition to traditional techniques to explore microbial worlds. Among all methods, SSU rRNA gene microbial diversity analysis has gained maximum attention. A number of ecosystems were explored for total microbial diversity. These studies have speculated that each ecosystem posses a huge microbial diversity and only a fraction was represented in the form of cultured representation. This majority of unexplored microbes was considered as uncultured/uncultivated. This uncultured group found to occupy a major portion of microbial diversity (90–99 %) in all studied ecosystems. This view was further strengthened by the concept of "Great Plate Count Anomaly" (Bik 2009), which indicates that there is a huge difference in the number of microbes observed under the microscope and on a culture medium. This small percentage of cultivable microorganism is due to many reasons like lack of essential nutrients, growth stimulus, desired growth conditions, symbionts, competition, communication, etc. Even due to lack of necessary information for the growth of a specific microorganism in specific requirements, majority of microbes were not cultured. It creates a hurdle to provide an optimum culturing condition for all microorganisms to grow them as a pure culture, so the majority of microorganisms remain dormant during classical culturing conditions (Rondon et al. 1999; Handelsman 2004; Riesenfeld et al. 2004a, b; Streit and Schmitz 2004; Schloss and Handelsman 2004; Steele and Streit 2005; Singh 2010).

All these results speculated that a majority of the microbial gene pool from any environmental or biological sample remains untapped in these uncultured microbes and cannot be explored by culture-based methods. This vast gene pool is a reservoir to identify novel genes for biocatalysts, bioactive compounds, and novel physiological pathways, which can be explored for various industrial, biomedical, and biotechnological potential (Schmeisser et al. 2003; Venter et al. 2004). It has generated requirement of a robust technology for the exploration of this untapped gene pool; neither all these potential genes/operons will remain hidden due to nonavailability of all microbes as pure culture. Accordingly, a culture-independent molecular technique was developed to explore a vast gene pool of a microbiome without the need of culturing. Later on, this technique was coined with the terminology of "Metagenomics" (Handelsman 2004; Shade and Handelsman 2012).

18.1.1 What Is Metagenomics?

Metagenomics is a culture-independent approach to gain admittance to the total gene pool of microorganisms residing in an ecosystem or metagenomics is a technique for analyzing genetic contents of a microbial community (Fig. 18.1). The metagenomics term was conceived for describing the study of similar but not identical subjects. This technique starts with isolation of genomic DNA from all microorgansims habitating in an environmental/biological, followed by either cloning or direct sequencing (Daniel 2005). This genomic DNA has a representation of genetic contents from all available microorganisms available at environmental niche; hereby, it is called "Metagenomic DNA" (Rondon et al. 1999; Handelsman 2004; Riesenfeld et al. 2004a, b; Streit and Schmitz 2004; Schloss and Handelsman 2004; Steele and Streit 2005). Now this genetic material could be explored directly without culturing the organisms under study. This is how metagenomics provides direct access to total gene pool of all available microorganisms from an environment, even in the absence of their culture representatives (Handelsman et al. 1998). This metagenomic DNA could be either used to

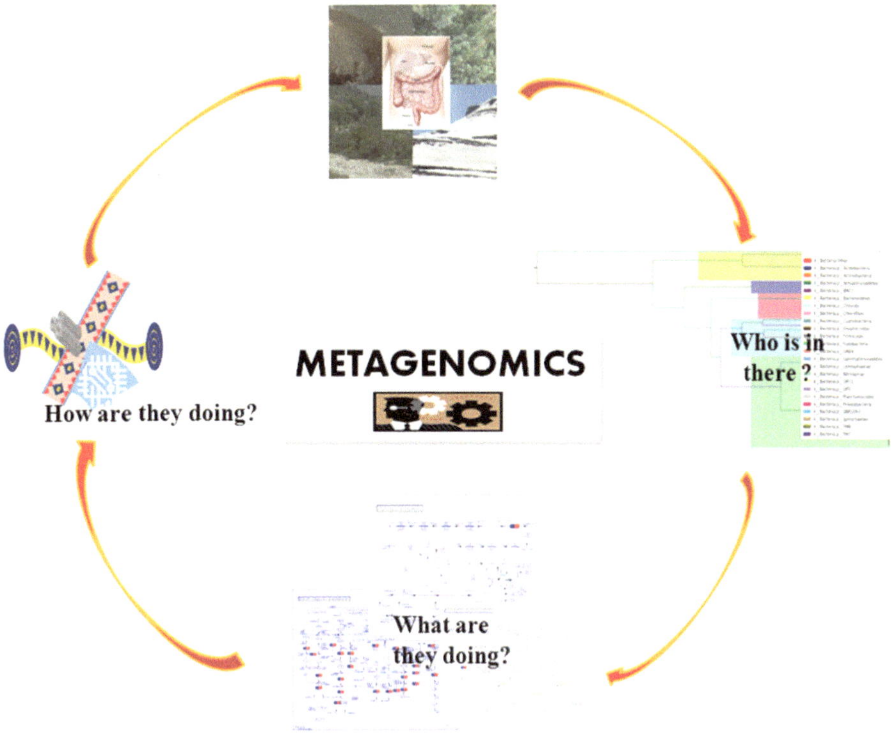

Fig. 18.1 An overview of metagenomic technology

amplify the rRNA gene (also known as 16S gene) to define its microbial diversity or sequenced directly to identify various microbial genetic contents for various physiological pathways (Tamakai et al. 2011). Overall, metagenomic technique tries to answer three basic questions associated with any sample, i.e., who are there?, what they are doing ?, and how they are performing a function? (Fig. 18.1).

Metagenomics helps us to define microbial community structure, genetic makeover, and their probable functions in a specific environment. Along with this, cloning of the environmental DNA into a cultured organism is also helpful to preserve them. Pace (1995) was the first to propose the concept of cloning the metagenomic DNA directly from the environmental sample (Pace 1995). It has created a revolution in the field of microbiology, and a number of breakthrough discoveries were made, which are otherwise impossible to perform with classical microbiological techniques.

18.1.2 Why Metagenomics?

Current small subunit rRNA (SSU rRNA)-based microbial diversity studies indicate that >99 % of the microbes inhabiting in various environmental niche are not easily cultivable. It makes them unavailable for biotechnology applications or basic research. A small fraction of cultivable microorganism from any niche provides a very limited information about growth conditions of majority of microorganisms, and when cultured under standard laboratory conditions, these microorganisms fail to grow because of lack of essential nutrients, growth stimulus, desired growth conditions, competition, communications, and symbionts. Along with this, a number of approaches used to explore the microbial diversity for basic research and industrial applications are biased due to the constrains of culturing methodologies. To conquer the limitations of cultivation techniques, a number of nucleic acid-based molecular methods were developed to get instant access to the microbial gene pool of an ecosystem. For example, SSU rRNA gene analysis of an environmental sample provides detailed information about the phylogenetic affiliation of all its inhabitating microbes (Tamakai et al. 2011). However, these types of analysis provide information only about the compositions of microbial communities; however, no information can be retrieved about their genetic contents and respective physiological role in microbial community. So, it created a need of a robust, culture-independent technology to study the genetic components of any microbial community. This technique should elude the need of microbial culturing for studying them. Among the various methods developed to establish an access to the ecology, and gene pool of uncultured organisms, metagenomics has emerged as a powerful attraction. Metagenomics is one of the young members of scientific arena, which has fetched a huge scientific attention within a few years of its establishment. Nowadays, it is used very frequently to decipher the complex metagenome composition of any microbial niche to decode various scientific mysteries.

18.1.3 Metagenomics Technology

The metagenomics technology is a culture-independent multistep molecular technology. This technique starts with a direct DNA isolation from any environmental sample, followed by either cloning in a suitable cultivable host followed by screening of genes or sequencing directly to elucidate genetic components of that microbial ecosystem (Fig. 18.2). Overall, metagenomics technology has two standard approaches to study the microbial communities to decipher various novel genetic elements, i.e., functional metagenomics and sequence metagenomics.

18.1.3.1 Functional Metagenomics

Functional metagenomics is a culture-independent molecular approach to identify a novel gene/operon based on the expression of specific character, such as biocatalytic activity or particular physiological function like resistance towards biotic/abiotic stress, etc. (Fig. 18.1). Functional metagenomics studies are usually started with isolation of environmental DNAs. A number of difficulties were found associated with isolation and purification of good-quality high molecular weight DNA. Among all, contamination of phenolic compounds, various polycyclic impurities in purified metagenomic DNA, very low yield, and DNA shearing are some of the major issues in metagenomic DNA isolation. The presence of phenolic/polycyclic compounds generally inhibits activities of various enzymes used in molecular cloning experiments. Even low-quality (sheared) metagenomic DNA is also not suitable for the construction of metagenomic library. In addition, in case of aqueous samples, a large volume of sample is required to obtain sufficient amount of sample for isolation of good-quality metagenomic DNA for downstream applications. As an outcome, the construction of large-sized metagenomic libraries (as this library represents cloned genomic fragments of various microorganisms) is quite challenging. In such circumstances, various research groups perform precultivations of microbial communities in laboratory conditions to attain good-quality high molecular weight metagenomic DNA. Now this metagenomic DNA is employed for construction of a metagenomic library. However, due to precultivation of microbial community in standard laboratory conditions, there is a strong possibility of enrichment of a specific microbial community in the whole sample followed by in metagenomic library. Nevertheless, this technique has been found to be more successful to isolate high molecular weight metagenomic DNA for cloning in suitable vector system to generate metagenomic libraries. Simultaneously, development of β-agarase-based gel purifications of metagenomic DNA has enabled to enrich high-quality, high molecular weight metagenomic DNA directly from the native sample without any prior precultivation. Isolation and purification of metagenomic DNA is followed by cloning of metagenomic DNA into a suitable cloning vectors and host strains for development of metagenomic libraries. In general, development of small-sized, plasmid-based metagenomic libraries (2–10 KB) in an *Escherichia coli* strain is a

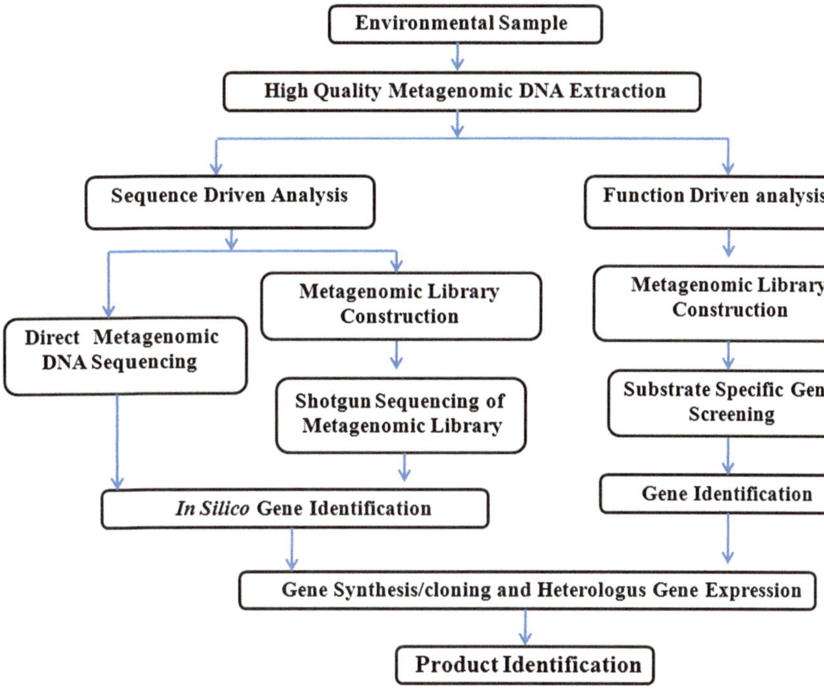

Fig. 18.2 General metagenomic approach for the discovery of novel gene from an environmental sources

classical approach (Brady et al. 2001; Ogawa and Maki 2003; Brady and Clardy 2004; Gabor et al. 2004; Grant et al. 2004; Riesenfeld et al. 2004a, b; Robertson et al. 2004; Yun et al. 2004; Ranjan et al. 2005; Rhee et al. 2005; LeCleir et al. 2007; Chauhan et al. 2009; Kapardar et al. 2010; Fu et al. 2013). However, due to the presence of small DNA inserts, the identification possibility of large gene clusters is very negligible. Accordingly, researchers have used other vector system like fosmid/cosmid to clone 25–35 Kb DNA insert and bacterial artificial chromosome (BAC) vectors to clone >100 Kb DNA inserts. Researches have constructed cosmid/fosmid and BAC-based metagenomic libraries in *Escherichia coli* host from various environments like compost soil, human gut, etc. A number of studies were performed to identify various novel genetic elements, either for any biocatalytic activity or any physiological functions (Tuffin et al. 2009). However, the discovery rate for these novel genetic elements is very poor. This poor rate of gene discovery is attributed to a number of issues like codon usage, incompatible host machinery, toxic gene product, cloned DNA stability, poor or low gene expression, improper screening methods, etc. Among all, heterologous gene expression in *E. coli* is a major hurdle to extract complete metagenomic information in the functional metagenomics analysis. Simultaneous development of novel vectors and usage of multiple host have provided solutions to some of these issues (Lim et al. 2005; Hedlund and Staley 2006; Wang et al. 2006; Jiang and Wu 2007; Chae et al. 2008; Chung et al. 2008; Lee et al. 2008; Kennedy et al. 2011).

18.1.3.2 Sequence Metagenomics

Sequenced-based metagenome analysis is the identification of various genetic elements of a microbial community either by directly sequencing the metagenomic DNA isolated from an environment or sequencing the clones from a metagenomic library. After the sequencing, sequences were analyzed for the gene of interest and phylogenetic anchors (Fig. 18.1). Sequence analysis with the identification of phylogenetic anchors is a systematic approach. This approach was first implemented by DeLong (2006), who has analyzed the first genomic sequence of an uncultured archaeon (Tamakai et al. 2011; Sul et al. 2011). In this approach, identification of the gene of interest is solely based on the information available for their respective database homologs. However, a majority of sequence data in various databases is still uncharacterized and annotated; in such scenario, majority of sequences generated with sequence metagenomics remained unidentified. Simultaneously, the size of any environmental metagenome is really large, so it requires a huge amount of finances to get it sequence. These issues limit applicability of sequence metagenomics to decipher novel genes/operons for various biotechnological applications.

Though both of these metagenomic approaches have strengths and limitations, but still a number of discoveries were made using metagenomic approaches.

18.2 Metagenomics and Its Industrial Application

As more than 99 % of total microorganism corresponds to non-cultured microbes in most of the microbial niches, metagenome searches with either functional metagenomics or sequence metagenomics will always lead to identification of novel genes and pathways. It can be easily understood that rate of identification of novel genes/appearance will be always higher with the metagenomic approach encompassing any approach based on culturing of microbes. With the onset of metagenomic era, a large number of novel genes/operons encoding biocatalysts or bioactive molecules were identified. Majority of them are having good applicability in industrial and pharmaceutical products. Even the metagenomic libraries from various environments have been searched for antibiotics, enzymes, and other bioactive compounds, which are a booming area of biotechnology. The industrial applications of metagenomics include identification of novel biocatalysts, discovery of new antibiotics for personalized medicine, and bioremediation. The huge amount of genomic information gathered by metagenomics has huge promises to provide new enzymes and molecules with diverse applications (Gilbert and Dupont 2011). There is a huge demand for novel enzymes and biocatalysts, and metagenomics is currently one of the most promising technologies to provide the desired molecules (Lorenz et al. 2002; Schloss and Handelsman 2003). Proteases, cellulases, esterases/lipases, xylanases, and various other industrially important enzymes have been produced through metagenomics (Fig. 18.2). The following are some of the enzymes that have been unlocked from genetically untapped resources.

Cellulases (EC 3.2.1.4.) are a group of enzymes known to catalyze the hydrolysis of cellulose. This group of enzymes was mainly identified from fungi, bacteria, and protozoans. Though cellulase has also been identified from even higher eukaryotes, recent studies indicated the microbial origin of these cellulases in these hosts. As an example, cellulases have been identified from wood-eating termites, but these cellulases are produced by the microbial intestinal symbionts of termites. Based on the type of catalyzed reaction, there are five general types of cellulases, i.e., exocellulase, endocellulase, cellobiase, oxidative cellulase, and cellulose phosphorylase (Lynd et al. 2002). They are used in a number of applications, like paper recycling, cotton processing, detergent enzymes, and so forth. Thus, cellulase is the third largest industrial enzyme worldwide, by dollar volume (Lynd et al. 2002). With a huge industrial significance, cellulases, along with hemicellulases, have occupied a remarkable position in the list of very important biocatalysts (Lynd et al. 2002; Collins et al. 2005; Steele et al. 2009; Wilson 2009). Metagenomics approach has been utilized to identify novel genes for cellulases from various environments like compost soils, soil from cold regions, rumen samples, etc. Ruminal cellulolytic microorganisms were widely known to digest cellulosic materials. Accordingly, this ecosystem was explored with metagenomic approach, and a number of novel genes encoding cellulases were identified (Krause and Combs 2003). A breakthrough study to identify cellulolytic genes for liquid biofuel production from biomass sugars was carried out by Hess et al. (2011) and identified 27,755 candidate genes for possible cellulolytic activity (Hess et al. 2011). They had used sequence metagenomic approach to accurately reveal cellulolytic genes at a massive scale. In addition, Alvarez et al. (2013) isolated and characterized a novel cellulase from a sugarcane soil metagenome (Alvarez et al. 2013), while Duan et al. (2009) isolated and characterized metagenomic gene encoding acidic cellulases from buffalo rumen metagenome (Voget et al. 2006; Duan et al. 2009) and also characterized a highly stable metagenome-derived halotolerant cellulase. The metagenome-derived cellulase is ideal for industrial applications (Voget et al. 2006). Given the amount of research that is underway in order to unlock novel cellulases from nature, it is expected that the vast gene mining for cellulase enzymes will become literally possible in the near future.

Lipases (E.C.3.1.1.3) are a significant member of hydrolase family. It performs hydrolysis and synthesis of long chain acyl glycerol with trioleyl glycerol substrate. A number of lipase-encoding DNA sequences were identified through recent metagenomic studies. Peng et al. (2014) isolated a unique lipase from a metagenomic library constructed from marine sediments. The lipase was found alkaline stable and has potential application into food industry (Peng et al. 2014). Ranjan et al. (2005) have identified 12 lipolytic clones from pond water metagenomic library. Majority of these clones harbor novel lipolytic genes, which have shared very low similarity with database homologs (Ranjan et al. 2005). Lee et al. (2006) isolated and characterized a new family of bacterial lipase from tidal flat sediments (Lee et al. 2006). Hardeman and Sjoling (2007) also isolated a cryoactive lipase from marine sediment metagenomic library (Hardeman and Sjoling 2007). Selvin et al. (2012) have identified a novel halotolerant lipase that was isolated from a marine sponge fosmid metagenomic library (Selvin et al. 2012). In the recent past, a number of lipases were isolated using metagenomic approach, and these lipases have showed novel characteristics, namely, thermal stability, alkaline stability, organic solvent tolerance, cold active nature, and so forth, making them potential candidates for industrial use (Glogauer et al. 2011; Chow et al. 2012; Ngo et al. 2013). Preeti et al. (2014) have explored human oral microbiome with functional metagenomic approach and identified an OMLip1 insensitive lipase. Identifications of novel lipases having unique characteristics like substrate specificity, enantio selectivity, pH and salt tolerance, etc. with culture-independent metagenomic approach have proved the potential of metagenomic approach for biotechnological advancement.

Xylanase catalyzes progressive breakdown β-1,4 glycosidic bonds of the xylan. Xylan is a highly abundant polysaccharide in plant materials. Hereby, xylanase has significant application in biofuel production from plant biomass. Accordingly, detailed studies are required to identify and utilize the xylanase. Verma et al. (2013) have identified a novel xylanase from compost-soil metagenome. This novel xylanase has showed alkali stability and thermostability (Verma et al. 2013). Gong et al. (2013) isolated and characterized a GH10 family xylanase through functional screening of cow rumen metagenomic BAC library (Gong et al. 2013).

Proteases are another largest groups of industrial important hydrolytic enzymes (Orhan et al. 2005). The proteases have wide range of demand in detergent industry, food industry, and leather industry with a number of medical applicabilities. During the past decades, many efforts have been made to identify new proteases with highly promising metagenomic approach. Biver et al. (2013) isolated an alkaline protease using metagenomic approach. These protease showed its possible applications in the detergent and bleaching industries (Biver et al. 2013). Two detergent-resistant serine proteases were discovered from the surface sand of deserts (Glogauer et al. 2011). A novel protease belonging to chymotrypsin-like S1 serine proteases was isolated by Niehaus et al. (2011). Neveu et al. (2011) have constructed metagenomic libraries of the Gobi and Death Valley deserts and screened two novel serine proteases (Neveu et al. 2011), while Pushpam et al. (2011) identified a metagenomic alkaline serine protease from goat skin surface metagenome (Pushpam et al. 2011).

Pectinases are glycosidic hydrolases to degrade pectin polymers present in various plant components. Singh et al. 2012 discovered a thermostable and thermoactive novel pectinase from the soil metagenome. Sathya et al. (2014) have identified a novel metagenomic pectinase clone from Western Ghats, India. This soil metagenomic pectinase-positive clone belongs to GH 28 family and shared only 52 % sequence similarity with the known PGs (polygalacturonases). Importantly, it has acted in a broad range of temperature and pH.

Amylases (EC 3.2.1.) are GH-13 family hyrdolases that catabolize hydrolytic cleavage of starch molecules into diversified products (Bordbar et al. 2005). Amylases are of three types: alpha (α) amylase, beta (β) amylase, and gamma (γ) amylase. The α-amylases (1,4-α-D-glucanglucanohydrolase) are calcium-dependent metalloenzymes. The α-amylase randomly cleaves the various long chain polysaccharides and generates varying size products like maltotriose, maltose, glucose, etc. β-amylase was first purified in the 1940s, which is commonly synthesized by microorganisms and higher eukaryotes. β-amylase catabolizes the hydrolysis of starch polymer from nonreducing end at penultimate α-1,4 glycosidic bond and releases two dextrose units. Gamma amylase cleaves at the last α (1–4) glycoside bond at the nonreducing end of starch, and dextrose is released. The γ-amylase works effectively in an acidic environment and can also cleave at α (1–6) glycosidic linkages. All these amylases are of industrial importance, and various metagenomic investigations have been performed to identify novel amylase genes. Vidya et al. (2011) isolated a thermostable and calcium-dependant amylase from a soil metagenome. They have also shown its potential application in baking and destarching (Sharma et al. 2010). In addition, Sharma et al. (2010) identified a metagenomic amylase from the soil derived from Northwestern Himalayas (Vidya et al. 2011).

Oxidoreductases are another group of extremely important catalysts identified through metagenome searches. These groups of enzymes are having high enantioselectivity and widely used for the synthesis of a number of pharmaceutically important compounds. This group of enzyme catalyzes synthesis of various chiral alcohols, carbonyl compounds, and hydroxy acids, whose chemical synthesis is quite challenging (Schloss and Handelsman 2003). Singh et al. (2010) have used functional metagenomic approach to identify two novel monoxygenase coding genes from an effluent treatment plant sludge microbiome (Singh et al. 2010). Jadeja et al. (2014) have identified a number of oxygenases involved in catabolism of phenolic compounds like naphthalene, anthracene, phenol, etc. (Jadeja et al. 2014).

Finally, isolation and identification of various genetic elements coding therapeutic compounds are always a core area of research. Schirmer et al. (2005) have used the metagenomic approach and identified and characterized a >100 Kb polyketide synthesis (PKS) gene cluster from a marine sponge *Discodermia dissoluta.* Silva et al. (2013) have identified six type I PKS groups in the sponge *Arenosclera brasiliensis* microbiome.

18.3 Metagenomics and Microbial Ecology

Sequence metagenomics provides a huge amount of sequence data for the understanding of the genetic makeover of a microbial community in an environmental niche. This information could provide significant details of various physiological pathways prevalent in the community, which can be harnessed to study ecological roles of microbes in a environmental niche. The significance of metagenomics to reveal microbial ecology in complex environments can be easily accessed.

18.3.1 Proteorhodopsin-Based Energy Fixation

Metagenomic survey of Sargasso sea leads to the discovery of the rhodopsin-like photoreceptors in marine bacteria. Bacteria, Archaea, and Eukaryota were found to reside in the ocean's photic zone under the constant exposure to sunlight. Bacteria residing in these unique environments contained photoproteins called proteorhodopsins. These photoproteins catalyze light-driven translocation of the proton across the membrane to generate a proton gradient. This proton gradient was utilized to generate cellular energy for microbial physiology. This discovery has helped to understand growth and survival mechanism among different marine microbial communities. Previously, rhodopsins were considered of only Archaeal origin. In 2000, Beja et al. sequenced a 130 Kb fragment of an uncultured γ-proteobacterium belonging to SAR 86 group and reported the presence of bacteriorhodopsin even in other microbial groups. Bacteriorhodopsins couple light-energy harvesting with carbon-cycling in the ocean via non-chlorophyll-based pathways.

18.3.2 Survival Strategies in Acidophilic Environment

In 2004, Tyson using a metagenomic approach elucidated community structure of acidophilic biofilms and their survivability in that ecosystem. They have identified that the biofilm sample contains approximately 108 *Leptospirillum* group II cells. The metagenomic analysis of these samples indicates the presence of the gene involved in nitrogen fixation and carbon fixation along with iron oxidation. The presence of a variety of proton efflux systems, ATPases, antiporters, and symporters indicated how these microbial communities maintain neutral pH in host cytoplasm. Simultaneously, it has been identified that a majority of microbial community members possess resistance genes for a number of toxic metals. All these genetic systems overall maintain growth of these microbial communities, even in acidic conditions (Tyson et al. 2004).

18.3.3 Role of Microorganisms in Honey Bee Colony Collapse Disorder

From many decades, many American countries, where honey production is the main source of income, were suffering from big losses, because of a large number of bee hives were collapsing. Beginning in 2006, beekeepers noticed an unusual disappearance of their honeybee colonies. There were no traces of disappearance, and even no dead bees were being found near the colonies. Since then, more than 30 % of the honeybee colonies disappeared each year. Due to the continual disappearance of worker bees, their colonies become weak which failed to function. Due to the collapse of the colonies, this phenomenon is properly named the Colony Collapse Disorder (CCD). There are many proposed causes were speculated for this phenomenon, including: the use of pesticides and insecticides, poor nutrition, habitat loss, and stress factors. Many research groups tried to identify the cause of honey bee colony collapse using culture isolation approach, but they failed to diagnose the causative agent. Later on, a study was performed by Cox-Foster in 2007 using metagenomic approach and analyzed the colony collapse disorder among infected hives compared to uninfected hives. He was the first person to reveal the role of microorganisms. According to this analysis, abundance of some specific group of gamma proteobacteria along with some fungal and viral members leads to diseases (Cox-Foster et al. 2007).

18.3.4 Aerobic Anoxygenic Photosynthesis

From many decades, it was a big mystery to understand how microbes residing in aerobic environment able to perform anaerobic photosynthesis. But in 2005, Waidner and Kirchman had performed a study using metagenomics approach, and a novel photosynthetic center for aerobic anoxygenic photosynthesis was discovered. This type of anoxygenic photosynthesis was the most ancient form of photosynthesis. In general, anoxygenic photosynthesis is the phototrophic process where light energy is captured and converted to ATP, without the production of oxygen. The light-harvesting complexes (LHC) encircle the reaction centers (RCs) to make an active photosynthetic core complexes in anoxygenic microbes. The *pufM* and *pufX* genes of the *puf* operon encode the structural proteins of

RC-LHC1. This study has identified diverse group of *pufM* and *pufX* which shows a limited homology *Rhodobacter* strains.

18.3.5 Reverse Methanogenesis

Reverse methanogenesis is a microbial process of methane consumption in anoxic sediments. This physiochemical process has huge impacts on the global environment. It has been observed that annually this unique microbial process consumes approximate 70 billion kilograms of methane in marine sediments. This process reduces the release of greenhouse gases from marine ecosystem to the surrounding atmosphere. Besides its ecological impact, the biological mechanisms involved in anaerobic methane oxidation were not studied in detail. Recent studies have indicated the role of archaea and sulfate-reducing bacteria (SRB). It was speculated that both these microbial groups can couple reverse methanogenesis to sulfate reduction. However, due to lack of their cultured representatives, there was no significant information available about the individual role of these microbial groups in anoxic methanogenesis. But the metagenomic study done by Hallam et al. (2006) concludes that during course of evolution, some of the methanogenic *Archaea* evolved genetic machinery to perform reverse methanogenesis for consumption of methane in order to produce the cellular components and energy. Simultaneous genomic analysis of methane oxidizers from deep-sea sediments has indicated that all genetic elements involved in reverse methanogenesis belong to an individual group (i.e., ANME-1 group) of archaeal methanotrophs.

18.3.6 Deepwater Horizon Oil Spill

The Deepwater Horizon (DWH) oil spill in 2010 represents a series of metagenomic studies to study impact of oil spills on the marine microbial ecology in the Gulf of Mexico. High-throughput 16S gene surveys have deciphered the enrichment of hydrocarbon-degrading bacteria taxa *Colwellia* and *Oceanospirillales* in the deepwater "oil plume" (Hazen et al. 2010; Baelum et al. 2012). Kimes et al. (2013) have used functional metagenomics and sequence metagenomics approach to identify the metabolic process for hydrocarbon degradation. They have identified anaerobic metabolism of hydrocarbon degradation in deep-sea sediments, compared to predominantly aerobic processes in surface oil degradation.

Simultaneously, a significant role of natural hydrocarbon seeps inhabitant, Deltaproteobacteria, was observed in degrading hydrocarbon compounds. In another investigation of deep-sea sediments, Mason et al. (2014) showed that sediment communities possessed specific degradation pathways for cyclohexane, aliphatics, and simple aromatics. Even they have speculated that genes for these xenobiotics degradations were already present in a natural ecosystem, but got enriched in marine sediments after the DWH oil spill (Montagna et al. 2013). In addition, metagenomic data were able to illustrate biases in the metabolism of various hydrocarbon compounds. Even they have deciphered the basic cause of slower biodegradation of certain hydrocarbon classes in some marine ecosystems (Mason et al. 2012).

18.3.7 Antibiotic Resistance Pathways

Infectious diseases have been a major cause of death all over the world, especially in developing countries. Along with this, the majority of these infectious agents have evolved antibiotic resistance mechanisms. However, till date there is debate about the origin of these antibiotic resistance mechanisms. In this field, cultured microorganisms were explored to decipher antibiotic resistance genes and antimicrobial compounds. Recently, functional and sequence-based metagenomic approaches were used to decipher the antibiotic resistome in the uncultured microbial world from various environments. Recently, there have been many investigations to identify antibiotic resistome in various natural environments like deep-sea sediments (Song et al. 2005), human oral cavity (Xie et al. 2010), human gut (Schjorring and Krogfelt 2011), pristine cave ecosystems (Bhullar et al. 2012), urban sewage, and wastewater (Novo et al. 2013). Riesenfeld et al. (2004) used functional metagenomics approach and identified nine unique antibiotic resistance clones against aminoglycoside and tetracycline antibiotics from Wisconsin remnant oak savannah soil (Riesenfeld et al. 2004). Simultaneously, undisturbed Alaskan soil was explored to identify antibiotic resistance genes towards β-lactam antibiotics. This study has identified novel β-lactamases genes (Allen et al. 2009). Marathe et al. (2013) have identified the enrichment of antibiotic resistance genes in microbial community of a treatment plant. This study has highlighted the genetic process involved in acquaintance of antibiotic resistance mechanisms in microbes (Marathe et al. 2013). In overall, metagenomic approach has helped to understand that resistance genes are highly diversified and abundant in various natural environments.

In addition to these studies, Chauhan et al. (2009) have used metagenomic approach to study arsenic detoxification mechanisms prevalent in various effluent treatment plant microbiome. It leads to identification of a novel gene *ArsN* involved in arsenate metabolism. Kapardar et al. (2010) have explored freshwater pond microbiome with functional metagenomic approach and identified two unique genes involved in salt tolerance (Kapardar et al. 2010). Sangwan et al. 2014a, b, have performed metagenomic analysis to identify an association between microbial community composition and differential gene expression with the change in the environment. They have identified the enrichment of a microbial group involved in hexachlorocyclohexane (HCH) degradation at the dumping site of HCH isomers (Sangwan et al. 2014a, b). More et al. (2014) have performed metagenomic analysis of activated sludge to reveal bioactive microbial communities and metabolic pathways involved in degradation of various aromatic compounds. This study leads to identification of the microbial groups representing *Synergistetes* and *Elusimicrobia* phylums. Even this study has identified various metabolic pathways involved in catabolism of phenolic compounds (More et al. 2014).

18.4 Metagenomics and Human Gut Microbiome

Humans live in constant association with microbes, and some of these microbes became permanent inhabitant on surfaces and inside of the human body. The number of our microbial companions outnumbers our own cells. In an approximation, a number of microbes are tenfold higher than the total number of cells. These microbial companions express approximately 100 times higher number of genes in comparison of our own genome, and due to these facts, human microbiome has been considered another body organ. Among all human microbiomes, the human intestinal microbiome is a dense and complex ecosystem residing in the distal part of our digestive tract. Its role in metabolizing dietary constituents (Sonnenburg et al. 2005; Flint et al. 2008; Ley et al. 2008) and in protecting the host against pathogens is crucial to human health (Macdonald and Monteleone 2005; Manichanh et al. 2006; Turnbaugh and Gordon 2009). It is mainly composed of commensal bacteria from the *Bacteroidetes, Firmicutes, Proteobacteria*, and *Actinobacteria* phyla, as well as several archaeal and eukaryotic species. With up to 10^{12} cells per gram of feces, the bacterial abundance is estimated to reach 1000 operational taxonomic units (OTUs) per individual, 70–80 % of the most dominant ones being subject specific (Tap et al. 2009; Tasse et al. 2010). Not only from the gut, but microbes have also been identified from various anatomical sites like oral cavity, dental plaque, ear, skin, armpits, vagina, etc. This human microbiota plays an important role in human physiology (Yatsunenko et al. 2012). In such scenario, information of the human microbiome along with human genome is crucial to understand the human biology (Mueller et al. 2012). However, all these studies have been conducted using culture-independent molecular tools. Among all these tools, metagenomics is one of the pioneer technology to understand and elucidate the molecular components of the human microbiome. Various metagenomics-based studies have shown profound effect of human microbiome on human physiology, nutrition, and immunity (Nicholson et al. 2012). Characterization of the microbiota for comparison of microbial communities and their contribution to human health and disease (Dave et al. 2012) have been carried out recently in the human microbiome project. These studies have provided insights into the composition of microbial community at various anatomical sites (Li et al. 2012; Human Microbiome project Consortium 2012; Parfrey and Knight 2012). A number of studies have shown that, after the establishment of a stable microbial community structure at an anatomical site (gut), it remains resilient until extreme selection pressure is applied (surgery, long-term antibiotic usage). In case if there are any disruptions in these human-associated microbial communities, It generally leads to unhealthy states of the body and develops diseases (Gordon 2012; Neuman and Nanau 2012; Sudo 2012). Recent studies have given indications that state of human body, i.e., either healthy or diseased, is influenced by its microbiome (Albenberg et al. 2012; Gordon 2012; Hunter 2012; Harris et al. 2012). Discovery of new antibiotic resistance genes in human microbiome is one of the new successes through new functional metagenomic techniques. Many studies report the antibiotic resistance genes present in human microbiome by using PCR methods. Functional metagenomic screens of 95 unique inserts, represent different known resistance genes within 10 novel beta-lactamase gene families (Moore et al. 2011). All these discoveries showing the role of human microbiomes in human health with culture-independent metagenomic approach have proved the potential of metagenomic approach for human healthcare. Using metagenomic approach, human microbiome project (HMP) has established a functional association between human microbiomes with host system. Advent of economical high-throughput sequencing technology and improved bioinformatics pipelines for metagenomic analysis has ensured easy access to the vast amount of knowledge of human microbes. This vast information of the human microbiome could result in development of novel, economical, sustainable, and efficient methods of clinical treatment of various physiological disorders to improve human healthcare.

18.5 Conclusions

Metagenomics has changed the prospective of the microbiologists and redefined the concept of a genome. Metagenomics has provided instant access to total gene pool of a microbiome from any living and nonliving environment. Culture-independent approach of metagenomics has provided a platform for accelerated rate of gene discovery. Metagenomics has created new avenues of research in the field of biotechnology. The functional metagenomics approach leads to identifications of novel enzymes and antibiotics from diverse environments. Metagenomic sequencing has open new avenues to understand microbial interactions, survival in complex environments like human gut, acid mine drainage, etc. Metagenomics provides the tools to decipher abundant information with an understanding of the uncultured microbial groups. A number of unique ecosystems exist on earth, which have not been studied and required a detailed analysis. Metagenomics can boost our understanding of various alluring and usual ecosystems like hyperthermal vents, soda lakes, volcanic springs, deserts, polar ice, wastewater treatment plants, deep soil sediments, and eukaryotic host systems. Metagenomic studies of such unique ecosystems will expand our understanding and continue to enhance our knowledge of microbial world.

Acknowledgement The author wish to thank CSIR scheme project 60(0099)/11/EMRII and UGC major research project SR1256(2012) for providing the necessary funds and facilities.

References

Albenberg LG, Lewis JD, Wu GD (2012) Food and the gut microbiota in inflammatory bowel diseases: a critical connection. Curr Opin Gastroenterol 28:314–320. doi:10.1097/MOG.0b013e328354586f

Allen HK, Moe LA, Rodbumrer J, Gaarder A, Handelsman J (2009) Functional metagenomics reveals diverse beta-lactamases in a remote Alaskan soil. ISME J 3:243–251. doi:10.1038/ismej.2008.86

Alvarez TM, Paiva JH, Ruiz DM, Cairo JPLF, Pereira IO, Paixao DAA, Almeida RF, Tonoli CCC, Ruller R, Santos CR, Squina FM, Murakami MT (2013) Structure and function of a novel cellulase 5 from sugarcane soil metagenome. PLoS One 8:e83635. doi:10.1371/journal.pone.0083635

Baelum J, Borglin S, Chakraborty R, Fortney JL, Mason RLOU, Auer M, Zemla M, Bill M, Conrad ME (2012) Deep-sea bacteria enriched by oil and dispersant from the deepwater horizon spill. Environ Microbiol 14:2405–2416. doi:10.1111/j.1462-2920.2012.02780.x

Beja O, Aravind L, Koonin EV, Suzuki MT, Hadd A, Nguyen LP, Jovanovich SB, Gates CM, Feldman RA, Spudich JL, Spudich EN, DeLong EF (2000) Bacterial Rhodopsin: evidence for a new type of phototrophy in the sea. Science 289:1902–1906. doi:10.1126/science.289.5486

Bhullar K, Waglechner N, Pawlowski A, Koteva K, Banks ED, Johnston MD, Barton HA, Wright GD (2012) Antibiotic resistance is prevalent in an isolated cave microbiome. PLoS One 7:e34953. doi:10.1371/journal.pone.0034953

Bik EM (2009) Composition and function of the human associated microbiota. Nutr Rev 67:164–171. doi:10.1111/j.1753-4887.2009.00237.x

Biver S, Portetelle D, Vandenbol M (2013) Characterization of a new oxidant-stable serine protease isolated by functional metagenomics. SpringerPlus 2:410. doi:10.1186/2193-1801-2-410

Bordbar A, Omidiyan K, Hosseinzadeh R (2005) Study on interaction of α -amylase from *Bacillus subtilis* with cetyl trimethylammonium bromide. Colloids Surf B Biointerfaces 40:67–71. doi:10.1016/j.colsurfb.2004.10.002

Brady SF, Clardy J (2004) Palmitoyl putrescine, an antibiotic isolated from the heterologous expression of DNA extracted from bromeliad tank water. J Nat Prod 67:1283–1286. doi:10.1021/np0499766

Brady SF, Chao CJ, Handelsman J, Clardy J (2001) Cloning and heterologous expression of a natural product biosynthetic gene cluster from eDNA. Org Lett 3:1981–1984. doi:10.1021/ol015949k

Chae JC, Song B, Zylstra GJ (2008) Identification of genes coding for hydrolytic dehalogenation in the metagenome derived from a denitrifying 4-chlorobenzoate degrading consortium. FEMS Microbiol Lett 281:203–209. doi:10.1111/j.1574-6968.2008.01106.x

Chauhan NS, Ranjan R, Purohit HJ, Kalia VC, Sharma R (2009) Identification of genes conferring arsenic resistance to *Escherichia coli* from an effluent treatment plant sludge metagenomic library. FEMS Microbiol Ecol 67:130–139. doi:10.1111/j.1574-6941.2008.00613.x

Chow J, Kovacic F, Antonia YD, Krauss U, Fersini F, Schmeisser C, Lauinger B, Bongen P, Pietruszka J, Schmidt M, Menyes I, Bornscheuer UT, Eckstein M, Thum O, Liese A, Dieckmann JM, Jaeger KE, Streit WR (2012) The metagenome derived enzymes LipS and LipT increase the diversity of known lipases. PLoS One 7:e47665. doi:10.1371/journal.pone.0047665

Chung EJ, Lim HK, Kim JC, Choi GJ, Park EJ, Lee MH, Chung YR, Lee SW (2008) Forest soil metagenome gene cluster involved in antifungal activity expression in *Escherichia coli*. Appl Environ Microbiol 74:723–730. doi:10.1128/AEM.01911-07

Collins T, Gerday C, Feller G (2005) Xylanases, xylanase families and extremophilic xylanases. FEMS Microbiol Rev 29:3–23. doi:10.1016/j.femsre.2004.06.005

Cox-Foster DL, Conlan S, Holmes EC, Palacios G, Evans JD, Moran NA, Quan PL, Briese T, Hornig M, Geiser DM, Martinson V, Engelsdorp DV, Kalkstein AL, Drysdale A, Hui J, Zhai J, Cui L, Hutchison SK, Simons JF, Egholm M, Pettis JS, Lipkin WI (2007) A metagenomic survey of microbes in honey bee colony collapse disorder. Science 318:283–287. doi:10.1126/science.1146498

Daniel R (2005) The metagenomics of soil. Nat Rev Microbiol 3:470–478. doi:10.1038/nrmicro1160

Dave M, Higgins PD, Middha S, Rioux KP (2012) The human gut microbiome: current knowledge, challenges, and future directions. Transl Res 160:246–257. doi:10.1016/j.trsl.2012.05.003

DeLong EF (2006) Archaeal mysteries of the deep revealed. Proc Natl Acad Sci U S A 103:6417–6418. doi:10.1073/pnas.0602079103

Duan CJ, Xian L, Zhao GC, Feng Y, Pang H, Bai XL, Tang JL, Ma QS, Feng JX (2009) Isolation and partial characterization of novel genes encoding acidic cellulases from metagenomes of buffalo rumens. J Appl Microbiol 107:245–256. doi:10.1111/j.1365-2672.2009.04202.x

Flint HJ, Bayer EA, Rincon MT, Lamed R, White BA (2008) Polysaccharide utilization by gut bacteria: potential for new insights from genomic analysis. Nat Rev Microbiol 6:121–131. doi:10.1038/nrmicro1817

Fu J, Leiros HKS, Pascale D, Johnson KA, Blencke HM, Landfald B (2013) Functional and structural studies of a novel cold-adapted esterase from an Arctic intertidal metagenomic library. Appl Microbiol Biotechnol 97:3965–3978. doi:10.1007/s00253-012-4276-9

Gabor EM, de Vries EJ, Janssen DB (2004) Construction, characterization, and use of small-insert gene banks of

DNA isolated from soil and enrichment cultures for the recovery of novel amidases. Environ Microbiol 6:948–958. doi:10.1111/j.1462-2920.2004.00643.x

Gilbert JA, Dupont CL (2011) Microbial metagenomics: beyond the genome. Ann Rev Mar Sci 3:347–371. doi:10.1146/annurev-marine-120709-142811

Glogauer A, Martini VP, Faoro H, Couto GH, Muller-Santos M, Monteiro RA, Mitchell DA, de Souza EM, Pedrosa FO, Krieger N (2011) Identification and characterization of a new true lipase isolated through metagenomic approach. Microb Cell Fact 10:54. doi:10.1186/1475-2859-10-54

Gong X, Gruninger RJ, Forster RJ, Teather RM, McAllister TA (2013) Biochemical analysis of a highly specific, pH stable xylanase gene identified from a bovine rumen-derived metagenomic library. Appl Microbiol Biotechnol 97:2423–2431. doi:10.1007/s00253-012-4088-y

Gordon JI (2012) Honor thy gut symbionts redux. Science 336:1251–1253. doi:10.1126/science.1224686

Grant S, Sorokin DY, Grant WD, Jones BE, Heaphy S (2004) A phylogenetic analysis of Wadi el Natrun soda lake cellulase enrichment cultures and identification of cellulase genes from these cultures. Extremophiles 8:421–429. doi:10.1007/s00792-004-0402-7

Hallam SJ, Mincer TJ, Schleper C, Preston CM, Roberts K, Richardson PM, DeLong EF (2006) Pathways of carbon assimilation and ammonia oxidation suggested by environmental genomic analyses of marine *Crenarchaeota*. PLoS Biol 4:e95. doi:10.1371/journal.pbio.0040095

Handelsman J (2004) Metagenomics: application of genomics to uncultured microorganisms. Microbiol Mol Biol Rev 68:669–685. doi:10.1128/MMBR.68.4.669-685.2004

Handelsman J, Rondon MR, Brady SF, Clardy J, Goodman RM (1998) Molecular biological access to the chemistry of unknown soil microbes: a new frontier for natural products. Chem Biol 5:245–249. doi:10.1016/S1074-5521(98)90108-9

Hardeman F, Sjoling S (2007) Metagenomic approach for the isolation of a novel low-temperature-active lipase from uncultured bacteria of marine sediment. FEMS Microbiol Ecol 59:524–534. doi:10.1111/j.1574-6941.2006.00206.x

Harris K, Kassis A, Major G, Chou CJ (2012) Is the gut microbiota a new factor contributing to obesity and its metabolic disorders? J Obes 2012:879151. doi:10.1155/2012/879151

Hazen TC, Dubinsky EA, DeSantis TZ, Andersen GL, Piceno YM, Singh N, Jansson JK, Probst A, Borglin SE, Fortney JL (2010) Deep-sea oil plume enriches indigenous oil-degrading bacteria. Science 330:204–208. doi:10.1126/science.1195979

Hedlund BP, Staley JT (2006) Isolation and characterization of *Pseudoalteromonas* strains with divergent polycyclic aromatic hydrocarbon catabolic properties. Environ Microbiol 8:178–182. doi:10.1111/j.1462-2920.2005.00871.x

Hess M, Sczyrba A, Egan R, Kim TW, Chokhawala H, Schroth G, Luo S, Clark DS, Chen F, Zhang T, Mackie RI, Pennacchio LA, Tringe SG, Visel A, Woyke T, Wang Z, Rubin EM (2011) Metagenomic discovery of biomass-degrading genes and genomes from cow rumen. Science 331:463–467. doi:10.1126/science.1200387

Human Microbiome project Consortium (2012) Structure, function and diversity of the healthy human microbiome. Nature 486:207–214. doi:10.1038/nature11234

Hunter P (2012) The changing hypothesis of the gut. The intestinal microbiome is increasingly seen as vital to human health. EMBO Rep 13:498–500. doi:10.1038/embor.2012.68

Jadeja NB, More RP, Purohit HJ, Kapley A (2014) Metagenomic analysis of oxygenases from activated sludge. Bioresour Technol 165:250–256. doi:10.1016/j.biortech.2014.02.045

Jiang C, Wu B (2007) Molecular cloning and functional characterization of a novel decarboxylase from uncultured microorganisms. Biochem Biophys Res Commun 357:421–426. doi:10.1016/j.bbrc.2007.03.159

Kapardar RK, Ranjan R, Grover A, Puri M, Sharma R (2010) Identification and characterization of genes conferring salt tolerance to *Escherichia coli* from pond water metagenome. Bioresour Technol 101:3917–3924. doi:10.1016/j.biortech.2010.01.017

Kennedy J, OLeary ND, Kiran GS, Morrissey JP, OGara F, Selvin J, Dobson AD (2011) Functional metagenomic strategies for the discovery of novel enzymes and biosurfactants with biotechnological applications from marine ecosystems. J Appl Microbiol 111:787–799. doi:10.1111/j.1365-2672.2011.05106.x

Kimes NE, Callaghan AV, Aktas DF, Smith WL, Sunner J, Golding B, Drozdowska M, Hazen TC, Suflita JM, Morris PJ (2013) Metagenomic analysis and metabolite profiling of deep-sea sediments from the Gulf of Mexico following the deepwater horizon oil spill. Front Microbiol 4:50. doi:10.3389/fmicb.2013.00050

Krause KM, Combs DK (2003) Effects of forage particle size, forage source, and grain fermentability on performance and ruminal pH in midlactation cows. J Dairy Sci 86:1382–1397. doi:10.3168/jds.S0022-0302(03)73722-9

LeCleir GR, Buchan A, Maurer J, Moran MA, Hollibaugh JT (2007) Comparison of chitinolytic enzymes from an alkaline, hypersaline lake and an estuary. Environ Microbiol 9:197–205. doi:10.1111/j.1462-2920.2006.01128.x

Lee M, Lee C, Oh T, Song JK, Yoon J (2006) Isolation and characterization of a novel lipase from a metagenomic library of tidal flat sediments: evidence for a new family of bacterial lipases. Appl Environ Microbiol 72:7406–7409. doi:10.1128/AEM.01157-06

Lee MY, Cheon JH, Hidaka T, Tsuno H (2008) The performance and microbial diversity of temperature-phased hyperthermophilic and thermophilic anaerobic digestion system fed with organic waste. Water Sci Technol 57:283–289. doi:10.2166/wst.2008.062

Ley RE, Hamady M, Lozupone C, Turnbaugh PJ, Ramey RR, Bircher JS, Schlegel ML, Tucker TA, Schrenzel MD, Knight R, Gordon JI (2008) Evolution of mammals and their gut microbes. Science 320:1647–1651. doi:10.1126/science.1155725

Li K, Bihan M, Yooseph S, Methe BA (2012) Analyses of the microbial diversity across the human microbiome. PLoS One 7:e32118. doi:10.1371/journal.pone.0032118

Lim HK, Chung EJ, Kim JC, Choi GJ, Jang KS, Chung YR, Cho KY, Lee SW (2005) Characterization of a forest soil metagenome clone that confers indirubin and indigo production on *Escherichia coli*. Appl Environ Microbiol 71:7768–7777. doi:10.1128/AEM.71.12.7768-7777

Lorenz P, Liebeton K, Niehaus F, Eck J (2002) Screening for novel enzymes for biocatalytic processes: accessing the metagenome as a resource of novel functional sequence space. Curr Opin Biotechnol 13:572–577. doi:10.1128/AEM.71.12.7768-7777

Lynd LR, Weimer PJ, van Zyl WH, Pretorius IS (2002) Microbial cellulose utilization: fundamentals and biotechnology. Microbiol Mol Biol Rev 66:506–577. doi:10.1128/MMBR.66.3.506-577

Macdonald TT, Monteleone G (2005) Immunity, inflammation, and allergy in the gut. Science 307:1920–1925. doi:10.1126/science.1106442

Manichanh C, Rigottier-Gois L, Bonnaud E, Gloux K, Pelletier E, Frangeul L, Nalin R, Jarrin C, Chardon P, Marteau P, Roca J, Dore J (2006) Reduced diversity of faecal microbiota in Crohn's disease revealed by a metagenomic approach. Gut 55:205–211. doi:10.1136/gut.2005.073817

Marathe NP, Regina VR, Walujkar SA, Charan SS, Moore ER, Larsson DG, Shouche YS (2013) A treatment plant receiving waste water from multiple bulk drug manufacturers is a reservoir for highly multi-drug resistant integron-bearing bacteria. PLoS One 8:e77310. doi:10.1371/journal.pone.0077310

Mason OU, Hazen TC, Borglin S, Chain PS, Dubinsky EA, Fortney JL, Han J, Holman HYN, Hultman J, Lamendella R (2012) Metagenome, metatranscriptome and single-cell sequencing reveal microbial response to deepwater horizon oil spill. ISME J 6:1715–1727. doi:10.1038/ismej.2012.59

Mason OU, Scott NM, Gonzalez A, Robbins-Pianka A, Baelum J, Kimbrel J, Bouskill NJ, Prestat E, Borglin S, Joyner DC (2014) Metagenomics reveals sediment microbial community response to deepwater horizon oil spill. ISME J 8:1464–1475. doi:10.1038/ismej.2013.254

Montagna PA, Baguley JG, Cooksey C, Hartwell I, Hyde LJ, Hyland JL, Kalke RD, Kracker LM, Reuscher M, Rhodes AC (2013) Deep-sea benthic footprint of the deepwater horizon blowout. PLoS One 8:e70540. doi:10.1371/journal.pone.0070540

Moore AM, Munck C, Sommer MOA, Dantas G (2011) Functional metagenomic investigations of the human intestinal microbiota. Front Microbiol 2:1–8. doi:10.3389/fmicb.2011.00188

More RP, Mitra S, Raju SC, Kapley A, Purohit HJ (2014) Mining and assessment of catabolic pathways in the metagenome of a common effluent treatment plant to induce the degradative capacity of biomass. Bioresour Technol 153:137–146. doi:10.1016/j.biortech.2013.11.065

Mueller K, Ash C, Pennisi E, Smith O (2012) The gut microbiota. Introduction. Science 336:1245. doi:10.1126/science.336.6086.1245

Neuman MG, Nanau RM (2012) Inflammatory bowel disease: role of diet, microbiota, life style. Transl Res 16:29–44. doi:10.1016/j.trsl.2011.09.001

Neveu TJ, Regeard C, Dubow MS (2011) Isolation and characterization of two serine proteases from metagenomic libraries of the Gobi and Death Valley deserts. Appl Microbiol Biotechnol 91:635–644. doi:10.1007/s00253-011-3256-9

Ngo TD, Ryu BH, Ju H, Jang E, Park K, Kim KK, Kim TD (2013) Structural and functional analyses of a bacterial homologue of hormone-sensitive lipase from a metagenomic library. Acta Crystallogr D Biol Crystallogr 69:1726–1737. doi:10.1107/S0907444913013425

Nicholson JK, Holmes E, Kinross J, Burcelin R, Gibson G, Jia W, Pettersson S (2012) Host-gut microbiota metabolic interactions. Science 336:1262–1267. doi:10.1126/science.1223813

Niehaus F, Gabor E, Wieland S, Siegert P, Maurer KH, Eck J (2011) Enzymes for the laundry industries: tapping the vast metagenomic pool of alkaline proteases. Microb Biotechnol 4:767–776. doi:10.1111/j.1751-7915.2011.00279

Novo A, Andre S, Viana P, Nunes OC, Manaia CM (2013) Antibiotic resistance, antimicrobial residues and bacterial community composition in urban wastewater. Water Res 47:1875–1887. doi: 10.1016/j.watres.2013.01.010

Ogawa K, Maki Y (2003) Cellulose as extracellular polysaccharide of hot spring sulfur-turf bacterial mat. Biosci Biotechnol Biochem 67:2652–2654. doi:10.1271/bbb.67.2652

Orhan E, Omay D, Guvenilir Y (2005) Partial purification and characterization of protease enzyme from *Bacillus subtilis* and *Bacillus cereus*. Appl Biochem Biotechnol 121:183–194. doi:10.1007/978-1-59259-991-2_16

Pace NR (1995) Opening the door onto the natural microbial world: molecular microbial ecology. Harvey Lect 91:59–78

Parfrey LW, Knight R (2012) Spatial and temporal variability of the human microbiota. Clin Microbiol Infect 18:8–11. doi:10.1111/j.1469-0691.2012.03861

Peng Q, Wang X, Shang M, Huang J, Guan G, Li Y, Shi B (2014) Isolation of a novel alkalinestable lipase from a metagenomic library and its specific application for milk fat flavor production. Microb Cell Fact 13:1. doi:10.1186/1475-2859-13-1

Preeti A, Hemalatha D, Rajendhran J, Mullany P, Gunasekaran P (2014) Cloning, expression and characterization of a lipase encoding gene from human oral metagenome. Indian J Microbiol 54:284–292. doi:10.1007/s12088-014-0455-y

Pushpam PL, Rajesh T, Gunasekaran P (2011) Identification and characterization of alkaline serine protease from goat skin surface metagenome. AMB Express 1:3. doi:10.1186/2191-0855-1-3

Ranjan R, Grover A, Kapardar RK, Sharma R (2005) Isolation of novel lipolytic genes from uncultured bacteria of pond water. Biochem Biophys Res Commun 335:57–65. doi:10.1016/j.bbrc.2005.07.046

Rhee JK, Ahn DG, Kim YG, Oh JW (2005) New thermophilic and thermostable esterase with sequence similarity to the hormone-sensitive lipase family, cloned from a metagenomic library. Appl Environ Microbiol 71:817–825. doi:10.1128/AEM.71.2.817-825

Riesenfeld CS, Goodman RM, Handelsman J (2004a) Uncultured soil bacteria are a reservoir of new antibiotic resistance genes. Environ Microbiol 6:981–989. doi:10.1111/j.1462-2920.2004.00664

Riesenfeld CS, Schloss PD, Handelsman J (2004b) Metagenomics: genomic analysis of microbial communities. Annu Rev Genet 38:525–52. doi:10.1146/annurev.genet.38.072902.091216

Robertson DE, Chaplin JA, DeSantis G, Podar M, Madden M, Chi E, Richardson T, Milan A, Miller M, Weiner DP, Wong K, McQuaid J, Farwell B, Preston LA, Tan X, Snead MA, Keller M, Mathur E, Kretz PL, Burk MJ, Short JM (2004) Exploring nitrilase sequence space for enantioselective catalysis. Appl Environ Microbiol 70:2429–2436. doi:10.1128/AEM.70.4.2429-2436

Rondon MR, Goodman RM, Handelsman J, Rondon MR, Goodman RM, Handelsman J (1999) The Earth's bounty: assessing and accessing soil microbial diversity. Trends Biotechnol 17:403–409. doi:10.1016/S0167-7799(99)01352-9

Sangwan N, Lata P, Dwivedi V, Singh A, Niharika N, Kaur J, Anand S, Malhotra J, Jindal S, Nigam A, Lal D, Dua A, Saxena A, Garg N, Verma M, Kaur J, Mukherjee U, Gilbert JA, Dowd SE, Raman R, Khurana P, Khurana JP, Lal R (2014a) Comparative metagenomic analysis of soil microbial communities across three hexachlorocyclohexane contamination levels. PLoS One 7:e46219. doi:10.1371/journal.pone.0046219

Sangwan N, Verma H, Kumar R, Negi V, Lax S, Khurana P, Khurana JP, Gilbert JA, Lal R (2014b) Reconstructing an ancestral genotype of two hexachlorocyclohexane-degrading *Sphingobium* species using metagenomic sequence data. ISME J8:398–408. doi:10.1038/ismej.2013.153

Sathya TA, Jacob AM, Khan M (2014) Cloning and molecular modelling of pectin degrading glycosyl hydrolase of family 28 from soil metagenomic library. Mol Biol Rep 41:2645–2656. doi:10.1007/s11033-014-3380-6

Schirmer A, Gadkari R, Reeves CD, Ibrahim F, DeLong EF, Hutchinson CR (2005) Metagenomic analysis reveals diverse polyketide synthase gene clusters in microorganisms associated with the marine sponge *Discodermia dissoluta*. Appl Environ Microbiol 71:4840–4849. doi:10.1128/AEM.71.8.4840-4849.2005

Schjorring S, Krogfelt KA (2011) Assessment of bacterial antibiotic resistance transfer in the gut. Int J Microbiol 2011:1–10. doi:10.1155/2011/312956

Schloss PD, Handelsman J (2003) Biotechnological prospects from metagenomics. Curr Opin Biotechnol 14:303–310. doi:10.1016/S0958-1669(03)00067-3

Schloss PD, Handelsman J (2004) Status of the microbial census. Microbiol Mol Biol Rev 68:686–691. doi:10.1128/MMBR.68.4

Schmeisser C, Stockigt C, Raasch C, Wingender J, Timmis KN, Wenderoth DF, Flemming HC, Liesegang H, Schmitz RA, Jaeger KE, Streit WR (2003) Metagenome survey of biofilms in drinking-water networks. Appl Environ Microbiol 69:7298–7309. doi:10.1128/AEM.69.12

Selvin J, Kennedy J, Lejon DPH, Kiran GS, Dobson ADW (2012) Isolation identification and biochemical characterization of a novel halo-tolerant lipase from the metagenome of the marine sponge *Haliclona simulans*. Microb Cell Fact 11:72. doi:10.1186/1475-2859

Shade A, Handelsman J (2012) Beyond the Venn diagram: the hunt for a core microbiome. Environ Microbiol 14:4–12. doi:10.1111/j.1462-2920.2011.02585.x

Sharma S, Khan FG, Qazi GN (2010) Molecular cloning and characterization of amylase from soil metagenomic library derived from northwestern Himalayas. Appl Microbiol Biotechnol 86:1821–1828. doi:10.1007/s00253-009-2404-y

Silva AET, Rua CPJ, Andrade BGN, Vicente ACP, Silva GGZ, Berlinck RGS, Thompson FL (2013) Polyketide synthase gene diversity within the microbiome of the sponge *Arenosclera brasiliensis*, endemic to the southern Atlantic ocean. Appl Environ Microbiol 79:1598–1605. doi:10.1128/AEM.03354-12

Singh BK (2010) Exploring microbial diversity for biotechnology: the way forward. Trends Biotechnol 28:111–116. doi:10.1016/j.tibtech.2009.11.006

Singh A, Chauhan NS, Thulasiram HV, Taneja V, Sharma R (2010) Identification of two flavin monooxygenases from an effluent treatment plant sludge metagenomic library. Bioresour Technol 101:8481–8484. doi:10.1016/j.biortech.2010.06.025

Singh R, Dhawan S, Singh K, Kaur J (2012) Cloning, expression and characterization of a metagenome derived thermoactive/thermostable pectinase. Mol Biol Rep 39:8353–8361. doi:10.1007/s11033-012-1685-x

Song JS, Jeon JH, Lee JH, Jeong SH, Jeong BC, Kim SJ, Lee JH, Lee SH (2005) Molecular characterization of

TEM-type beta-lactamases identified in cold-seep sediments of Edison Seamount (South of Lihir Island, Papua New Guinea). J Microbiol 43:172–178. doi:10.1002/9781118010549.ch51

Sonnenburg JL, Xu J, Leip DD, Chen CH, Westover BP, Weatherford J, Buhler JD, Gordon JI (2005) Glycan foraging in vivo by an intestine-adapted bacterial symbiont. Science 307:1955–9. doi:10.1126/science.1109051

Steele HL, Streit WR (2005) Metagenomics: advances in ecology and biotechnology. FEMS Microbiol Lett 247:105–111. doi:10.1016/j.femsle.2005.05.011

Steele HL, Jaeger KE, Daniel R, Streit WR (2009) Advances in recovery of novel biocatalysts from metagenomes. J Mol Microbiol Biotechnol 16:25–37. doi:10.1159/000142892

Streit WR, Schmitz RA (2004) Metagenomics the key to the uncultured microbes. Curr Opin Microbiol 7:492–498. doi:10.1016/j.mib.2004.08.002

Sudo N (2012) Role of microbiome in regulating the HPA axis and its relevance to allergy. Chem Immunol Allergy 98:163–75. doi:10.1159/000336510

Sul WJ, Cole JR, Jesus EC, Wang Q, Farris RJ, Fish JA, Tiedjea JM (2011) Bacterial community comparisons by taxonomy-supervised analysis independent of sequence alignment and clustering. Proc Natl Acad Sci U S A 108:4637–14642. doi:10.1073/pnas.1111435108

Tamakai H, Wright CL, Li X, Lin Q, Hwang C, Wang S, Thimmapurma J, Kamagata Y, Liu WT (2011) Analysis of 16S rRNA amplicon sequencing option on the Roche/454 next-generation titanium sequencing platform. PLoS One 6:e25263. doi:10.1371/journal.pone.0025263

Tap J, Mondot S, Levenez F, Pelletier E, Caron C, Furet JP, Ugarte E, Tamayo RM, Paslier DL, Nalin R, Dore J, Leclerc M (2009) Towards the human intestinal microbiota phylogenetic core. Environ Microbiol 11: 2574–2584. doi:10.1111/j.1462-2920.2009.01982.x

Tasse L, Bercovici J, Pizzut-Serin S, Robe P, Tap J, Klopp C, Cantarel BL, Coutinho PM, Henrissat B, Leclerc M, Doré J, Monsan P, Remaud-Simeon M, Potocki-Veronese G (2010) Functional metagenomics to mine the human gut microbiome for dietary fiber catabolic enzymes. Genome Res 20:1605–1612. doi:10.1101/gr.108332.110

Tuffin M, Anderson D, Heath C, Cowan DA (2009) Metagenomic gene discovery: how far have we moved into novel sequence space? Biotechnol J 4(12):1671–1683. doi:10.1002/biot.200900235

Turnbaugh PJ, Gordon JI (2009) The core gut microbiome, energy balance and obesity. J Physiol 587:4153–4158. doi:10.1113/jphysiol.2009.174136

Tyson GW, Chapman J, Hugenholtz P, Allen EE, Ram RJ, Richardson PM, Solovyev VV, Rubin EM, Rokhsar DS, Banfield JF (2004) Community structure and metabolism through reconstruction of microbial genomes from the environment. Nature 428:37–43

Venter JC, Remington K, Heidelberg JF, Halpern AL, Rusch D, Eisen JA, Wu D, Paulsen I, Nelson KE, Nelson W, Fouts DE, Levy S, Knap AH, Lomas MW, Nealson K, White O, Peterson J, Hoffman J, Parsons R, Baden-Tillson H, Pfannkoch C, Rogers YH, Smith HO (2004) Environmental genome shotgun sequencing of the Sargasso Sea. Science 304:66–74. doi:10.1126/science.1093857

Verma D, Kawarabayasi Y, Miyazaki K, Satyanarayana T (2013) Cloning, expression and characteristics of a novel alkalistable and thermostablexylanase encoding gene (Mxyl) retrieved from compost-soil metagenome. PLoS One 8:e52459. doi:10.1371/journal.pone.0052459

Vidya J, Swaroop S, Singh SK, Alex D, Sukumaran RK, Pandey A (2011) Isolation and characterization of a novel α -amylase from a metagenomic library of western Ghats of Kerala, India. Biologia 66:939–944. doi:10.2478/s11756-011-0126-y

Voget S, Steele HL, Streit WR (2006) Characterization of a metagenome-derived halotolerant cellulase. J Biotechnol 126:26–36. doi:10.1016/j.jbiotec.2006.02.011

Waidner LA, Kirchman DL (2005) Aerobic anoxygenic photosynthesis genes and operons in uncultured bacteria in the Delaware River. Environ Microbiol 7:1896–1908. doi:10.1111/j.1462-2920.2005.00883.x

Wang L, Chen S, Xiao X, Huang X, You D, Zhou X, Deng Z (2006) arsRBOCT arsenic resistance system encoded by linear plasmid pHZ227 in *Streptomyces sp.* strain FR-008. Appl Environ Microbiol 72:3738–3742. doi:10.1128/AEM.72.5.3738-3742

Wilson DB (2009) Cellulases and biofuels. Curr Opin Biotechnol 20:295–299. doi:10.1016/j.copbio.2009.05.007

Xie G, Chain PS, Lo CC, Liu KL, Gans J, Merritt J, Qi F (2010) Community and gene composition of a human dental plaque microbiota obtained by metagenomic sequencing. Mol Oral Microbiol 25:391–405. doi:10.1111/j.2041-1014.2010.00587.x

Yatsunenko T, Rey FE, Manary MJ, Trehan I, Dominguez-Bello MG, Contreras M, Magris M, Hidalgo G, Baldassano RN, Anokhin AP, Heath AC, Warner B, Reeder J, Kuczynski J, Caporaso JG, Lozupone CA, Lauber C, Clemente JC, Knights D, Knight R, Gordon JI (2012) Human gut microbiome viewed across age and geography. Nature 486:222–227. doi:10.1038/nature11053

Yun J, Kang S, Park S, Yoon H, Kim MJ, Heu S, Ryu S (2004) Characterization of a novel amylolytic enzyme encoded by a gene from a soil-derived metagenomic library. Appl Environ Microbiol 70:7229–7235. doi:10.1128/AEM.70.12.7229-7235

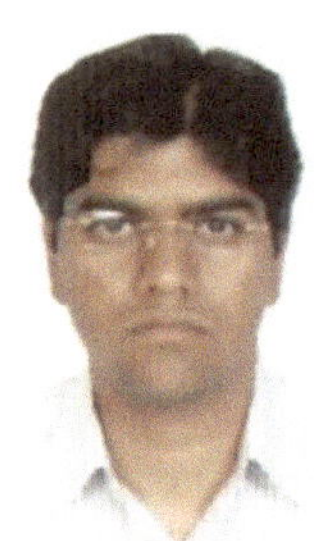

Manoj Kumar has attained the doctorate degree in Environmental Science from BBAU Lucknow in 2009. His current area of research is metagenomics. Currently, he is working as a DST Scientist in Department of Biochemistry, M.D. University, Rohtak, Haryana.

Jitendra Kumar has completed a Master's Degree in Philosophy from the Department of Zoology, University of Delhi, Delhi. Presently, he is pursuing Ph.D. from Department of Biochemistry, M.D. University, Rohtak, Haryana, in the area of metagenomics.

Nar Singh Chauhan has received his Ph.D. degree in Biotechnology from University of Pune, Pune Maharashtra. He is presently the Professor (Asst.) at the Department of Biochemistry, Maharshi Dayanand University, Rohtak, Haryana. His current research area is exploration of microbial ecosystems with systems biology approach.

In Silico Reconstitution of Novel Routes for Microbial Plastic

19

Vipin Chandra Kalia, Sadhana Lal, Rashmi, Ashwini Chauhan, and Goutam Bhattacharyya

Abstract

Polyhydroxyalkanoate (PHA) biosynthesis is a phenotype expressed under environmental stress. PHA synthesis relies on important intermediates tricarboxylic acid (TCA) cycle and fatty acid metabolism especially acetyl CoA. Acetyl CoA is in demand by a host of metabolisms and is not easily available under normal "unstressed" conditions. This work was done to find out alternative routes for copolymer polyhydroxybutyrate (PHB), which may exist in nature but not explored so far. Using metabolic pathway database (KEGG), we have devised in silico novel metabolic pathways, which are independent of acetyl CoA, i.e., may operate under different conditions. Moreover, organisms possessing enzymes for these simulated PHB production routes have also been identified. In this comparative analysis, we found some potential organisms such as *Brucella*, *Deinococcus*, *Homo sapiens*, *Mus musculus*, *Rattus*, *Thermoanaerobacter*, and *Xanthomonas* which have the necessary genetic machinery for one or more of these novel pathways. Incidentally, they have not been reported previously as PHB producers. On the other hand, these novel routes have also been observed in certain known PHB producers such as *Clostridium*, *Ralstonia*, *Pseudomonas,* and *Mesorhizobium.* The advantage of these novel routes in known PHB producers is expanding the source of starting material for PHB synthesis.

19.1 Introduction

Increasing dependency on chemically synthesized polymers has resulted in environmental- and human health-related issues (Kalia et al. 2003a). Polyhydroxyalkanoates (PHAs) are biopolymers, which are biodegradable as well. These offer exciting possibilities since these can be produced from renewable resources in con-

V.C. Kalia (✉) • Rashmi • A. Chauhan
G. Bhattacharyya
Microbial Biotechnology and Genomics,
CSIR-Institute of Genomics and Integrative Biology,
Delhi University Campus, Mall Road,
Delhi 110007, India
e-mail: vckalia@igib.res.in; vc_kalia@yahoo.co.in

S. Lal
CSIR-Institute of Genomics and Integrative Biology,
Delhi University Campus, Delhi, India
e-mail: sadhana.lal@gmail.com;
Sadhana.Lal@umanitoba.ca

V.C. Kalia (ed.), *Microbial Factories*, DOI 10.1007/978-81-322-2595-9_19

trast to petrochemical-derived plastics. Among the various types of naturally occurring biopolymers, PHAs accumulate as insoluble inclusions in the cytoplasm for storage of carbon and energy in microbes (Park et al. 2002; Steinbüchel and Lutke-Eversloh 2003; Verlinden et al. 2007; Rehm 2010). These polymers have generated commercial interest as these are biocompatible as well (Lee 1996), but it is essential to improve their material properties and to reduce the production cost (Agus et al. 2006; Tomizawa et al. 2011; Kumar et al. 2014, 2015a, b; Patel et al. 2015; Singh et al. 2015). Microbes have been found to produce poly-3-hydroxybutyrate (P3HB) – the most common type of PHA, polymers of 4-hydroxybutyrate (P4HB), hydroxyvalerate (PHV), hydroxyhexanoate (PHH), hydroxyoctanoate (PHO), and their copolymers (Anil-Kumar et al. 2007; Chookietwattana and Khonsarn 2011; Khiyami et al. 2011; Masood et al. 2012; Sangkharak and Prasertsan 2012; Santimano et al. 2009; Shamala et al. 2012; Shrivastav et al. 2010; Yan et al. 2010). PHB has been the most interesting as it has properties similar to plastics like polypropylene (PP). The biosynthesis of PHB depends heavily upon tricarboxylic acid (TCA) cycle, fatty acid biosynthesis, and degradation (β-oxidation) products especially acetyl CoA. Thus, there is a need to look for alternative metabolism for PHB synthesis, which can utilize different substrates as starting materials. In this article, in silico reconstitution of the four "novel" metabolic routes for PHB biosynthesis may exist in nature but not explored so far. This offers an opportunity to synthesize novel polymers, which can be exploited for desirable properties and can be produced from a broad substrate range (Aldor and Keasling 2003).

19.2 Metabolic Pathways for PHA Biosynthesis

PHAs comprise a large group of biopolymers, the most widely studied being poly(3-hydroxybutyric acid) (P(3HB)). P(3HB) biosynthesis is catalyzed by three enzymes (Fig. 19.1). The biosynthesis initiates by transforming acetyl CoA into acetoacetyl-CoA by action of β-ketothiolase (encoded by the gene *pha*A), which in turn lead to the production of 3-hydroxybutyryl-CoA with the help of an NADPH-dependent enzyme – acetoacetyl-CoA dehydrogenase (which is encoded by the gene *pha*B). In methylotrophs, this reaction is NADH dependent (Breuer et al. 1995). These monomers get polymerized to PHB by a polymerase (encoded by *pha*C) (Reddy et al. 2003; Rehm 2010; Singh et al. 2009). Although P(3HB) accumulation is widely distributed among prokaryotes (Table 19.1), most investigations related to the enzymatic mechanisms involved in this pathway have focused on a few of the natural PHA producers (Madison and Huisman 1999; Singh et al. 2009).

19.3 Secondary Pathways for PHA Biosynthesis

In addition to the P(3HB) biosynthetic pathway described above, a few variations have been recorded (Aldor and Keasling 2003; Steinbüchel and Lutke-Eversloh 2003). They also vary in their source of origin and its further metabolism (Fig. 19.1).

(i) In *Rhodospirillum rubrum* PHB synthesis involves five steps. S(+)-3-hydroxybutyryl-CoA synthesized by an NADH-dependent acetoacetyl-CoA reductase is transformed to R(−) form by two 2-enoyl-CoA hydratases, which is then polymerized (Schubert et al. 1988).
(ii) P(3HB-HV) production through methylmalonyl-CoA metabolic pathway: Here, the conversion of succinyl-CoA to (2R)-methylmalonyl-CoA is mediated by the enzyme methylmalonyl-CoA mutase (Aldor et al. 2002). In the next stage, the (2R)-methylmalonyl-CoA form is changed to (2S)-methylmalonyl-CoA form, the reaction is mediated by the enzyme methylmalonyl-CoA epimerase. The product of the previous reaction is transformed to propionyl-CoA, by two alternative enzymes as observed in *Propionigenium modestum* (Dimroth and Schink 1998) and

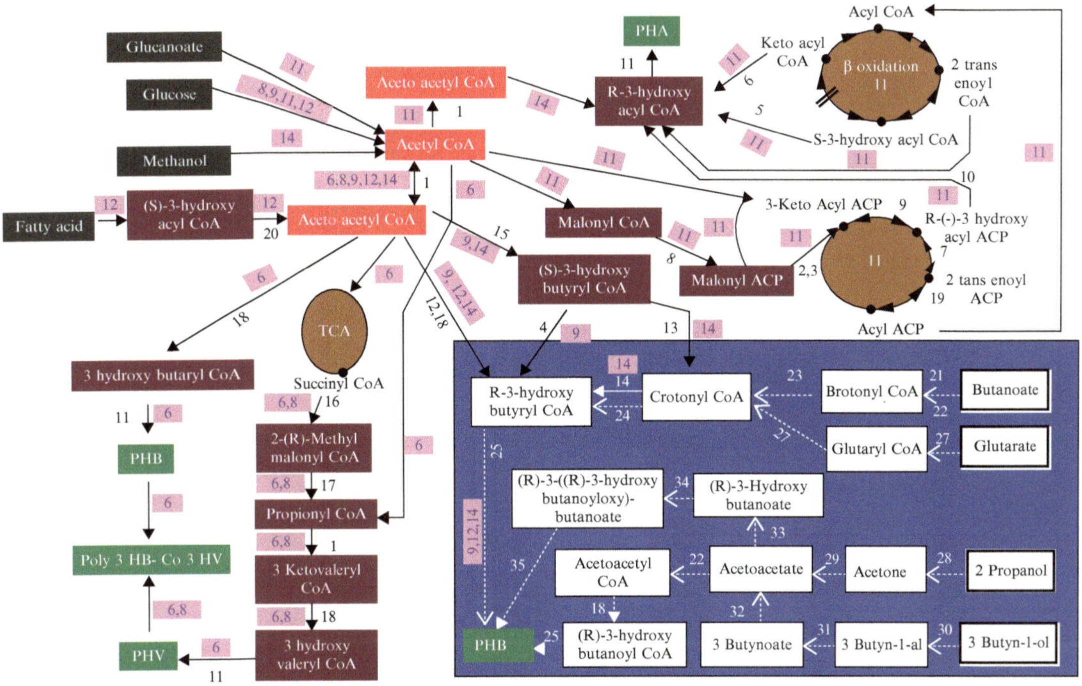

Fig. 19.1 Classical, alternative, and novel (←) metabolic pathways for PHA production. Numbers [1–36] represent different enzymes. Numbers written in *gray* color box represent references. Name of enzymes: *1* Beta ketothiolase, *2* Beta-ketoacyl-ACP synthase I, *3* Beta-ketoacyl-ACP synthase II, *4* (R)-specific enoyl-CoA reductase, *5* Epimerase, *6* 3-Ketoacyl-CoA reductase, *7* 3-Hydroxydecanoyl-ACP dehydrogenase, *8* Malonyl-CoA-ACP transacylase, *9* Beta-ketoacyl-ACP reductase, *10* 3-Hydroxydecanoyl-ACP-CoA transacylase, *11* PHA synthase, *12* NADPH-linked acetoacetyl-CoA reductase, *13* S(+)-Specific crotonyl-CoA hydratase, *14* R(−)-Specific crotonyl-CoA hydratase, *15* NADH-linked acetoacetyl-CoA reductase, *16* Methylmalonyl-CoA mutase (sleeping beauty mutase), *17* Methylmalonyl-CoA decarboxylase, *18* Acetoacetyl-CoA reductase, *19* Enoyl-ACP reductase, *20* L-(+)-Beta-hydroxyacyl-CoA dehydrogenase, *21* Butyrate-CoA ligase, *22* Acetate-CoA transferase, *23* Butyryl-CoA dehydrogenase, *24* 3-Hydroxybutyryl-CoA dehydratase, *25* PHB synthase, *26* Glutarate-CoA ligase, *27* Glutaryl-CoA dehydrogenase, *28* Isopropanol dehydrogenase, *29* Acetoacetate decarboxylase, *30* Alcohol dehydrogenase, *31* Aldehyde dehydrogenase, *32* Acetylene carboxylate hydratase, *33* 3-Hydroxybutyrate dehydrogenase, *34* Hydroxybutyrate-dimer hydrolase, *35* Carboxylic-ester hydrolases

Propionibacterium shermanii (Aldor et al. 2002). A few bacteria such as *Rhodococcus ruber* (Williams et al. 1994) and *Nocardia corallina* (Braunegg et al. 1998) are unique in their abilities to accumulate copolymers of PHAs having 3HV in spite of the fact that the precursors for HV: propionate and /or valerate (Ackermann and Babel 1997), were not provided as supplements.

(iii) De novo fatty acid synthetic pathway: PHA production by intermediates of β-oxidation or fatty acid metabolism were reported from Pseudomonads (Lee and Choi 1999). These precursors for PHA synthesis are produced via de novo fatty acid metabolic routes, which are critical for bacteria growth in the presence of simple carbon compounds (Yu 2001). The crucial enzyme linking these two pathways was identified to be a 3-hydroxyacyl-CoA-ACP (acyl carrier protein) transferase (Rehm et al. 2001). The PHA biosynthesis through this route in *Pseudomonas aeruginosa* consists of 3-hydroxydecanoate (3HD). *P. oleovorans* accumulates 3-hydroxyoctanoate (3HO) and 3-hydroxyhexanoate (3HHx) containing PHA copolymers, in the presence of octane, octanol, or octanoate in the medium. The evolution has thus diverged at the level of monomer-CoA supply. PHA synthase

Table 19.1 Polyhydroxyalkanoate (PHA)-producing organisms (review of literature)

Archaea		
Halobacterium mediterranei	*Haloferax mediterranei*	*Methanomonas*
Actinobacteria		
Actinomycetes	*Corynebacterium*	*Nocardia corallina*
Rhodococcus sphaeroides	*R. ruber*	*R. ruber* PP2
Streptomyces	*Micrococcus*	
Cyanobacteria		
Aphanothece	*Gloeothece* sp. PCC 6909	*Microcoleus*
Microcystis	*Spirulina platensis*	*S. platensis* K4
Synechococcus sp. MA 19	*Synechocystis*	
Firmicutes		
Bacillus megaterium QMB 1551	*Caryophanon*	*Clostridium*
Syntrophomonas sp.		
Proteobacteria (Alpha)		
Agrobacterium sp.	*Azorhizobium caulinodans*	*Azospirillum*
Beijerinckia	*Bradyrhizobium japonicum*	*Caulobacter crescentus*
Hyphomicrobium	*Methylobacterium extorquens* 1BT6	*M. rhodesianum* MB 126
Methylocystis parvus	*Methylosinus trichosporium*	*Mycoplana rubra* B 346
Nitrobacter	*Paracoccus denitrificans*	*Pedomicrobium*
Rhizobium etli	*R. meliloti*	*Rhodobacter*
Rhodopseudomonas	*Rhodospirillum rubrum*	*R. rubrum* ATCC 25903
R. rubrum Ha	*R. sphaeroides*	*Sinorhizobium meliloti* 41
Stella	*Xanthobacter*	
Proteobacteria (Beta)		
Alcaligenes sp. SH69	*A. latus*	*A. rubrum*
Aquaspirillum	*Chromobacterium violaceum*	*Comamonas acidovorans* DS17
Leptothrix	*Ralstonia eutropha*	*R. metallidurans*
Sphaerotilus natans	*Spirillum*	*Thiobacillus*
Zoogloea ramigera		
Proteobacteria (Gamma)		
Acinetobacter sp. RA 3849	*Actinobacillus* sp. EL9	*Aeromonas caviae*
A. hydrophila 4AK4	*Allochromatium vinosum*	*Azomonas*
Azotobacter beijerinckii	*A. chroococcum* H23	*A. paspali*
A. vinelandii	*Beggiatoa*	*Beneckea*
Burkholderia cepacia	*B. sacchari* 1PT 101	*Candida tropicalis*
Chlorogloea	*Chromatium purpuratum*	*C. violaceum*
C. vinosum	*Escherichia coli* JM 83	*E. coli* K-12
Ectothiorhodospira saposhnikovi	*Haemophilus*	*Klebsiella aerogenes*
Lamprocystis	*Lampropedia*	*Moraxella*
Nitrococcus	*Oceanospirillum*	*Photobacterium*
Protomonas extorquens	*Pseudomonas* sp. 61-3	*P. acidophila*
P. aeruginosa	*P. cepacia*	*P. citronellolis* MTCC 1191
P. fluorescens BM 07	*P. lemoignei*	*P. oleovorans*
P. pseudoflava	*P. putida*	*P. putida* 01
P. putida GPp 104	*P. resinovorans*	*P. syringae*
P. stutzeri	*P. testosteroni*	*Salmonella enterica*

(continued)

Table 19.1 (continued)

Thiocapsa	*Thiocystis violacea* 2311	*Thiosphaera*
Vibrio harveyi		
Eukaryota		
Umbellularia californica		

Organisms, 124; genera, 90; species, 118 (+6 strains); {*E. coli* 2, *P. putida* 3, *R. ruber* 2, and *R. rubrum* 3.}

evolved in a manner whereby it could polymerize short- or medium- chain monomers. In the case of *Azotobacter vinelandii*, UWD grown on glucose medium, copolymer poly(3-HB-co-3-HV) was supported by valerate and other short-chain, uneven length fatty acids (Page and Manchak 1995). Another route for bioconversion of succinyl-CoA to propionyl-CoA leading to copolymer production through recombinant technology has also been reported (Aldor et al. 2002).

(iv) In yet another development, it has been found that PHB synthesis can be achieved through by a pathway which involves five steps. In a deviation from the widely reported pathway, in the present case, NADH-dependent enzyme acetoacetyl-CoA reductase is responsible for the transformation reaction leading to S(+)-3-hydroxybutyryl-CoA. The S to R form transformation is executed by two stereospecific 2-enoyl-CoA hydratases, which is then polymerized, e.g., in *Methylobacterium rhodesianum* (Mothes and Babel 1995; Lee and Choi 1999). Another interesting feature is the constitutive expression of the two acetoacetyl-CoA reductases – NADH and NADPH dependent. These multiple features may prove effective in the synthesis of 3-hydroxybutyryl-CoA in spite of a low transhydrogenase activity. Here, P(3HB) synthesis depends upon the growth phase (Mothes et al. 1998).

Catabolism of propionic acid is observed to be carried out through parallel routes: 2-methylcitric acid cycle operative in yeast and mold, *Escherichia coli*, and other non-PHA producers (Bramer et al. 2002). It may be the reason why copolymers of PHA have low 3HV content. It becomes interesting to search pathways which can specifically aid P(3HB-co-HV) production on industrial scale.

19.4 Metabolic Engineering for Novel PHA-Producing Pathways and Novel Organisms

Metabolic engineering is envisaged as an important tool for allowing bacteria to utilize a wider range of feed substrates, with the ultimate objective of improving the PHA yields and their quality. Recombination strategy for introducing PHA biosynthetic genes in to the non-PHA producers seems more promising (Lee 1996; Agus et al. 2010; Hyakutake et al. 2011; Tomizawa et al. 2011; Hiroe et al. 2012; Kumar et al. 2013). The key intermediates in almost all the pathways involved in the production of PHA are acetyl CoA and acetoacetyl-CoA. Pyruvate, the major precursor to acetyl CoA, is always in great demand. Acetyl CoA competes with amino acid biosynthesis, oxaloacetate, and acetyl P for pyruvate. Subsequently, acetyl CoA is involved in a host of pathways, e.g., TCA cycle, butanoate and propanoate metabolisms, fatty acid biosynthesis, amino acid biosynthesis, etc. (Fig. 19.2). Acetyl CoA is an important intermediate of PHA producers such as *Ralstonia eutropha* (Lee and Choi 1999). It has the possibility of either flowing into the TCA cycle or may be diverted to PHA synthesis. The most important mechanism which regulates the diversion of acetyl CoA is the ratio of NAD(P)H/NAD(P). During limiting nitrogen availability, this ratio increases and in turn leads to the inhibition of two enzymes – citrate synthase and isocitrate dehydrogenase of the TCA

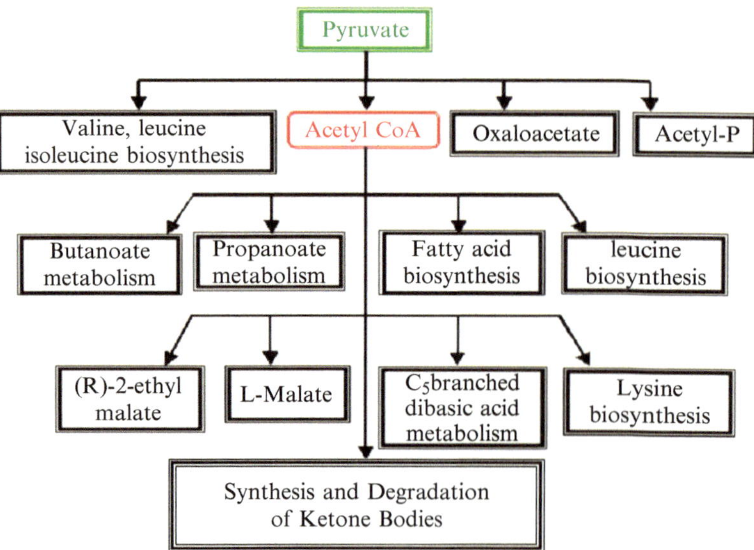

Fig. 19.2 Metabolism of pyruvate and acetyl CoA in different metabolic pathways (Source: KEGG)

cycle. Consequently, the flux of acetyl CoA into TCA cycle gets reduced, which allow acetyl CoA diversion to the synthesis of acetoacetyl-CoA with the help of the enzyme β-ketothiolase. On the contrary, if nitrogen is in sufficient quantities, the quantum of CoA goes up with the entry of acetyl CoA into TCA cycle. High concentration of CoA inhibits β-ketothiolase activity and consequently the biosynthesis of PHA stops as well.

A variation in the feed material also affects PHA composition. In the presence of octane and 1-octene, *P. oleovorans* produces PHAs where monomers with an unsaturated bond may vary from as low as nil to as high as 50 %. It depends upon the formation of 1-octene as an intermediate (Mergaert et al. 1993). In the presence of crotonyl CoA or hexenoyl CoA, *Aeromonas caviae* uses the enzyme enoyl-CoA hydratase to form (R)-3-hydroxy monomer. This happens primarily when the enzymes thiolase and reductase are absent. In yet another scenario, bacteria may metabolize sugars to P(2HB-3HV) copolymers through 3-HV generated via methylmalonyl-CoA route (Ackermann and Babel 1997; Dimroth and Schink 1998; Aldor et al. 2002). *R. eutropha* preferred butyric acid along with acetic acid for high PHA formation rate. On the other hand, propionic acid and acetic acid mixture results in low PHA formation rate. Acetic acid reduced propionic acid metabolism and resulted in hydroxyvalerate formation (Pozo et al. 2002). These metabolic pathways are quite complicated and operate on acetyl CoA, which is efficiently diverted to PHA production only under environmental stress.

In such a scenario, the question being raised is whether PHA biosynthesis can operate via an alternative route, which may bypass acetyl CoA and work under different ecological stress. An enormous microbial diversity including sequenced genomes has motivated the effort to determine novel metabolic capabilities of the organisms and link their enzymes with specific functions (Moreira and López-García 2002; Schloss and Handelsman 2004).

In order to improve the PHA production mechanisms, it become imperative to understand their metabolic regulatory properties. This process can be aided by the availability of metabolic databases, which can come handy in developing organism-specific metabolic network (Kalia et al. 2003a; Ma and Zeng 2003). A survey of KEGG database for 120 metabolic pathways revealed nine metabolic pathways that lead to the formation of PHA. Based on this information, we have constructed in silico four novel routes for PHB synthesis (Fig. 19.1) which may function independently of acetyl CoA. These four metabolisms involve butanoate, glutarate, 2-propanol, and 3-butyn-1-ol as starting substrates.

A search for the presence of enzymes of these four alternative pathways was made using methods detailed in our previous works (Kalia et al. 2003a, b, 2007). The aim of this analysis was to find out those microbes, which possess the enzymes for these novel alternative pathways.

The search for PHA biosynthetic routes, which may function independently of acetyl CoA, resulted in four novel routes such as (1) butanoate and (2) glutarate leading to the synthesis of crotonyl CoA via butanoyl CoA and glutaryl CoA, respectively. Crotonyl CoA is then converted into PHB via R-3-hydroxybutyryl-CoA (Tables 19.2 and 19.3). On the other hand, during the metabolisms of (3) 2-propanol and (4) 3-butyn-1-ol to PHB, acetoacetate is a common intermediate. These reactions are carried out by a set of enzymes (Tables 19.4 and 19.5).

In the butanoate to PHB biotransformation pathway, the primary reaction of butanoate to butanoyl CoA can be catalyzed by either butyrate-CoA ligase or acetoacetate-CoA transferase, and the subsequent reactions involve one enzyme each. The ability for this biotransformation extends from *Firmicutes* such as *Thermoanaerobacter*; *Clostridium* sp.; *Proteobacteria* such as *Brucella*, *Ralstonia*, *Pseudomonas*, and *Xanthomonas*; and *Deinococcaceae* member such as *Deinococcus* to eukaryotes such as *Homo sapiens*, *Mus musculus*, and *Rattus norvegicus*. Most other organisms with ability to metabolize butanoate invariably lack butyryl-CoA dehydrogenase. Since *Clostridium*, *Ralstonia*, and *Pseudomonas* are already known to produce PHB, under stress conditions, the presence of alternative route implies the possibilities of PHB production being operative even under different environmental conditions. Glutarate metabolic pathway for PHB production involves glutarate-CoA ligase, glutaryl-CoA dehydrogenase, enoyl-CoA hydratase, and PHB synthase. A large number of organisms have genes for three of the four enzymes of this route. However, these organisms need to be complemented by glutarate-CoA ligase gene from eukaryotic organisms such as *Rattus*. Among these *Agrobacterium*, *Arabidopsis*, *Bacillus*, *Caulobacter*, *Mesorhizobium* (Kalia et al. 2003a), *Pseudomonas*, *Ralstonia*, *Sinorhizobium*, and *Streptomyces* have already been shown to have the ability for PHB synthesis and hence can be easily employed after necessary modification(s) for alternative route as well. *Mycobacterium* and *Drosophila*, which are not known PHB producers, need complementation for glutarate-CoA ligase for exploitation of alternative routes of PHB biosynthesis. For the metabolism of 2-propanol to PHB, the survey revealed that only two organisms such as *M. loti* and *R. solanacearum* have the complete genetic makeup for this pathway. The primary difficulty is the non-availability of acetoacetate decarboxylase, which has been reported rarely. In addition, the other potential candidates for engineering this pathway are *Agrobacterium* sp., *B. melitensis*, *E. coli*, and *R. metallidurans* where they need to be complemented with acetoacetate decarboxylase. Utilization of 3-butyn-1-ol as substrate for PHB production involves a host of enzymes distributed over two different routes with acetoacetate as a common intermediate. Of these, a majority can be found only among α- and β-proteobacteria. The major limiting factors for exploiting these routes are the enzyme acetylene carboxylase hydratase and hydroxybutyrate-dimer hydrolase, which could not be detected among the organisms in the available databases. We may thus need to search organism possessing either or both of these enzymes. The likely candidates, who can be looked for these enzymes, can be those which have alcohol dehydrogenase and aldehyde dehydrogenase (Table 19.5).

Among the various potential organisms possessing genes for one or more of these novel pathways, *Brucella*, *Deinococcus*, *H. sapiens*, *M. musculus*, *Rattus*, *Thermoanaerobacter*, and *Xanthomonas* have the necessary genetic machinery, e.g., for butanoate to PHB route. Incidentally, they have neither been reported previously as PHB producers nor have the genes for PHB synthesis (Kalia et al. 2003a). On the other hand, certain known PHA producers such as *Clostridium*, *Ralstonia*, *Pseudomonas*, and *Mesorhizobium*

Table 19.2 Occurrence of enzymes for first metabolic pathway (starting substrate – butanoate) for PHA production in different organisms

Organisms	BCL/AACT[a]	BCD[b]	3-HBCD[c]	PHAs[d]
Firmicutes				
Thermoanaerobacter tengcongensis	+/+	+	+	+
Clostridium acetobutylicum	–/+	+	+	+
C. perfringens	–/+	+	+	+
C. tetani E88	–/+	+	+	–
C. thermosaccharolyticum	–/–	+	–	–
Streptococcus pyogenes	–/+	–	–	+
S. pyogenes M3	–/+	–	–	+
S. pyogenes M18	–/+	–	–	+
Fusobacteria				
Fusobacterium nucleatum	–/–	+	+	+
F. nucleatum subsp. *nucleatum* ATCC 25586	–/+	+	+	–
F. nucleatum subsp. *vincentii* ATCC 49256	–/+	–	+	–
Proteobacteria				
Alpha				
Brucella melitensis	+/+	+	+	+
Beta				
Ralstonia solanacearum	+/+	+	+	+
R. metallidurans	+/+	+	+	+
Gamma				
Escherichia coli O157:H7 EDL933	+/+	–	+	+
E. coli CFT073	+/+	–	+	+
E. coli strain O157:H7, substrain RIMD 0509952	+/+	–	+	–
E. coli K12	+/+	–	+	–
E. coli K12 W3110	–/–	–	–	+
E. coli K12 MG1655	–/–	–	–	+
Haemophilus influenzae Rd KW20	+/+	–	+	–
H. influenzae	–/+	–	+	–
H. ducreyi 35000HP	–/–	–	+	–
Pseudomonas aeruginosa PA01	+/+	+	+	+
P. aeruginosa UCBPP-PA14	+/+	+	–	+
P. putida KT2440	+/+	–	+	+
P. fluorescens PfO-1	+/+	–	+	+
P. syringae pv. *tomato str.* DC3000	+/+	–	+	+
P. aeruginosa	–/–	+	+	+
P. syringae pv. *syringae* B728a	+/+	–	+	–
Xanthomonas campestris pv. *campestris str.* ATCC 33913	+/+	+	+	+
X. campestris	+/–	–	+	+
X. axonopodis	–/–	–	+	+
Deinococcus				
Deinococcus radiodurans	+/+	+	+	+

(continued)

Table 19.2 (continued)

Organisms	BCL/AACT[a]	BCD[b]	3-HBCD[c]	PHAs[d]
Eukaryotes				
Metazoa				
Homo sapiens	+/+	+	+	+
Rattus norvegicus	+/+	+	+	+
Mus musculus	+/+	+	+	+

[a]Butyrate-CoA ligase (BCL: 6.2.1.2)/Acetate-CoA transferase (AACT: 2.8.3.8)
[b]Butyryl-CoA dehydrogenase (BCD: 1.3.99.2)
[c]3-Hydroxybutyryl-CoA dehydratase (3-HBCD: 4.2.1.55)
[d]PHA synthase (PHAS: 2.3.1.-)

Table 19.3 Occurrence of enzymes for second metabolic pathway (starting substrate – glutarate) for PHA production in different organisms

Organisms	GCL[a]	GCD[b]	ECH[c]	PHAS[d]
Archaea				
Archaeoglobus fulgidus	–	+	+	–
Halobacterium sp.	–	+	+	–
Actinobacteria				
Mycobacterium tuberculosis CDC 1551	–	+	+	+
M. tuberculosis H37Rv	–	+	+	+
Streptomyces coelicolor	–	+	+	+
Firmicutes				
Bacillus subtilis	–	+	+	+
B. halodurans	–	+	+	+
Staphylococcus aureus N315	–	+	+	+
S. aureus MU50	–	+	+	+
S. aureus MW2	–	+	+	+
Streptococcus pyogenes	–	–	+	+
S. pyogenes M3	–	–	+	+
S. pyogenes M18	–	–	+	+
Fusobacteria				
Fusobacterium nucleatum	–	–	+	+
Proteobacteria				
Alpha				
Agrobacterium tumefaciens C58 UWASH	–	+	+	+
A. tumefaciens C58 CEREON	–	+	+	+
Brucella melitensis	–	+	+	+
Caulobacter crescentus	–	+	+	+
Mesorhizobium loti	–	+	+	+
Rickettsia prowazekii	–	–	+	+
R. conorii	–	–	+	–
Sinorhizobium meliloti	–	+	+	+
Beta				
Ralstonia solanacearum	–	+	+	+

(continued)

Table 19.3 (continued)

Organisms	GCL[a]	GCD[b]	ECH[c]	PHAS[d]
Gamma				
Escherichia coli K12W3110	–	–	+	+
E. coli K12 MG1655	–	–	+	+
Pseudomonas aeruginosa	–	+	+	+
Salmonella typhi	–	–	+	+
S. typhimurium	–	–	+	+
Vibrio cholerae	–	–	+	+
Xanthomonas axonopodis	–	+	+	+
X. campestris	–	+	+	+
Yersinia pestis	–	–	+	+
Deinococcus				
Deinococcus radiodurans	–	+	+	+
Eukaryotes				
Metazoa				
Caenorhabditis elegans	–	+	+	–
Canis familiaris	+	–	–	–
Columba livia	+	–	–	–
Drosophila melanogaster	–	+	+	+
Homo sapiens	–	+	+	+
Mus musculus	–	+	+	+
Rattus norvegicus	+	–	+	+
Viridiplantae				
Arabidopsis thaliana	–	+	+	+

[a]Glutarate-CoA ligase (GCL: 6.2.1.6)
[b]Glutaryl-CoA dehydrogenase (GCD: 1.3.99.7)
[c]Enoyl-CoA hydratase (ECH: 4.2.1.17)
[d]PHA synthase (PHAS: 2.3.1.-)

Table 19.4 Occurrence of enzymes for third metabolic pathway (starting substrate – 2-propanol) for PHA production in different organisms

Organisms	IPD[a]	AADC[b]	AACT[c]	AACR[d]	PHAS[e]
Firmicutes					
Clostridium acetobutylicum	+	+	–	+	+
Thermoanaerobacter tengcongensis	+	–	+	–	+
Streptococcus agalactiae 2603 V/R	+	–	+	+	–
S. pyogenes MGAS8232	+	–	+	+	–
S. pyogenes M3	–	–	–	–	+
S. pyogenes M18	–	–	–	–	+
Proteobacteria					
Alpha					
Mesorhizobium loti	+	+	+	+	+
Agrobacterium tumefaciens C58 CEREON	+	–	+	+	+
A. tumefaciens C58 UWASH	+	–	+	+	+

(continued)

Table 19.4 (continued)

Organisms	IPD[a]	AADC[b]	AACT[c]	AACR[d]	PHAS[e]
A. tumefaciens	–	–	+	+	+
Brucella melitensis	+	–	+	+	+
B. suis 1330	+	–	+	+	–
Caulobacter crescentus	+	–	–	+	+
C. crescentus CB15		–	+	+	+
Rickettsia prowazekii	–	–	–	+	+
R. conorii	–	–	–	+	–
Beta					
Ralstonia solanacearum	+	+	+	+	+
R. metallidurans	+	–	+	+	+
Gamma					
Escherichia coli O157:H7 EDL933	+	–	+	+	+
E. coli CFT073	+	–	+	+	+
E. coli K12	+	–	+	+	–
E. coli K12 W3110	–	–	–	–	+
E. coli K12MG1655	–	–	–	–	+
Vibrio cholerae O1 *biovar eltor* str. N16961	+	–	–	+	+
V. parahaemolyticus RIMD 2210633	+	–	–	+	+
V. vulnificus YJ016	+	–	–	+	+
V. vulnificus CMCP6	+	–	–	+	+
V. cholerae	–	–	–	+	+
Xylella fastidiosa 9a5c	+	–	–	+	–
Haemophilus somnus 129PT	+	–	–	+	–
H. influenzae Rd KW20	+	–	–	+	–
Eukaryotes					
Euglenozoa					
Phytomonas sp.	+	–	–	–	–

[a]Isopropanol dehydrogenase (IPD: 1.1.1.80)
[b]Acetoacetate decarboxylase (AADC: 4.1.1.4)
[c]Acetoacetyl-CoA transferase (AACT: 2.8.3.8)
[d]Acetoacetyl-CoA reductase (AACR: 1.1.1.36)
[e]PHA synthase (PHAS: 2.3.1.-)

Table 19.5 Occurrence of enzymes for fourth metabolic pathway (starting substrate – 3-butyn 1-al) for PHA production in different organisms

Organisms	ACDH[a]	ADDH[b]	ACH[c]	HBDH[d]	HBH[e]	CEH[f]	AACT[g]	AACR[h]	PHAS[i]
Archaea									
Thermoplasma acidophilum	–	+	–	+	–	–	+	+	–
T. volcanium	–	+	–	+	–	–	–	–	–
Methanosarcina acetivorans	–	+	–	+	–	–	–	–	+
M. mazei	–	+	–	+	–	–	–	–	+
Halobacterium sp.	–	+	–	+	–	–	–	–	–
Actinobacteria									
Mycobacterium tuberculosis H37Rv	–	+	–	–	–	+	+	+	+
M. tuberculosis CDC1551	–	+	–	–	–	–	+	+	+
M. leprae	–	+	–	–	–	+	–	+	+
Streptomyces coelicolor	–	+	–	+	–	+	–	+	+
Thermobifida fusca	–	+	–	+	–	+	–	+	–
Aquificales									
Aquifex aeolicus	–	+	–	+	–	–	–	+	–
Green sulfur bacteria									
Chlorobium tepidum	–	+	–	+	–	–	–	+	+
Cyanobacteria									
Nostoc punctiforme	–	+	–	+	–	+	–	+	–
Synechocystis sp.	–	+	–	+	–	–	–	–	+
Anabaena sp.	–	+	–	–	–	–	–	–	+
Firmicutes									
Bacillus subtilis	–	+	–	+	–	–	+	+	+
B. halodurans	–	+	–	+	–	–	+	+	+
B. anthracis A2012	–	+	–	+	–	–	+	+	+
Clostridium perfringens	–	+	–	+	–	–	+	+	+
C. tetani	–	–	–	–	–	–	–	+	–
Staphylococcus aureus Mu50	–	+	–	+	–	–	–	+	+
S. aureus N315	–	+	–	+	–	–	–	+	+
S. aureus MW2	–	+	–	+	–	–	–	+	+
Oceanobacillus iheyensis	–	+		+	–	–	–	–	+

Lactobacillus plantarum WCFS1	–	+	–	+	–	–	–	+	–
Proteobacteria									
Alpha									
Sinorhizobium meliloti	+	+	–	+	–	+	+	+	+
Bradyrhizobium japonicum	+	+	–	+	–	+	+	+	+
Mesorhizobium loti	+	+	–	+	–	+	+	+	+
Rhodobacter sphaeroides	+	+	–	+	–	+	+	–	+
Novosphingobium aromaticivorans	+	+	–	+	–	–	+	+	+
Rhodopseudomonas palustris	+	+	–	+	–	–	+	+	+
Caulobacter crescentus CB15	–	+	–	+	–	+	+	+	+
Agrobacterium tumefaciens	+	+	–	–	–	–	+	+	+
A. tumefaciens C58 CEREON	–	+	–	–	–	–	+	+	+
A. tumefaciens C58 UWASH	–	+	–	+	–	–	+	+	+
Rhodospirillum rubrum	–	+	–	+	–	–	+	+	+
Beta									
Ralstonia solanacearum	–	+	–	+	–	+	+	+	+
R. metallidurans	–	+	–	+	–	+	+	+	+
Gamma									
Azotobacter vinelandii	+	+	–	+	–	+	+	+	+
Escherichia coli O157 EDL 933	+	+	–	+	–	–	+	+	+
E. coli CFT073	+	+	–	+	–	–	–	+	+
E. coli K12	+	+	–	+	–	–	+	+	–
E. coli K12 MG1655	–	+	–	–	–	–	–	–	+
E. coli K12 W3110	–	+	–	–	–	–	–	–	+
E. coli O157 Sakai	–	+	–	–	–	–	–	–	+
Pseudomonas aeruginosa PA01	+	+	–	+	–	–	+	+	+
P. putida KT2440	+	+	–	+	–	–	+	+	+
P. syringae pv. *tomato str.* DC3000	+	+	–	+	–	–	+	+	+
P. aeruginosa UCBPP-PA14	–	+	–	–	–	–	+	+	+
P. fluorescens PfO-1	–	+	–	–	–	–	+	+	+
P. aeruginosa	+	+	–	–	–	–	–	+	+

(continued)

Table 19.5 (continued)

Organisms	ACDH[a]	ADDH[b]	ACH[c]	HBDH[d]	HBH[e]	CEH[f]	AACT[g]	AACR[h]	PHAS[i]
Burkholderia fungorum	+	+	–	–	–	+	+	+	+
Xanthomonas axonopodis	+	+	–	+	–	–	–	+	+
X. campestris	+	+	–	+	–	–	–	+	+
Salmonella typhimurium LT2	+	+	–	+	–	–	+	+	–
S. enterica subsp. *enterica serovar Typhi*	+	+	–	+	–	–	–	+	–
Vibrio cholerae O1 biovar *eltor str.* N16961	–	+	–	+	–	–	–	+	+
V. cholerae	–	+	–	–	–	–	–	+	+
Shigella flexneri 2a str. 301	+	+	–	+	–	–	–	+	–
Deinococcus									
Deinococcus radiodurans	–	+	–	+	–	–	+	+	+
Eukaryotes									
Metazoa									
Drosophila melanogaster	–	+	–	+	–	–	+	+	+
Homo sapiens	–	+	–	+	–	–	+	+	+
Rattus norvegicus	–	+	–	+	–	–	+	+	+
Mus musculus	–	+	–	–	–	–	+	+	+
Caenorhabditis elegans	–	+	–	–	–	–	+	–	–
Fungi									
Saccharomyces cerevisiae	–	+	–	–	–	–	+	–	+
Schizosaccharomyces pombe	–	+	–	–	–	–	–	–	+
Viridiplantae									
Arabidopsis thaliana	–	+	–	–	–	–	–	+	+

[a]Alcohol dehydrogenase (ACDH: 1.1.99.8)
[b]Aldehyde dehydrogenase (ADDH: 1.2.1.3, 1.2.99.3)
[c]Acetylene carboxylase hydratase (ACH: 4.2.1.71)
[d]3-hydroxybutyrate dehydrogenase (HBDH: 1.1.1.30)
[e]Hydroxybutyrate-dimer hydrolase (HBH: 3.1.1.22)
[f]Carboxylic-ester hydrolases (CEH: 3.1.1.-)
[g]Acetate-CoA transferase (AACT: 2.8.3.8)
[h]Acetoacetyl-CoA reductase (AACR: 1.1.1.36)
[i]PHA synthase (PHAS: 2.3.1.-)

can thus be made to produce PHB by alternative routes under different environmental conditions.

Different approaches have been attempted or suggested, which can lead to the reduction in the PHA production costs. These suggestions include methods such as mixing the polymers with cheap materials, use of cheaper raw materials, synthesis optimization, selection and genetic improvement of bacterial strains, production of transgenic plants, and even clubbing it with hydrogen production (Daniell and Dhingra 2002; Snell and Peoples 2002; Aldor and Keasling 2003; Ma and Zeng 2003; Kumar et al. (2009); Patel et al. 2011, 2012, 2015; Singh et al. 2013; Kumar et al. (2014), 2015a, b, c). Among the organisms possessing machinery for PHA production through alternative routes, *Aquifex aeolicus*, *Burkholderia fungorum*, *Novosphingobium aromaticivorans* (Kalia et al. 2003a), *Rhodopseudomonas palustris* (Kalia et al. 2003a), *Clostridium*, and *Pseudomonas* assume higher significance because of their potential to produce hydrogen, a clean fuel from a wide range of wastes. In conclusion, in silico analysis can prove helpful in reducing cost by saving time and money, which is otherwise required to develop metabolically engineered strains for developing various biopolymers through the genetic modifications.

Acknowledgments We are thankful to the Director of CSIR-Institute of Genomics and Integrative Biology, and CSIR project WUM (ESC0108) for providing the necessary funds, facilities, and moral support.

References

Ackermann JU, Babel W (1997) Growth-associated synthesis of poly (hydroxybutyric acid) in *Methylobacterium rhodesianum* as an expression of an internal bottleneck. Appl Microbiol Biotechnol 47:144–149

Agus J, Kahar P, Abe H, Doi Y, Tsuge T (2006) Molecular weight characterization of poly(R)-3-hydroxybutyrate synthesized by genetically engineered strains of *Escherichia coli*. Polym Degrad Stab 91:1138–1146. doi:10.1016/j.polymdegradstab.2005.07.006

Agus J, Kahar P, Hyakutake M, Tomizawa S, Abe H, Tsuge T, Satoh Y, Tajima K (2010) Unusual change in molecular weight of polyhydroxyalkanoate (PHA) during cultivation of PHA-accumulating *Escherichia coli*. Polym Degrad Stab 95:2250–2254. doi:10.1016/j.polymdegradstab.2010.09.009

Aldor IS, Keasling JD (2003) Process design for microbial plastic factories: metabolic engineering of polyhydroxyalkanoates. Curr Opin Biotechnol 14:475–483. doi:10.1016/j.copbio.2003.09.002

Aldor HS, Kim SW, Prather KLJ, Keasling JD (2002) Metabolic engineering of a novel propionate-independent pathway for the production of poly(3-hydroxybutyrate-co-3-hydroxyvalerate) in recombinant *Salmonella enterica serovar typhimurium*. Appl Environ Microbiol 68:3848–3854. doi:10.1128/AEM.68.8.3848-3854.2002

Anil-Kumar PK, Shamala TR, Kshama L, Prakash MH, Joshi GJ, Chandrashekar A, Latha Kumari KS, Divyashree MS (2007) Bacterial synthesis of poly(hydroxybutyrate-co-hydroxyvalerate) using carbohydrate-rich mahua (*Madhuca* sp.) flowers. J Appl Microbiol 103:204–209. doi:10.1111/j.1365-2672.2006.03221.x

Bramer CO, Silva LF, Gomez JGC, Priefert H, Steinbuchel A (2002) Identification of the 2-methylcitrate pathway involved in the catabolism of propionate in the polyhydroxyalkanoate-producing strain *Burkholderia sacchari* IPT101^T and analysis of a mutant accumulating a copolyester with higher 3-hydroxyvalerate content. Appl Environ Microbiol 68:271–279. doi:10.1128/AEM.68.1.271-279.2002

Braunegg G, Lefebvre G, Genser KF (1998) Polyhydroxyalkanoates, biopolyesters from renewable resources: physiological and engineering aspects. J Biotechnol 65:127–161

Breuer U, Ackermann JU, Babel W (1995) Accumulation of poly (3-hydroxybutyric acid) and over production of exopolysaccharides in a mutant of a methylotrophic bacterium. Can J Microbiol 41(1):55–59

Chookietwattana K, Khonsarn N (2011) Biotechnological conversion of wastewater to polyhydroxyalkanoate by *Bacillus* in a sequencing batch reactor. World Appl Sci J 15:1425–1434

Daniell H, Dhingra A (2002) Multigene engineering: dawn of an exciting new era in biotechnology. Curr Opin Biotechnol 13:136–141. doi:10.1016/S0958-1669(02)00297-5

Dimroth P, Schink B (1998) Energy conservation in the decarboxylation of dicarboxylic acid by fermenting bacteria. Arch Microbiol 170:69–77

Hiroe A, Tsuge K, Nomura CT, Itaya M, Tsuge T (2012) Rearrangement of gene order in the *pha*CAB operon leads to effective production of ultrahigh-molecular-weight poly [(R)-3-Hydroxybutyrate] in genetically engineered *Escherichia coli*. Appl Environ Microbiol 78:3177–3184. doi:10.1128/AEM.07715-11

Hyakutake M, Saito Y, Tomizawa S, Mizuno K, Tsuge T (2011) Polyhydroxyalkanoate (PHA) synthesis by Class IV PHA synthases employing *Ralstonia eutropha* PHB$^-$4 as host strain. Biosci Biotechnol Biochem 75:1615–1617

Kalia VC, Chauhan A, Bhattacharyya G, Rashmi (2003a) Genomic databases yield novel bioplastic producers.

Nat Biotechnol 21:845–846. doi:10.1038/nbt0803-845

Kalia VC, Lal S, Ghai R, Mandal M, Chauhan A (2003b) Mining genomic databases to identify novel hydrogen producers. Trends Biotechnol 21:152–156. doi:10.1016/S0167-7799(03)00028-3

Kalia VC, Lal S, Cheema S (2007) Insight in to the phylogeny of polyhydroxyalkanoate biosynthesis: horizontal gene transfer. Gene 389:19–26. doi:10.1016/j.gene.2006.09.010

Khiyami MA, Al-Fadual SM, Bahklia AH (2011) Polyhydroxyalkanoates production via *Bacillus* plastic composite support (PCS) biofilm and date palm syrup. J Med Plant Res 5:3312–3320

Kumar T, Singh M, Purohit HJ, Kalia VC (2009) Potential of *Bacillus* sp. to produce polyhydroxybutyrate from biowaste. J Appl Microbiol 106:2017–2023. doi:10.1111/j.1365-2672.2009.04160.x

Kumar P, Patel SKS, Lee JK, Kalia VC (2013) Extending the limits of *Bacillus* for novel biotechnological applications. Biotechnol Adv 31:1543–1561. doi:10.1016/j.biotechadv.2013.08.007

Kumar P, Singh M, Mehariya S, Patel SKS, Lee JK, Kalia VC (2014) Ecobiotechnological approach for exploiting the abilities of *Bacillus* to produce co-polymer of polyhydroxyalkanoate. Indian J Microbiol 54:151–157. doi:10.1007/s12088-014-0457-9

Kumar P, Mehariya S, Ray S, Mishra A, Kalia VC (2015a) Biodiesel industry waste: a potential source of bioenergy and biopolymers. Indian J Microbiol 55:1–7. doi:10.1007/s12088-014-0509-1

Kumar P, Ray S, Patel SKS, Lee JK, Kalia VC (2015b) Bioconversion of crude glycerol to PHA by *Bacillus thuringiensis* under non-limiting nitrogen conditions. Int J Biol Macromol 78:9–16. doi:10.1016/j.ijbiomac.2015.03.046

Kumar P, Sharma R, Ray S, Mehariya S, Patel SKS, Lee JK, Kalia VC (2015c) Dark fermentative bioconversion of glycerol to hydrogen by *Bacillus thuringiensis*. Bioresour Technol 182:383–388. doi:10.1016/j.biortech.2015.01.138

Lee SY (1996) Plastic bacteria? Progress and prospects for polyhydroxyalkanoates production in bacteria. Trends Biotechnol 14:431–438. doi:10.1016/0167-7799(96)10061-5

Lee SY, Choi JI (1999) Production and degradation of polyhydroxyalkanoates in waste environment. Waste Manag 19:133–139

Ma H, Zeng AP (2003) Reconstruction of metabolic networks from genome data and analysis of their global structure for various organisms. Bioinformatics 19:270–277. doi:10.1093/bioinformatics/19.2.270

Madison LL, Huisman GW (1999) Metabolic engineering of poly (3-hydroxyalkanoates): from DNA to plastic. Microbiol Mol Biol Rev 63:21–53

Masood F, Hasan F, Ahmed S, Chen P, Hameed A (2012) Biosynthesis and characterization of poly-(3-hydroxybutyrate-co-3-hydroxyvalerate) from *Bacillus cereus* S10. J Polym Environ 20:865–871. doi:10.1007/s10924-012-0457-y

Mergaert J, Webb A, Anderson C, Wouters A, Swings J (1993) Microbial degradation of poly (3-hydroxybutyrate) and poly (3-hydroxybutyrate-co-3-hydroxyvalerate) in soils. Appl Environ Microbiol 59:3233–3238

Moreira D, López-García P (2002) The molecular ecology of microbial eukaryotes unveils a hidden world. Trends Microbiol 10:31–38

Mothes G, Babel W (1995) *Methylobacterium rhodesianum* MB 126 possesses two stereospecific crotonyl-CoA hydratases. Can J Microbiol 41(1):68–72

Mothes G, Ackermann JU, Babel W (1998) Regulation of poly(3-hydroxybutyrate) synthesis in *Methylobacterium rhodesianum* MB 126 growing on methanol of fructose. Arch Microbiol 169:360–363

Page WJ, Manchak J (1995) The role of β-oxidation of short-chain alkanoates in polyhydroxyalkanoate copolymer synthesis in *Azotobacter vinelandii* UWD. Can J Microbiol 41(1):106–114

Park SJ, Park JP, Lee SY (2002) Metabolic engineering of *Escherichia coli* for the production of medium-chain-length polyhydroxyalkanoates rich in specific monomers. FEMS Microbiol Lett 214:217–222

Patel SKS, Singh M, Kalia VC (2011) Hydrogen and polyhydroxybutyrate producing abilities of *Bacillus* spp. from glucose in two stage system. Indian J Microbiol 51:418–423. doi:10.1007/s12088-011-0236-9

Patel SKS, Singh M, Kumar P, Purohit HJ, Kalia VC (2012) Exploitation of defined bacterial cultures for production of hydrogen and polyhydroxybutyrate from pea-shells. Biomass Bioenergy 36:218–225. doi:10.1016/j.biombioe.2011.10.027

Patel SKS, Kumar P, Singh M, Lee JK, Kalia VC (2015) Integrative approach to produce hydrogen and polyhydroxybutyrate from biowaste using defined bacterial cultures. Bioresour Technol 176:136–141. doi:10.1016/j.biortech.2014.11.029

Pozo C, Martinez-Toledo MV, Rodelas B, González-López J (2002) Effect of culture conditions on the production of polyhydroxyalkanoates by *Azotobacter chroococcum* H23 in media containing a high concentration of alpechin (wastewater from olive oil mills) as primary carbon source. J Biotechnol 97:125–131. doi:10.1016/S0168-1656(02)00056-1

Reddy SK, Ghai R, Rashmi KVC (2003) Polyhydroxyalknoates: an overview. Bioresour Technol 87:137–146. doi:10.1016/S0960-8524(02)00212-2

Rehm BHA (2010) Bacterial polymers: biosynthesis, modifications and applications. Nat Rev Microbiol 8:578–592. doi:10.1038/nrmicro2354

Rehm BHA, Mitsky TA, Steinbüchel A (2001) Role of fatty acid de novo biosynthesis in polyhydroxyalkanoic acid (PHA) and rhamnolipid synthesis by *Pseudomonads*: establishment of the transacylase (PhaG)-mediated pathway for PHA biosynthesis in *Escherichia coli*. Appl Environ Microbiol 67:3102–3109. doi:10.1128/AEM.67.7.3102-3109.2001

Sangkharak K, Prasertsan P (2012) Screening and identification of polyhydroxyalkanoates producing bacteria and biochemical characterization of their possible application. J Gen Appl Microbiol 58:173–182. doi:10.2323/jgam.58.173

Santimano MC, Prabhu NN, Garg S (2009) PHA production using low-cost agro-industrial wastes by *Bacillus* sp. strain COL1/A6. Res J Microbiol 4:89–96. doi:10.3923/jm.2009.89.96

Schloss PD, Handelsman J (2004) Status of the microbial census. Microbiol Mol Biol Rev 68:686–691. doi:10.1128/MMBR.68.4.686-691.2004

Schubert P, Steinbüchel A, Schlegel HG (1988) Cloning of the *Alcaligenes eutrophus* genes for synthesis of poly-β-hydroxybutyric acid (PHB) and synthesis of PHB in *Escherichia coli*. J Bacteriol 170:5837–5847

Shamala TR, Vijayendra SVN, Joshi GJ (2012) Agro-industrial residues and starch for growth and co-production of polyhydroxyalkanoate copolymer and α-amylase by *Bacillus* sp. CFR-67. Braz J Microbiol 43:1094–1102. doi:10.1590/S1517-838220120003000036

Shrivastav A, Mishra SK, Shethi B, Pancha I, Jain D, Mishra S (2010) Isolation of promising bacterial strains from soil and marine environment for polyhydroxyalkanoates (PHAs) production utilizing *Jatropha* biodiesel byproduct. Int J Biol Macromol 47:283–287. doi:10.1016/j.ijbiomac.2010.04.007

Singh M, Patel SK, Kalia VC (2009) Bacillus subtilis as potential producer for polyhydroxyalkanoates. Microb Cell Fact 8:38. doi:10.1186/1475-2859-8-38

Singh M, Kumar P, Patel SKS, Kalia VC (2013) Production of polyhydroxyalkanoate co-polymer by *Bacillus thuringiensis*. Indian J Microbiol 53:77–83. doi:10.1007/s12088-012-0294-7

Singh M, Kumar P, Ray S, Kalia VC (2015) Challenges and opportunities for customizing polyhydroxyalkanoates. Indian J Microbiol 55:235–249. doi:10.1007/s12088-015-0528-6

Snell KD, Peoples OP (2002) Polyhydroxyalkanoate polymers and their production in transgenic plants. Metab Eng 4:29–40. doi:10.1006/mben.2001.0214

Steinbüchel A, Lutke-Eversloh T (2003) Metabolic engineering and pathway construction for biotechnological production of relevant polyhydroxyalkanoates in microorganisms. Biochem Eng J 3734:1–16

Tomizawa S, Hyakutake M, Saito Y, Agus J, Mizuno K, Abe H, Tsuge T (2011) Molecular weight change of polyhydroxyalkanoate (PHA) caused by the PhaC subunit of PHA synthase from *Bacillus cereus* YB-4 in recombinant *Escherichia coli*. Biomacromolecules 12:2660–2666. doi:10.1021/bm2004687

Verlinden RAJ, Hill DJ, Kenward MA, Williams CD, Radecka I (2007) Bacterial synthesis of biodegradable polyhydroxyalkanoates. J Appl Microbiol 102:1437–1449. doi:10.1111/j.1365-2672.2007.03335.x

Williams DR, Anderson AJ, Dawes EA, Ewing DF (1994) Production of a co-polyester of 3-hydroxybutyric acid and 3-hydroxyvaleric acid from succinic acid by *Rhodococcus ruber*: biosynthetic considerations. Appl Microbiol Biotechnol 40:717–723

Yan Q, Zhao M, Miao H, Ruana W, Song R (2010) Coupling of the hydrogen and polyhydroxyalkanoates (PHA) production through anaerobic digestion from Taihu blue algae. Bioresour Technol 101:4508–4512. doi:10.1016/j.biortech.2010.01.073

Yu J (2001) Production of PHA from starchy wastewater via organic acids. J Biotechnol 86:105–112. doi:10.1016/S0168-1656(00)00405-3

Vipin Chandra Kalia received his M.Sc. and Ph.D. degree in Genetics from IARI, New Delhi. He is presently the Chief Scientist at CSIR-IGIB and Professor of AcSIR. He is a Fellow of Association of Microbiologists of India and Fellow of National Academy of Sciences (FNASc). He is the Editor in Chief of the Indian Journal of Microbiology (India). His current interests are bioenergy, biopolymers, and quorum sensing. Google Scholar: http://scholar.google.co.in/citations?hl=en&user=XaUw-VIAAAAJ, Website: http://www.igib.res.in/?q=V.C.Kalia

Sadhana Lal received his M.Sc. degree in Bioscience from IIT Roorkee, Uttarakhand, India, and Ph.D. in Biochemistry from Jiwaji Gwalior University, MP, India, and CSIR-IGIB, Delhi, India. She is presently working as a Research Associate in Biosystems Engineering Department at University of Manitoba, Winnipeg, Canada. She is a member of the American Society for Microbiology (ASM). Her current research interests are bacterial genome annotation, in silico comparative analysis of bacterial metabolisms, genetic engineering, and bioenergy generation.

Investigating the Phylogeny of Hydrogen Metabolism by Comparative Genomics: Horizontal Gene Transfer

20

Sadhana Lal, Dhananjay V. Raje, Simrita Cheema, Atya Kapley, Hemant J. Purohit, and Vipin Chandra Kalia

Abstract

The phylogenetic analysis based on molecular characteristics indicates that lithotrophic metabolism was followed by phototrophy. Hydrogen (H_2) metabolism is a signature of such environments. This property is prominent among organisms found in geothermal conditions and in deep aquifers. H_2 is generated readily by abiotic mechanisms where the terminal electron acceptor is likely to be the limiting factor. In the post-fossil fuel era, H_2 has in fact emerged as a strong contender for future fuel. It is thus important to understand the molecular mechanisms which lead to H_2 production and associated biological systems. These can help to comprehend issues such as sustainability, environmental emissions and energy security. Comparative genomic analysis reveals events of horizontal transfer of genes of H_2 metabolism among taxonomically diverse organisms. This offers an opportunity to identify those genomes which can be tailored for transforming presently 'non'-H_2 producers into producers. This also suggests that naturally occurring events can be mimicked to provide future fuel H_2.

S. Lal
CSIR-Institute of Genomics and Integrative Biology, Delhi University Campus, Delhi, India

S. Cheema • V.C. Kalia (✉)
Microbial Biotechnology and Genomics, CSIR-Institute of Genomics and Integrative Biology, 110007 Delhi, India
e-mail: vckalia@igib.res.in; vc_kalia@yahoo.co.in

D.V. Raje • A. Kapley
Environmental Genomics Division, CSIR-National Environmental Engineering Research Institute, 440020 Nagpur, Maharastra, India
e-mail: a_kapley@neeri.res.in

H.J. Purohit
Environmental Genomics Division, CSIR-National Environmental Engineering Research Institute (NEERI), Nehru Marg, 440015 Nagpur, India
e-mail: hj_purohit@neeri.res.in; hemantdrd@hotmail.com

V.C. Kalia (ed.), *Microbial Factories*, DOI 10.1007/978-81-322-2595-9_20

20.1 Introduction

Phylogenetic trees based on small subunit of ribosomal DNA gene sequences provide intriguing insights into microbial evolution. The phylogenetic analysis based on molecular characteristics indicates that lithotrophic metabolism was followed by phototrophy and that it is relatively more widespread. As the lithotrophic biosphere is spread quite deep into the Earth (Ghiorse 1997), it indicates higher contribution to global mass than is presently understood. Hydrogen (H_2) metabolism is a signature of such environments, where methane is the final product. It is significantly an important characteristic among organisms which live in geothermal conditions and in deep aquifers (Stevens and McKinley 1995). H_2 is generated by abiotic systems, which are limited by the terminal electron acceptor. In the post-fossil fuel era, H_2 has in fact emerged as a strong contender for future fuel (Veziroglu 1995). It can address issues of sustainability, environmental emissions and energy security (Turner 2004). The interest in H_2 metabolism lies in technologies where organisms and enzymes might be exploited for commercial H_2 production and especially for use in fuel cells (Fraser 2004). In view of the diverse applications of H_2 in various fields, its large scale and economic production have always been a challenging research area.

A deep comprehension of the molecular fundamentals underlying H_2 production and the biological systems responsible for it is really important for future ventures. Biologically H_2 involves microbial growth and metabolism. H_2 production per se is a mechanism for disposal of excess electrons generated during respiration. The functions of hydrogenases (H_2ase) are often associated with cellular localization, e.g., H_2 evolution is most often cytosolic, whereas H_2 uptake is usually periplasmic or membrane localized. Anaerobic organisms rely on H_2ase enzymes, whereas facultative and heterotrophic anaerobes depend up on soluble formate dehydrogenase (FDH) and its association with an insoluble H_2ase enzyme. Fermentative bacteria metabolize pyruvate and formate to H_2 by employing a combination of FDH and H_2ase enzymes.

Fermentative H_2 production system (Vignais et al. 2001) involves enzyme complex: three subunits of FDH (-α, -β and -γ) coded by *fdh*A, *fdh*B and *fdh*C genes and two subunits of H_2ase (small subunit and large subunit) coded by *hyd*A and *hyd*B genes. Mining of genomic database for these enzymatic subunits has revealed organisms which had the potential to produce H_2 (Kalia et al. 2003b). It also showed a large genetic variable, where organisms could not evolve H_2 because of lack of one of the enzymes or their subunits. The genetic variability in these organisms has led us to explore the role of different genes and their phylogenetic status in H_2 metabolism. Recombinant DNA technology for transferring genes from certain organisms into others, including non-H_2 producers, has met little success. It may thus be interesting and necessary to evaluate the extent to which evolutionary processes have driven genome modifications (Delmotte et al. 2006), the variation in the codon usage pattern, the extent of gene expression within a genome (Codon Adaptation Index, CAI) and their integration within the host organisms (Sharp and Li 1987). Low CAI values and atypical composition of certain genes such as hemolysin genes in *Escherichia coli* (Felmlee et al. 1985) and polyhydroxybutyrate biosynthesis gene *phaC* in *Azotobacter vinelandii* and *Pseudomonas oleovorans* (Kalia et al. 2007) suggest that these genes are not well adapted and are of recent acquisitions. However, similarities in CAI values of the foreign genes and those of the highly expressed genes of the host indicate adaptation to equivalent extents and can be used to predict levels of gene expression (Carbone et al. 2003; Kalia et al. 2007). In this study, we have presented the results of phylogenetic analyses, Chi square (χ^2) values, G+C content and CAI of genes of the H_2 metabolic pathway (Ermolaeva et al. 1998; Mount 2001; Garcia-Vallve et al. 2003; Kalia et al. 2007). These indicators of horizontal gene transfer (HGT) events during evolution have enabled us to identify certain organisms, which can be exploited to transform 'non'-H_2 producers into producers.

20.2 Phylogenetic Analysis

The phylogenetic distribution of five enzymatic subunits of FDH and H_2ase among 183 organisms reflects on the naturally occurring genetic variation. The wide phylogenetic distribution of each subunit of FDH and H_2ase reflected the significant events in their evolution. Only 12 genera had all the five subunits of FDH and H_2ase. The prevalence of all the five subunits was highest among γ-Proteobacteria (Table 20.1). The phylogenetic trees of the different enzymatic subunits do not correspond to the 16S rDNA-based microbial phylogeny. This suggests that HGT might have played a catalytic role in the 'rapid' evolution of H_2 metabolism. It also revealed that some organisms have the ability to acquire genes through HGT (Lal et al. 2008).

20.2.1 Phylogenetic Analysis of *fdh*A Gene

The wide diversity in the Fdh-α subunit encoded by *fdh*A gene is reflected by its paraphyletic distribution in organisms belonging to Archaea, Actinobacteria, Aquificales, CFB, Firmicutes, various Proteobacteria and extends even up to Fungi and Planctomycetes. A comparison of the phylogenetic tree of complete amino acid sequences of Fdh-α indicates that the *fdh*A phylogeny is inconsistent with that of 16S rDNA phylogeny (Fig. 20.1). Such inconsistencies were widespread among different taxonomical groups. Among these strong inconsistencies (BV ≥500), only a few could be categorized as HGT events based on significant χ^2 values and deviations in GC contents.

An archaeal member, *Haloferax mediterranei* is clubbed among Actinobacteria, Firmicutes and γ-Proteobacteria on one hand and Cyanobacteria, α-, β-, γ-Proteobacteria, Aquificales and CFB on the other. Its wide separation from all other archaeal members in the phylogenetic tree for *fdh*A is supported by higher GC content of its gene (65.91 %) and genome (61.05 %), compared to low GC content, 35–54 % for the *fdh*A gene and 34.5–51.9 % for the genomes of other archaeal members.

Among Firmicutes, three organisms showed significant deviations in their *fdh*A gene compared to their total genome: *Bacillus cereus*, *B. subtilis* and *Clostridium perfringens* str.13 (χ^2: 204, 303 and 342, respectively). The strong but abnormal association of *C. perfringens* str.13 with *Pirellula* sp. 1 (Planctomycetes) (BV 1000) is further supported by their association with *Aspergillus nidulans* FGSC A4 (Eukaryote) (BV 533). On a major clade, *Bacillus* spp., *Listeria* and *Staphylococcus* had a strong association with *Sulfolobus* species (Archaea) on one hand and *Xanthomonas campestris* (γ-Proteobacteria) on the other hand. Very interestingly, *X. campestris* seems to have received *fdh*A through HGT events (χ^2: 1909.430). This close association of distantly related organism indicates that *fdh*A has been shared among organisms under conditions where energy generation became of paramount importance. In spite of a large number of γ-proteobacterial members possessing *fdh*A, only two of them have shown highly significant deviation from their genomes: *X. campestris* and *Shigella flexneri*. *X. campestris* seems to have acquired its *fdh*A gene more recently than *S. flexneri* since there is a wide deviation in the CAI value in the former in comparison to its genome. *Xanthomonas axonopodis* pv. citri. str. 306 have been recently shown to have acquired the genes of polyhydroxyalkanoate (PHA) biosynthesis through HGT (Kalia et al. 2007). In *S. flexneri*, the gene seems to have got well adapted to its host genome after its transfer from a distantly related organism (Carbone et al. 2003; Kalia et al. 2007).

The three actinobacterial members possessing *fdh*A are widely separated. Their association with γ-Proteobacteria is quite strong (BV 717 and 1000). On the clade harbouring *Streptomyces coelicolor*, *Gloeobacter violaceus* (γ-Proteobacteria) and *Chloroflexus aurantiacus* (Chloroflexales), high BV and significant deviation in the gene characteristics of *S. coelicolor* strongly suggest HGT event. The separation of the two species of *Streptomyces* is justified by the highly significant deviation, χ^2: 4661.298 in the *fdh*A gene of *S. coelicolor* from its genome. Although GC content of the gene and genome of

Table 20.1 Distribution[a] of FDH and H_2ase subunits in different organisms

S no.	Organisms	Formate dehydrogenase (FDH)			Hydrogenase (H_2ase)	
		Alpha subunit	Beta subunit	Gamma subunit	Large subunit	Small subunit
1	*Actinobacillus pleuropneumoniae serovar* 1 str. 4074	+	+	+	+	+
2	*Aquifex aeolicus* VF5	+	+	+	+	+
3	*Azotobacter vinelandii*	+	+	+	+	+
4	*Bradyrhizobium japonicum* USDA 110	+	+	+	+	+
5	*Campylobacter jejuni* subsp. *jejuni* NCTC 11168	+	+	+	+	+
6	*Desulfitobacterium hafniense*	+	+	+	+	+
7	*Helicobacter hepaticus* ATCC 51449	+	+	+	+	+
8	*Magnetospirillum magnetotacticum*	+	+	+	+	+
9	*Escherichia coli* CFT073	+	+	+	+	+
10	*E. coli* K12	+	+	+	+	+
11	*E. coli* O157:H7 EDL933	+	+	+	+	+
12	*Salmonella enterica* subsp. *enterica serovar typhi*	+	+	+	+	+
13	*S. typhimurium* LT2	+	+	+	+	+
14	*Shewanella oneidensis* MR-1	+	+	+	+	+
15	*Wolinella succinogenes*	+	+	+	+	+
16	*Archaeoglobus fulgidus* DSM 4304	+	+	–	+	+
17	*Desulfovibrio desulfuricans* G20	+	+	–	+	+
18	*Escherichia coli*	+	+	–	+	+
19	*Geobacter metallireducens*	+	+	–	+	+
20	*Magnetococcus* sp. MC–1	+	+	–	+	+
21	*Rhodospirillum rubrum*	+	+	–	+	+
22	*Ralstonia metallidurans*	+	+	+	–	+
23	*Shigella flexneri* 2a str. 301	+	+	+	–	+
24	*Burkholderia fungorum*	+	+	+	–	–
25	*Chromobacterium violaceum* ATCC 12472	+	+	+	–	–
26	*Pseudomonas aeruginosa* PAO1	+	+	+	–	–
27	*P. fluorescens* PfO-1	+	+	+	–	–
28	*P. putida* KT2440	+	+	+	–	–
29	*Bordetella parapertussis*	+	+	+	–	–
30	*Sinorhizobium meliloti* (Rhizobium meliloti)	+	+	+	–	–
31	*Vibrio parahaemolyticus* RIMD 2210633	+	+	+	–	–
32	*V. vulnificus* CMCP6	+	+	+	–	–
33	*V. vulnificus* YJ016	+	+	+	–	–
34	*Yersinia pestis* CO92	+	+	+	–	–
35	*Dechloromonas aromatica* RCB	+	+	–	+	–

(continued)

Table 20.1 (continued)

S no.	Organisms	Formate dehydrogenase (FDH)			Hydrogenase (H_2ase)	
		Alpha subunit	Beta subunit	Gamma subunit	Large subunit	Small subunit
36	*Streptomyces avermitilis* MA-4680	+	–	+	+	–
37	*Rhodobacter sphaeroides*	+	–	–	+	+
38	*Rhodopseudomonas palustris* CGA009	+	–	–	+	+
39	*Ralstonia eutropha*	+	–	–	+	+
40	*Desulfovibrio desulfuricans*	–	+	–	+	+
41	*D. gigas*	–	+	–	+	+
42	*Geobacter sulfurreducens* PCA	–	+	–	+	+
43	*Methanosarcina acetivorans* C2A	–	+	–	+	+
44	*Rubrivivax gelatinosus*	–	+	–	+	+
45	*Bordetella bronchiseptica* RB50	+	+	–	–	–
46	*Chloroflexus aurantiacus*	+	+	–	–	–
47	*Cytophaga hutchinsonii*	+	+	–	–	–
48	*Escherichia coli* O157:H7	+	+	–	–	–
49	*Haloferax mediterranei*	+	+	–	–	–
50	*Mesorhizobium loti*	+	+	–	–	–
51	*Methanococcus maripaludis*	+	+	–	–	–
52	*Methanopyrus kandleri* AV19	+	+	–	–	–
53	*Methanothermobacter thermautotrophicus* str. Delta H	+	+	–	–	–
54	*Microbulbifer degradans* 2–40	+	+	–	–	–
55	*Novosphingobium aromaticivorans*	+	+	–	–	–
56	*Pirellula* sp. 1	+	+	–	–	–
57	*Pyrobaculum aerophilum* str. IM2	+	+	–	–	–
58	*Pyrococcus abyssi*	+	+	–	–	–
59	*P. horikoshii*	+	+	–	–	–
60	*Sulfolobus solfataricus*	+	+	–	–	–
61	*S. tokodaii*	+	+	–	–	–
62	*Methylobacterium extorquens*	+	–	+	–	–
63	*Pseudomonas syringae* pv. *syringae* B728a	+	–	+	–	–
64	*P. syringae* pv. *tomato* str. DC3000	+	–	+	–	–
65	*Ralstonia solanacearum*	+	–	+	–	–
66	*Sulfurospirillum multivorans*	+	–	+	–	–
67	*Methanosarcina barkeri*	+	–	–	–	+
68	*Haemophilus ducreyi* 35000HP	–	+	+	–	–
69	*Haemophilus influenzae*	–	+	+	–	–
70	*H. influenzae* Rd KW20	–	+	+	–	–
71	*Pasteurella multocida*		+	+	–	–
72	*Photorhabdus luminescens* subsp. *laumondii* TTO1	–	+	+	–	–

(continued)

Table 20.1 (continued)

S no.	Organisms	Formate dehydrogenase (FDH)			Hydrogenase (H_2ase)	
		Alpha subunit	Beta subunit	Gamma subunit	Large subunit	Small subunit
73	*Vibrio cholerae O1* biovar eltor str. N16961	–	+	+	–	–
74	*Chlorobium tepidum* TLS	–	+	–	+	–
75	*Desulfovibrio vulgaris*	–	+	–	+	–
76	*Methanosarcina mazei* Goe1	–	+	–	–	+
77	*Alcaligenes hydrogenophilus*	–	–	–	+	+
78	*Azotobacter chroococcum*	–	–	–	+	+
79	*A. chroococcum* (strain mcd 1)	–	–	–	+	+
80	*Citrobacter freundii*	–	–	–	+	+
81	*Clostridium acetobutylicum*	–	–	–	+	+
82	*Corynebacterium diphtheriae*	–	–	–	+	+
83	*Desulfitobacterium dehalogenans*	–	–	–	+	+
84	*Desulfomicrobium baculatum*	–	–	–	+	+
85	*Desulfovibrio fructosovorans*	–	–	–	+	+
86	*D. vulgaris* (strain Miyazaki)	–	–	–	+	+
87	*Helicobacter pylori* 26695	–	–	–	+	+
88	*H. pylori* J99	–	–	–	+	+
89	*Methylococcus capsulatus* str. Bath	–	–	–	+	+
90	*Oligotropha carboxidovorans*	–	–	–	+	+
91	*Pseudomonas hydrogenovora*	–	–	–	+	+
92	*Thiocapsa roseopersicina*	–	–	–	+	+
93	*Rhizobium leguminosarum*	–	–	–	+	+
94	*R. leguminosarum* bv. *viciae*	–	–	–	+	+
95	*Rhodobacter capsulatus*	–	–	–	+	+
96	*Agrobacterium tumefaciens* str. C58 (Cereon)	+	–	–	–	–
97	*A. tumefaciens* str. C58 (U. Washington)	+	–	–	–	–
98	*Amycolatopsis mediterranei*	+	–	–	–	–
99	*Aquifex aeolicus*	+	–	–	–	–
100	*Aspergillus nidulans* FGSC A4	+	–	–	–	–
101	*Azospirillum brasilense*	+	–	–	–	–
102	*Bacillus anthracis* str. A2012	+	–	–	–	–
103	*B. anthracis* str. Ames	+	–	–	–	–
104	*B. cereus* ATCC 14579	+	–	–	–	–
105	*B. halodurans*	+	–	–	–	–
106	*B. subtilis*	+	–	–	–	–
107	*B. subtilis* subsp. *subtilis* str. 168	+	–	–	–	–
108	*Campylobacter jejuni*	+	–	–	–	–
109	*Caulobacter crescentus* CB15	+	–	–	–	–
110	*Clostridium perfringens*	+	–	–	–	–
111	*C. perfringens* str. 13	+	–	–	–	–
112	*Cyanothece* sp. PCC 8801	+	–	–	–	–

(continued)

Table 20.1 (continued)

		Formate dehydrogenase (FDH)			Hydrogenase (H_2ase)	
S no.	Organisms	Alpha subunit	Beta subunit	Gamma subunit	Large subunit	Small subunit
113	*Deinococcus radiodurans* R1	+	–	–	–	–
114	*Enterococcus faecalis* V583	+	–	–	–	–
115	*Eubacterium acidaminophilum*	+	–	–	–	–
116	*Ferroplasma acidarmanus*	+	–	–	–	–
117	*Gloeobacter violaceus*	+	–	–	–	–
118	*Klebsiella oxytoca*	+	–	–	–	–
119	*Klebsiella pneumoniae*	+	–	–	–	–
120	*Listeria innocua*	+	–	–	–	–
121	*L. monocytogenes* EGD–e	+	–	–	–	–
122	*Methanothermobacter thermautotrophicus*	+	–	–	–	–
123	*Methanococcus voltae*	+	–	–	–	–
124	*Methanobacterium formicicum*	+	–	–	–	–
125	*Methanocaldococcus jannaschii*	+	–	–	–	–
126	*Moorella thermoacetica*	+	–	–	–	–
127	*Mycobacterium avium* subsp. paratuberculosis str. k10	+	–	–	–	–
128	*Nostoc punctiforme*	+	–	–	–	–
129	*Nostoc* sp. PCC 7120	+	–	–	–	–
130	*Pseudomonas putida*	+	–	–	–	–
131	*Pseudomonas* sp. G–179	+	–	–	–	–
132	*Pyrococcus furiosus* DSM 3638	+	–	–	–	–
133	*Salmonella enterica* subsp. *enterica serovar typhi* (strain CT18)	+	–	–	–	–
134	*Sphingomonas elodea*	+	–	–	–	–
135	*Staphylococcus aureus* subsp. *aureus* Mu50	+	–	–	–	–
136	*S. aureus* subsp. *aureus* N315	+	–	–	–	–
137	*Streptomyces coelicolor* A3(2)	+	–	–	–	–
138	*Synechococcus* sp. PCC 7002	+	–	–	–	–
139	*Synechococcus* sp. PCC 7942	+	–	–	–	–
140	*Synechococcus* sp. WH 8102	+	–	–	–	–
141	*Thermococcus litoralis*	+	–	–	–	–
142	*Thermosynechococcus elongatus* BP–1	+	–	–	–	–
143	*Thermoplasma acidophilum*	+	–	–	–	–
144	*Trichodesmium erythraeum* IMS101	+	–	–	–	–
145	*Xanthomonas campestris* pv. *campestris* str. ATCC 33913	+	–	–	–	–
146	*Acidianus ambivalens*	–	+	–	–	–
147	*Aeropyrum pernix*	–	+	–	–	–
148	*Allochromatium vinosum*	–	+	–	–	–
149	*Anopheles gambiae* str. PEST	–	+	–	–	–

(continued)

Table 20.1 (continued)

S no.	Organisms	Formate dehydrogenase (FDH) Alpha subunit	Beta subunit	Gamma subunit	Hydrogenase (H_2ase) Large subunit	Small subunit
150	*Azoarcus evansii*	–	+	–	–	–
151	*Bdellovibrio bacteriovorus*	–	+	–	–	–
152	*Brucella melitensis* 16 M	–	+	–	–	–
153	*Brucella suis* 1330	–	+	–	–	–
154	*Carboxydothermus hydrogenoformans*	–	+	–	–	–
155	*Desulfovibrio vulgaris* subsp. *vulgaris* str. Hildenborough	–	+	–	–	–
156	*Haemophilus somnus* 129PT	–	+	–	–	–
157	*Haloarcula marismortui* subsp. *marismortui*	–	+	–	–	–
158	*Halobacterium* sp. NRC–1	–	+	–	–	–
159	*Ideonella dechloratans*	–	+	–	–	–
160	*Leptospira interrogans serovar* Lai str. 56601	–	+	–	–	–
161	*Paracoccus pantotrophus*	–	+	–	–	–
162	*Pelobacter acidigallici*	–	+	–	–	–
163	*Pseudomonas stutzeri*	–	+	–	–	–
164	*Rhodovulum sulfidophilum*	–	+	–	–	–
165	*Selenomonas ruminantium*	–	+	–	–	–
166	*Salmonella typhimurium*	–	+	–	–	–
167	*Shewanella* sp. ANA–3	–	+	–	–	–
168	*Shigella flexneri* 2a str. 2457 T	–	+	–	–	–
169	*Bradyrhizobium japonicum*	–	–	+	–	–
170	*Bordetella pertussis* Tohama I	–	–	+	–	–
171	*Photobacterium damselae*	–	–	–	+	–
172	*Aquifex pyrophilus*	–	–	–	–	+
173	*Azorhizobium* sp. UPM1161	–	–	–	–	+
174	*A. caulinodans*	–	–	–	–	+
175	*Bradyrhizobium* sp. UPM1029	–	–	–	–	+
176	*Bradyrhizobium* sp. UPM1166	–	–	–	–	+
177	*Bradyrhizobium* sp. UPM1167	–	–	–	–	+
178	*Bradyrhizobium* sp. UPM1171	–	–	–	–	+
179	*Bradyrhizobium* sp. UPM1172	–	–	–	–	+
180	*Bradyrhizobium* sp. UPM860	–	–	–	–	+
181	*Methanosarcina mazei*	–	–	–	–	+
182	*Rhizobium tropici*	–	–	–	–	+
183	*Shigella flexneri*	–	–	–	–	+

[a]Data has been arranged in the descending order of the number of subunits present in different organisms

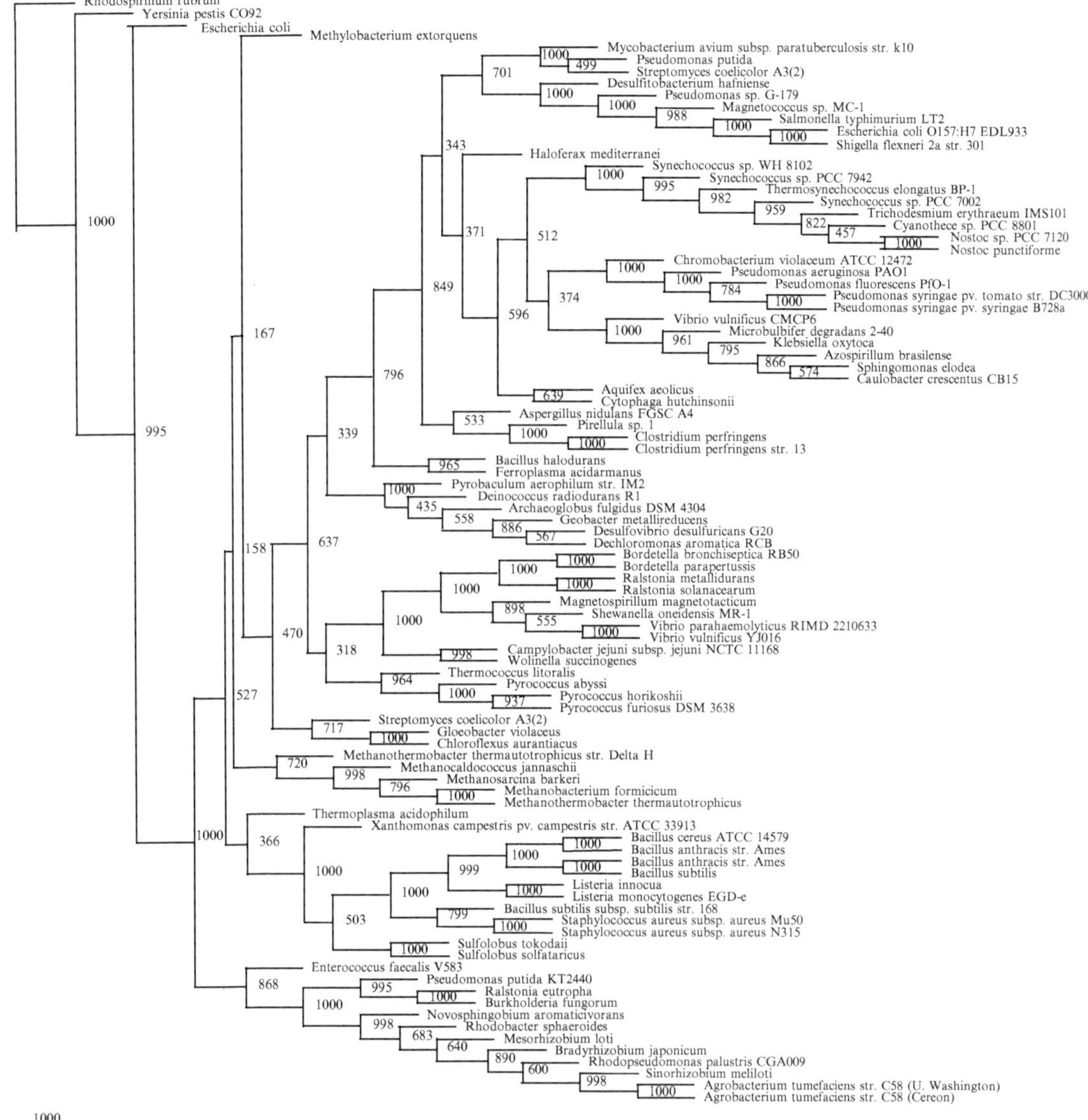

Fig. 20.1 Phylogenetic tree based on the amino acid sequences of α-subunit of FDH

these two species is quite similar (71 %), however, there are wide differences in the CAI values: 0.409–0.411 for *S. coelicolor* and 0.754–0.758 for *Streptomyces avermitilis*.

A large number of α-proteobacterial members with strong associations (BV 600–1000) are on a single clade. On this particular clade, *Bradyrhizobium japonicum* and *Rhodopseudomonas palustris* show significant deviation for their *fdh*A gene (χ^2 = 2586 and 1576, respectively). Among the β-proteobacterial members, *fdh*A gene of only *Chromobacterium violaceum* ATCC 12472 shows significant deviations, χ^2 = 2931.54, +5.68 % higher GC content compared to that of its genome and notable differences from the highly expressed genes (CAI). These parameters taken together can support HGT. *C. violaceum* ATCC 12472 has been observed to have received its *pha*B gene of PHA biosynthesis through HGT and was similarly associated on the phylogenetic tree with γ-proteobacterial member (*Azotobacter* sp. FA8)

(11). *Campylobacter jejuni* subsp. *jejuni* NCTC 11168 (ε-Proteobacteria) has shown significant deviation in its *fdh*A gene (χ^2: 498.807). Although it is clubbed along with other ε-Proteobacteria, however, the nearest associate from a distant taxon is *Ralstonia metallidurans* (β-Proteobacteria) (BV 1000), which incidentally also shows close homology for its *fdh*B gene (BV 974).

In quite a few strong associations, such as those between *Ferroplasma acidarmanus* and *Bacillus halodurans* (BV 965) and *Desulfitobacterium hafniense* among γ-Proteobacteria and Actinobacteria (BV 701–1000), no other evidences could be generated to support them as cases of HGT. *Archaeoglobus fulgidus* shows a significant deviation for its *fdh*A gene (χ^2: 817.381); however, its association with *Deinococcus radiodurans* is weaker (BV 435) than with *Geobacter metallireducens* (BV 558) (ε-Proteobacteria), although DNA pattern analysis does show significant similarities among these three organisms, which lends support to possibility of gene movement across taxa. These evidences could not be further corroborated by other parameters of HGT. On the branch harbouring *Thermococcus litoralis* and *Pyrococcus* species although *Pyrococcus furiosus* does have a significantly high χ^2 of 143.046 other statistical analyses are not significantly deviating. Similarly, the clade of methanogens does have *Methanobacterium* with χ^2 for *fdh*A, 107.897, but no other significant deviations could be recorded.

There is a minor discrepancy in the deep placing of *Burkholderia fungorum* (α-Proteobacteria) on a clade harbouring *Pseudomonas putida* (γ-Proteobacteria) and *Ralstonia eutropha* (β-Proteobacteria). Similarly, *Magnetospirillum magnetotacticum* (α-Proteobacteria) is placed among β- and γ-proteobacterial members with some deviations. The deeply seated α-proteobacterial members, *Azospirillum*, *Sphingomonas* and *Caulobacter* sp. on the clade harbouring β-, γ-Proteobacteria and Cyanobacteria, are strongly associated among themselves as evident by high BV. Such a grouping could not be supported by other evidences. Although most γ-proteobacterial members have similar GC contents for their *fdh*A genes and their genomes, there are wide differences of up to 11 % within each group. *Klebsiella oxytoca*, *Vibrio vulnificus* and *Yersinia pestis* are among those which have the largest deviations; however, these are not supported by other parameters.

In brief, *S. coelicolor*, *X. campestris*, *C. violaceum*, *B. cereus*, *B. subtilis* and *C. perfringens* str.13, *B. japonicum*, *R. palustris* and *S. flexneri* seem to have acquired their *fdh*A gene through HGT and some of them have got well adapted to their host genome.

20.2.2 Phylogenetic Analysis of *fdh*B Gene

Formate dehydrogenase β-subunit, an important component of the enzyme, is distributed among a wide range of taxonomical classes. A comparison of the phylogenetic trees based on Fdh-β protein sequences with 16S rDNA gene reveals inconsistencies (based on BV ≥500), which are widespread among different taxonomical groups (Fig. 20.2).

The distribution of *fdh*B gene extends from Archaea to Metazoa; however, it is concentrated largely among Archaea and Proteobacteria. Fourteen organisms (12 genera, 14 species) of archaeal group have paraphyletic distribution on the phylogenetic tree. Methanogens and *Pyrococcus* are largely on a single branch. *Pyrococcus horikoshii* showed homology with its archaeal members such as *Pyrococcus abyssi*, but it seems more closely linked to *Novosphingobium aromaticivorans* (α-Proteobacteria), since there is large deviation in the GC contents (6.17 % higher). The association of *N. aromaticivorans* among archaeal members, however, could find support from other evidences. On yet another branch, quite prominent grouping of archaeal members along with β-, γ-Proteobacteria, Firmicutes and *Bacteroides* gives rise to the possibility of an HGT event. Incidentally, only *A. fulgidus* DSM 4304 among the four archaeal members has a highly significant χ^2 value for its Fdh-β gene. *Selenomonas ruminantium* (Firmicute), *Chlorobium tepidum* TLS

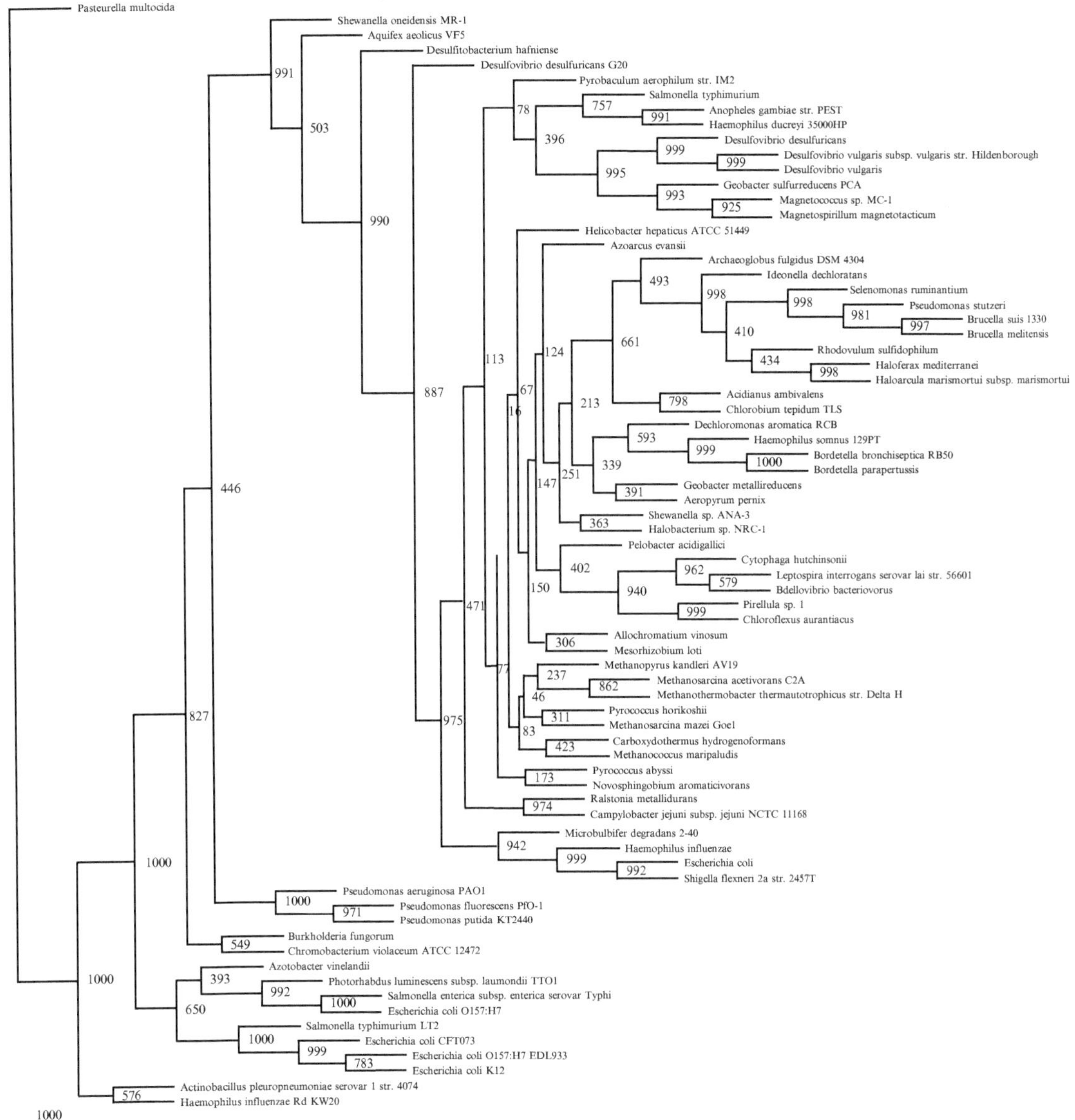

Fig. 20.2 Phylogenetic tree based on the amino acid sequences of β-subunit of FDH

(*Bacteroides*) and *Brucella* species show significant homologies (BV 981–998). *Brucella suis*1330 on this branch also has a significant χ^2 value. These are strong cases of HGT, where the *fdh*B gene seems to have got well adapted to its hosts as evident by CAI value, which is quite similar to those of the highly expressed genes. The GC content of *fdh*B gene in *A. fulgidus* DSM 4304 deviates (5.65 % higher) from that of its host. However, *fdh*B gene of *B. suis*1330 does not show such deviation. Such a difference can perhaps be assigned to the length of the *fdh*B gene, 549 bp in *A. fulgidus* DSM 4304 and 1539 bp in the case of *B. suis* 1330.

Among Proteobacteria, the representatives are largely paraphyletic. The seven genera of α-Proteobacteria show close homologies to δ-Proteobacteria and other Proteobacteria. *Brucella* and *Rhodovulum sulfidophilum* among other α-Proteobacteria have shown homology to β- and γ-Proteobacteria, Firmicutes and even Archaea. *fdh*B gene of *B. suis* and *A. fulgidus* DSM 4304, on this branch, seems to have been involved in HGT event, as discussed above. Incidentally, the pairing of *B. fungorum* (α-Proteobacteria) with *C. violaceum* ATCC 12472 (β-Proteobacteria) (BV 549) is also supported by significant χ^2 values. *B. fungorum* might have transmitted its gene to *C. violaceum* ATCC 12472 by an HGT event since they have quite similar GC content for their *fdh*B gene in spite of large differences in the GC content of their genomes, 59.05 and 65.71 %, respectively.

Among β-Proteobacteria, eight organisms (seven genera) in which the *fdh*B gene has been recorded are closely placed on the phylogenetic tree. The paraphyletic nature of *fdh*B gene of β-Proteobacteria is quite evident. *fdh*B gene of *Azoarcus evansii* shows significant χ^2 value (135.063), 2.15 % lower GC content compared to the whole genome (67.14 %). Incidentally, it is placed on an independent branch, quite close to *A. fulgidus* DSM 4304, another candidate supporting HGT event. *R. metallidurans* (β-Proteobacteria) showed high homology to *C. jejuni* subsp. *jejuni* NCTC 11168 (ε-Proteobacteria) (BV 974). However, only the latter is supported to be involved in HGT as evident by high χ^2 value (822.723) and differences in CAI and GC contents.

Among δ-Proteobacteria, *fdh*B gene is represented by ten organisms belonging to six genera. Interestingly, like *fdh*B of Archaea, α- and β-Proteobacteria, this group is also paraphyletic in nature. *Pelobacter acidigallici* and *Bdellovibrio bacteriovorus* are grouped on the phylogenetic tree along with members of *Bacteroides*, Spirochetes, Planctomycetes and *Chloroflexus*. This branch has some phylogenetically close relatives like *Cytophaga hutchinsonii* (*Bacteroides*), *Chloroflexus aurantiacus* (Chloroflexi) and *Pirellula* sp. 1 (Planctomycetes) clubbed with *B. bacteriovorus* (δ-Proteobacteria) and *Leptospira interrogans* (Spirochetes). The GC contents of their respective genomes vary widely, which range from 36.42 % to 57.69 %. The GC content of the *fdh*B gene of these organisms also varies a lot among themselves (40.07 % in *C. hutchinsonii* to 56.03 % in *Pirellula* sp. 1), although GC contents of *fdh*B gene of *L. interrogans serovar* Lai str. 56601 and *Pirellula* sp. 1 varied a lot compared to their respective genomes, 6.28 and 11.87 %. However, only *fdh*B gene of *Pirellula* sp. 1 had a significant χ^2 value of 544.70. The high homology between *Pirellula* sp. 1 and *C. aurantiacus* (BV 999) is also evident by the similarity in the GC contents of the *fdh*B gene in spite of the variations in their genomic GC contents, 44.16 and 57.69 %, respectively. The strong phylogenetic associations among these five organisms (BV 579–999) are also evident by significant similarities in their DNA pattern for *fdh*B gene. Among these, Chloroflexus was recently reported to be apparently involved in HGT for its *pha*B gene (Kalia et al. 2007).

Proteobacteria-ε are distributed in two branches. *C. jejuni* subsp. *jejuni* NCTC 11168 – *R. metallidurans* association (BV 974) is supported by high χ^2 value and other parameters for the former associate. *Helicobacter hepaticus* ATCC 51449 is another ε-Proteobacteria with high χ^2 value.

The paraphyletic nature of γ-Proteobacteria for their *fdh*B gene is also quite evident. *fdh*B gene of *Actinobacillus pleuropneumoniae serovar* 1 str. 4074 is the only case, which can be grouped under HGT events, as it is supported by significantly high χ^2 value, 11.07 % (higher) deviation in its GC content from that of its host genome (35.15 %). It forms a distant group with *Haemophilus influenzae* Rd KW20. *Haemophilus* genera also have very interesting information to reveal. *Haemophilus* species have quite similar GC contents for their *fdh*B gene (39–41 %) and their genomes (37–38 %). However, all the three species are widely separated from one another, such as *Haemophilus ducreyi* 35000HP with *Anopheles gambiae* str. PEST (Metazoa) (BV 991) or with *Brucella* (β-Proteobacteria) (BV 999) or with γ-Proteobacteria (BV 999). There is a great likelihood of *Haemophilus fdh*B gene to have originated from different donors. Alternatively, *fdh*B gene in these organisms might have diverged after duplication (Calteau et al. 2005). These evidences lead to the belief that *H. influenzae* biotype aegyptius might have evolved through HGT events (Achtman and Hakenbeck 1992; Hacker et al. 1997).

E. coli and *Salmonella* species of γ-Proteobacteria are known to share 90 % of their genes. *Salmonella* has 10 % unique genes. *fdh*B gene of these two organisms seems to share quite a lot of their characteristics such as CAI and GC contents (54–55 % GC for the gene and 51–53 % for the genome). However, in spite of low overall differences, they are quite widely separated on the phylogenetic tree. *Salmonella enterica* subsp. *enterica serovar typhi* and *E. coli* are independently placed, whereas *Salmonella typhimurium* LT2 and other *E. coli* strains are all grouped together.

The *fdh*B gene of *Shewanella* species are also widely separated phylogenetically. The association of Shewanella *oneidensis* MR-1 and *Aquifex aeolicus* seems to be rather strong (BV 991). There is high homology with *D. hafniense* (Firmicutes) and *Desulfovibrio desulfuricans* G20 (δ-Proteobacteria). *Shewanella* sp. ANA-3 clubs poorly with *Halobacterium* sp. NRC-1 (Archaea) (BV 363) but is branched among β- and γ-Proteobacteria. The association between *Shewanella* sp. ANA3 and *Halobacterium* sp. NRC-1 is quite weak, although the GC content of their *fdh*B gene is quite different. The deviation between the two species of *Shewanella*, *Shewanella* sp. ANA-3 and *S. oneidensis* MR-1, is justified by the GC content of 48.02 and 66.88 %, respectively, for their *fdh*B gene, whereas their genomes differ little with respect to their GC contents (46.81 % and 48.02 %).

Pyrobaculum aerophilum str. IM2 is branched along with Proteobacteria (α-, γ-, δ-, etc.) and Metazoa (*Anopheles*) (BV 78). Similarly, high homology between *Anopheles* (Metazoa) and *H. ducreyi* (γ-Proteobacteria) is also evident by BV991. However, none of these three could be supported by χ^2 analysis. *fdh*B gene of *H. ducreyi* 35000HP indicates the possibility of it being involved in an HGT event because of its higher GC content compared to the rest of the genome. The presence of *Carboxydothermus hydrogenoformans* (ε-Proteobacteria) among archaeal members (BV423) is poorly supported by low χ^2 value. However, the GC content of its *fdh*B gene has deviated quite a lot, i.e., 7.05 % lower compared to its genome. *H. hepaticus* ATCC 51449 (ε-Proteobacteria) has a significant χ^2 value of 120.476 for its *fdh*B; however, a close associate with strong BV could not be traced in these analyses. Another archaeal member, *Aeropyrum pernix* is deeply placed among β-, γ- and δ-Proteobacterial members on the phylogenetic tree for *fdh*B gene. The relatively close associations of (i) *A. pernix* and *Geobacter* (δ-Proteobacteria) (BV 391) and (ii) *Halobacterium sp.* NRC-1 (Archaea) and *Shewanella* sp. ANA-3 (γ-Proteobacteria) seem to be apparent cases of HGT, which do not withstand closer scrutiny. The association of *fdh*B gene sequences of *C. hydrogenoformans* with *Methanococcus maripaludis* (Archaea), *M. magnetotacticum* (α-Proteobacteria) and its closest associate *Magnetococcus* sp. MC-1 (other Proteobacteria) seems to be apparent cases of HGT.

In conclusion, *fdh*B gene seems to have also been transmitted by horizontal inheritance, as these cases have withstood the closer scrutiny of statistical methods. Incidentally, the HGT could be detected among Archaea, Planctomycetes and α-, β-, γ- and ε-Proteobacteria. These give rise to the idea that this system must have spread rapidly to provide energy-generating system to the organisms.

20.2.3 Phylogenetic Analysis of *fdh*C Gene

Unlike the presence of *fdh*A and *fdh*B genes among Archaea, *fdh*C gene could not be detected among the sequenced archaeal genomes (Fig. 20.3). The only representative of Actinobacteria on the *fdh*C phylogenetic tree, *S. avermitilis* MA-4680 is closely associated with *D. hafniense* (Firmicute) (BV 1000). Similarly, *A. aeolicus* VF5 representing Aquificales is weakly associated with β- and γ-Proteobacteria. The few α-Proteobacteria members are widely distributed as far as their *fdh*C gene is concerned. In all the cases, they are grouped along with Proteobacteria. *B. japonicum* USDA 110 has a highly significant χ^2 value for *fdh*C gene. This gene seems to have

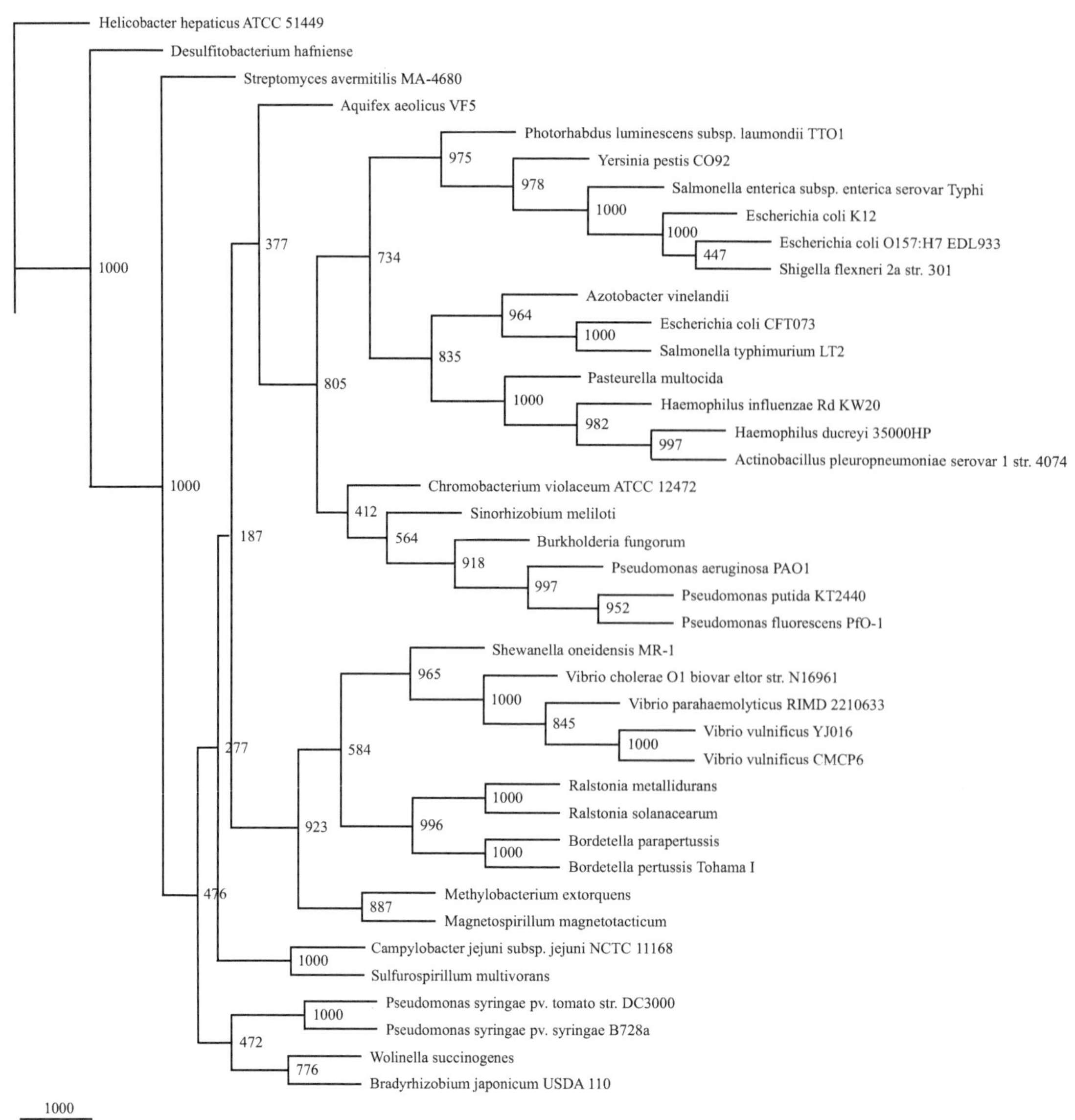

Fig. 20.3 Phylogenetic tree based on the amino acid sequences of γ-subunit of FDH

got well adapted to its host since there are small deviations in CAI and GC contents of the gene and the genome. The strong association of *B. japonicum* (α-Proteobacteria) with *Wolinella succinogenes* (ε-Proteobacteria) is reflected by high BV of 776.

Among β-Proteobacteria, the different organisms possessing *fdh*C gene are bifurcated, *C. violaceum* ATCC 12472 and *B. fungorum* branching along with α- and γ-Proteobacterial members. In *C. violaceum* ATCC 12472, *fdh*C gene has an extremely significant χ^2 value. The CAI values and GC content indicate that this gene got well adapted to its host after acquisition, since the deviations are rather small. Its association with *Sinorhizobium meliloti* (α-Proteobacteria) is quite weak (BV 412). On the other branch harbouring β-Proteobacteria, none of them show any abnormal gene characteristics and seem to follow vertical inheritance.

Among the ε-Proteobacteria, only *C. jejuni* subsp. *jejuni* NCTC 11168 showed significant deviation in its *fdh*C gene characteristics. It has a strong association with *Sulfurospirillum multivorans* (BV 1000), and there are differences in their GC contents. *fdh*C gene seems to have evolved sufficiently to get adapted to its host within the ε-Proteobacterial group.

The distribution of *fdh*C gene among the various organisms of γ-Proteobacteria on the phylogenetic tree is paraphyletic. There is clear-cut segregation among enteric and non-enteric γ-Proteobacteria. Although most of the organisms represented in this tree are closely related to other groups of Proteobacteria such as α-, β- and ε-Proteobacteria, none of them is closely related with δ-Proteobacteria. This reflects the vertical inheritance of *fdh*C gene among γ-Proteobacteria. However, two cases showing wide deviation have also been recorded. The distribution of *S. flexneri* 2a str. 301 and *Vibrio parahaemolyticus* RIMD 2210633 on two widely separated branches can be assigned to the characteristics of the *fdh*C gene and their respective genomes. *S. flexneri* 2a str. 301 is closely placed along with *E. coli* and *Salmonella* with a GC content of 51–53 % for their *fdh*C gene and genome, whereas *V. parahaemolyticus* RIMD 2210633 has a GC content in the range of 46–47 %. In the case of *S. flexneri* 2a str. 301, the differences in CAI values for the gene and the highly expressed gene indicate that the *fdh*C gene is not well adapted to its host. On the other hand, *fdh*C gene in *V. parahaemolyticus* RIMD 2210633 seems to have got well adapted to its host. Among the *Vibrio* species, the major difference lies only in the CAI, 0.76 for its gene and 0.81 for the genome of *V. vulnificus*, whereas *V. parahaemolyticus* RIMD 2210633 has CAI value in the range of 0.458–0.468.

In conclusion, just like *fdh*B gene, there are certain cases of HGT which involve *fdh*C gene. Incidentally, unlike *fdh*B gene, where the HGT cases were distributed over a wide range of taxonomical classes, the events in the case of *fdh*C gene are limited only to proteobacterial members among the organisms which have been sequenced for their genomes. A perusal of all the cases of major inconsistencies in the evolutionary trees of genes – *fdh*A, *fdh*B and *fdh*C – reveal two organisms in which all these are involved in the HGT, i.e., *C. violaceum* ATCC 12472 (β-Proteobacteria) and *C. jejuni* subsp. *jejuni* NCTC 11168 (ε- Proteobacteria).

20.2.4 Phylogenetic Analysis of *hyd*A Gene

Hydrogenase small subunit (H_2ase-SSU encoded by hydA gene) is an important component of the energy-generating system. It is distributed among a wide range of taxonomical classes. An overview of the phylogenetic tree indicates certain discrepancies, which may be explained by HGT events (Fig. 20.4).

Archaeal members, *A. fulgidus* DSM4304 and *Methanosarcina acetivorans* C2A are strongly associated among themselves (BV 973–1000). *A. fulgidus* DSM4304 among Archaea shows a significant deviation for its H_2ase-SSU gene contents (χ^2 value: 422.9). The gene seems to have adapted well to its host genome. Its GC content of 52.73 % is +3.37 % higher than that of its genome and deviates also from that of *Methanosarcina* species (47–48 %). These features lend strong support to the acquisition of *hyd*A gene by *A. fulgidus* DSM4304 through HGT mechanism. This archaeal group is strongly associated with Firmicutes (BV 925). The segregation of Firmicutes into two branches can be assigned to wide differences between their GC contents: *hyd*A gene and genomes of *Clostridium acetobutylicum*, 36.30 and 33.16 %; and *D. hafniense*, 53.33 and 49.21 %, respectively. The distribution of *hyd*A among Actinobacteria and Aquificales indicates it to have followed a vertical inheritance. The similarities in the CAI values of the *hyd*A and the highly expressed genes of these organisms indicate that it has adapted well during evolution.

Out of a group of five ε-Proteobacteria, three have shown significant χ^2 values for their *hyd*A gene. The deviation is also supported by large differences of 6–7 % in the GC contents of their *hyd*A genes and their respective genomes. Although *W. succinogenes* and *H. hepaticus*

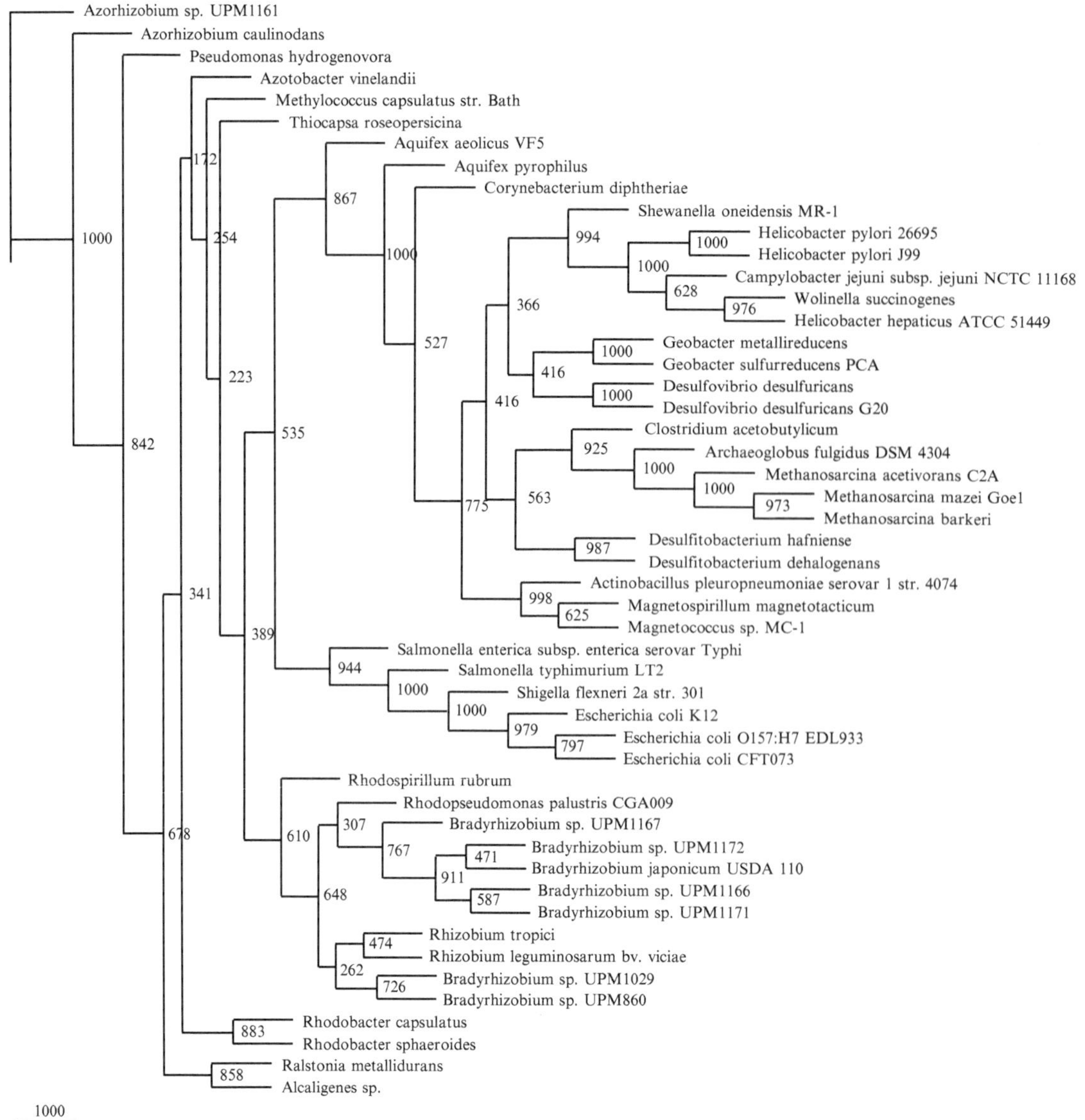

Fig. 20.4 Phylogenetic tree based on the amino acid sequences of large subunit of H_2ase

ATCC 51449 are strongly associated (BV 976), however, this close association does not find further support from their supplementary information on CAI values or GC contents. *H. hepaticus* ATCC 51449 may have received this gene through HGT (χ^2 value: 102.053), but there are no compelling evidences to designate *Wolinella* as the true donor. The *hyd*A gene of *Helicobacter pylori* 26695 has a significant value of 209.825; however, it is associated largely with other ε-Proteobacteria such as *C. jejuni* subsp. *jejuni* NCTC 11168 and *H. hepaticus* 51449, which have also acquired this gene through HGT. The deeply seated branches of γ-, δ- and ε-Proteobacteria among Firmicutes, Aquificales and Actinobacteria appear to be an interesting event. The occurrence of HGT events among ε-Proteobacteria justifies the clubbing. However, studying these organisms for other genes as well might reveal more information.

The two members of α-Proteobacteria, *R. palustris* CGA009 and *B. japonicum* USDA110 show deviation in their *hyd*A gene as evident from the highly significant χ^2 values. The adaptation *hyd*A gene to their respective hosts during evolution is quite clear from their CAI values, which differ marginally from those of the highly expressed genes. Here, we may designate *R. palustris* CGA009 and *B. japonicum* USDA110 as the recipients of *hyd*A by HGT event. However, with the available genome database, we found it difficult to trace the donor(s).

The association of *M. magnetotacticum* among γ-Proteobacteria and *Magnetococcus* sp. MC-1 (other Proteobacteria) is supported by a strong BV (625–998). *A. pleuropneumoniae serovar*1 str. 4074 on this branch has significant χ^2 value along with a large difference of 10.17 % in the GC content of its gene (45.32 %) compared to that of its genome (35.15 %). It may thus be designated as a recipient of *hyd*A through HGT, but *M. magnetotacticum* and *Magnetococcus* sp. MC-1 cannot be designated among donors, since they are more deeply seated on this phylogenetic branch.

Among γ-Proteobacteria, only two members have shown wide deviations in their *hyd*A gene characteristics. The significant χ^2 values of *S. flexneri* 2a str. 301 and *A. pleuropneumoniae serovar* 1 str. 4074 are also strongly supported by the 10–11 % deviation in the GC contents of the gene and genomes. A very interesting observation is that concerning *S. flexneri* 2a str. 301, where the GC contents of its genome (44.63 %) vary significantly from all other members of γ-Proteobacteria (51–53 %). All these features support the acquisition of *hyd*A by *S. flexneri* 2a str. 301 through HGT event.

The significant deviations in the *hyd*A gene of archaeal and α-, γ- and ε-Proteobacteria emphasize the role of HGT in the evolution of energy metabolism. *A. fulgidus* DSM4304, *A. pleuropneumoniae serovar* 1 str. 4074, *C. jejuni* subsp. *jejuni* NCTC 11168, *H. pylori* J99, *H. hepaticus* ATCC 514449, *R. palustris* CGA009 and *S. flexneri* 2a str. 301 have also shown the acquisition of other genes of H_2 generation through HGT. In this study, none of the members of Aquificales and green sulphur bacteria have shown the involvement of their genes for H_2 evolution through HGT. Although, it is relatively easy to designate the recipients, search for donors will have to wait for more sequenced genomes.

20.2.5 Phylogenetic Analysis of *hyd*B Gene

The phylogenetic distribution of large subunit of H_2ase (encoded by *hyd*B gene) among the sequenced genomes of a wide range of organisms, Archaea to Proteobacteria, is presented in Fig. 20.5. Incidentally, the classes are the same as those where distribution of *hyd*A gene was recorded. An overview of the phylogenetic tree indicates vertical inheritance of the *hyd*B gene. However, certain discrepancies can be recorded and may be explained by HGT events, particularly those that withstand closer statistical scrutiny. Among the organisms whose genomes have been sequenced, two Archaea have the *hyd*B gene for H_2ase-LSU enzyme. *A. fulgidus* and *M. acetivorans* have a strong association between them (BV 665). *A. fulgidus* has a significant χ^2 value, although there are not much deviation in CAI and GC content of its *hyd*B gene. The association of *A. fulgidus* and *Desulfovibrio desulfuricans* G20 (δ-Proteobacteria) is strongly supported by high BV (1000). The two members of Actinobacteria *S. avermitilis* MA 4680 and *Corynebacterium diphtheriae* are widely separated and are grouped among Archaea and Firmicutes on one hand and with Aquificales on the other. The separation is supported by the differences in the CAI values of their highly expressed genes and the GC content of their genome (*S. avermitilis* 71.14 % and *C. diphtheriae* 54.09 %) and even *hyd*B gene, 53.33–65.55 %. Even the association of *C. diphtheriae* (Actinobacteria) and *A. aeolicus* (Aquificales) has little homology (BV 427) and no deviation in their gene characteristics.

Firmicutes possessing *hyd*B gene are distributed largely along with Archaea and Actinobacteria. However, *D. hafniense* is deeply rooted along with *C. tepidum* (*Bacteroides*) among a large group of diverse Proteobacteria. There does not appear any HGT event to have taken place among these organisms for *hyd*B.

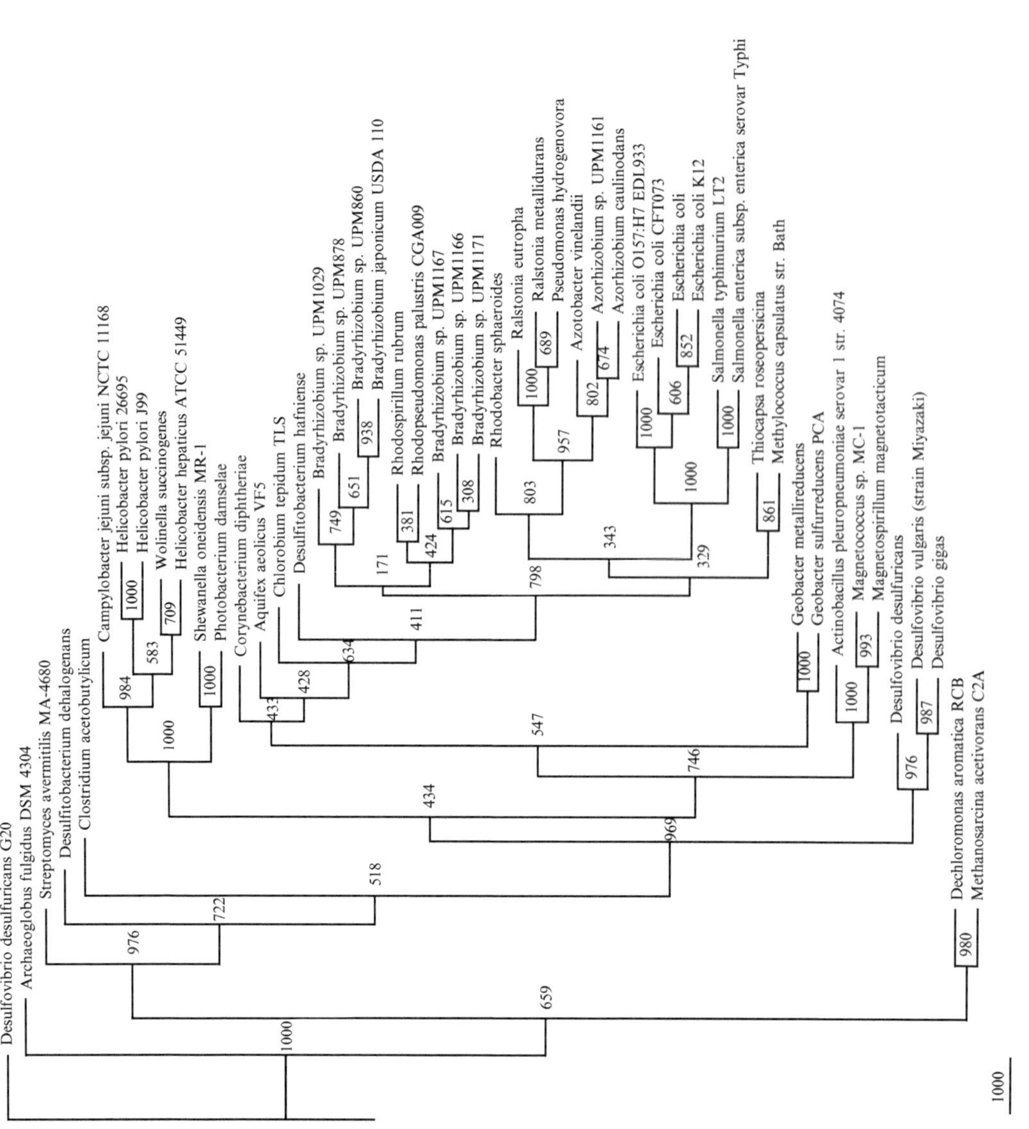

Fig. 20.5 Phylogenetic tree based on the amino acid sequences of small subunit of H_2ase

A large number of α-Proteobacteria have been observed to posses *hyd*B gene. Most of these are branched along with members of β- and γ-Proteobacteria. However, they are still segregated on two different branches on the phylogenetic tree. There are no major atypical sequence characteristics as diagnosed by the GC contents of the *hyd*B gene and the respective genomes of most of these α-Proteobacterial members. However, there is little variation in the CAI values of *B. japonicum* and *R. palustris* (CAI: 0.55–0.734).

The three representatives of β-Proteobacteria are well separated from each other. *Ralstonia* species are clubbed among α- and γ-Proteobacteria. The strong association of *M. acetivorans* (Archaea) and *Dechloromonas aromatica* RCB (β-Proteobacteria) (BV 980) seems to be limited only to their hydrogenase gene. Although it is difficult to justify their linkage to archaeal members, however, their segregation into two branches can be assigned to the wide variations in the GC composition of their *hyd*B genes and that of the whole genomes (*R. eutropha*, 58.27 and 66.17 %; *D. aromatica*, 62.98 and 54.71 %). The separation of *D. desulfuricans* from other species of *Desulfovibrio* is perhaps justified by the observation that the deviation in the GC content of *hyd*B gene ranges from 54.83 % to 65.08 %, whereas that of their genomes ranges from 58.08 % to 63.78 %. *Desulfovibrio gigas* has a significant χ^2 value for its *hyd*B gene and is associated closely with *Desulfovibrio vulgaris*. Phylogenetic tree showing evolutionary relationship of iron hydrogenase showed that the two species of *Desulfovibrio*, *D. fructosovorans* and *D. vulgaris*, did not cluster together. It was viewed as a strong evidence of gene duplication and/or potential gene transfer (Embley et al. 2003).

Among ε-Proteobacteria, there are two major branches. The branch carrying *Campylobacter*, *Helicobacter* and *Wolinella* species has the largest number of organisms, which have highly significant χ^2 value for their *hyd*B genes. Among these organisms, *Wolinella* does not show any significant χ^2 value. It does have a high BV 709 with *H. hepaticus*. There are wide differences in their GC contents. *hyd*B gene of *Helicobacter* and *Campylobacter* seems to be well adapted to its host, as the CAI values are quite similar to those of their highly expressed genes. One of the branches harbouring *Geobacter* species does not show any close association with any other bacteria. However, its wide separation from other ε-Proteobacteria is supported by the observation that its GC content of 61.62 % compared to a GC content of 30.83–48.90 % of those organisms on the other branch.

The distribution of *hyd*B gene among γ-Proteobacteria is recorded on three major branches, largely among members from other proteobacterial groups. For *hyd*B gene, none of the γ-Proteobacteria showed any significant χ^2 value. The segregation of γ-Proteobacteria on various branches on the phylogenetic tree is a reflection of the wide variation in the GC content of *hyd*B gene (encoding for the hydrogenase) and the genome.

In conclusion, we may record the observation that the *hyd*B gene is well adapted to its host genome, as evident by the CAI values and GC contents. *hyd*B gene is widely distributed among Archaea, Actinobacteria, *Bacteroides*, Firmicutes and α-, β-, ε- and γ-Proteobacteria. There are a few members of ε-Proteobacteria and one each of Archaea, α- and δ-Proteobacteria, which had significant χ^2 values. Among these, only *A. fulgidus* DSM 43904 was associated with *C. acetobutylicum* (Firmicutes). Once more, *C. jejuni* has shown that just like *fdh*A, *fdh*B, *fdh*C and *hydA* gene, *hyd*B gene also seems to have been transferred by an HGT event. *A. fulgidus* and *H. hepaticus* have significant deviation in their *fdh*B and *hyd*B genes, indicating HGT events.

20.2.6 Tailoring Genomes for Novel H_2 Producers

Out of 183 genomes analysed for the subunits of FDH and H_2ases, 26 organisms have been observed to show significant deviations in their gene characteristics, which implies HGT events (Table 20.2). The movements of the various subunits of FDH and H_2ases in natural microbial iso-

Table 20.2 Genetic characterization of a typical associations of organisms indicating horizontal gene transfer events. nd: The associated organism could not be designated because either it was not detectable among sequenced organisms or both belonged to the same taxa

Organisms	Gene	Chi square statistic	Bootstrap value	Associated organisms
Archaea				
Archaeoglobus fulgidus DSM 4304	*fdh*A	817.381	558	*Geobacter metallireducens* (δ-Proteobacteria)
A. fulgidus DSM 4304	*fdh*B	836.808	661	*Chlorobium tepidum* TLS (*Bacteroidetes*)
A. fulgidus DSM 4304	*hyd*A	708.704	925	*Clostridium acetobutylicum* (Firmicutes)
A. fulgidus DSM 4304	*hyd*B	422.969	1000	*Desulfovibrio desulfuricans* G20 (δ-Proteobacteria)
Methanobacterium formicicum	*fdh*A	107.897	–	nd
Pyrococcus furiosus DSM 3638	*fdh*A	143.046	–	nd
Actinobacteria				
Streptomyces coelicolor A3(2)	*fdh*A	4461.298	717	*Gloeobacter violaceus* (Cyanobacteria)
Cyanobacteria				
Cyanothece sp. PCC 8801	*fdh*A	160.537	–	nd
Nostoc sp. PCC 7120	*fdh*A	157.741	–	nd
Synechococcus sp. PCC 7002	*fdh*A	214.799	–	nd
Synechococcus sp. PCC 7942	*fdh*A	182.408	–	nd
Synechococcus sp. WH 8102	*fdh*A	727.740	–	nd
Firmicutes				
Bacillus cereus ATCC 14579	*fdh*A	204.277	–	nd
B. subtilis subsp. subtilis str. 168	*fdh*A	303.300	–	nd
Clostridium perfringens str 13	*fdh*A	342.633	1000	*Pirellula* sp. (Planctomycetes)
Planctomycetes				
Pirellula sp. 1	*fdh*B	544.705	940	*Cytophaga hutchinsonii* (*Bacteroidetes*)
α-Proteobacteria				
Bradyrhizobium japonicum USDA 110	*fdh*A	2586.194	–	nd
B. japonicum USDA 110	*fdh*C	2190.875	776	*Wolinella succinogenes* (ε-Proteobacteria)
B. japonicum USDA 110	*hyd*A	1602.637	–	nd
Brucella suis 1330	*fdh*B	1468.820	981	*Pseudomonas stutzeri* (γ-Proteobacteria)
Rhodopseudomonas palustris CGA009	*fdh*A	1576.594	–	nd
R. palustris CGA009	*hyd*B	1614.523	–	nd
R. palustris CGA009	*hyd*A	2827.092	–	nd
β-Proteobacteria				
Azoarcus evansii	*fdh*B	135.062	124	*Halobacterium* sp. NRC-1 (Archaea)

(continued)

Table 20.2 (continued)

Organisms	Gene	Chi square statistic	Bootstrap value	Associated organisms
Chromobacterium violaceum ATCC 12472	*fdh*A	2931.541	1000	*Pseudomonas aeruginosa* PAO1 (γ-Proteobacteria)
C. violaceum ATCC 12472	*fdh*B	3130.499	549	*Burkholderia fungorum* (α-Proteobacteria)
C. violaceum ATCC 12472	*fdh*C	3235.607	412	*Sinorhizobium meliloti* (α-Proteobacteria)
δ-Proteobacteria				
Desulfovibrio gigas	*hyd*B	83.801	–	nd
ε-Proteobacteria				
Campylobacter jejuni subsp. *jejuni* NCTC 11168	*fdh*A	498.807	1000	*Ralstonia metallidurans* (β-Proteobacteria)
C. jejuni subsp. *jejuni* NCTC 11168	*fdh*B	822.723	974	*R. metallidurans* (β-Proteobacteria)
C. jejuni subsp. *jejuni* NCTC 11168	*fdh*C	361.407	–	nd
C. jejuni subsp. *jejuni* NCTC 11168	*hyd*B	519.091	–	nd
C. jejuni subsp. *jejuni* NCTC 11168	*hyd*A	84.912	–	nd
Helicobacter hepaticus ATCC 51449	*fdh*B	120.476	–	nd
H. hepaticus ATCC 51449	*hyd*B	153.126	–	nd
H. hepaticus ATCC 51449	*hyd*A	102.053	–	nd
Helicobacter pylori 26695	*hyd*B	273.493	–	nd
H. pylori 26695	*hyd*A	209.825	–	nd
γ-Proteobacteria				
Actinobacillus pleuropneumoniae serovar 1 str. 4074	*fdh*B	1187.532	–	nd
A. pleuropneumoniae serovar 1 str. 4074	*hyd*A	510.915	998	*Magnetospirillum magnetotacticum* (α-Proteobacteria)
Shigella flexneri 2a str. 301	*fdh*A	515.476	988	*Magnetococcus* sp. MC-1 (unclassified Proteobacteria)
S. flexneri 2a str. 301	*fdh*C	362.934	–	nd
S. flexneri 2a str. 301	*hyd*A	271.378	–	nd
Vibrio parahaemolyticus RIMD 2210633	*fdh*C	379.009	–	nd
Xanthomonas campestris pv. *campestris* str. ATCC 33913	*fdh*A	1909.430	–	Firmicutes group

lates reiterate that energy generation system might have evolved (perhaps in response to envi ronmental changes) much faster than can be expected by vertical inheritance (Jain et al. 2003). The most prominent participants in these genetic transmissions are *A. fulgidus* DSM4304 (Archaea); *B. japonicum* USDA 110 and *R. palustris* CGA009 (α-Proteobacteria); *C. violaceum* ATCC 12472 (β-Proteobacteria); *C. jejuni*, *H. hepaticus* and *H. pylori* (ε-Proteobacteria); and *A. pleuropneumoniae serovar*1 str. 4074 and *S. flexneri* 2a str. 301 (γ-Proteobacteria). These events offer an opportunity to identify those genomes which can be tailored for transforming presently non-H_2 producers into producers.

Out of 26 organisms, *A. pleuropneumoniae*, *B. japonicum*, *C. jejuni* and *H. hepaticus* have all the five subunits of FDH and H_2ase. However, only *C. jejuni* has significant deviations in all its genes compared to the rest of the genome. *A. fulgidus*, *B. suis* 1330, *C. violaceum* ATCC 12472, *A. evansii*, *H. pylori* 26695, *S. flexneri* and *X. campestris* possess one to four subunits, and incidentally, each show significant deviations (χ^2) in their gene contents (Table 20.1). Since these organisms have a potential ability to acquire genes even from distantly related organisms, the missing gene(s) can be brought in either from such organisms, which have acquired these genes through HGT events, or from closely related ones. A closer scrutiny of the phylogenetic trees and homology between the nearest neighbours provides clues to the origin of the genotype of certain organisms such as *A. fulgidus* DSM4304, *B. japonicum* USDA 110, *R. palustris* CGA009, *C. jejuni*, *H. hepaticus*, *H. pylori* and *S. flexneri* 2a str. 301 (Figs. 20.6, 20.7, 20.8 and 20.9). It is also possible to detect those organisms, which may be exploited to supplement the missing gene.

A. fulgidus DSM4304 has two subunits each of FDH and H_2ase, which deviate significantly from the host genome characteristics. Critical analyses of phylogenetic trees provide an insight into the possible donors for different subunits (Fig. 20.6). *A. fulgidus* shows a high BV of 1000 with *fdh*A gene of *P. aerophilum* str. IM2 and *hydA* gene of *M. acetivorans* C2A on their respective phylogenetic trees. The association of *A. fulgidus* and *P. aerophilum* is also supported by DNA pattern analysis (class 3 of α-subunit of FDH). Among the organism distantly related to *A. fulgidus* for its two subunits are *C. acetobutylicum* (Firmicute) having a BV 925 (for *hyd*A) and *D. desulfuricans* G20 (γ-Proteobacterium) with a BV 587 for *fdh*A. A weak association of *A. fulgidus* DSM4304 was also seen for *fdh*B gene of *Ideonella dechlorotans*. However, the donor organism for *fdh*C of *A. fulgidus* could not be traced.

The possible origin of *fdh*A, *fdh*B, *fdh*C, *hyd*A and *hyd*B genes in *B. japonicum* USDA110 and *R. palustris* CGA009 has been depicted in Fig. 20.7. *B. japonicum* USDA110 genotype showed significant deviation in the characteristics for its *fdh*A, *fdh*C and *hyd*A genes (Fig. 20.7). Its two other genes *fdh*B and *hyd*B did not show any deviations. On the basis of phylogenetic trees for the different genes of FDH and H_2ase, *fdh*A of *B. japonicum* USDA 110 and *R. palustris* CGA009 showed high BV with *Mesorhizobium lot*,, *S. meliloti* and *Agrobacterium tumefaciens* str. C58 (all α-Proteobacteria). *fdh*C of *B. japonicum* USDA110 seems to have been contributed by *W. succinogenes* (ε-Proteobacteria) (BV 776 and significant DNA pattern similarities) where as *fdh*C of *Pseudomonas syringae* pv. *syringae* B728a (γ-Proteobacteria) is weakly associated (BV 472). Incidentally, *B. japonicum* USDA110 and *R. palustris* CGA009 show significant deviations in their *hyd*B genes. A very low BV of 107 for *hyd*B gene between these two organisms strongly indicates their divergence after a common origin. Since *R. palustris* CGA009 phylogenetically belongs to the same taxon as *B. japonicum* USDA110, it can perhaps be supplemented for its *fdh*B subunit from the latter and for *fdh*C subunit from either *B. japonicum* USDA110 or *S. meliloti* and can be transformed into H_2-producer. The possibility of *B. japonicum* USDA110 to donate its genes was recorded also for *pha*C gene of PHA biosynthesis pathway, where HGT was observed between this organism and *R. metallidurans* (β-Proteobacteria) (Kalia et al. 2007).

A unique case of HGT (χ^2, CAI, GC content, rare patterns) is represented by ε-proteobacterial member *C. jejuni*, where all the five genes encoding for three subunits of FDH and two subunits of H_2ase had significantly deviated χ^2 values. It was the only organism which seems to be genotypically highly flexible as a recipient of foreign DNA (Fig. 20.8). It has previously been shown to have the abilities to get transformed. Based on high BV, *fdh*A seems to have originated from either *R. metallidurans* (β-Proteobacteria) or *W. succinogenes* (ε-Proteobacteria), *fdh*B from

Proteobacteria
Archaea
Actinobacteria
Aquificales
GS Bacteria
Firmicutes
Chloroflexi
Alpha
Beta
Delta
Epsilon
Gamma
Pyrobaculum aerophilumstr. IM2
[α, β]
α
P
BV 1000
Clostridium Acetobutylicum
[L, S]
S
BV 925
Archaeoglobus fulgidus DSM 4304
[α, β, L, S] -γ
α
BV 587
Desulfovibrio desulfuricans G20
[α, β, L, S]
β
BV 493
Ideonella Dechloratans
[β]
S
BV 1000
Methanosarcina acetivorans C2A
[β, L, S]

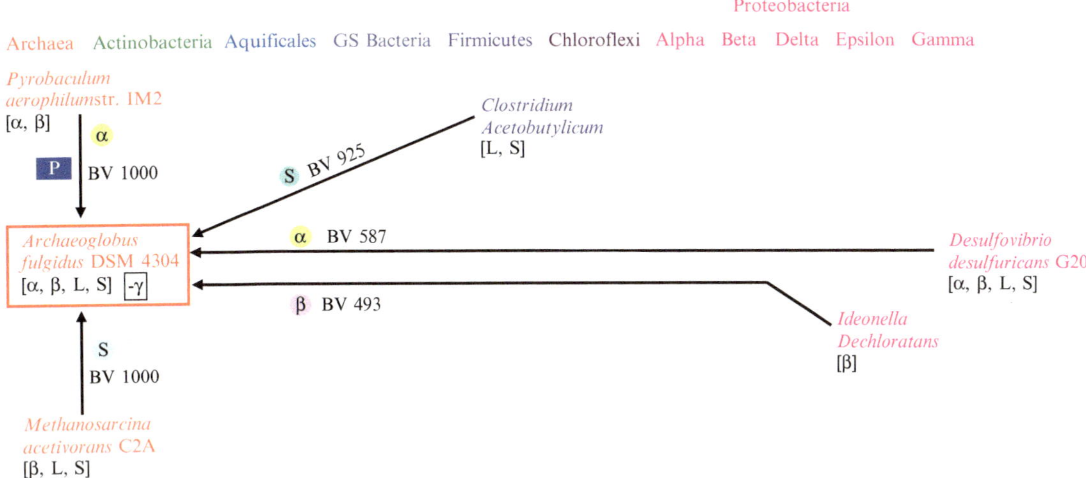

Fig. 20.6 Possible origin of formate dehydrogenase and hydrogenase genes in *Archaeoglobus fulgidus* DSM 4304

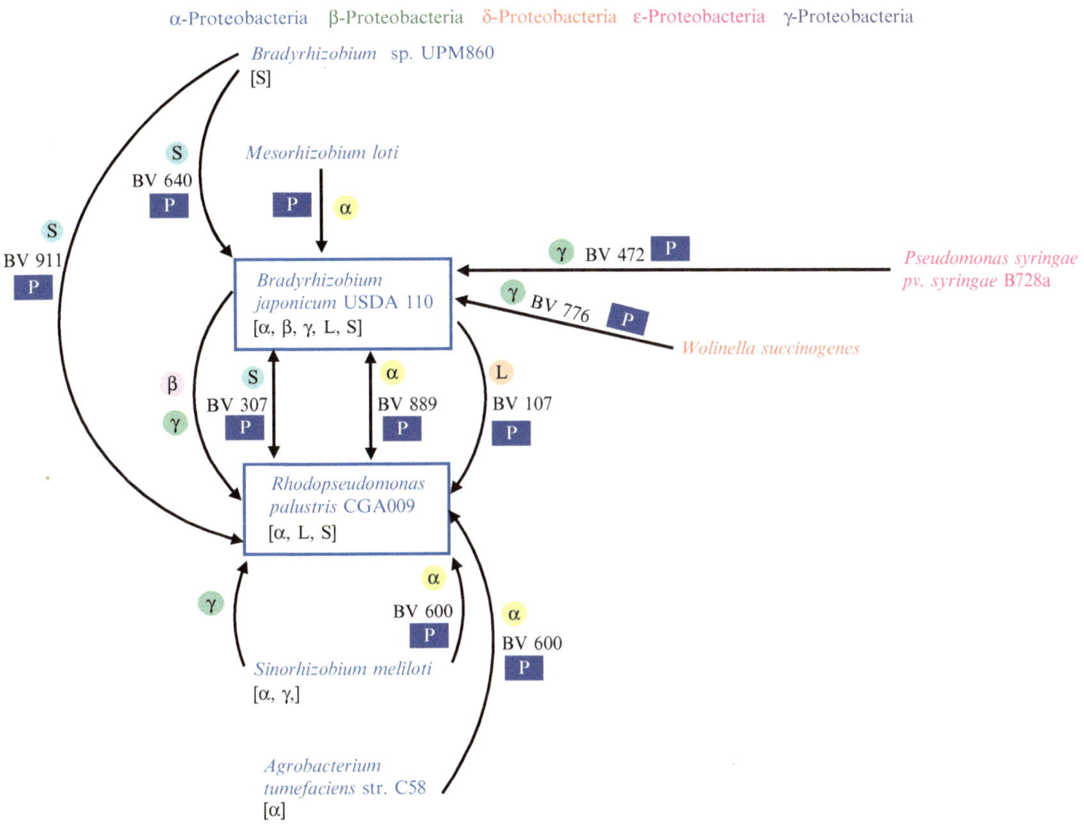

α-Proteobacteria
β-Proteobacteria
δ-Proteobacteria
ε-Proteobacteria
γ-Proteobacteria
Bradyrhizobium sp. UPM860
[S]
Mesorhizobium loti
S
BV 640
P
S
BV 911
P
P
α
γ
BV 472
P
Pseudomonas syringae pv. syringae B728a
Bradyrhizobium japonicum USDA 110
[α, β, γ, L, S]
γ
BV 776
P
Wolinella succinogenes
β
γ
S
BV 307
P
α
BV 889
P
L
BV 107
P
Rhodopseudomonas palustris CGA009
[α, L, S]
γ
α
BV 600
P
α
BV 600
P
Sinorhizobium meliloti
[α, γ,]
Agrobacterium tumefaciens str. C58
[α]

Fig. 20.7 Possible origin of formate dehydrogenase and hydrogenase genes in *Bradyrhizobium japonicum* USDA110 and *Rhodopseudomonas palustris* CGA009

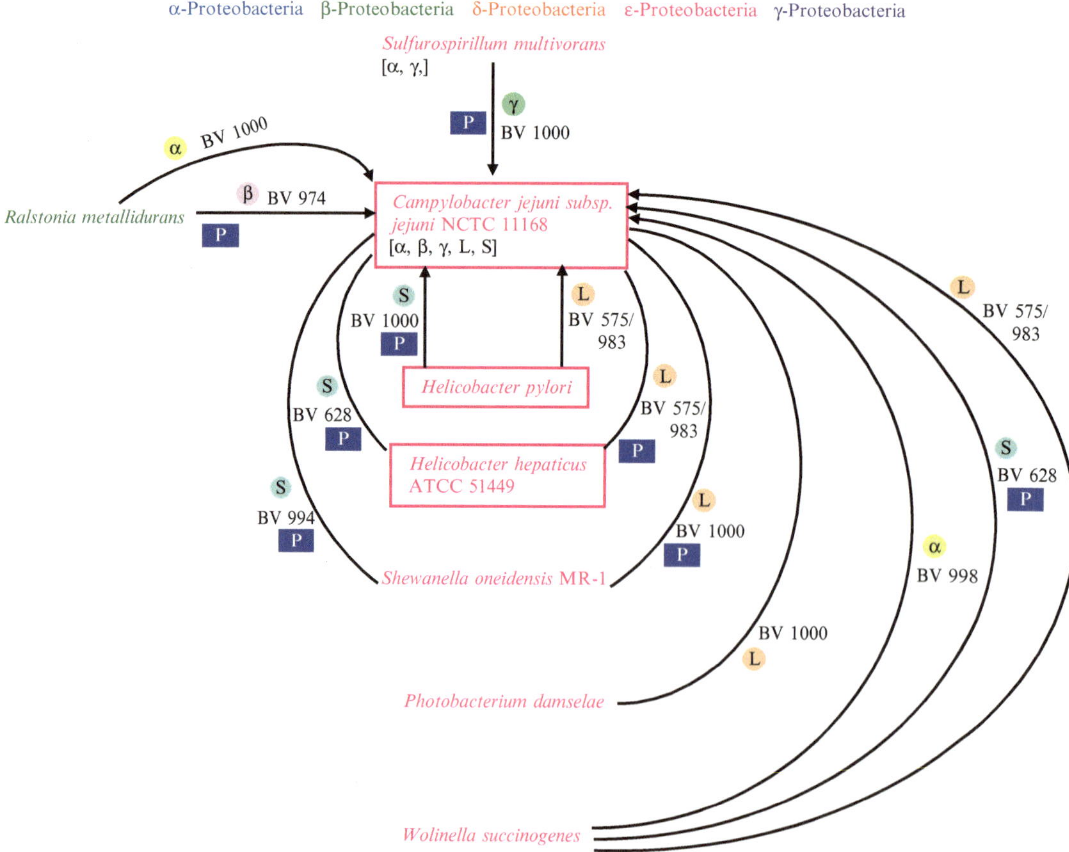

Fig. 20.8 Possible origin of formate dehydrogenase and hydrogenase genes in *Campylobacter jejuni* subsp. *jejuni* NCTC11168

R. metallidurans, *fdh*C from *S. multivorans* and the two subunits of H_2ase from *H. hepaticus* ATCC 51449, *H. pylori*, *S. oneidensis* MR-1, *Photobacterium damselae*, or even *W, succinogenes*. Very interestingly, both the *Helicobacter* species have been involved in HGT, and *H. hepaticus* has all the five genes of FDH and H_2ase.

Another potential candidate for genome shuffling is *S. flexneri*. It is presently not among the known H_2-producers perhaps because of lack of H_2ase LSU. Out of the four other genes (*fdh*A, *fdh*B, *fdh*C and *hyd*A), three have shown significant deviations in their characteristics supporting HGT. By analysing the phylogenetic trees for these four genes, a strong association between *S. flexneri* 2a str. 301 could be observed with other γ-proteobacterial members (Fig. 20.9). The association between *S. flexneri* and *E. coli* for *fdh*A and *fdh*C could be further supported by high BV of 1000. On the other hand, the strong association between *S. flexneri* and *S. typhimurium* LT2 was based on high BV of 1000. For *fdh*B, the only close associate of *S. flexneri* was *E. coli* (BV 992). The *hyd*A of *S. flexneri* showed almost similar homology with *E. coli* (BV 979) and *S. typhimurium* (BV 1000). There were other members of γ-Proteobacteria such as *S. enterica* subsp. *enterica serovar typhi*, *Photorhabdus luminescens* subsp. laumondii TT01 and *Microbulbifer degradans* 2–40, which showed high BV 972–1000 for their *fdh*C. *Magnetococcus* MC-1 was the only organism which did not belong to γ-Proteobacteria but still had very close association with *S. flexneri* for its *fdh*A (BV 988). Based on the strong associations on different

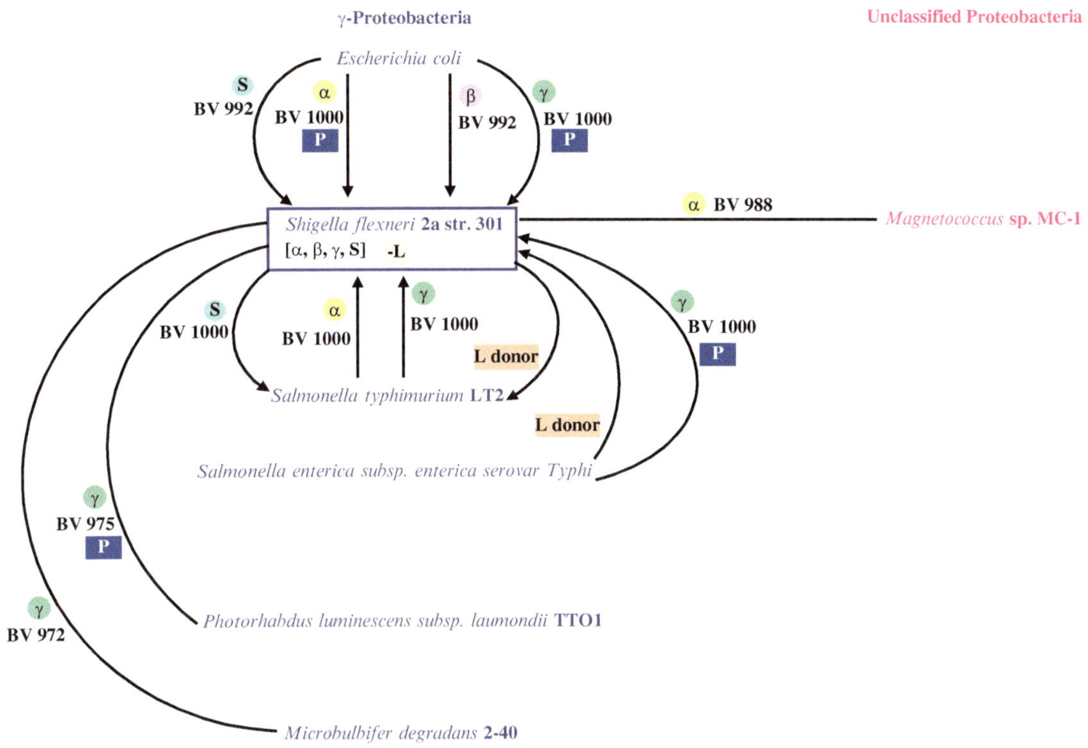

Fig. 20.9 Possible origin of formate dehydrogenase and hydrogenase genes in *Shigella flexneri* 2a str. 301

phylogenetic trees, the potential donors for supplementing *hyd*B of *S. flexneri* genome could be traced to *S. typhimurium* LT2 and *S. enterica* subsp. *enterica serovar typhi*. Taking into consideration all the associates of *S. flexneri*, it seems that since it received the genes *fdh*A, *fdh*C and *hyd*A through HGT from a donor(s), which does not seem to be among those organisms, whose genomes have been sequenced so far, it is difficult to trace the donor. It, however, points toward the possibility of the further transmission of these genes after acquisition among its associates by vertical inheritance.

The evidences for the movement of genes across wide taxonomic classes provide us with an opportunity to transform certain other non-H_2 producers such as *R. rubrum*, *D. desulfuricans* G20, *Magnetococcus* sp. MC-1 and *G. metallireducens* by supplementing their genomes with *fdh*C gene and that of *R. metallidurans* with *hyd*B gene.

20.3 Conclusions

Acquisition of novel metabolic properties allows the recipient organism to inhabit diverse environmental niches. HGT helps in modifying bacterial genomes by allowing rapid acquisition of capabilities necessary for meeting new challenges (Caro-Quintero and Konstantinidis 2014). The role of HGT in evolution is evident by exchange of genes between eukaryotes and prokaryotes such as *D. radiodurans* (Koonin et al. 2001). Comparative genomic analysis hyperthermophiles belonging to Archaea and Bacteria (Aravind et al. 1998) or the intracellular pathogenic bacteria, for example, *Chlamydia* and *Rickettsia* (Wolf et al. 1998) have revealed common genes, which reflect their independent adaptation to similar environmental conditions and in other cases adaptation to stress (Foflonker et al. 2015). Although acquisitions and disseminations is enhanced by different vehicles such as mobile elements, plasmids, gene cassettes, transposons, or even bacteriophages (Di Gioia et al. 1998; Campbell et al. 1999; Dubnau 1999), however, the analyses of flanking regions of the five genes presented here did not show their presence.

Many organisms seem to have come together during evolution, and the transmission of genes seems to have happened as a network. In these networks, although theoretically all the genes should have followed a vertical gene transfer system, however, it has not happened so. Either the organisms have been selective in receiving and incorporating foreign genes or the prevailing environment forced some to part with its genome in a move to protect its identity (by preserving it in other organisms) and others to receive/borrow in order to develop/complement/maintain its own energy metabolism (Brown 2003). Analyses of genomes of 183 organisms show that these five genes involved in this H_2 evolution pathway appeared in 12 genera. The high prevalence of all the five subunits was among Proteobacteria is perhaps an act of the acquiring novel characteristics through HGT, an important feature responsible for adaptive evolution (Lawrence and Ochman 1998; Doolittle 1999). Of the various analysed genomes, *A fulgidus* (Archaea) and *S. flexneri* (γ-Proteobacteria) have four of the five subunits of FDH and H_2ase. There can be an opportunity to supplement the H_2 production system for these bacteria if their HGT events are taken in to consideration. This proposes that nature might be looking for an alternative/new pathway, which might handle the electrons with different option or maybe a new electron acceptor. The various associations of H_2 producing bacteria and certain eukaryotes come out of these studies to support the H_2 theory. Incongruence in phylogenetic trees and statistically significant deviations in gene characteristics of metabolic gene sequences reflects HGT events among genomes. These support the widely established notion that the eukaryotic organelles such as chloroplasts mitochondria have bacterial contribution leading to their origin. Comparative genomics reveals mitochondria to have evolved from Proteobacteria (such as *Agrobacterium* and *Escherichia*), and chloroplasts might have originated from Cyanobacteria (*Gloeobacter* and *Synechococcus*) (Norman 1997). The highly divergent eukaryotes such as *Vairimorpha*, *Trichomonas* and *Giardia* lack mitochondria (Cavalier-Smith 1993) but still contain few bacterial-type genes (Cavalier-Smith 1993; Bui et al. 1996; Germot et al. 1996; Roger et al. 1996). γ-Proteobacteria represents a model bacterial diversification, and HGT is known to be extensive in this group (Lerat et al. 2003). Cloning of genes across taxa such as hydrogenase gene (*hyd*) from *Clostridium butyricum* (Firmicute) in *E. coli* (γ-Proteobacteria) in fact resulted in 3.1–3.5 times higher activity (Karube et al. 1983). The possibility of such transfers among widely related organisms has even resulted in the construction of new pathways for PHA synthesis. Here, genes of Firmicutes such as *C. acetobutylicum* (butyrate kinase (*Buk*) and phosphotransbutyrylase (*Ptb*)) were expressed along with genes *pha*E and *pha*C encoding for PHA synthase belonging to γ-Proteobacteria and *Thiocapsa pfennigi* in γ-Proteobacteria (*E. coli*) resulted in PHA synthesis (Liu and Steinbüchel 2000). Similar transfers and expressions have been reported in cotton (John and Keller 1996) and *Arabidopsis* (Poirier et al. 1992). Of the three putative HGT events between Bacteria (*Thermoanaerobacter tengcongensis* and *D. gigas* (δ-Proteobacteria)) and Archaea species, which are relatively closer to *Methanosarcina*, influenced different operons, which also included *ech*, coding for NiFe hydrogenases (Calteau et al. 2005). Ech hydrogenase of *T. tengcongensis* favours H_2 evolution rather than H_2 uptake (Soboh et al. 2004). These gene transfers between the diverse evolutionary domains reflect on potential genetic evidences, which were lost over time. These evidences support organisms such as *Haloferax* (Archaea), *Gloeobacter* (Cyanobacteria), *Anopheles* (Metazoa) and *Aspergillus* (Eukaryote) to act as the reservoirs for individual genes being presented here. Nonetheless, these trends reflect on the ancestral nature of these three domains of life (Kandler 1993). The most perplexing thing is the acquisition of *fdh*A and *fdh*B by eukaryotes through unspecified mechanisms (Rivera and Lake 2004). With the present status of sequenced genomes and understanding of microbial ecosystems, it is difficult to reconstruct the future path of this evolutionary process (Martin and Embley 2004).

HGTs are largely viewed as events which lead to discrepancies in the phylogenetic tree (Ge et al. 2005; Rochette et al. 2014) compared to those established on the basis of small subunit of rDNA (Woese 1987). Since 16S rDNA trees are expected to reflect phylogeny, the possible explanations for these inconsistencies could be the existence of paralogs which might have diverged by different genetic mechanisms, such as gene recruitment, gene duplication, fusion and shuffling. These mechanisms have been proposed to be instrumental in the evolution of photosynthetic machinery (Raymond et al. 2002) and *arsC* gene encoding for arsenate reductase (Jackson and Dugas 2003). On the contrary, we may look at HGT as a means for rapid evolution, which otherwise may take much longer to happen. From this perspective, our study shows certain organisms where HGT events have lead to greater genetic variability during their evolution. This study has shown the extensive amount of HGT in H_2 metabolism, providing supporting evidence for its strong contribution for its evolution (Koumandou and Kossida 2014). Such deviations may look insignificant at present, but may prove important in some other context like recombinant gene technology and gene expression. Among the identified organisms that look most promising for H_2 production, some assume higher significance for their ability to produce PHAs as well, such as *B. fungorum*, *N. aromaticivorans*, *R. palustris* (Kalia et al. 2003a, b) and *B. japonicum*, which have the ability to degrade a range of carbon compounds with low nitrogen content, have magnetotactic ability, dehalogenate and degrade environmental pollutants, or even fix biological nitrogen. This raises the intriguing possibility that they could be exploited to breakdown wastes, generate H_2 produce value added chemicals such as PHAs, and to further enhance the efficiency of the bioconversion process (Angenent et al. 2004).

Acknowledgments We are thankful to Directors of CSIR-Institute of Genomics and Integrative Biology, CSIR- National Environmental Engineering Research Institute and CSIR and CSIR project WUM (ESC0108) for providing the necessary funds, facilities and moral support.

References

Achtman M, Hakenbeck R (1992) Recent developments regarding the evolution of pathogenic bacteria. In: Hormaeche CE (ed) Molecular biology of bacterial infection: current status and future perspective. Cambridge University Press, New York, pp 13–31

Angenent LT, Karim K, Al-Dahhan MH, Wrenn BA, Domiguez-Espinosa R (2004) Production of bioenergy and biochemicals from industrial and agricultural wastewater. Trends Biotechnol 22:477–485. doi:10.1016/j.tibtech.2004.07.001

Aravind L, Tatusov RL, Wolf YI, Walker DR, Koonin EV (1998) Evidence for massive gene exchange between archaeal and bacterial hyperthermophiles. Trends Genet 14:442–444. doi:10.1016/S0168-9525(98)01553-4

Brown RJ (2003) Ancient horizontal gene transfer. Nat Rev Genet 4:121–132. doi:10.1038/nrg1000

Bui ET, Bradley PJ, Johnson PJ (1996) A possible mitochondrial gene in the early-branching amitochondriate protist *Trichomonas vaginalis*. Proc Natl Acad Sci U S A 93:9651–9656

Calteau A, Gouy M, Perriere G (2005) Horizontal transfer of two operons coding for hydrogenases between bacteria and archaea. J Mol Evol 60:557–565. doi:10.1007/s00239-004-0094-8

Campbell A, Mrazek J, Karlin S (1999) Genome signature comparisons among prokaryote, plasmid, and mitochondrial DNA. Proc Natl Acad Sci U S A 96:9184–9189. doi:10.1073/pnas.96.16.9184

Carbone A, Zinovyev A, Kepes F (2003) Codon adaptation index as a measure of dominating codon bias. Bioinform 19:2005–2015. doi:10.1093/bioinformatics/btg272

Caro-Quintero A, Konstantinidis KT (2014) Inter-phylum HGT has shaped the metabolism of many mesophilic and anaerobic bacteria. ISME J. doi:10.1038/ismej.2014.193

Cavalier-Smith T (1993) Kingdom protozoa and its 18 phyla. Microbiol Rev 57:953–994

Delmotte F, Rispe C, Schaber J, Silva FJ, Moya A (2006) Tempo and mode of early gene loss in endosymbiotic bacteria from insects. BMC Evol Biol 6:56. doi:10.1186/1471-2148-6-56

Di Gioia D, Peel M, Fava F, Wyndham RC (1998) Structures of homologous composite transposons carrying *cba*ABC genes from Europe and North America. Appl Environ Microbiol 64:1940–1946

Doolittle WF (1999) Phylogenetic classification and the universal tree. Science 284:2124–2128. doi:10.1126/science.284.5423.2124

Dubnau D (1999) DNA uptake in bacteria. Annu Rev Microbiol 53:217–244. doi:10.1146/annurev.micro.53.1.217

Embley TM, van der Giezen M, Horner DS, Dyal PL, Bell S, Foster PG (2003) Hydrogenosomes, mitochondria and early eukaryotic evolution. IUBMB Life 55:387–395. doi:10.1080/1521654031000159283 4

Ermolaeva O, Rastogi M, Pruitt KD, Schuler GD, Bittner ML, Chen Y, Simon R, Meltzer P, Trent JM, Boguski MS (1998) Data management and analysis for gene expression arrays. Nat Genet 20:19–23. doi:10.1038/1670

Felmlee T, Pellett S, Welch RA (1985) Nucleotide sequence of an *Escherichia coli* chromosomal hemolysin. J Bacteriol 163:94–105

Foflonker F, Price DC, Qiu H, Palenik B, Wang S, Bhattacharya D (2015) Genome of the halotolerant green alga *Picochlorum* sp. Reveals strategies for thriving under fluctuating environmental conditions. Environ Microbiol 17:412–426. doi:10.1111/1462-2920.12541

Fraser AA (2004) Hydrogenases: active site puzzles and progress. Curr Opin Chem Biol 8:133–140. doi:10.1016/j.cbpa.2004.02.004

Garcia-Vallve S, Guzman E, Montero MA, Romeu A (2003) Horizontal gene transfer in bacterial and archaeal complete genomes. Nucleic Acids Res 31:187–189. doi:10.1093/nar/gkg004

Ge F, Wang LS, Kim J (2005) The cobweb of life revealed by genome-scale estimates of horizontal gene transfer. PLoS Biol 3, e316. doi:10.1371/journal.pbio.0030316

Germot A, Philippe H, Guyader HL (1996) Presence of a mitochondrial-type 70 kDa heat shock protein in *Trichomonas vaginalis* suggests a very early mitochondrial endosymbiosis in eukaryotes. Proc Natl Acad Sci U S A 93:14614–14617

Ghiorse WC (1997) Subterranean life. Science 275:789–790

Hacker J, Blum-Oehler G, Mühldorfer I, Tschäpe H (1997) Pathogenicity islands of virulent bacteria: structure, function and impact on microbial evolution. Mol Microbiol 23:1089–1097. doi:10.1046/j.1365-2958.1997.3101672.x

Jackson CR, Dugas SL (2003) Phylogenetic analysis of bacterial and archaeal *arsC* gene sequences suggests an ancient, common origin arsenate reductase. BMC Evol Biol 3:18. doi:10.1186/1471-2148-3-18

Jain R, Rivera MC, Moore JE, Lake JA (2003) Horizontal gene transfer accelerates genome innovation and evolution. Mol Biol Evol 20:1598–1602. doi:10.1093/molbev/msg154

John ME, Keller G (1996) Metabolic pathway engineering in cotton: biosynthesis of polyhydroxybutyrate in fiber cells. Proc Natl Acad Sci U S A 93:12768–12773

Kalia VC, Chauhan A, Bhattacharyya G, Rashmi (2003a) Genomic databases yield novel bioplastic producers. Nature Biotechnol 21:845–846. doi:10.1038/nbt0803-845

Kalia VC, Lal S, Ghai R, Mandal M, Chauhan A (2003b) Mining genomic databases to identify novel hydrogen producers. Trends Biotechnol 21:152–156. doi:10.1016/S0167-7799(03)00028-3

Kalia VC, Lal S, Cheema S (2007) Insight in to the phylogeny of polyhydroxyalkanoate biosynthesis: horizontal gene transfer. Gene 389:19–26. doi:10.1016/j.gene.2006.09.010

Kandler O (1993) The early diversification of life. In: Bengtson S (ed) Early life on earth. Nobel symp 84. Columbia Univ Press, New York, pp 152–160

Karube I, Urano N, Yamada T, Hirochika H, Sakaguchi K (1983) Cloning and expression of the hydrogenase gene from *Clostridium butyricum* in *Escherichia coli*. FEBS Lett 158:119–122

Koonin EV, Makarova KS, Aravind L (2001) Horizontal gene transfer in prokaryotes: quantification and classification. Annu Rev Microbiol 55:709–742. doi:10.1146/annurev.micro.55.1.709

Koumandou VL, Kossida S (2014) Evolution of the F_0F_1 ATP synthase complex in light of the patchy distribution of different bioenergetic pathways across prokaryotes. PLoS Comput Biol 10, e1003821. doi:10.1371/journal.pcbi.1003821

Lal S, Cheema S, Kalia VC (2008) Phylogeny vs genome reshuffling: horizontal gene transfer. Indian J Microbiol 48:228–242. doi:10.1007/s12088-008-0034-1

Lawrence JG, Ochman H (1998) Molecular archaeology of the *Escherichia coli* genome. Proc Natl Acad Sci U S A 95:9413–9417

Lerat E, Daubin V, Moran NA (2003) From gene trees to organismal phylogeny in prokaryotes: the case of the γ-proteobacteria. PLoS Biol 1, e19. doi:10.1371/journal.pbio.0000019

Liu S-J, Steinbüchel A (2000) A novel genetically engineered pathway for synthesis of poly (hydroxyalkanoic acids) in *Escherichia coli*. Appl Environ Microbiol 66:739–743. doi:10.1128/AEM.66.2.739-743.2000

Martin W, Embley TM (2004) Early evolution comes full circle. Nature 431:134–137. doi:10.1038/431134a

Mount DW (2001) Phylogenetic prediction. In: Argentine J (ed) Bioinformatics: sequence and genome analysis, 2nd edn. Cold Spring Harbour Laboratory Press, New York, pp 247–248

Norman RP (1997) A molecular view of microbial diversity and the biosphere. Science 276:734–740

Poirier Y, Dennis DE, Klomparens K, Somerville C (1992) Polyhydroxybutyrate, a biodegradable thermoplastic, produced in transgenic plants. Science 256:520–523

Raymond J, Zhaxybayeva O, Gogarten JP, Gerdes SY, Blankenship RE (2002) Whole-genome analysis of photosynthetic prokaryotes. Science 298:1616–1620. doi:10.1126/science.1075558

Rivera MC, Lake JA (2004) The ring of life provides evidence for a genome fusion origin of eukaryotes. Nature 431:152–155. doi:10.1038/nature02848

Rochette NC, Brochier-Armanet C, Gouy M (2014) Phylogenomic test of the hypotheses for the evolutionary origin of eukaryotes. Mol Biol Evol 31:832–845. doi:10.1093/molbev/mst272

Roger AJ, Clark CG, Doolittle WF (1996) A possible mitochondrial gene in the amitochondriate protist *Trichomonas vaginalis*. Proc Natl Acad Sci U S A 93:14618–14622

Sharp PM, Li WH (1987) The codon adaptation index- a measure of directional synonymous codon usage bias, and its potential applications. Nucleic Acids Res 15:1281–1295

Soboh B, Linder D, Hedderich R (2004) A multisubunit membrane-bound (NiFe) hydrogenase and a NADH-dependent Fe-only hydrogenase in the fermenting bacterium *Thermoanaerobacter tengcongensis*. Microbiology 150:2451–2463. doi:10.1099/mic.0.27159-0

Stevens TO, McKinley JP (1995) Lithoautotrophic microbial ecosystems in deep basalt aquifers. Science 270:450–455

Turner JA (2004) Sustainable hydrogen production. Science 305:972–974. doi:10.1126/science.1103197

Veziroglu TN (1995) Twenty years of hydrogen movement 1974–1994. Int J Hydrog Energ 20:1–7

Vignais PM, Billoud B, Meyer J (2001) Classification and phylogeny of hydrogenases. FEMS Microbiol Rev 25:455–501. doi:10.1111/j.1574-6976.2001.tb00587.x

Woese CR (1987) Bacterial evolution. Microbiol Rev 51:221–271

Wolf YI, Aravind L, Koonin EV (1998) *Rickettsiae and Chlamydia*: evidence of horizontal gene transfer and gene exchange. Trends Genet 14:442–444. doi:10.1016/S0168-9525(98)01553-4

Sadhana Lal received her M.Sc. degree in Bioscience from IIT Roorkee, Uttarakhand, India, and Ph.D. degree in Biochemistry from Jiwaji Gwalior University, MP, India, and CSIR-IGIB, Delhi, India. She is presently working as a Research Associate in Biosystems Engineering Department at University of Manitoba, Winnipeg, Canada. She is a member of the American Society for Microbiology (ASM). Her current research interests are bacterial genome annotation, in silico comparative analysis of bacterial metabolisms, genetic engineering, and bioenergy generation.

Dr. (Mrs.) Atya Kapley, Ph.D. from University of Hyderabad, Hyderabad, Principal Scientist at the National Environmental Engineering Research Institute, Nagpur, works in the field of environmental genomics. She uses a multidisciplinary approach, combining conventional microbiology tools with bioinformatics and molecular tools to address rising levels of environmental contamination. She has worked on the bioremediation of pesticide-contaminated soil and uses metagenomics approach to study microbial communities in activated biomass of wastewater treatment plants with the aim to understand and improve biological treatment capacity.

Hemant J. Purohit is a microbiologist with training in molecular biology working as Head of Environmental Genomics Division, CSIR-NEERI. His current interests include developing deep insight through gene expression and their networks into biofuel production, bioremediation, and stress tolerance through application of omics tools. He is Editor of Indian Journal of Microbiology, Associate Editor of Applied Biochemistry and Biotechnology, and Academic Editor of Plos One.

Vipin Chandra Kalia received his M.Sc. and Ph.D. degree in Genetics from IARI, New Delhi. He is presently the Chief Scientist at CSIR-IGIB and Professor of AcSIR. He is a Fellow of Association of Microbiologists of India and Fellow of National Academy of Sciences (FNASc). He is the Editor in Chief of the Indian Journal of Microbiology (India). His current interests are bioenergy, biopolymers, and quorum sensing. Google Scholar: http://scholar.google.co.in/citations?hl=en&user=XaUw-VIAAAAJ. Website: http://www.igib.res.in/?q=V.C.Kalia

Prokaryotic Contributions Towards Eukaryotic Powerhouse

21

Vipin Chandra Kalia

Abstract

The origin and evolution of prokaryotes and eukaryotes is a scientific curiosity which always appeared to be obscure and complicated. Evolution is primarily an adaptive step towards survival against adverse environmental conditions, leading to the development of multi-characteristic organisms. The evidences for transition of prokaryotes to eukaryotes came from the discovery of DNA in two most important eukaryotic cell organelles: chloroplasts and mitochondria. This nexus then appears to have relocated quite a bit of its genetic material to the nucleus and in turn receives proteins for their functioning through a dedicated reimport system. This article presents the contributions of different prokaryotic lineages towards evolution of eukaryotic powerhouse, the mitochondria.

21.1 Introduction

The scientific curiosities on the origin and evolution or merger of prokaryotes and eukaryotes have been topics of debate and rife explorations (Rochette et al. 2014). The discovery of the presence of DNA in two important eukaryotic cell organelles (chloroplasts and mitochondria) raised more queries than it answered on their origin and evolution, and many of them continue to be unanswered even today (Tekle et al. 2009). The endosymbiotic nature of chloroplasts and mitochondria was originally proposed by Konstantin Mereschkowsky in 1905 and by Ivan Wallin in the 1920s, respectively [http://infao5501.ag5.mpi-sb.mpg.de:8080/topx/archive?link=Wikipedia-Lip6-2/970821.xml&style]. It was only towards the end of the twentieth century that the endosymbiotic hypothesis was popularized by different groups (Gray 1992; Margulis 1996; Gray and Spencer 1996; Blanchard and Lynch 2000; Brinkman et al. 2002). Endosymbiont theory emerged as the most favorable explanation for the origin of eukaryotic cell organelles (Gray 1992; Gray and Spencer 1996; Tekle et al. 2009; Koumandou and Kossida 2014). These nexus then appear to have relocated quite a bit of their genetic material to the nucleus (Douglas 1998; Koonin 2010; Pina et al. 2011; Lombard et al. 2012a; Koonin and

V.C. Kalia (✉)
Microbial Biotechnology and Genomics,
CSIR-Institute of Genomics and Integrative Biology,
Delhi University Campus, Mall Road,
Delhi 110007, India
e-mail: vckalia@igib.res.in; vc_kalia@yahoo.co.in

V.C. Kalia (ed.), *Microbial Factories*, DOI 10.1007/978-81-322-2595-9_21

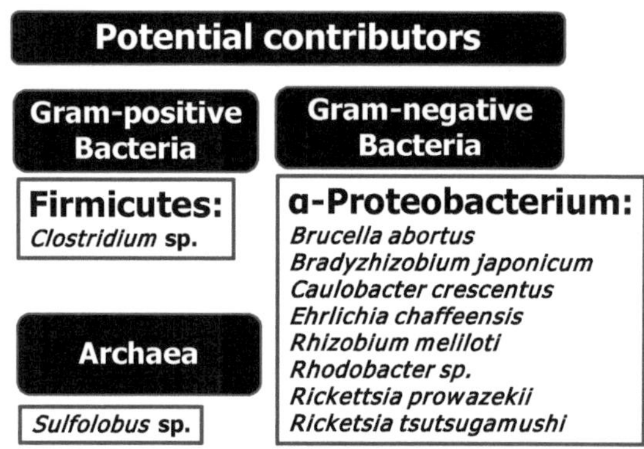

Fig. 21.1 Bacterial contribution to the eukaryotic energy power house – the mitochondria

Wolf 2012; Martijn and Ettema 2013). In this article, we have presented the contributions of different prokaryotic lineages towards evolution of the energy generation system of eukaryotes (Fig. 21.1).

21.2 Adaptive Step Towards Survival and Chimeric Evolution

Evolutionary relationships based on phylogenetic analyses of diverse genes allowed scientist to propose two models for justifying the origin of eukaryotic cells (Gray et al. 1999; Archibald 2008). According to the first model of progressive evolution, archaebacteria evolved as eukaryotic cells, and they share ancestral relations with *Crenarchaeota* – eocytes (Lake et al. 1984; Brown et al. 2001; Cox et al. 2008; Forterre 2013). The chimeric model states that eukaryotic cell nucleus evolved via symbiogenesis of metabolically dependent consortia – an archaebacterium (a thermoacidophile) and a motile eubacterium (Gupta 1995; Tekle et al. 2009) as a chimera. This emergence of partners might be under selective environmental pressures such as a threat of oxygen, limitation of carbon compounds, and acceptors of electrons. Archaeal nucleocytoplasm acquired swimming motility from eubacteria to become mastigotes (the early Eukarya). Some of these mastigotes became aerobes by incorporating purple eubacteria, which can be considered as potential mitochondrial precursors, a potential stage for the evolution of protoctists, animals, and fungi. On the other hand, phototrophic protoctists like algae originated by the association of aerobes with plastid precursor cyanobacteria (Margulis 1996). The transient host to the mitochondria has been the amitochondriate eukaryotes (termed Archezoa) (Lang et al. 1999). During that era, alternative symbiotic scenarios appeared in the eukaryotic cell evolution (Martin and Müller 1998; Moreira and López-García 1998). These invoked the notion of syntrophy, where the end product of one of the partners – H_2 – served as an essential nutrient for the other. The "ox-tox" hypothesizes that the symbiont is responsible for O_2 detoxification and subsequently provides ATP to the host (Gray et al. 1999). It may be envisaged that the eukaryotic ancestor got transformed in "chimeras" as a result of genome fusion of different prokaryotes (Golding and Gupta 1995; Gupta 1998; Margulis et al. 2000; Kurland et al. 2006; De Duve 2007; Brochier-Armanet et al. 2008; Poole and Neumann 2011; Maguire and Richards 2014).

21.3 Origin of Mitochondria: The Gram-Positive Bacterial Route

Eukaryotic origins seem to have taken place by archaeal-bacterial partnerships. The primary basis of hydrogen hypothesis is the symbiotic relation of "methanogen" with "α-proteobacterium" (Martin and Müller 1998; Karlin et al. 1999). These studies required compatibility of the genetic systems of the archaebacterium and the eubacterium for heterologous expression of gene(s), which can be expected from similarities in their genomic signatures (Karlin et al. 1999). Genome DNA signatures of *Clostridium* spp. (Firmicutes) and *Sulfolobus* spp. (Archaea) are sufficiently similar to meet the abovementioned criteria (Fig. 21.1). They have significantly more similarity between them than with all other available prokaryotes, including classical α-proteobacterium. Based on the preservation of DNA features of *Clostridium*- and *Sulfolobus*-like organisms in animal mitochondria and several amitochondrial eukaryotes, the close relationship between *Clostridium* and *Sulfolobus* nexus has been hypothesized for the origin of a stable fusion ancestor. It finds support from the observation that *Clostridium*-like organisms contribute their unique characteristic of generating H_2 and acetate by transforming organic matter. The endospores of *Clostridium* provide a strong platform for establishing cytoplasm and nucleus within the eukaryotic cell (Karlin et al. 1999). On the other hand, the *Sulfolobus*-like organism have a unique ability to utilize H_2 and link the electrons to a respiratory chain components: quinones and cytochrome cofactors. It thus helps to establish an organismic partnership. This nexus may lose certain genes, and some may even be relocated to the nucleus followed by acquisition of genes from other organisms such as α-proteobacterium and/or other eubacteria (Douglas 1998). The genomic reasons for *Sulfolobus-/Clostridium*-like ancestors are that all animal mitochondria are significantly underrepresented in their GC content (21–46%) (Karlin et al. 1999), which is quite similar to Gram-positive prokaryotes: *Sulfolobus* and *Clostridium* (Karlin 1998; Campbell et al. 1999) unlike α-proteobacteria where it is >60%. These two organisms also have a high CC/GG relative abundance just like animal mitochondria in contrast to Gram-negative bacteria – *Rickettsia prowazekii*, which is normal for CC/GG. Similar patterns of dinucleotide TA frequency provide interesting evidences supporting the contributions of *Sulfolobus* spp., *Clostridium* spp., and *R. prowazekii* (Campbell et al. 1999) in mitochondrial genomes. It is, however, drastically underrepresented in α- and β-proteobacteria (Karlin et al. 1999). On the contrary, the frequency of occurrence of dinucleotide (GC) is relatively high in proteobacteria especially β- and γ-groups, comparatively balanced in the group of α-proteobacteria, in animal mitochondria, and in *Sulfolobus* spp. and *Clostridium* spp. but significantly high in the pathogen – *R. prowazekii*. These observations support *Sulfolobus*- and *Clostridium*-like combination (Campbell et al. 1999).

Although *Clostridium* lacks an ability to oxidize acetate and a respiratory chain for energy conservation, however, it possesses enzymes, which are characteristic of hydrogenosome, an H_2-producing organelles present in certain anaerobic eukaryotes (Karlin et al. 1999). Trichomonads such as *Trichomonas vaginalis* and *Histomonas meleagridis* possess key enzymes present in hydrogenosome organelle, where pyruvate gets oxidized to CO_2 and acetyl-CoA and in the process leads to the generation of acetate and ATP (Mazet et al. 2008). Prokaryotic and mitochondrial genomic comparison revealed that mitochondria (animal and fungal) are moderately similar to *Sulfolobus* and *Clostridium* and distantly related to *Chlamydomonas reinhardtii* and protists. On the other hand, mitochondrial (plant and green algae) genomes are quite similar to *Anabaena* species and close to red algae, *Chondrus crispus* (Karlin et al. 1999). On the basis of these evidences, it can be proposed that mitochondrial genome might have originated from that of (eu)bacteria (Gray 1999; Lang et al. 1999) and not the archaeal (Karlin et al. 1999) domain of life.

21.4 Origin of Mitochondria: The Gram-Negative Bacterial Route

Comparisons of the biochemical and morphological features of mitochondria, chloroplasts, and bacteria belonging to cyanobacteria and proteobacteria have been taken into account while proposing the evolution of eukaryotic cells by endosymbiosis. These proposals have been supported by molecular sequence data of a wide range of genes (Gray 1992; Gray et al. 2001; Deschamps 2014). Phylogenetic trees based on the bacterial and mitochondrial genes which encode for proteins involved in energy metabolism (NADH dehydrogenase) and genetic processes (ribosomal proteins) have shown that *R. prowazekii* is more close to mitochondria than bacteria (Fig. 21.1) (Gray and Spencer 1996; Andersson et al. 1998; Kitada et al. 2007). There is a lack of genes responsible for metabolizing sugars, synthesis of amino acids, and nucleotides (Gray 1998). Peptidase from *Rickettsia* shows more resemblance to α and β subunits of mitochondrial processing peptidase (Kitada et al. 2007). The closer relationship between α-proteobacterium and mitochondria is also supported by phylogenetic studies based on heat shock proteins (HSP60 and HSP10) sequences from α-proteobacteria (*Rickettsia tsutsugamushi* and *Ehrlichia chaffeensis*) and mitochondria (Gupta 1995; 1998).

A specific phylogenetic analysis of HSP70s showed highest sequence identity between mitochondria and α-proteobacterial members such as *Rhizobium meliloti*, *Brucella abortus* and *Caulobacter crescentus* (Falah and Gupta 1994). This relationship was further supported by 4-amino acid insert signature sequence lacking in α-proteobacterium and mitochondria but present in β- or γ-subdivisions of proteobacteria (Gupta 1998). Specific signature sequences (4-amino acid deletion and 1-amino acid insert) are present in the cytosolic and endoplasmic reticulum but not in any Hsp70s of prokaryotic or organellar origin. These signatures are specific to eukaryotes and might have been introduced into a common ancestor (Gupta 1998).

Other proteins whose signature sequences have been uniquely shared by eukaryotes and Gram-negative bacterium are aspartate amino transferase, alanyl-tRNA synthetase, glutamate dehydrogenase, glutamate-1-semialdehyde, glutamine synthase, 2-1-aminomutase, and phosphoribosylformylglycineamide amidotransferase protein (Gupta and Golding 1996; Gupta 1998). The signatures of these proteins provide additional evidences of their eubacterial rather than the archaebacterial origin (Gupta 1998). There is a wide variation in the mitochondrial gene content, 67 in *Reclinomonas americana* and only 3 in the Apicomplexans (Gray et al. 1999; Lang et al. 1999). Here *R. prowazekii* (Andersson et al. 1998; Gray et al. 2001) could be categorized closest to mitochondria. Among other α-proteobacteria, the genes of *C. crescentus*, *Bradyrhizobium japonicum*, and *Rhodobacter* have homologs in mitochondrial genomes (Karlberg et al. 2000; Gray et al. 2001).

21.5 The Interim Eukaryote: Hydrogenosome

Evolutionary analysis of 16S rRNA gene, widely employed for elucidating phylogenetic relationships (Porwal et al. 2009), had demonstrated that facultative anaerobic protists, trichomonads, diverged from eukaryotic evolution before the establishment of mitochondria (Viscoglioso et al. 1993; Bui et al. 1996). Trichomonads lack two eukaryotic organelles, mitochondria and peroxisomes. Incidentally, an unusual organelle hydrogenosome involved in carbohydrate metabolism is present in them (Bui et al. 1996). Hydrogenosome shares the double-membrane characteristic of mitochondria to produce ATP from pyruvate and show the presence of cristae (Martin 2005). These can be distinguished from mitochondria by the presence of enzymes commonly found in anaerobes: pyruvate/ferredoxin oxidoreductases and hydrogenases. These enzymes enable them to evolve H_2 (Bui et al. 1996). Hydrogenosome is also found in taxonomically diverse organisms like fungi and ciliates (Tan et al. 2002). The linkage between hydrogenosomes and mitochondria has been based on the presence of genes for mitochondria-like proteins in eukaryotes possessing hydrogenosomes (Gray et al. 1999; Lang et al. 1999; Van der Giezen et al. 2002). The presence of signature sequences in heat shock proteins (Hsp10, Hsp60 and Hsp70), in hydrogenosome, and in mitochondria and α-Gram-negative purple bacteria further supports the linkages (Gupta 1995; 1998; Bui et al. 1996). In eukaryotic cells, specific signature sequences are used to distinguish Hsp70 homologs in different compartments (Gupta 1998).

21.6 Evolution via Horizontal Gene Transfer

The presence of certain classes of conserved proteins across taxonomic and phylogenetic range like a heat shock family of proteins (a 70 KDa – Hsp70) (Gupta 1995) present an attractive system for elucidating evolutionary events. The indication that α-proteobacteria are close to endosymbionts which might have given rise to mitochondria is based on phylogenetic analyses of the rRNA and cytochrome C sequences (Gray 1992; Bertini et al. 2007). However, these inferences are limited by the wide range of size variations of rRNAs in eukaryotes and their phylogenetic incongruencies with other metabolic gene sequences, i.e., it might be a case of horizontal gene transfer (HGT) (Kalia et al. 2007; Lal et al. 2008). Of all prokaryotic genes, HGT appears to have affected between 2 and 60% of them with a frequency of about 1.1 HGT event in a gene family life span (Dagan and Martin 2007). This rate of HGT is sufficient to allow integration of ancestral genomes into present-day genomes (Koonin and Wolf 2006; Whitaker et al. 2009).

Gene transfer events have been reported from prokaryotes to eukaryotes (Koonin et al. 2001). These transfers confer metabolic advantage to the recipient (Andersson 2005). Support to such proposal comes from the transfer of small subunit of glutamate synthase, from a Gram-positive bacteria with a low GC content to an eukaryotic ancestor (Andersson and Roger 2002). Cyanobacterial HSP70 shows around 70% identity for chloroplast homologs (Falah and Gupta 1994), indicating a close relationship between them. It seems that around 18% of the *Arabidopsis* genome has its origin in a cyanobacterial ancestor of the eukaryotic organelle – chloroplast (Martin et al. 2002; Kleffmann et al. 2004). On the other hand, the origin of nuclear genes is likely to be from α-proteobacterial ancestor of mitochondria (Esser et al. 2004). Blanchard and Lynch (2000) in an article on why do organellar genes end up in the nucleus have discussed the theories on transfer of genes from mitochondria and chloroplast into the nucleus. They have also provided details of the steps, which are needed to complete this process (Blanchard and Lynch 2000).

Bioinformatic tools to identify the phylogenetic distribution of mitochondrial proteins have revealed that nucleus encoded proteins fall into three groups: (a) prokaryotic, (b) eukaryotic, and (c) organism specific (Marcotte et al. 2000). In *Saccharomyces cerevisiae*, 10% of the nuclear genome or 630 genes is linked to mitochondrial activities. In *Caenorhabditis elegans*, around 400 nucleus encoded mitochondrial genes seem to be responsible for prokaryotic mitochondrial ancestor (Marcotte et al. 2000). Prokaryotic heritage inferred from organeller morphology (Gray 1999) reveals that many of the proteins resemble more to prokaryotes than of eukaryotes (Tatusov et al. 2000). Functioning of mitochondrial genes depends on their phylogenetic origin, such that those derived from prokaryotes are predominantly involved in metabolic activities, producing energy, protein synthesis, and mitochondrial organization (Caro-Quintero and Konstantinidis 2014). On the other hand, eukaryotic mitochondrial proteins are devoted to transport and protein targeting metabolism and mitochondrial organization (Marcotte et al. 2000).

Horizontal acquisition of mitochondrial genes by angiosperms (Bergthorsson et al. 2003) and their transmission onto plants such as *Amborella trichopoda* (Bergthorsson et al. 2004) have been limited by poor rate of nucleotide changes in plant mitochondrial DNA (Laroche et al. 1997). HGT is now established for mitochondrial DNA and relatively less known in chloroplasts (Bergthorsson et al. 2004). In a phylogenetic study, 24 genes of *Cryptosporidium parvum* were reported to have a eubacterial origin (Huang et al. 2004). Phylogenetic study of sulfide dehydrogenase in prokaryotes and eukaryotes showed that diplomonad sequences may cluster with prokaryotes (Andersson and Roger 2002). The phylogenetic trees of different classes of alcohol dehydrogenases have different prokaryotic affinities. It is indicative that this gene may be a strong HGT event in both pro- and eukaryotes (Andersson 2005). Recent studies reveal that chloroplast and mitochondrial genomes share sequences, which indicate the presence of promiscuous DNA of plastid origin in the angiosperm mtDNA genome. It also speculates unknown gene transfer route, which might be responsible for transfer from mitochondrion to the chloroplast (Goremykin et al. 2009; Straub et al. 2013). They, however, did not find HGT to be as rampant as reported previously by others (Bergthorsson et al. 2003; 2004; Goremykin et al. 2009). The recent view has been that bacterial genes get into archaea through HGT as they stay together in communities (Cohan and Koeppel 2008; Sato and Atomi 2011; Lombard et al. 2012b)

21.7 Conclusion

The pioneering and painstaking works of Charles Darwin on the evolution of living beings laid the foundation for evolutionary studies. It very vividly demonstrated that to understand the historical processes, a comparison of different advanced stages of a series of transformation is quite critical (Rodríguez-Ezpeleta and Philippe 2006). In nature, whole organisms are selected under specific environmental pressures (Foflonker et al. 2014). Although such selections can be traced down to certain molecular changes, however, natural selection process takes place throughout the life history of living organisms (Margulis 1996). Evolutionary genomics of microbes, microbial eukaryotes, eukaryotes, and their organelles showcases the amazing genomic diversity found in these organisms. It also provides a glimpse of the status of our little knowledge about microbial and eukaryotic associations.

21.8 The Debates Continue

The debate(s) continues as to why mitochondria and plastids continue to have their own genomes, and possible answers are still relevant and continuously sought (Martin 2003; Rodríguez-Ezpeleta et al. 2005; Theissen and Martin 2006; Yoon et al. 2006; Bodył et al. 2007; Non et al. 2007; Forterre 2013).

Acknowledgements We are thankful to Director of CSIR-Institute of Genomics and Integrative Biology and WUM (ESC0108) for providing necessary facilities and support.

References

Andersson JO (2005) Lateral gene transfer in eukaryotes. Cell Mol Life Sci 62:1182–1197. doi:10.1007/s00018-005-4539-z

Andersson JO, Roger AJ (2002) Evolutionary analyses of the small subunit of glutamate synthase: gene order conservation, gene fusions and prokaryote-to-eukaryote lateral gene transfers. Eukaryot Cell 1:304–310. doi:10.1128/EC.1.2.304-310.2002

Andersson SGE, Zomorodipour A, Andersson JO, Sicheritz-Ponten T, Alsmark UC, Podowski RM, Naslund AK, Eriksson AS, Winkler HH, Kurland CG (1998) The genome sequence of *Rickettsia prowazekii* and the origin of mitochondria. Nature 396:133–140. doi:10.1038/24094

Archibald JM (2008) The eocyte hypothesis and the origin of eukaryotic cells. Proc Natl Acad Sci USA 105:20049–20050. doi:10.1073/pnas.0811118106

Bergthorsson U, Adams KL, Thomason B, Palmer JD (2003) Widespread horizontal transfer of mitochondrial genes in flowering plants. Nature 424:197–201. doi:10.1038/nature01743

Bergthorsson U, Richardson AO, Young GJ, Goertzen LR, Palmer JD (2004) Massive horizontal transfer of mitochondrial genes from diverse land plant donors to the basal angiosperm *Amborella*. Proc Natl Acad Sci USA 101:17747–17752. doi:10.1073/pnas.0408336102

Bertini I, Cavallaro G, Rosato A (2007) Evolution of mitochondrial-type cytochrome *c* domains and of the protein machinery for their assembly. J Inorg Biochem 101:1798–1811. doi:10.1016/j.jinorgbio.2007.02.001

Blanchard JL, Lynch M (2000) Organellar genes: why do they end up in the nucleus? Trends Genet 16:315–320. doi:10.1016/S0168-9525(00)02053-9

Bodył A, Mackiewicz P, Stiller JW (2007) The intracellular cyanobacteria of *Paulinella chromatophora*: endosymbionts or organelles? Trends Microbiol 15:295–296. doi:10.1016/j.tim.2007.05.002

Brinkman FSL, Blanchard JL, Cherkasov A, Av-Gay Y, Brunham RC, Fernandez RC, Finlay BB, Otto SP, Ouellette BFF, Keeling PJ, Rose AM, Hancock REW, Jones SJM (2002) Evidence that plant-like genes in chlamydia species reflect an ancestral relationship between chlamydiaceae, cyanobacteria, and the chloroplast. Genome Res 12:1159–1167. doi:10.1101/gr.341802

Brochier-Armanet C, Gribaldo S, Forterre P (2008) A DNA topoisomerase IB in thaumarchaeota testifies for the presence of this enzyme in the last common ancestor of archaea and eucarya. Biol Direct 3:54. doi:10.1186/1745-6150-3-54

Brown JR, Douady CJ, Italia MJ, Marshall WE, Stanhope MJ (2001) Universal trees based on large combined protein sequence data sets. Nature Genet 28:281–285. doi:10.1038/90129

Bui ETN, Bradley PJ, Johnson PJ (1996) A common evolutionary origin for mitochondria and hydrogenomes. Proc Natl Acad Sci USA 93:9651–9656

Campbell A, Mrázek J, Karlin S (1999) Genome signature comparisons among prokaryote, plasmid, and mitochondrial DNA. Proc Natl Acad Sci USA 96:9184–9189

Caro-Quintero A, Konstantinidis KT (2014) Inter-phylum HGT has shaped the metabolism of many mesophilic and anaerobic bacteria. The ISME Journal. doi:10.1038/ismej.2014.193

Cohan FM, Koeppel AF (2008) The origins of ecological diversity in prokaryotes. Curr Biol 18:R1024–R1034. doi:10.1016/j.cub.2008.09.014

Cox CJ, Foster PG, Hirt RP, Harris SR, Embley TM (2008) The archaebacterial origin of eukaryotes. Proc Natl Acad Sci USA 105:20356–20361. doi:10.1073/pnas.96.16.9184

Dagan T, Martin W (2007) Ancestral genome sizes specify the minimum rate of lateral gene transfer during prokaryote evolution. Proc Natl Acad Sci USA 104:870–875

De Duve C (2007) The origin of eukaryotes: a reappraisal. Nat Rev Genetics 8:395–403. doi:10.1038/nrg2071

Deschamps P (2014) Primary endosymbiosis: have cyanobacteria and chlamydiae ever been roommates? Acta Societatis Botanicorum Poloniae. doi:10.5586/asbp.2014.048

Douglas SE (1998) Plastid evolution: origins, diversity, trends. Curr Opin Genet Dev 8:655–661. doi:10.1016/S0959-437X(98)80033-6

Esser C, Ahmadinejad N, Wiegand C, Rotte C, Sebastiani F, Gelius-Dietrich G, Henze K, Kretschmann E, Richly E, Leister D, Bryant D, Steel MA, Lockhart PJ, Penny D, Martin W (2004) A genome phylogeny for mitochondria among α-proteobacteria and a predominantly eubacterial ancestry of yeast nuclear genes. Mol Biol Evol 21:1643–1660. doi:10.1093/molbev/msh160

Falah M, Gupta RS (1994) Cloning of the hsp70 (*dnaK*) genes from *Rhizobium meliloti* and *Pseudomonas cepacia*: phylogenetic analyses of mitochondrial origin based on a highly conserved protein sequence. J Bacteriol 176:7748–7753

Foflonker F, Price DC, Qiu H, Palenik B, Wang S (2014) Bhattacharya D. Genome of the halotolerant green alga Picochlorum sp. reveals strategies for thriving under fluctuating environmental conditions, Environ Microbiol. doi:10.1111/1462-2920.12541

Forterre P (2013) The common ancestor of archaea and eukarya was not an archaeon. Archaea 2013:Article ID 372396. doi:10.1155/2013/372396

Golding GB, Gupta RS (1995) Protein-based phylogenies support a chimeric origin of the eukaryotic genome. Mol Biol Evol 12:1–6. doi:10.1093/oxfordjournals.molbev.a040178

Goremykin VV, Salamini F, Velasco R, Viola R (2009) Mitochondrial DNA of *Vitis vinifera* and the issue of rampant horizontal gene transfer. Mol Biol Evol 26:99–110. doi:10.1093/molbev/msn226

Gray MW (1992) The endosymbiont hypothesis revisited. Int Rev Cytol 141:233–357

Gray MW (1998) Rickettsia, typhus and the mitochondrial connection. Nature 396:109–110. doi:10.1038/24030

Gray MW (1999) Evolution of organellar genomes. Curr Opin Genet Dev 9:678–687. doi:10.1016/S0959-437X(99)00030-1

Gray MW, Burger G, Lang BF (2001) The origin and early evolution of mitochondria. Genome Biol 2:1018.1–1018.5. doi:10.1186/gb-2001-2-6-reviews1018

Gray MW, Burger G, Lang BF (1999) Mitochondrial evolution. Science 283:1476–1481. doi:10.1126/science.283.5407.1476

Gray MW, Spencer DF (1996) Organellar evolution. In: Roberts DM, Sharp P, Alderson G, Collins M (eds) Evolution of Microbial Life. Cambridge University Press, Cambridge, MA, pp 109–126

Gupta RS (1995) Evolution of the chaperonin families (Hsp60, Hsp10 and Tcp-1) proteins and the origin of eukaryotic cells. Mol Microbiol 15:1–11. doi:10.1111/j.1365-2958.1995.tb02216.x

Gupta RS (1998) Protein phylogenies and signature sequences: a reappraisal of evolutionary relationships among archaebacteria, eubacteria and eukaryotes. Microbiol Mol Biol Rev 62:1435–1491

Gupta RS, Golding GB (1996) The origin of the eukaryotic cell. Trends Biochem Sci 21:166–171

Huang J, Mullapudi N, Lancto CA, Scott M, Abrahamsen MS, Kissinger JC (2004) Phylogenomic evidence supports past endosymbiosis, intracellular and horizontal gene transfer in *Cryptosporidium parvum*. Genome Biol 5:R88. doi:10.1186/gb-2004-5-11-r88

Kalia VC, Lal S, Cheema S (2007) Insight in to the phylogeny of polyhydroxyalkanoate biosynthesis: horizontal gene transfer. Gene 389:19–26. doi:10.1016/j.gene.2006.09.010

Karlberg O, Canbäk B, Kurland CG, Andersson SGE (2000) The dual origin of the yeast mitochondrial proteome. Yeast 17:170–187. doi:10.1002/1097-0061(20000930)17:3<170::AID-YEA25>3.0.CO;2-V

Karlin S (1998) Global dinucleotide signatures and analysis of genomic heterogeneity. Curr Opin Microbiol 1:598–610. doi:10.1016/S1369-5274(98)80095-7

Karlin S, Brocchieri L, Mrázek J, Campbell AM, Spormann AM (1999) A chimeric prokaryotic ancestry of mitochondria and primitive eukaryotes. Proc Natl Acad Sci USA 96:9190–9195

Kitada S, Uchiyama T, Funatsu T, Kitada Y, Ogishima T, Ito A (2007) A protein from a parasitic microorganism, *Rickettsia prowazekii*, can cleave the signal

sequences of proteins targeting mitochondria. J Bacteriol 189:844–850. doi:10.1128/JB.01261-06
Kleffmann T, Russenberger D, von Zychlinski A, Christopher W, Sjölander K, Gruissem W, Baginsky S (2004) The *Arabidopsis thaliana* chloroplast proteome reveals pathway abundance and novel protein functions. Curr Biol 14:354–362
Koonin EV (2010) The origin and early evolution of eukaryotes in the light of phylogenomics. Genome Biol 11:209. doi:10.1186/gb-2010-11-5-209
Koonin EV, Makarova KS, Aravind L (2001) Horizontal gene transfer in prokaryotes: quantification and classification. Ann Rev Microbiol 55:709–742. doi:10.1146/annurev.micro.55.1.709
Koonin EV, Wolf YI (2006) Evolutionary systems biology: links between gene evolution and function. Curr Opin Biotechnol 17:481–487. doi:10.1016/j.copbio.2006.08.003
Koonin EV, Wolf YI (2012) Evolution of microbes and viruses: a paradigm shift in evolutionary biology? Front Cell Infect Microbiol 2:119, doi:eCollection 2012
Koumandou VL, Kossida S (2014) Evolution of the F_0F_1 ATP synthase complex in light of the patchy distribution of different bioenergetic pathways across prokaryotes. PLoS Comput Biol 10, e1003821. doi:10.1371/journal.pcbi.1003821
Kurland CG, Collins LJ, Penny D (2006) Genomics and the irreducible nature of eukaryote cells. Science 312:1011–1014. doi:10.1126/science.1121674
Lake JA, Henderson E, Oakes M, Clark MW (1984) Eocytes: A new ribosome structure indicates a kingdom with a close relationship to eukaryotes. Proc Natl Acad Sci USA 81:3786–3790
Lal S, Cheema S, Kalia VC (2008) Phylogeny vs genome reshuffling: horizontal gene transfer. Indian J Microbiol 48:228–242. doi:10.1007/s12088-008-0034-1
Lang BF, Gray MW, Burger G (1999) Mitochondrial genome evolution and the origin of eukaryotes. Ann Rev Genet 33:351–397
Laroche J, Li P, Maggia L, Bousquet J (1997) Molecular evolution of angiosperm mitochondrial introns and exons. Proc Natl Acad Sci USA 94:5722–5727
Lombard J, López-García P, Moreira D (2012a) The early evolution of lipid membranes and the three domains of life. Nat Rev Microbiol 10:507–515. doi:10.1038/nrmicro2815
Lombard J, López-García P, Moreira D (2012) Phylogenomic investigation of phospholipid synthesis in archaea. Archaea 2012:Article ID 630910. doi:10.1155/2012/630910
Maguire F, Richards TA (2014) Organelle evolution: a mosaic of 'mitochondrial' functions. Curr Biol 24:R518–R520. doi:10.1016/j.cub.2014.03.075
Marcotte EM, Xenarios I, van Der Bliek AM, Eisenberg D (2000) Localizing proteins in the cell from their phylogenetic profiles. Proc Natl Acad Sci USA 97:12115–12120. doi:10.1073/pnas.220399497
Margulis L (1996) Archael-eubacterial mergers in the origin of eukarya: phylogenetic classification of life. Proc Natl Acad Sci USA 93:1071–1076
Margulis L, Dolan MF, Guerrero R (2000) The chimeric eukaryote: origin of the nucleus from the karyomastigont in amitochondriate protists. Proc Natl Acad Sci USA 97:6954–6959. doi:10.1073/pnas.97.13.6954
Martijn J, Ettema TJ (2013) From archaeon to eukaryote: the evolutionary dark ages of the eukaryotic cell. Biochem Soc Trans 41:451–457. doi:10.1042/BST20120292
Martin W (2003) Gene transfer from organelles to the nucleus: frequent and in big chunks. Proc Natl Acad Sci USA 100:8612–8614. doi:10.1073/pnas.1633606100
Martin W (2005) The missing link between hydrogenosomes and mitochondria. Trends Microbiol 13:457–459. doi:10.1016/j.tim.2005.08.005
Martin W, Müller M (1998) The hydrogen hypothesis for the first eukaryote. Nature 392:37–41. doi:10.1038/32096
Martin W, Rujan T, Richly E, Hansen A, Cornelsen S, Lins T, Leister D, Stoebe B, Hasegawa M, Penny D (2002) Evolutionary analysis of *Arabidopsis*, cyanobacterial and chloroplast genomes reveals plastid phylogeny and thousands of cyanobacterial genes in the nucleus. Proc Natl Acad Sci USA 99:12246–12251. doi:10.1073/pnas.182432999
Mazet M, Diogon M, Alderete JF, Vivarès CP, Delbac F (2008) First molecular characterization of hydrogenosomes in the protozoan parasite *Histomonas meleagridis*. Int J Parasitol 38:177–190. doi:10.1016/j.ijpara.2007.06.006
Moreira D, López-García P (1998) Symbiosis between methanogenic Archaea and α-Proteobacteria as the origin of eukaryotes: the syntrophic hypothesis. J Mol Evol 47:517–530
Non AL, Kitchen A, Mulligan CJ (2007) Identification of the most informative regions of the mitochondrial genome for phylogenetic and coalescent analyses. Mol Phylogenet Evol 44:1164–1171. doi:10.1016/j.ympev.2006.12.020
Pina M, Bize A, Forterre P, Prangishvili D (2011) The archeoviruses. FEMS Microbiol Rev 35:1035–1054. doi:10.1111/j.1574-6976.2011.00280.x
Poole M, Neumann N (2011) "Reconciling an archaeal origin of eukaryotes with engulfment: a biologically plausible update of the Eocyte hypothesis,". Res Microbiol 162:71–76. doi:10.1016/j.resmic.2010.10.002
Porwal S, Lal S, Cheema S, Kalia VC (2009) Phylogeny in aid of the present and novel microbial lineages: diversity in *Bacillus*. PLoS ONE 4, e4438. doi:10.1371/journal.pone.0004438
Rochette NC, Brochier-Armanet C, Gouy M (2014) Phylogenomic test of the hypotheses for the evolutionary origin of eukaryotes. Mol Biol Evol 31:832–845. doi:10.1093/molbev/mst272

Rodríguez-Ezpeleta N, Brinkmann H, Burey SC, Roure B, Burger G, Roger AJ, Gray MW, Philippe H, Lang BF (2005) Monophyly of primary photosynthetic eukaryotes: green plants, red algae, and glaucophytes. Curr Biol 15:1325–1330. doi:10.1016/j.cub.2005.06.040

Rodríguez-Ezpeleta N, Philippe H (2006) Plastid origin: replaying the tape. Curr Biol 16:R53–R56. doi:10.1016/j.cub.2006.01.006

Sato T, Atomi H (2011) Novel metabolic pathways in Archaea. Curr Opin Microbiol 14:307–314. doi:10.1016/j.mib.2011.04.014

Straub SC, Cronn RC, Edwards C, Fishbein M, Liston A (2013) Horizontal transfer of DNA from the mitochondrial to the plastid genome and its subsequent evolution in milkweeds (apocynaceae). Genome Biol Evol 5:1872–1885. doi:10.1093/gbe/evt140

Tan KSW, Singh M, Yap EH (2002) Recent advances in *Blastocystis hominis* research: hot spots in terra incognita. Int J Parasitol 32:789–804. doi:10.1016/S0020-7519(02)00005-X

Tatusov RL, Galperin MY, Natale DA, Koonin EV (2000) The COG database: a tool for genome-scale analysis of protein functions and evolution. Nucleic Acids Res 28:33–36. doi:10.1093/nar/28.1.33

Tekle YI, Parfrey LW, Katz LA (2009) Molecular data are transforming hypotheses on the origin and diversification of eukaryotes. BioScience 59:471–481. doi:10.1525/bio.2009.59.6.5

Theissen U, Martin W (2006) The difference between organelles and endosymbionts. Curr Biol 16:R1016–R1017. doi:10.1016/j.cub.2006.11.020

Van der Giezen M, Slotboom DJ, Horner DS, Dyal PL, Harding M, Xue G-P, Embley TM, Kunji ERS (2002) Conserved properties of hydrogenosomal and mitochondrial ADP/ATP carriers: a common origin for both organelles. EMBO J 21:572–579. doi:10.1093/emboj/21.4.572

Viscoglioso E, Philippe H, Baroin A, Perasso R, Brugerolle G (1993) Phylogeny of trichomonads based on partial sequences of large subunit rRNA and on cladistic analysis of morphological data. J Eukaryot Microbiol 40:411–421

Whitaker JW, McConkey GA, Westhead DR (2009) The transferome of metabolic genes explored: analysis of the horizontal transfer of enzyme encoding genes in unicellular eukaryotes. Genome Biol 10:R36. doi:10.1186/gb-2009-10-4-r36

Yoon HS, Reyes-Prieto A, Melkonian M, Bhattacharya D (2006) Minimal plastid genome evolution in the *Paulinella* endosymbiont. Curr Biol 16:R670–R672. doi:10.1016/j.cub.2006.08.018

http://infao5501.ag5.mpi-sb.mpg.de:8080/topx/archive?link=Wikipedia-Lip6-2/970821.xml&style

Vipin Chandra Kalia received his M.Sc. and Ph.D. degree in Genetics from IARI, New Delhi. He is presently the Chief Scientist at CSIR-IGIB and Professor of AcSIR. He is a Fellow of Association of Microbiologists of India and Fellow of National Academy of Sciences (FNASc). He is the Editor in Chief of the Indian Journal of Microbiology (India). His current interests are bioenergy, biopolymers, and quorum sensing. Google Scholar: http://scholar.google.co.in/citations?hl=en&user=XaUw-VIAAAAJ. Website: http://www.igib.res.in/?q=V.C.Kalia